Advanced Nanomaterials

Edited by
Kurt E. Geckeler and
Hiroyuki Nishide

Further Reading

Tjong, Sie Chin

Carbon Nanotube Reinforced Composites

Metal and Ceramic Matrices

2009
ISBN: 978-3-527-40892-4

Zehetbauer, M. J., Zhu, Y. T. (eds.)

Bulk Nanostructured Materials

2009
ISBN: 978-3-527-31524-6

Vollath, D.

Nanomaterials

An Introduction to Synthesis, Properties and Applications

2008
ISBN: 978-3-527-31531-4

Astruc, D. (ed.)

Nanoparticles and Catalysis

2008
ISBN: 978-3-527-31572-7

Lee, Yoon S.

Self-Assembly and Nanotechnology

A Force Balance Approach

2008
ISBN: 978-0-470-24883-6

Eftekhari, Ali (Ed.)

Nanostructured Materials in Electrochemistry

2008
ISBN: 978-3-527-31876-6

Lazzari, M. / Liu, G. / Lecommandoux, S. (eds.)

Block Copolymers in Nanoscience

2006
ISBN: 978-3-527-31309-9

Kumar, Challa S. S. R. (Ed.)

Nanomaterials for the Life Sciences

10 Volume Set

2010
ISBN: 978-3-527-32261-9

Kumar, Challa S. S. R. (Ed.)

Nanotechnologies for the Life Sciences

10 Volume Set

2007
ISBN: 978-3-527-31301-3

Rao, C. N. R., Müller, A., Cheetham, A. K. (eds.)

Nanomaterials Chemistry

Recent Developments and New Directions

2007
ISBN: 978-3-527-31664-9

Advanced Nanomaterials

Edited by
Kurt E. Geckeler and Hiroyuki Nishide

Volume 2

WILEY-VCH

WILEY-VCH Verlag GmbH & Co. KGaA

The Editors

Prof. Dr. Kurt E. Geckeler
Department of Nanobio Materials and
Electronics
World-Class University (WCU)
and
Department of Materials Science and
Engineering
Gwangju Institute of Science and
Technology (GIST)
1 Oryong-dong, Buk-gu
Gwangju 500-712
South Korea
E-mail: keg@gist.ac.kr

Prof. Hiroyuki Nishide
Department of Applied Chemistry
Waseda University
Ohkubo 3, Shinjuku
Tokyo 169-8555
Japan
E-mail: nishide@waseda.jp
and
Department of Nanobio Materials and
Electronics
World-Class University (WCU)
Gwangju Institute of Science and
Technology (GIST)
1 Oryong-dong, Buk-gu
Gwangju 500-712
South Korea

Library of Congress Card No.: applied for

British Library Cataloguing-in-Publication Data
A catalogue record for this book is available from
the British Library.

**Bibliographic information published by the Deutsche
Nationalbibliothek**
The Deutsche Nationalbibliothek lists this
publication in the Deutsche Nationalbibliografie;
detailed bibliographic data are available on the
Internet at <http://dnb.d-nb.de>.

Composition   Toppan Best-set Premedia Limited
Printing and Bookbinding   Strauss GmbH,
Mörlenbach
Cover Design   Schulz Grafik-Design, Fußgönheim

Printed in the Federal Republic of Germany
Printed on acid-free paper

ISBN: 978-3-527-31794-3

Contents

Advanced Nanomaterials. Edited by Kurt E. Geckeler and Hiroyuki Nishide
Copyright © 2010 WILEY-VCH Verlag GmbH & Co. KGaA, Weinheim
ISBN: 978-3-527-31794-3

Volume 1

Preface

Nanotechnology has found an incredible resonance and a vast number of applications in many areas during the past two decades. The resulting deep paradigm shift has opened up new horizons in materials science, and has led to exciting new developments. Fundamentally, nanotechnology is dependent on the existence or the supply of new nanomaterials that form the prerequisite for any further progress in this new and interdisciplinary area of science and technology. Evidently, nanomaterials feature specific properties that are characteristic of this class of materials, and which are based on surface and quantum effects.

Clearly, the control of composition, size, shape, and morphology of nanomaterials is an essential cornerstone for the development and application of nanomaterials and nanoscale devices. The complex functions of nanomaterials in devices and systems require further advancement in the preparation and modification of nanomaterials. Such advanced nanomaterials have attracted tremendous interest during recent years, and will form the basis for further progress in this area. Thus, the major classes of novel materials are described in the twenty-eight chapters of this two-volume monograph.

The initializing concept of this book was developed at the *3rd IUPAC International Symposium on Macro- and Supramolecular Architectures and Materials (MAM-06): Practical Nanochemistry and Novel Approaches,* held in in Tokyo, Japan, 2006, within the framework of the biannual MAM symposium series. This monograph provides a detailed account of the present status of nanomaterials, and highlights the recent developments made by leading research groups. A compilation of state-of-the-art review chapters, written by over sixty contributors and well-known experts in their field from all over the world, covers the novel and important aspects of these materials, and their applications.

The different classes of advanced nanomaterials, such as block copolymer systems including block copolymer nanofibers and nanotubes, smart nanoassemblies of block copolymers for drug and gene delivery, aligned and ordered block copolymers, helical polymer-based supramolecular films, as well as novel composite materials based on gold nanoparticles and carbon nanotubes, are covered in the book. Other topics include the synthesis of inorganic nanotubes, metal nanoparticle-attached electrodes, radical polymers in organic polymer batteries, oxidation catalysis by nanoscale gold, silver, copper, self-assembling

Advanced Nanomaterials. Edited by Kurt E. Geckeler and Hiroyuki Nishide
Copyright © 2010 WILEY-VCH Verlag GmbH & Co. KGaA, Weinheim
ISBN: 978-3-527-31794-3

nanoclusters, optically responsive polymer nanocomposites, renewable thermoplastic matrices based on phyllosilicate nanocomposites, amphiphilic polymer–clay intercalation and applications, the synthesis and catalysis of mesoporous alumina, and nanoceramics for medical applications.

In addition, this book highlights the recent progress in the research and applications of structural ceramics, the ecological toxicology of engineered carbon nanoparticles, carbon nanotubes as adsorbents for the removal of surface water contaminants, molecular imprinting with nanomaterials, near-field Raman imaging of nanostructures and devices, fullerene-rich nanostructures, nanoparticle-cored dendrimers and hyperbranched polymers, as well as the interactions of carbon nanotubes with biomolecules. The book is completed with a series of chapters featuring concepts in self-assembly, nanostructured organogels via molecular self-assembly, the self-assembly of linear polypeptide-based block copolymers, and information-guided self-assembly by structural DNA nanotechnology.

The variety of topics covered in this book make it an interesting and valuable reference source for those professionals engaged in the fundamental and applied research of nanotechnology. Thus, scientists, students, postdoctoral fellows, engineers, and industrial researchers, who are working in the fields of nanomaterials and nanotechnology at the interface of materials science, chemistry, physics, polymer science, engineering, and biosciences, would all benefit from this monograph.

The advanced nanomaterials presented in this book are expected to result in commercial applications in many areas. As the science and technology of nanomaterials is still in its infancy, further research will be required not only to develop this new area of materials science, but also to explore the utilization of these novel materials. All new developments impart risks, and here also it is important to evaluate the risks and benefits associated with the introduction of such materials into the biosphere and ecosphere.

On behalf of all contributors to we thank the publishers and authors on behalf of all contributors for granting copyright permissions to use their illustrations in this book. It is also very much appreciated that the authors devoted their time and efforts to contribute to this monograph. Last, but not least, the major prerequisite for the success of this comprehensive book project was the cooperation, support, and understanding of our families, which is greatly acknowledged.

The Editors

List of Contributors

Kotoji Ando
Yokohama National University
Department of Material Science
and Engineering
79-1 Tokiwadai
Hohogaya-ku
Yokohama 240-8501
Japan

Besim Ben-Nissan
University of Technology
Faculty of Science
Broadway
P.O. Box 123
Sydney
NSW 2007
Australia

Simona Bronco
CNR-INFM-PolyLab c/o
Dipartimento di Chimica e
Chimica Industriale
Università di Pisa
Via Risorgimento 35
56126 Pisa
Italy

Horacio Cabral
The University of Tokyo
Department of Materials Engineering
Graduate School of Engineering
7-3-1 Hongo, Bunkyo-ku
Tokyo 113-8656
Japan

Ying-Nan Chan
National Taiwan University
Institute of Polymer Science and
Engineering
Taipei 10617
Taiwan

and

National Chung Hsing University
Department of Chemical Engineering
Taichung 40227
Taiwan

Wen-Hsin Chang
National Taiwan University
Institute of Polymer Science and
Engineering
Taipei 10617
Taiwan

Advanced Nanomaterials. Edited by Kurt E. Geckeler and Hiroyuki Nishide
Copyright © 2010 WILEY-VCH Verlag GmbH & Co. KGaA, Weinheim
ISBN: 978-3-527-31794-3

Jing Cheng
The University of Western
Ontario
Department of Civil and
Environmental Engineering
London, ON N6A 5B9
Canada

Andy H. Choi
University of Technology
Faculty of Science
Broadway
P.O. Box 123
Sydney
NSW 2007
Australia

Francesco Ciardelli
University of Pisa
CNR-INFM-PolyLab
c/o Department of Chemistry,
and Industrial Chemistry
Via Risorgimento 35
56126 Pisa
Italy

Serena Coiai
Centro Italiano Packaging and
Dipartimento di Chimica e
Chimica Industriale
Università di Pisa
Via Risorgimento 35
56126 Pisa
Italy

Maria-Beatrice Coltelli
Centro Italiano Packaging and
Dipartimento di Chimica e
Chimica Industriale
Università di Pisa
Via Risorgimento 35
56126 Pisa
Italy

Claudio De Rosa
University of Napoli "Federico II"
Department of Chemistry
Complesso Monte S. Angelo
Via Cintia
80126 Napoli
Italy

Soorly G. Divakara
Colorado School of Mines
Department of Chemistry and
Geochemistry
1500 Illinois St.
Golden, CO 80401
USA

Kevin Flavin
Queen Mary University of London
School of Biological and Chemical
Sciences
Mile End Road
London E1 4NS
UK

Michiya Fujiki
Nara Institute of Science and
Technology
Graduate School of Materials Science
8916-5 Takayama
Ikoma
Nara 630-0101
Japan

Kurt E. Geckeler
Gwangju Institute of Science and
Technology (GIST)
Department of Materials Science and
Engineering
1 Oryong-dong, Buk-gu
Gwangju 500-712
South Korea

Achutharao Govindaraj
International Centre for
Materials Science
New Chemistry Unit and CSIR
Centre of Excellence in
Chemistry
Jawaharlal Nehru Centre for
Advanced Scientific Research
Jakkur P. O.
Bangalore 560 064
India

and

Solid State and Structural
Chemistry Unit
Indian Institute of Science
Bangalore 560 012
India

Jose E. Herrera
The University of Western
Ontario
Department of Civil and
Environmental Engineering
London, ON N6A 5B9
Canada

Ishenkumba A. Kahwa
The University of the West
Indies
Chemistry Department
Mona Campus
Kingston 7
Mona
Jamaica

Johnson Kasim
Nanyang Technological
University
School of Physical and
Mathematical Sciences
Division of Physics and Applied
Physics
Singapore 637371
Singapore

Kazunori Kataoka
The University of Tokyo
Department of Materials Engineering
Graduate School of Engineering
7-3-1 Hongo
Bunkyo-ku
Tokyo 113-8656
Japan

and

The University of Tokyo
Center for Disease Biology and
Integrative Medicine
Graduate School of Medicine
7-3-1 Hongo
Bunkyo-ku
Tokyo 113-0033
Japan

and

The University of Tokyo
Center for NanoBio Integration
7-3-1 Hongo
Bunkyo-ku
Tokyo 113-8656
Japan

Yonggang Ke
Arizona State University
Department of Chemistry and
Biochemistry & The Biodesign
Institute
Tempe, AZ 85287
USA

Harm-Anton Klok
Ecole Polytechnique Fédérale de
Lausanne (EPFL)
Institut des Matériaux, Laboratoire des
Polymères
STI-IMX-LP
MXD 112 (Bâtiment MXD), Station 12
1015 Lausanne
Switzerland

Arjun S. Krishnan
North Carolina State University
Department of Chemical &
Biomolecular Engineering
Raleigh, NC 27695
USA

Masashi Kunitake
Kumamoto University
Department of Applied
Chemistry and Biochemistry
2-39-1 Kurokami
Kumamoto 860-8555
Japan

Fernando Langa
Universidad de Castilla-La
Mancha
Facultad de Ciencias del
Medio Ambiente
45071 Toledo
Spain

Massimo Lazzari
University of Santiago de
Compostela
Department of Physical
Chemistry
Faculty of Chemistry and
Institute of Technological
Investigations
15782 Santiago de Compostela
Spain

Sebastien Lecommandoux
University of Bordeaux
Laboratoire de Chimie des
Polymères Organiques (LCPO)
UMR CNRS 5629
Institut Polytechnique de
Bordeaux
16 Avenue Pey Berland
33607 Pessac
France

Zhi Li
Colorado School of Mines
Department of Chemistry and
Geochemistry
1500 Illinois St.
Golden, CO 80401
USA

Jiang-Jen Lin
National Taiwan University
Institute of Polymer Science and
Engineering
Taipei 10617
Taiwan

Guojun Liu
Queens University
Department of Chemistry
50 Bader Lane
Kingston Ontario K7L 3N6
Canada

Yan Liu
Arizona State University
Department of Chemistry and
Biochemistry & The Biodesign
Institute
Tempe, AZ 85287
USA

Watoru Nakao
Yokohama National University
Department of Energy and Safety
Engineering
79-5 Tokiwadai
Hodogaya-ku
Yokohama 240-8501
Japan

Dhriti Nepal
Gwangju Institute of Science
and Technology (GIST)
Department of Materials Science
and Engineering
1 Oryong-dong, Buk-gu
Gwangju 500-712
South Korea

and

School of Polymer
Textile and Fiber Engineering
Georgia Institute of Technology
Atlanta, GA 30332
USA

Jean-François Nierengarten
Université de Strasbourg
Laboratoire de Chimie des
Matériaux Moléculaires
(UMR 7509)
Ecole Européenne de Chimie
Polymères et Matériaux
25 rue Becquerel
67087 Strasbourg, Cedex 2
France

Hiroyuki Nishide
Waseda University
Department of Applied
Chemistry
Tokyo 169-8555
Japan

Christopher K. Ober
Cornell University
Department of Materials Science
and Engineering
Ithaca, NY 14853
USA

Akihiro Ohira
National Institute of Advanced
Industrial Science and Technology
(AIST)
Polymer Electrolyte Fuel Cell Cutting-
Edge Research Center (FC-Cubic)
2-41-6 Aomi, Koto-ku
Tokyo 135-0064
Japan

Makoto Onaka
The University of Tokyo
Department of Chemistry
Graduate School of Arts and Sciences
Komaba, Meguro-ku
Tokyo 153-8902
Japan

Ryan R. Otter
Middle Tennessee State University
Department of Biology
Murfreesboro, TN 37132
USA

Kenichi Oyaizu
Waseda University
Department of Applied Chemistry
Tokyo 169-8555
Japan

Munetaka Oyama
Kyoto University
Graduate School of Engineering
Department of Material Chemistry
Nishikyo-ku
Kyoto 615-8520
Japan

Elisa Passaglia
University of Pisa
Department of Chemistry and
Industrial Chemistry
Via Risorgimento 35
56126 Pisa
Italy

Thathan Premkumar
Department of Materials Science
and Engineering
Gwangju Institute of Science
and Technology (GIST)
1 Oryong-dong, Buk-gu
Gwangju 500-712
South Korea

Andrea Pucci
University of Pisa
Department of Chemistry and
Industrial Chemistry
Via Risorgimento 35
56126 Pisa
Italy

Jeremy J. Ramsden
Cranfield University
Bedfordshire MK43 0AL
UK

and

Cranfield University at
Kitakyushu
2-5-4F Hibikino
Wakamatsu-ku
Kitakyushu 808-0135
Japan

C.N.R. Rao
International Centre for Materials
Science,
New Chemistry Unit and CSIR Centre
of Excellence in Chemistry
Jawaharlal Nehru Centre for Advanced
Scientific Research
Jakkur P. O.
Bangalore 560 064
India

and

Solid State and Structural Chemistry
Unit
Indian Institute of Science
Bangalore 560 012
India

Marina Resmini
Queen Mary University of London
School of Biological and
Chemical Sciences
Mile End Road
London E1 4NS
UK

Ryan M. Richards
Colorado School of Mines
Department of Chemistry and
Geochemistry
1500 Illinois St.
Golden, CO 80401
USA

Aaron P. Roberts
University of North Texas
Department of Biological Sciences &
Institute of Applied Sciences
Denton, TX 76203
USA

Kristen E. Roskov
North Carolina State University
Department of Chemical &
Biomolecular Engineering
Raleigh, NC 27695
USA

Giacomo Ruggeri
University of Pisa
CNR-INFM-PolyLab
c/o Department of Chemistry
and Industrial Chemistry
Via Risorgimento 35
56126 Pisa
Italy

Helmut Schlaad
Max Planck Institute of Colloids
and Interfaces
MPI KGF Golm
14424 Potsdam
Germany

Evan L. Schwartz
Cornell University
Department of Materials Science
and Engineering
Ithaca, NY 14853
USA

Tsunetake Seki
The University of Tokyo
Department of Chemistry
Graduate School of Arts and
Sciences
Komaba, Meguro-ku
Tokyo 153-8902
Japan

Ze Xiang Shen
Nanyang Technological University
School of Physical and Mathematical
Sciences
Division of Physics and Applied
Physics
Singapore 637371
Singapore

Young-Seok Shon
California State University, Long Beach
Department of Chemistry and
Biochemistry
1250 Bellflower Blvd
Long Beach, CA 90840
USA

Richard J. Spontak
North Carolina State University
Department of Chemical &
Biomolecular Engineering
Raleigh, NC 27695
USA

and

North Carolina State University
Department of Materials Science &
Engineering
Raleigh, NC 27695
USA

Koji Takahashi
Kyushu University
Hakozaki
Higashi-ku
Fukuoka 812-8581
Japan

Akrajas Ali Umar
Universiti Kebangsaan Malaysia
Institute of Microengineering and
Nanoelectronics
43600 UKM Bangi Selangor
Malaysia

Hao Yan
Arizona State University
Department of Chemistry and
Biochemistry & The Biodesign
Institute
Tempe, AZ 85287
USA

Ting Yu
Nanyang Technological
University
School of Physical and
Mathematical Sciences
Division of Physics and Applied
Physics
Singapore 637371
Singapore

Jingdong Zhang
Huazhong University of Science and
Technology
College of Chemistry and Chemical
Engineering
Wuhan 430074
China

14

Amphiphilic Poly(Oxyalkylene)-Amines Interacting with Layered Clays: Intercalation, Exfoliation, and New Applications

Jiang-Jen Lin, Ying-Nan Chan, and Wen-Hsin Chang

14.1
Introduction

During recent years, organic–inorganic hybrid materials have attracted a great deal of research interest due to their promising industrial applications [1–3]. The successful development of Nylon-6/montmorillonite (MMT) nanocomposites by the Toyota research group [4, 5], reflects the importance of utilizing mineral clays for improving polymer properties [6–8], including fire retardation and gas barrier. One of the crucial issues for preparing these advanced polymers is to overcome the problem of incompatibility between the hydrophilic smectite-clays and hydrophobic polymers. Organic modification is the common method to convert the hydrophilic clay into an organophile [9–13]. Fine dispersion of the silicate particles in polymer matrices was then achieved for a variety of hydrophobic nanocomposites including polypropylene [14], polystyrene (PS) [15], polyurethane [16], polyester [17], epoxy [18–20], and polyimide [21].

In the past, many reviews have been devoted to the subject of clay nanocomposites [1–5, 21–25]. Significant progress in polymer engineering and the limitations of polymer compatibility to the nanoscale clays in intercalated and exfoliated forms are already well reviewed. Recent research developments on utilizing the layered silicate clays have also diversified into the areas of biotechnology [26], such as the use of anionic clays for gene therapy. Actually, recent advances in biomedical applications involve different types of nanomaterial, which can be generally classified into three categories according to their geometric dimensions: spherical (e.g., silver and titanium oxide particles) [27, 28]; fibril-shape (e.g., carbon nanotube) [29]; and sheet-like (e.g., natural and synthetic clays) nanomaterials. With at least one dimension that falls into the nanometer scale, all of these inorganic materials possess a high-aspect ratio surface area relative to their weight. For the sheet-like clays, each thin layer platelet with a thickness of approximately 1 nm (almost at the molecular level) and the lateral dimension of approximately 50 nm to several micrometers provides an extremely large specific surface ($>500 \, m^2 \, g^{-1}$). Hence, the primary structure of the thin-layer clays is unique with respect to its

Advanced Nanomaterials. Edited by Kurt E. Geckeler and Hiroyuki Nishide
Copyright © 2010 WILEY-VCH Verlag GmbH & Co. KGaA, Weinheim
ISBN: 978-3-527-31794-3

geometric shape as well as surface ionic charge, which could be important when interacting with biological materials such as proteins, nucleic acids, and microorganisms.

In this chapter, we review the synthetic aspects of organic modifications of smectite clays, with particular attention being paid to reports on the uses of various intercalation agents for the polymer–clay composite applications, without emphasizing their composite performance. Recent efforts on using high-molecular-weight poly(oxyalkylene)-diamines to tailor the layered-silicate spacing is particularly addressed [30–33]. The reaction profile of the intercalation via ion exchange by different equivalents of organic salt to clay exchange capacity, as well as the unconventional mechanisms involving metal-ion chelation and hydrogen bonding, are also reviewed. With the incorporation of hydrophobic amine-salts, the resulting organoclays may exhibit unusual colloidal properties [34, 35] as well as an ability to self-assemble into rigid-rod nanoarrays [36–38]. Furthermore, the method of exfoliating the layered-silicate clays into random platelets has been recently reported [39–41]. This process involved the use of polymeric amines via zigzag conformation or phase inversion mechanisms for the platelet randomization. The tailored organoclays are found to be suitable for embedding biomaterials such as protein [42, 43]. The aim of this chapter is to summarize the literature activities in clay utilization, with emphasis placed on the organic modifications and the emerging research in the areas of biomedical applications.

14.2
Chemical Structures of Clays and Organic-Salt Modifications

14.2.1
Natural Clays and Synthetic Layered-Double-Hydroxide (LDH)

Smectic clays are naturally abundant, with a well-characterized lamellar structure of multiple inorganic plates, a high surface area, and ionic charges on the surface [9–11]. The phyllosilicate clays of the 2 : 1 type, or smectites such as MMT, bentonite, saponite, and hectorite, are conventionally utilized as catalysts [44–46], adsorbents [47, 48], metal-chelating agents [49], fillers for polymer composites [1, 3, 21], and so on. The generic structure is composed of multiple layers of silicate/aluminum oxide, for example, with layers of two tetrahedron sheets sandwiching an edge-shared octahedral sheet [21]. Counter metal ions populate the composition with variation in isomorphic substitution of silicon or aluminum by divalent metal ions such as Mg^{2+}, Ca^{2+}, or Fe^{2+}. These ionic charges are potentially exchangeable through further ionic exchange with alkali metal ions such as Na^+ or Li^+, as well as with organic ions. Although these layered silicates are hydrophilic and swell in water, they often exist as aggregates in micrometer sizes from their primary stack units. In the case of MMT, the primary stack structure possesses multiple aluminosilicate plates of irregular polygonal shapes at average dimension of ca. $100\,nm \times 100\,nm \times 1\,nm$ for individual platelets [50]. For the ionic-exchange reaction, the divalent counter-cations in most natural clays could be exchanged into different ions, including Na^+, Cu^{2+}, Zn^{2+}, Mg^{2+}, Ca^{2+} and an acidified H^+ form [51]. The replacement priority for these cations are: $Al^{3+} > Ca^{2+} > Mg^{2+} > K^+ = NH_4^+ > Na^+$ [46, 52]. According to this exchange order, organic quaternary ammonium salts can replace Na^+ ions, but not with divalent cations in Mg^{2+}-MMT and Ca^{2+}-MMT. Therefore, for most natural clays, the sodium ion exchange is necessary to facilitate the subsequent organic ion intercalation.

Synthetic fluorine mica, which structurally is similar to sodium tetrasilicic micas, is prepared from the Na_2SiF_6 treatment of talc at high temperature [53, 54]. The synthetic mica (Na^+-Mica) is water-dispersible and generally used as an inorganic thickener. This synthetic fluorinated mica has an average dimension of 300–1000 nm in 80–100 nm for MMT. Another class of synthetic clays, layered-double-hydroxided (LDHs), can be prepared from the coprecipitation of inorganic salts. The chemical structure is described as $[Mg_6Al_2(OH)_{16}]CO_3 \cdot 4H_2O]$ in the example of magnesium/aluminum hydroxides. Various metal hydroxides, including Ni, Cu, or Zn for divalent and Al, Cr, Fe, V, or Ga for trivalent metal ions, and anions such as CO_3^{2-}, Cl^-, SO_4^{2-}, NO_3^-, or other various organic anions, have been reported [55–58]. These LDHs are classified as anionic clays that can be organically modified through an anionic-exchange reaction, using substances such as carboxylic acids, anionic polymers, organic phosphoric acids, and so on [58]. These synthetic clays may have various applications, including heterogeneous catalysts, optical materials, biomimetic catalysts, separation agents, and DNA reservoirs [59–63]. Recently, Mg–Al LDHs were incorporated with poly(oxypropylene)-bis-amindocarboxylic acid salts (POP-acid) to result in a wide basal spacing of 92 Å [64]. This wide spacing, as well as the introduction of a hydrophobic POP backbone, may open up new applications for this class of anionic clays.

14.2.2
Low-Molecular-Weight Intercalating Agents and X-Ray Diffraction *d*-Spacing

The common strategy for utilizing smectite clays is to alter their inherent hydrophilic nature so that they become hydrophobic and organically compatible with polymers. For the synthesis of polymer–clay nanocomposites, organic oniums such as alkyl ammonium salts [22] are commonly used to intercalate the layered minerals. The resultant organoclays are then suitable for the consequent process of melt-blending with polymers and *in situ* polymerization. For example, sodium montmorillonite (Na^+-MMT), consisting of sodium ions on the silicate surface ($\equiv Si$–O^-Na^+), can be intercalated with organic onium salts. The quaternary alkyl ammonium ($R_4N^+X^-$) or alkyl phosphonium ($R_4P^+X^-$) salts are the common intercalating agents because of their commercial availability. The incorporation of organic intercalating agents also resulted in a silicate gallery expansion. For example, the C_{18}-alkyl quaternary salts may intercalate Na^+-MMT, causing a layer space expansion to 20–30 Å basal spacing from the pristine 12 Å clay gallery. It is noteworthy that the same organic quaternary salt may not exchange with natural clays with divalent counter ions such as M^{2+}-MMT [65] (where $M^{2+} = Mg^{2+}$ or Ca^{2+}).

A large number of intercalating agents for modifying the cationic smectite clays have been described, and it is possible to classify these according to their chemical structures and organic functionalities (see Table 14.1). Low-molecular-weight alkyl ammonium and phosphonium salts are the common intercalating agents, and normally these may result in a widening of the basal spacing in the range of 13 to 50 Å. Thermally stable and reactive surfactant types of imidazolium salts (with C_{12}, C_{16} and C_{18} alkyl groups) were reported for the preparation of polystyrene/ MMT nanocomposites [93]. The general idea here is that the presence of alkyl imidazolium may contribute to the thermal stability of the composites. Other reactive cationic surfactants, such as vinylbenzyl dimethyldodecyl ammonium chloride and surfactants with 2-methacryloyl functionalities, have been reported for poly(methylmethacrylate) (PMMA) and PS–clay nanocomposites. The purpose of introducing a reactive functionality to the intercalating agent is to facilitate not only a layer exfoliation but also fine dispersion in the polymer matrices.

14.3
Poly(Oxyalkylene)-Polyamine Salts as Intercalating Agents, and Their Reaction Profiles

14.3.1
Poly(Oxyalkylene)-Polyamine Salts as Intercalating Agents

The uses of poly(oxyalkylene)-amines of M_w 2000–4000 g mol^{-1} for the intercalation of Na$^+$-MMT to prepare organoclays with a large d spacing was reported in 2001 [30]. Poly(oxyalkylene)-diamines (POA-diamines) are commercially available polyether amines that are produced from the amination of polyols such as polyethylene glycols, polypropylene glycols, and their mixed poly(oxyethylene-oxypropylene) [99, 100]. Both, hydrophobic and hydrophilic types of POA-diamines, poly(oxypropylene)-amines (POP-amines) and poly(oxyethylene)-amines (POEamines), respectively, are available. By using the hydrophobic POP-diamines of 230, 400, 2000, and 4000 g mol^{-1}, Na$^+$-MMT was intercalated into organoclays at varied X-ray diffraction (XRD) basal spacings (15.0, 19.4, 58.0, and 92.0 Å, respectively). The lamellar expansion was found to be proportionally correlated with the POP molecular weights, through the POP tethering with silicate surfaces or the ionic-exchange reaction of the quaternary ammonium ions. Within the silicate confinement, the POP backbones may aggregate through a hydrophobic phase separation and consequently stretch out the basal spacing of the silicate interlayer gallery (Figure 14.1).

In contrast, the hydrophilic POE-diamine of M_w 2000 g mol^{-1} (POE2000) resulted in a spacing of 19.4 Å. The conclusion has been drawn that the intercalating agent requires a hydrophobic nature in order to expand the ionically charged platelets. The hydrophilic POE backbones or the $-(CH_2CH_2O)_x-$ structure, resulted in the organics associating tightly with the silicate surface. Only the hydrophobic POP backbone $-(CH_2CH(CH_3)O)_x-$ has shown an ability to generate a new "supporting" phase for widening the gallery space.

Table 14.1 Representatives of the various intercalating agents described in the literature.

Functionality	Intercalating agent	Clay	XRD (Å)	Reference(s)
POA-diamine salt	Poly(oxypropylene)-diamine (2000 M_w), Poly(oxypropylene)-diamine (POP-amines, 4000 M_w), Poly(oxypropylene)-diamine (POP-amines, 5000 M_w)	Li$^+$-fluorohectorite, Na$^+$-MMT, Na$^+$-Mica	46–92	[12, 30, 35]
Polymeric amine	Amine-terminated PS surfactants, Amine-terminated butadiene acrylonitrile copolymers	Na$^+$-MMT	10–155	[66–68]
Alkyl-amine salt	1-Hexadecylamine and octadecylamine	Na$^+$-MMT	16–37	[69–76]
Amino acid	6-Aminohexanoic acid, 12-Aminododecainoic acid	Na$^+$-MMT	17–49	[12, 77]
Acid	C_{18} carboxylic acids and other acids, 2-Acrylamido-2-methyl-1-propanesulfonic acid	Mg^{2+}- or Ca^{2+}-MMT, Na$^+$-MMT	35, Exfoliation	[65, 78–80]
Ammonium salt	Hexadecyltrimethylammonium bromide, Dioctadecyldimethyl ammonium bromide, Vinylbenzyldimethyldodecylammonium chloride, Dialkyldimethylammonium salt from hydrogenated tallow	Na$^+$-MMT	19–40	[69, 70, 81–86]
Phosphonium salts	10-[3,5-bis(methoxycarbonyl)phenoxy]decyltriphenylphosphonium bromide Dodecyltriphenyl phosphonium bromide, Hexadecyltributyl phosphonium bromide, Tetraoctyl phosphonium bromide	Na$^+$-Mica, Li$^+$-fluorohectorite, Na$^+$-MMT	24–32	[72, 87–92]
Imidazolium salt	1,2-Dimethyl-3-(benzyl ethyl isobutyl polyhedral oligomeric silsesquioxane) imidazolium chloride (DMIPOSS), Dioctadecyl imidazolium, 1,2-Dimethyl-3-hexadecyl imidazolium, vinyl-alkyl-imidazolium (C_{12}, C_{16} and C_{18})	Na$^+$-MMT	36	[93–96]
Stibonium salt	Triphenylhexadecyl stibonium trifluoromethyl sulfonate	Na$^+$-MMT	20	[97]
Poly(ethylene glycol)	PEG400 vs. POP-amine[a]	Na$^+$-MMT	17.7	[98]

a   POP-amines: poly(oxyalkylene)-amines of M_w 230–4000.

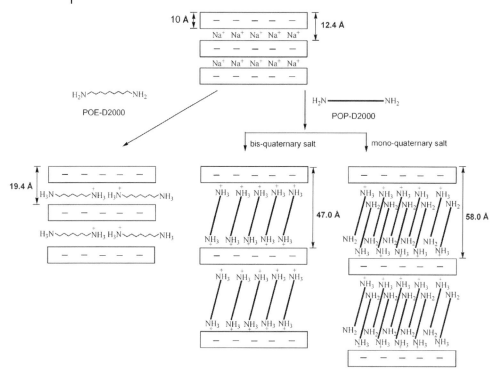

Figure 14.1 Schematic representation of Na⁺-MMT intercalation by hydrophilic and hydrophobic poly(oxyalkylene)-diamine salts of the same molecular weight [31].

It was noted that formation of the hydrophobic POP phase in the silicate gallery could be understood by estimating the theoretical length of the fully stretched POP backbone. Based on calculations of the bond lengths (1.54 Å for C–C and 1.43 Å for C–O) and bond angles (109.6° and 112°, respectively), the theoretical length (77 Å) for POP2000 was longer than the value of 58.0 Å observed for those molecules in confinement. Apparently, the POPs were aggregated into hydrophobic phase but not fully extended at the molecular level, nor arranged in a tilting orientation in the gallery confinement.

14.3.2
Critical Conformational Change in Confinement During the Intercalating Profile [31]

The intercalation profile was studied by varying the equivalent ratios (from 0.2 to 1.0) of POP2000 salt to the clay CEC (120 mEq 100 g⁻¹). Critical intercalation points were apparent, as shown in Figure 14.2, for the clay expansion with respect to the POP intercalants. Before the critical point, the basal spacing expansion was similar (19.0–20.2 Å) and in the range of amine addition, from 0.2 to 0.8 CEC equivalent.

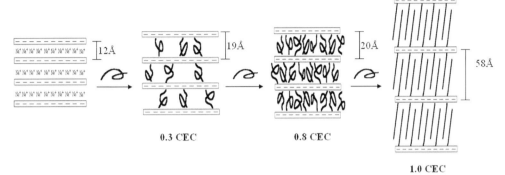

Figure 14.2 Schematic representation of POP-diamine forming a hydrophobic phase in a clay gallery [31].

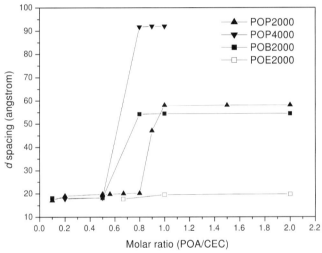

Figure 14.3 Critical points of poly(oxyalkylene)-diamine salts intercalating montmorillonite at different CEC ratios [31].

At 0.8 CEC equivalent, the POP addition expanded the d spacing suddenly to 58 Å in the case of POP2000 intercalation.

The critical concentration for the basal spacing expansion is generalized for several hydrophobic polyether-amines, as summarized in Figure 14.3. The POP4000 and poly(oxybutylene)diamine of 2000 g mol^{-1} M_w (POB2000) showed a critical point for the intercalation (Figure 14.3, ▼ and ■, respectively). The increases in basal spacing, from 20 Å to 92 Å for POP4000, and from 20 Å to 54 Å for POB2000, were observed and attributed to the molecular length, POP4000 > POP2000 > POB2000. The existence of a critical concentration was explained by the POP hydrophobic phase generated in the layer gallery. For the POE-backbone amine salt, there was no change in the critical basal spacing.

14.3.3
Correlation between MMT *d*-Spacing and Intercalated Organics [32, 33]

The basal spacing enlargement can only be achieved by a hydrophobic aggregation of the intercalating agents in the gallery. By taking in consideration the molecular end-to-end length in the confinement, a linear correlation was found for the basal spacing expansion. As the incorporated organics occupy the same volume as the measured silicate spacing, the relationship between the interlayer spacing (*D*) and the organic weight fraction (*R* or w/w of POP-amine/MMT by weight) is expressed by the following equation [32]:

$$D = R \times \frac{1}{A \times \rho} + t \tag{14.1}$$

where *A* is the surface area of MMT, *t* is the plate thickness, and ρ is the density of the intercalant in the gallery. The XRD measurement is a function of the organics occupied in the gallery, and the relationship of gallery distance, volume, and organic fraction is predictable.

14.4
Amphiphilic Copolymers as Intercalating Agents

14.4.1
Various Structures of the Amphiphilic Copolymers [34, 40, 101, 102]

In order to tailor the hydrophobic nature for organoclays, several research groups have used amphiphilic copolymers as intercalating agents. For example, the copolymers of POP-amine grafting on poly(propylene) (PP) [34, 40], poly(styrene-ethylene/butadiene-styrene) (SEBS) [101] and poly(styrene-maleic anhydride) (SMA) [102] copolymers were reported. These copolymers, which comprised multiple amines as the pendant groups, were comb-like in shape and amphiphilic in nature (Figure 14.4). The copolymers, after treating with HCl, were able to form stable emulsions in water and to intercalate with Na⁺-MMT. The generation of an emulsion at 670 nm diameter for the amine-grafted PP copolymer at ambient temperature, and at 560 nm at 75 °C in toluene/water, has also been reported [34]. The fine emulsion rendered the copolymers able to intercalate with Na⁺-MMT. The pendant quaternary ammonium ions of the copolymers may undergo an ionic-exchange reaction to afford the silicates of 19.5 Å spacing. The layered silicate platelet gallery was widened and surrounded with hydrophobic copolymer backbones. The resultant silicate/copolymer hybrids resembled a single micelle structure containing a hydrophilic rigid silicate core and a hydrophobic organic corona. The hybrids were dispersible in toluene and were found to be approximately 500 nm in size in the example of PP-POP2000/MMT.

Hydrophilic amine

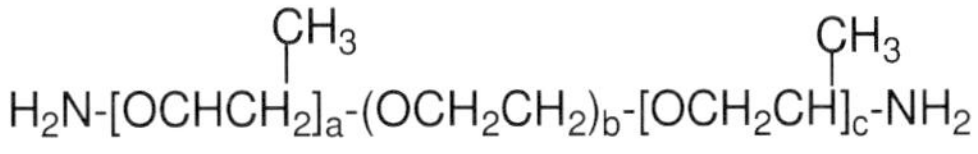

$$\underset{CH_3}{\qquad} \qquad \underset{CH_3}{\qquad}$$

$$H_2N\text{-}[OCHCH_2]_a\text{-}(OCH_2CH_2)_b\text{-}[OCH_2CH]_c\text{-}NH_2$$

a + c = 5, b = 39.5 (M_w = 2000; POE2000)

Hydrophobic amines

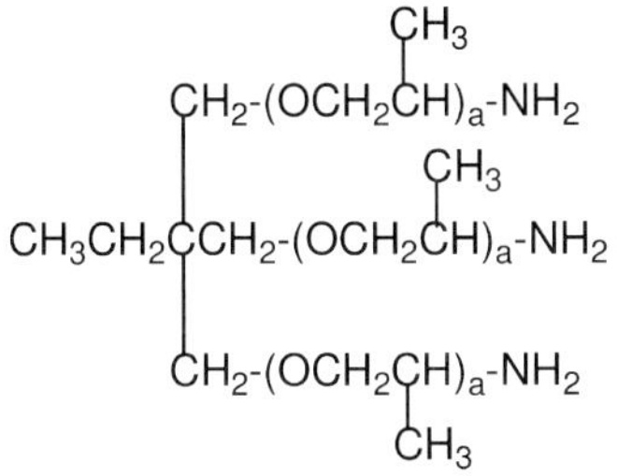

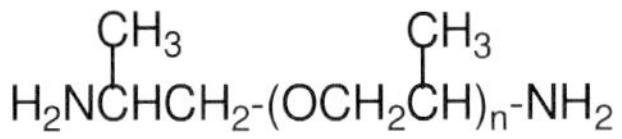

CH$_2$-(OCH$_2$CH)$_a$-NH$_2$ with CH$_3$

CH$_3$CH$_2$CCH$_2$-(OCH$_2$CH)$_a$-NH$_2$ with CH$_3$

CH$_2$-(OCH$_2$CH)$_a$-NH$_2$ with CH$_3$

$$H_2NCHCH_2\text{-}(OCH_2CH)_n\text{-}NH_2$$ with CH$_3$, CH$_3$

a = 4–5 (M_w = 400; POPT400)

a = 16–17 (M_w = 3000; POPT3000)

a = 27–28 (M_w = 5000; POPT5000)

n = 5–6 (M_w = 400; POPD400)

n = 33 (M_w = 2000; POPD2000)

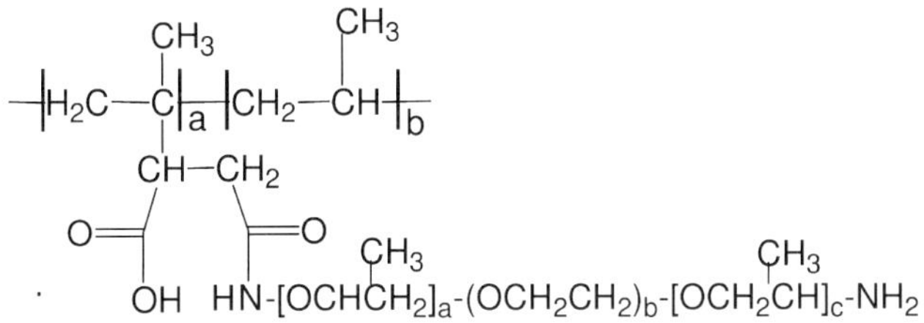

$$\left[H_2C\text{-}\underset{CH_3}{\overset{CH_3}{C}}\right]_a \left[CH_2\text{-}CH\right]_b$$

CH-CH$_2$

O= =O CH$_3$ CH$_3$

OH HN-[OCHCH$_2$]$_a$-(OCH$_2$CH$_2$)$_b$-[OCH$_2$CH]$_c$-NH$_2$

Figure 14.4 Chemical structures of POA-diamines and the grafted PP copolymers [34, 40].

The POA-diamine grafting onto SEBS-g-MA generated another class of comb-like and amphiphilic polyamines. The particular SEBS–POA copolymer, with an average of nine amine pendants, is capable of forming a fine emulsion and inter-calating with Na$^+$-MMT to afford a wide range of XRD *d* spacing, from 17 Å to 52 Å. The high *d* spacing was due to SEBS backbone intercalation, while the POA intercalation could result in only a low spacing. Thus, it was concluded that two different intercalating modes were present, the conceptual description of which is shown in Figure 14.5.

The amphiphilic properties of the copolymers can be better controlled by using poly(styrene-*co*-maleic anhydrides) (SMA) grafted with the POA-diamines. The

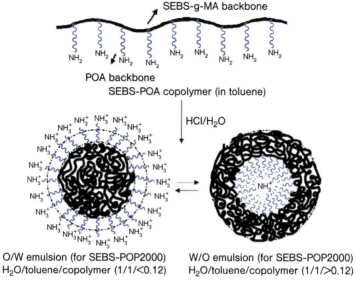

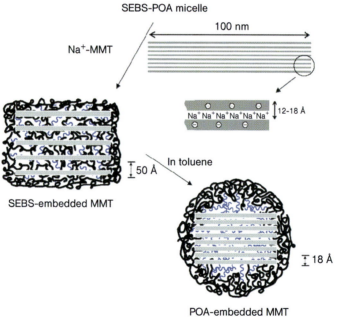

Figure 14.5 Schematic representation of copolymer/clay hybrids in different forms of intercalation [101].

copolymers with different ratios of styrene/MA monomers provided a wide range of hydrophobic properties. The SMA-POP copolymers generated the intercalated organoclays with 12.9 Å to 78.0 Å basal spacing. Two types of intercalation were also observed for the SMA-POP intercalation.

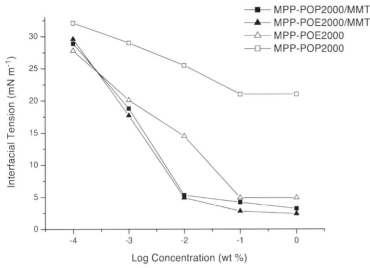

Figure 14.6 Interfacial tensions of the hybrids prepared from MMT intercalation with POP- or POE-grafted PP copolymers in toluene/water [34].

14.4.2
Colloidal Properties

The silicate–copolymer hybrids, intercalated with hydrophobic organics, are dispersible in toluene and exhibit a surfactant property [34, 35, 101]. In Figure 14.6, the hybrids can be seen to exhibit an ability to reduce the toluene/water interfacial tension, from $28\,\mathrm{mN\,m^{-1}}$ to $5\,\mathrm{mN\,m^{-1}}$. The critical micelle concentration (CMC) or critical aggregation concentration (CAC) was estimated as 0.01 wt%, by extrapolating from the interfacial tension versus concentration curve.

14.5
New Intercalation Mechanism Other than the Ionic-Exchange Reaction

14.5.1
Amidoacid and Carboxylic Acid Chelating [65, 103]

The conventional method of modifying $\mathrm{Na^+}$-MMT requires the use of organic onium salts, such as ammonium and phosphonium salts, for the ionic-exchanging intercalation. The driving force for the organic incorporation is via the ionic exchange reaction, the replacement of ammonium salts ($-\mathrm{NH_3^+Cl^-}$) for sodium ions ($\equiv\mathrm{Si-O^-Na^+}$). In the case of using poly(oxyalkylene)-amidoacids in the form of zwitterions as the intercalating agents, the intercalation of $\mathrm{Na^+}$-MMT occurred [103]. The mechanism via a chelating intermediate between the amidoacid and the

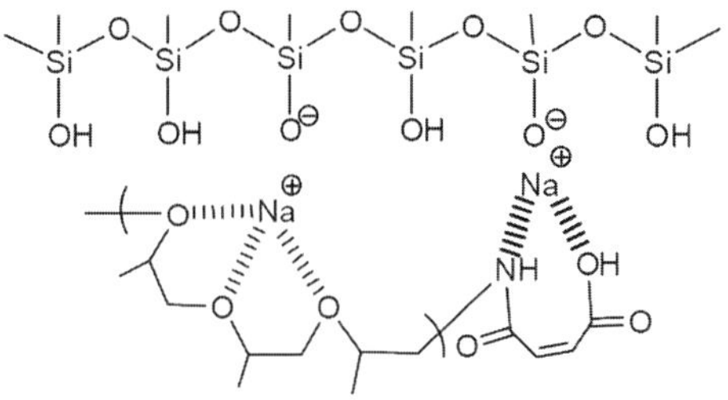

Figure 14.7 Schematic representation of sodium ion chelation via an oxypropylene-amidoacid [103].

counter sodium ion in the clay gallery was proposed. The chelating intermediate is supposedly stable due to the formation of a cyclic, seven-membered ring structure (Figure 14.7).

Another mechanism [65] was reported for metal chelating in the clay gallery. Different alkyl lengths of alkyl-carboxylic acids, $CH_3(CH_2)_nCOOH$, were allowed to intercalate into the clays that contained divalent metal counter ions. Conventionally, it was believed that the C_{12-18} carboxylic acids [e.g., lauric acid ($n = 10$) and stearic acid ($n = 16$)] would be unable to incorporate into the silicate clays, such as the naturally occurring MMT. However, it was reported that the sodium or potassium salts of alkyl carboxylates could be incorporated into Na^+-MMT to afford a low organic embedment of 10–15 wt% organics, and an XRD basal spacing of 15 Å. However, the same intercalating carboxylate salts in fact intercalated the divalent M^{2+}-MMT (XRD = 10.1 Å) to achieve a wide spacing of up to 30 or 43 Å, in the example of the naturally occurring smectite silicates (K10), which comprised layered aluminosilicates and mostly divalent counter cations. The difference in intercalation between the M^{2+}-MMT and Na^+-MMT was explained by the complexes of divalent M^{2+} with carboxylate ($-COO^-$) and SiO^- anions (Figure 14.8). Formation of the thermally stable complex served as the driving force for the organics to incorporate into the gallery.

14.5.2
Intercalation Involving Intermolecular Hydrogen Bonding [104]

Besides the chelating mechanism, hydrogen bonding of the intercalated organics (structures shown in Scheme 14.1) may serve as additional forces for the interaction, in similar fashion to the hydrophobic interaction described above. With the structure consisting of $-(CH_2)_8-$ and amide functionalities, the intercalating agents may be further accumulated in the clay gallery. In the examples of the layered aluminosilicates of Na^+ synthetic mica and MMT, in both cases intercalation with the POP-amidoamine quaternary salts afforded an enlarged basal spacing of up to 78 Å, with an organic embedment up to 70 wt% in the hybrids. The driving forces applied by the hydrogen bonding interaction were evidenced by using a stepwise intercalating method. Continuing addition of the POP-amidoamine salts after the

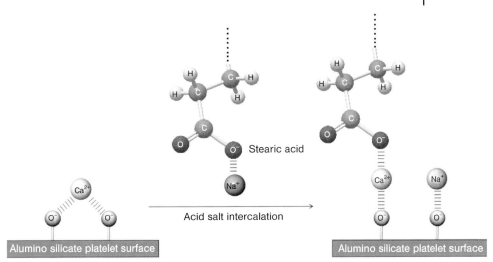

Figure 14.8 Diagrammatic representation of a carboxylic acid salt complex through divalent cations on silicate K10 [65].

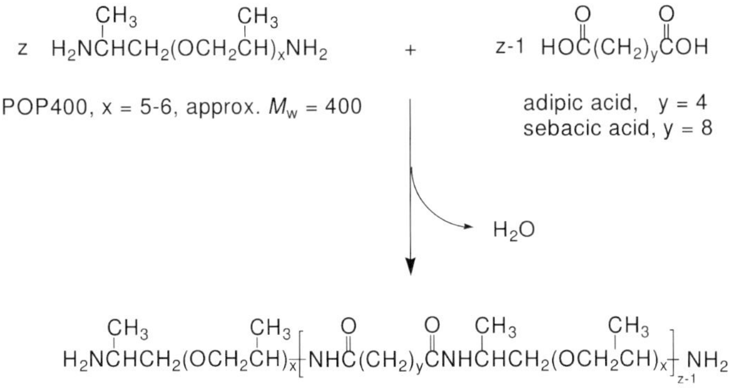

z H₂NĊHCH₂(OCH₂ĊH)ₓNH₂   +   z-1 HOĊ(CH₂)ᵧĊOH

$$z \quad H_2N\overset{\underset{|}{CH_3}}{C}HCH_2(OCH_2\overset{\underset{|}{CH_3}}{C}H)_xNH_2 \quad + \quad z\text{-}1 \quad HO\overset{\overset{O}{\parallel}}{C}(CH_2)_y\overset{\overset{O}{\parallel}}{C}OH$$

POP400, x = 5-6, approx. M_w = 400

adipic acid, y = 4
sebacic acid, y = 8

→ H_2O

$$H_2N\overset{\underset{|}{CH_3}}{C}HCH_2(OCH_2\overset{\underset{|}{CH_3}}{C}H)_{\overline{x}}\left[N\overset{\overset{O}{\parallel}}{H}\overset{\overset{O}{\parallel}}{C}(CH_2)_y\overset{\overset{O}{\parallel}}{C}N\overset{\underset{|}{CH_3}}{H}CH_2(OCH_2\overset{\underset{|}{CH_3}}{C}H)_{\overline{x}}\right]_{z\text{-}1}NH_2$$

amidoamines (z = 2, 3, and 4)

Scheme 14.1 Representative structure of POP-amidoamines with hydrogen-bonding functionalities [104].

CEC equivalent amount led to the generation of hybrids, while the basal spacing was increased from 58 Å to 70–78 Å.

14.6
Self-Assembling Properties of Organoclays [36, 37]

With the incorporation of hydrophobic POP, the MMT hybrids were amphiphilic in nature. For example, MMT/POP2000 at 58.0 Å basal spacing and 63 wt%

Figure 14.9 (a, b) Scanning electron microscopy images of the self-assembly of a POP-modified clay [36].

organics was dispersible in toluene, and exhibited an ability to lower the toluene/water interfacial surface tension. The surfactant property was shown to depend on the hydrophobic–hydrophilic balance between the hydrophobic POP portion and the ionic character of the platelet. With varied amounts of organics in the clay, the amphiphilic nature of the hybrids can be tailored. Those hybrids with an organic content in the range of 22 to 63 wt%, and a corresponding d spacing of 19–58 Å, can easily be prepared. Among these hybrids, the POP/MMT hybrid with a balanced hydrophobic–hydrophilic ratio (19 Å d spacing, 26 wt% organics) enabled self-assembly into rod-like microstructures which were up to 40 μm in length [36] (Figure 14.9). These unique self-assemblies were observed using scanning electron microscopy (SEM). The POP-modified MMT hybrid was first dispersed in a water/toluene mixture at a concentration of 1.0 wt%, and then subjected to evaporation at 80 °C before the SEM observations were made. The formation of rod-like arrays [37] was uniform, with individual rod dimensions ranging from 100 to 800 nm in width, and from 2 to 10 μm as the average length. The mechanism for the formation of lengthy rods was proposed to be a two-directional growth of horizontal and vertical self-alignment, as illustrated conceptually in Figure 14.10. The correct balance between the POP hydrophobic aggregation and the silicate ionic charge attraction is important for forming orderly microstructures.

14.7
Exfoliation Mechanism and the Isolation of Random Silicate Platelets

14.7.1
Thermodynamically Favored Exfoliation of Na⁺-MMT by the PP-POP Copolymers [40]

The hybrids prepared from the intercalation of PP-POP2000 copolymers with Na⁺-MMT at 19.5 Å [34] could be further exfoliated under the conditions of 120 °C and 500 p.s.i. N_2 in a closed system, such that the layered structure of Na⁺-MMT was transformed into random platelets. The process required the copolymers to form

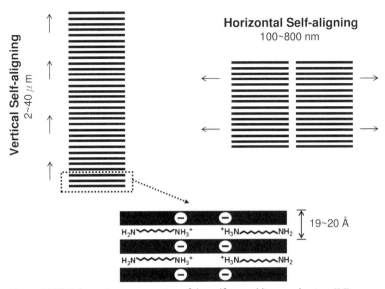

Figure 14.10 Schematic representation of the self-assembling mechanism [36].

a fine emulsion before interacting with the hydrophilic clays. Two types of inter-calating hybrid may be involved, as shown conceptually in Figure 14.11. Transformation of the initial copolymer/layered silicate hybrid into the exfoliation could be explained by the relative stability of the water-in-oil (W/O) and oil-in-water (O/W) hybrid structures. An inversion from the W/O to the O/W hybrid was favored thermodynamically such that, ultimately, exfoliation of the clay-layered structure would occur.

14.7.2
Zigzag Mechanism for Exfoliating Na⁺-MMT

An alternative mechanism of exfoliation was reported which used Mannich polyamines as the intercalating agents [39]; these were prepared from the reaction of *p*-cresol, formaldehyde, and POA-diamines. After being converted into the corresponding amine salts, the copolymers proved to be effective for exfoliating the layered silicates. Due to the presence of multiple amine sites in the polymer backbone for anchoring onto the silicate surface, a zigzag conformation of the copolymers within the confinement was proposed as the exfoliation mechanism. This process is illustrated schematically in Figure 14.12.

14.8
Isolation of the Randomized Silicate Platelets in Water [41]

The process of exfoliation led to the next step of isolating random silicate platelets. As shown in Scheme 14.2), the layered Na⁺-MMTs were exfoliated through

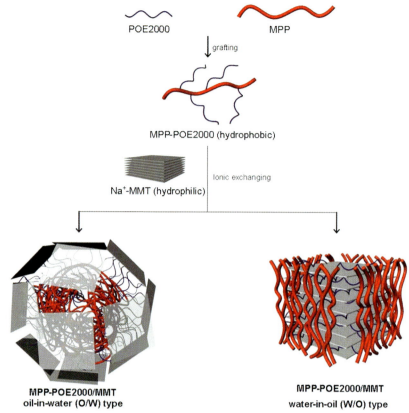

Figure 14.11 Emulsion of the copolymers affecting Na$^+$-MMT intercalation and exfoliation [40].

ionic-exchange reactions, while further Na$^+$ exchange permitted recovery of the organic amines.

The nature of the material of the random silicate platelets was confirmed by using XRD analysis, transmission electron microscopy (TEM), and atomic force microscopy (AFM). In particular, the material was shown to have a unique ionic character in water. Based on parameters of 1.2 mEq g^{-1} CEC for Na$^+$-MMT and a surface area of 720 m^2 g^{-1}, the platelets were calculated to contain 20000 ions per platelet, with an area per ion of 0.9 nm^2, and with 4×10^{16} platelets per gram. The intensive charges rendered the silicate plates with a high affinity for polar organics such as poly(ethylene glycol) (PEG) and water molecules. During measurement of the zeta potential (see Figure 14.13), a Na$^+$-MMT suspension in water demonstrated a relatively constant charge character at -18 mV, over a pH range from 2 to 11. However, the zeta potential of the exfoliated platelets showed an isoelectric point of pH 6.4, and varied from 21 mV to -25 mV over the pH range from 5 to 8. It appeared that the platelet surface possessed different ionic potentials that were originally buried within the MMT layered structure.

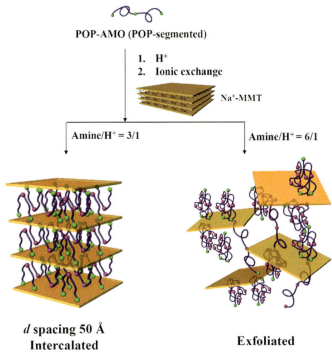

POP-AMO (POP-segmented)

1. H⁺
2. Ionic exchange

Na⁺-MMT

Amine/H⁺ = 3/1   Amine/H⁺ = 6/1

d spacing 50 Å
Intercalated   **Exfoliated**

Figure 14.12 Schematic representation of silicate stack and exfoliation by the presence of polyamine salts [39].

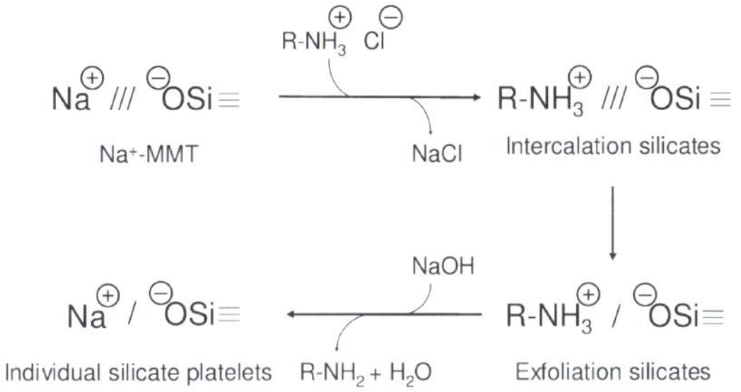

$$\overset{\oplus}{Na} /// \overset{\ominus}{O}Si \equiv \quad \xrightarrow[NaCl]{R-\overset{\oplus}{N}H_3 \ \overset{\ominus}{Cl}} \quad R-\overset{\oplus}{N}H_3 /// \overset{\ominus}{O}Si \equiv$$

Na⁺-MMT   Intercalation silicates

$$\overset{\oplus}{Na} / \overset{\ominus}{O}Si \equiv \quad \xleftarrow[R-NH_2 + H_2O]{NaOH} \quad R-\overset{\oplus}{N}H_3 / \overset{\ominus}{O}Si \equiv$$

Individual silicate platelets   Exfoliation silicates

Scheme 14.2 The scheme of stepwise intercalation, exfoliation, and isolation of random silicate platelets [41].

14.9
Emerging Applications in Biomedical Research

Both types of ionic clay, namely the anionic LDH and the cationic Na⁺-MMT, have been recognized for their abilities to adsorb biomaterials [105], the key properties for this application being the high surface area, the ionic character, and the layered structure [106–108]. Previously reported examples have used clays as the support

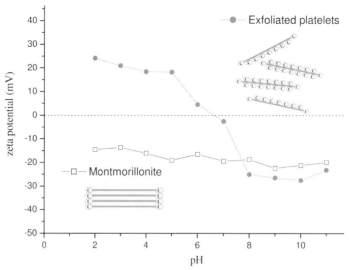

Figure 14.13 Zeta potentials of montmorillonite (□) and exfoliated silicate platelets (●) at different pH values [41].

for biomaterial adsorption [109], and DNA encapsulation for gene therapy [26]. In general, the key elements in this area of research are to specify the functions of encapsulating large molecules within the layered confinement. Earlier studies have revealed several examples of proteins complexing with naturally occurring clays, simply by their adsorption directly onto the clay surface [110, 111]. Such adsorption between proteins and silicates is affected by the properties of the clay, the ionic charges and the type of species, as well as the size of the protein and its isoelectric properties [111].

Recent improvements in this area have focused on the stability of the biomaterial, the controlled release profile, and targeting. Indeed, it has been reported that the incorporation of bovine serum albumin (BSA) into the layered spacing of MMT and synthetic fluorinated mica is possible via a step-wise ionic exchange reaction [42, 43], as shown in Figure 14.14. Specifically, the layer spacing of the synthetic mica was first expanded with a POA-amine salt of 2000 or 4000 M_w, from the pristine 12 Å to the final 93 Å, and this was followed by BSA incorporation. By comparison, the direct incorporation of BSA into a clay interlayer was found to be relatively difficult, particularly for those micas with a larger platelet size. This stepwise process was performed at a suitable pH (less than the isoelectric point of 4.8 for BSA), at which point the protein structure was cationic ($-NH_3^+/-COOH$); this proved to be ideal for the ionic-exchange intercalation with the counter ions in the silicate galleries. The stepwise intercalation can be generalized for embedding large biomolecules into the clay layer confinement. This synthetic methodology is considered to have practical applications, including biosensing for target molecular detection [112], as well as interface modification for drug delivery [113, 114] and tissue engineering [62, 115].

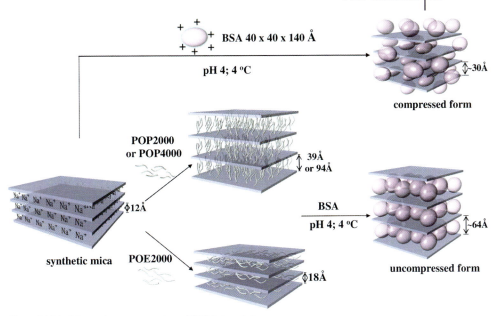

Figure 14.14 Schematic representation of BSA intercalation into mica, via direct or step-wise processes [42].

14.10
Conclusions

The naturally abundant clays have been long recognized for their wide applications as catalysts, absorbents, metal-chelating agents, colloid stabilizers, and fillers for polymer composites. Recent developments in nanocomposite designs have required the preparation of organophilic clays, which may then become compatible with the ionic smectite clays. Since 2001, the present authors' group has explored tailoring of the basal spacing of the layered silicates, as well as the reaction profile and amphiphilic properties of these hybrids. By using a series of hydrophobic and hydrophilic POA-salts, it was possible to control the XRD basal spacing within the range of 20–92 Å and, for the first time, to observe a unique self-assembled morphology. During this time, the use of anionic LDH clays for a variety of applications, including gene therapy, has been suggested. With these spatially enlarged silicates, biomolecules such as proteins could be embedded in the clay gallery, which would retain its original size. It was also found that several comb-branched polyamines could randomize the layered structure into individual platelets. More recently, innovative families of polymer–clay hybrids have been identified with new properties of interfacial tension activity, compatibility to hydrophobic materials, and intensive affinity for biomacromolecules. It appears that the fundamental exploration of clay surface chemistry has in fact directed us towards a new horizon of biomedical research opportunities, including enzyme/drug encapsulation, wound surface anti-adhesives, and biosensors.

References

1 Giannelis, E.P. (1996) *Adv. Mater.*, **8**, 29–35.

2 Utracki, L.A., Sepehr, M. and Boccaleri, E. (2007) *Polym. Adv. Technol.*, **18**, 1–37.

3 Alexandre, M. and Dubois, P. (2000) *Mater. Sci. Eng.*, **28**, 1–63.

4 Okada, A. and Usuki, A. (1995) *Mater. Sci. Eng. C*, **3**, 109–115.

5 Okada, A. and Usuki, A. (2006) *Macromol. Mater. Eng.*, **291**, 1449–1476.

6 Porter, D., Metcalfe, E. and Thomas, M.J.K. (2000) *Fire Mater.*, **24**, 45–52.

7 Wang, Z. and Pinnavaia, T.J. (1998) *Chem. Mater.*, **10**, 1820–1826.

8 Gilman, J.W. (1999) *Appl. Clay Sci.*, **15**, 31–49.

9 Theng, B.K.G. (1974) *The Chemistry of Clay-Organic Reactions*, 2nd edn, John Wiley & Sons, Inc., New York.

10 Van Olphen, H. (1977) *An Introduction to Clay Colloid Chemistry*, 2nd edn, John Wiley & Sons, Inc., New York.

11 Theng, B.K.G. (1979) *Formation and Properties of Clay-Polymer Complexes*, Elsevier, New York.

12 Triantafillidis, C.S., LeBaron, P.C. and Pinnavaia, T.J. (2002) *Chem. Mater.*, **14**, 4088–4095.

13 Muzny, C.D., Butler, B.D., Hanley, H.J.M., Tsvetkov, F. and Peiffer, D.G. (1996) *Mater. Lett.*, **28**, 379–384.

14 Bohning, M., Goering, H., Fritz, A., Brzezinka, K.W., Turky, G., Scholnhals, A. and Schartel, B. (2005) *Macromolecules*, **38**, 2764–2774.

15 Robello, D.R., Yamaguchi, N., Blanton, T. and Barnes, C. (2004) *J. Am. Chem. Soc.*, **126**, 8118–8119.

16 Tien, Y.I. and Wei, K.H. (2002) *J. Appl. Polym. Sci.*, **86**, 1741–1748.

17 Lepoittevin, B., Pantoustier, N., Alexandre, M., Calberg, C., Jerome, R. and Dubois, P. (2002) *J. Mater. Chem.*, **12**, 3528–3532.

18 Myskova, M.Z., Zelenka, J., Spacek, V. and Socha, F. (2003) *Macromol. Symp.*, **200**, 291–296.

19 Wang, M.S. and Pinnavaia, T.J. (1994) *Chem. Mater.*, **6**, 468–474.

20 Lan, T., Kaviratna, P.D. and Pinnavaia, T.J. (1996) *J. Phys. Chem. Sol.*, **57**, 1005–1010.

21 Ray, S.S. and Okamoto, M. (2003) *Prog. Polym. Sci.*, **28**, 1539–1641.

22 LeBaron, P.C., Wang, Z. and Pinnavaia, T.J. (1999) *Appl. Clay Sci.*, **15**, 11–29.

23 Giannelis, E.P. (1998) *Appl. Organomet. Chem.*, **12**, 675–680.

24 Zanetti, M., Lomakin, S. and Camino, G. (2000) *Macromol. Mater. Eng.*, **279**, 1–9.

25 Ishida, H., Campbell, S. and Blackwell, J. (2000) *Chem. Mater.*, **12**, 1260–1267.

26 Choy, J.H., Kwak, S.Y., Jeong, Y.J. and Park, J.S. (2000) *Angew. Chem. Int. Ed. Engl.*, **39**, 4041–4045.

27 Yi, Y., Wang, Y. and Liu, H. (2003) *Carbohyd. Polym.*, **53**, 425–430.

28 Fu, G., Vary, P.S. and Lin, C.T. (2005) *J. Phys. Chem. B*, **109**, 8889–8898.

29 Gu, L., Elkin, T., Jiang, X., Li, H., Lin, Y., Qu, L., Tzeng, T.R.J., Joseph, R. and Sun, Y.P. (2005) *Chem. Commun.*, 874–876.

30 Lin, J.J., Cheng, I.J., Wang, R. and Lee, R.J. (2001) *Macromolecules*, **34**, 8832–8834.

31 Lin, J.J., Cheng, I.J. and Chou, C.C. (2003) *Macromol. Rapid Commun.*, **24**, 492–495.

32 Chou, C.C., Shieu, F.S. and Lin, J.J. (2003) *Macromolecules*, **36**, 2187–2189.

33 Chou, C.C., Chang, Y.C., Chiang, M.L. and Lin, J.J. (2004) *Macromolecules*, **37**, 473–477.

34 Lin, J.J., Hsu, Y.C. and Chou, C.C. (2003) *Langmuir*, **19**, 5184–5187.

35 Lin, J.J. and Chen, Y.M. (2004) *Langmuir*, **20**, 4261–4264.

36 Lin, J.J., Chou, C.C. and Lin, J.L. (2004) *Macromol. Rapid Commun.*, **25**, 1109–1112.

37 Lin, J.J., Chu, C.C., Chou, C.C. and Shieu, F.S. (2005) *Adv. Mater.*, **17**, 301–304.

38 Lin, J.J., Chu, C.C., Chiang, M.L. and Tsai, W.C. (2006) *Adv. Mater.*, **18**, 3248–3252.

39 Chu, C.C., Chiang, M.L., Tsai, C.M. and Lin, J.J. (2005) *Macromolecules*, **38**, 6240–6243.

40 Chou, C.C. and Lin, J.J. (2005) *Macromolecules*, **38**, 230–233.

41 Lin, J.J., Chu, C.C., Chiang, M.L. and Tsai, W.C. (2006) *J. Phys. Chem. B*, **110**, 18115–18120.

42 Lin, J.J., Wei, J.C. and Tsai, W.C. (2007) *J. Phys. Chem. B*, **111**, 10275–10280.

43 Lin, J.J., Wei, J.C., Juang, T.Y. and Tsai, W.C. (2007) *Langmuir*, **23**, 1995–1999.

44 Cseri, T., Bekassy, S., Figueras, F. and Rizner, S. (1995) *J. Mol. Catal. A: Chem.*, **98**, 101–107.

45 Pinnavaia, T.J. (1983) *Science*, **220**, 365–371.

46 Ajjou, A.N., Harouna, D., Detellier, C. and Alper, H. (1997) *J. Mol. Catal. A: Chem.*, **126**, 55–60.

47 Celis, R., Hermosin, M.C., Carrizosa, M.J. and Cornejo, J. (2002) *J. Agric. Food Chem.*, **50**, 2324–2330.

48 Rawajfih, Z. and Nsour, N. (2006) *J. Colloid Interface Sci.*, **298**, 39–49.

49 Kiraly, Z., Veisz, B., Mastalir, A. and Kofarago, G. (2001) *Langmuir*, **17**, 5381–5387.

50 Usuki, A., Hasegawa, N., Kadoura, H. and Okamoto, T. (2001) *Nano. Lett.*, **1**, 271–272.

51 Sparks, D.L. (2003) *Environmental Soil Chemistry*, John Wiley & Sons, Ltd.

52 Wang, H., Zhao, T., Zhi, L., Yan, Y. and Yu, Y. (2002) *Macromol. Rapid Commun.*, **23**, 44–48.

53 Tateyama, H., Nishimura, S., Tsunematsu, K., Jinnai, K., Adachi, Y. and Kimura, M. (1992) *Clays Clay Miner.*, **40**, 180–185.

54 Kodama, T., Higuchi, T., Shimizu, T., Shimizu, K., Komarneni, S., Hoffbauer, W. and Schneider, H. (2001) *J. Mater. Chem.*, **11**, 2072–2077.

55 Rives, V. and Ulibarri, M.A. (1999) *Coord. Chem. Rev.*, **181**, 61–120.

56 Rives, V. (2002) *Mater. Chem. Phys.*, **75**, 19–25.

57 Khan, A.I. and O'Hare, D. (2002) *J. Mater. Chem.*, **12**, 3191–3198.

58 Aisawa, S., Takahashi, S., Ogasawara, W., Umetsu, Y. and Narita, E. (2001) *J. Solid State Chem.*, **162**, 52–62.

59 Iyi, N., Kurachima, K. and Fujita, T. (2002) *Chem. Mater.*, **14**, 583–589.

60 Choy, J.H., Kwak, S.Y., Park, J.S., Jeong, Y.J. and Portier, J. (1999) *J. Am. Chem. Soc.*, **121**, 1399–1400.

61 Leroux, F. and Besse, J.P. (2001) *Chem. Mater.*, **13**, 3507–3515.

62 Sels, B., Vos, D.D., Buntinx, M., Pierard, F., Mesmaeker, A.K.D. and Jacobs, P. (1999) *Nature*, **400**, 855–857.

63 Whilton, N.T., Vickers, P.J. and Mann, S. (1997) *J. Mater. Chem.*, **7**, 1623–1629.

64 Lin, J.J. and Juang, T.Y. (2004) *Polymer*, **45**, 7887–7893.

65 Chou, C.C., Chiang, M.L. and Lin, J.J. (2005) *Macromol. Rapid Commun.*, **26**, 1841–1845.

66 Beyer, F.L., Tan, N.C.B., Dasgupta, A. and Galvin, M.E. (2002) *Chem. Mater.*, **14**, 2983–2988.

67 Ha, Y.H., Kwon, Y., Breiner, T., Chan, E.P., Tzianetopoulou, T., Cohen, R.E., Boyce, M.C. and Thomas, E.L. (2005) *Macromolecules*, **38**, 5170–5179.

68 Kurian, M., Dasgupta, A., Galvin, M.E., Ziegler, C.R. and Beyer, F.L. (2006) *Macromolecules*, **39**, 1864–1871.

69 Yang, Y., Zhu, Z.K., Yin, J., Wan, X.Y. and Qi, Z.E. (1999) *Polymer*, **40**, 4407–4414.

70 Maiti, M., Bandyopadhyay, A. and Bhowmick, A.K. (2006) *J. Appl. Polym. Sci.*, **99**, 1645–1656.

71 Subramani, S., Lee, J.Y., Choi, S.W. and Kim, J.H. (2007) *J. Polym. Sci. B: Polym. Phys.*, **45**, 2747–2761.

72 Bourlinos, A.B., Jiang, D.D. and Giannelis, E.P. (2004) *Chem. Mater.*, **16**, 2404–2410.

73 Hotta, Y., Taniguchi, M., Inukai, K. and Yamagishi, A. (1996) *Langmuir*, **12**, 5195–5201.

74 Wang, K., Wang, L., Wu, J., Chen, L. and He, C. (2005) *Langmuir*, **21**, 3613–3618.

75 Szabo, A., Gournis, D., Karakassides, M.A. and Petridis, D. (1998) *Chem. Mater.*, **10**, 639–645.

76 Hotta, Y., Inukai, K., Taniguchi, M., Nakata, M. and Yamagishi, A. (1997) *Langmuir*, **13**, 6697–6703.

77 Byun, H.Y., Choi, M.H. and Chung, I.J. (2001) *Chem. Mater.*, **13**, 4221–4226.

78 Beatty, A.M., Granger, K.E. and Simpson, A.E. (2002) *Chem. Eur. J.*, **8**, 3254–3259.

79 Malakul, P., Srinivasan, K.R. and Wang, H.Y. (1998) *Ind. Eng. Chem. Res.*, **37**, 4296–4301.

80 Choi, Y.S., Choi, M.H., Wang, K.H., Kim, S.O., Kim, Y.K. and Chung, I.J. (2001) *Macromolecules*, **34**, 88978–88985.

81 Mahadevaiah, N., Venkataramani, B. and Prakash, B.S.J. (2007) *Chem. Mater.*, **19**, 4606–4612.

82 Vaia, R.A., Teukolsky, R.K. and Giannelis, E.P. (1994) *Chem. Mater.*, **6**, 1017–1022.

83 Hotta, S. and Paul, D.R. (2004) *Polymer*, **45**, 7639–7654.

84 Zeng, C. and Lee, L.J. (2001) *Macromolecules*, **34**, 4098–4103.

85 Fu, X. and Qutubuddin, S. (2001) *Polymer*, **42**, 807–813.

86 Fu, X. and Qutubuddin, S. (2000) *Mater. Lett.*, **42**, 12–15.

87 Imai, Y., Nishimura, S., Abe, E., Tateyama, H., Abiko, A., Yamaguchi, A., Aoyama, T. and Taguchi, H. (2002) *Chem. Mater.*, **14**, 477–479.

88 Chang, J.H., Jang, T.G., Ihn, K.J., Lee, W.K. and Sur, G.S. (2003) *J. Appl. Polym. Sci.*, **90**, 3208–3214.

89 Kim, M.H., Park, C.I., Choi, W.M., Lee, J.W., Lim, J.G., Park, O.O. and Kim, J.M. (2004) *J. Appl. Polym. Sci.*, **92**, 2144–2150.

90 Ijdo, W.L. and Pinnavaia, T.J. (1999) *Chem. Mater.*, **11**, 3227–3231.

91 Maiti, P., Yamada, K., Okamoto, M., Ueda, K. and Okamoto, K. (2002) *Chem. Mater.*, **14**, 4654–4661.

92 Xie, W., Xie, R., Pan, W.P., Hunter, D., Koene, B., Tan, L.S. and Vaia, R. (2002) *Chem. Mater.*, **14**, 4837–4845.

93 Bottino, F.A., Fabbri, E., Fragala, I.L., Malandrino, G., Orestano, A., Pilati, F. and Pollicino, A. (2003) *Macromol. Rapid Commun.*, **24**, 1079–1084.

94 Fox, D.M., Maupin, P.H., Harris, R.H. Jr., Gilman, J.W., Eldred, D.V., Katsoulis, D., Trulove, P.C. and DeLong H.C. (2007) *Langmuir*, **23**, 7707–7714.

95 Wang, Z.M., Chung, T.C., Gilman, J.W. and Manias, E. (2003) *J. Polym. Sci., B: Polym. Phys.*, **41**, 3173–3187.

96 Gilman, J.W., Awad, W.H., Davis, R.D., Shields, J., Harris, R.H. Jr., Davis, C., Morgan, A.B., Sutto, T.E., Callahan, J., Trulove, P.C. and DeLong. H.C. (2002) *Chem. Mater.*, **14**, 3776–3785.

97 Wang, D. and Wilkie, C.A. (2003) *Polym. Degrad. Stab.*, **82**, 309–315.

98 Greenwell, H.C., Harvey, M.J., Boulet, P., Bowden, A.A., Coveney, P.V. and Whiting, A. (2005) *Macromolecules*, **38**, 6189–6200.

99 Moss, P.H. (1964) Nickel-Copper-Chromia catalyst and the preparation thereof. US Patent 3, 152, 998.

100 Yeakey, E.L. (1972) Process for preparing polyoxyalkylene polyamines. US Patent 3, 654, 370.

101 Chang, Y.C., Chou, C.C. and Lin, J.J. (2005) *Langmuir*, **21**, 7023–7028.

102 Lin, J.J., Hsu, Y.C. and Wei, K.L. (2007) *Macromolecules*, **40**, 1579–1584.

103 Lin, J.J., Chang, Y.C. and Cheng, I.J. (2004) *Macromol. Rapid Commun.*, **25**, 508–512.

104 Lin, J.J., Chen, Y.M. and Yu, M.H. (2007) *Colloids Surf. A: Physicochem. Eng. Aspects*, **302**, 162–167.

105 Shan, D., Yao, W. and Xue, H. (2006) *Electroanalysis*, **18**, 1485–1491.

106 Zhou, Y., Hu, N., Zeng, Y. and Rusling, J.F. (2002) *Langmuir*, **18**, 211–219.

107 Kelleher, B.P., Oppenheimer, S.F., Han, F.X., Willeford, K.O., Simpson, M.J., Simpson, A.J. and Kingery, W.L. (2003) *Langmuir*, **19**, 9411–9417.

108 Baron, M.H., Revault, M., Servagent-Noinville, S., Abadie, J. and Quiquampoix, H. (1999) *J. Colloid Interface Sci.*, **214**, 319–332.

109 Naidja, A., Huang, P.M. and Bollag, J.M. (1997) *J. Mol. Catal. A: Chem.*, **115**, 305–316.

110 Causserand, C., Kara, Y. and Aimar, P. (2001) *J. Membr. Sci.*, **186**, 165–181.

111 Cristofaro, A.D. and Violante, A. (2001) *Appl. Clay Sci.*, **19**, 59–67.

112 Lvov, Y., Ariga, K., Ichinose, I. and Kunitake, T. (1995) *J. Am. Chem. Soc.*, **117**, 6117–6123.

113 Liu, Y.L., Hsu, C.Y., Su, Y.H. and Lai, J.Y. (2005) *Biomacromolecules*, **6**, 368–373.

114 Meziani, M.J. and Sun, Y.P. (2003) *J. Am. Chem. Soc.*, **125**, 8015–8018.

115 Wang, Q., Gao, Q. and Shi, J. (2004) *J. Am. Chem. Soc.*, **126**, 14346–14347.

15
Mesoporous Alumina: Synthesis, Characterization, and Catalysis

Tsunetake Seki and Makoto Onaka

15.1
Introduction

The pore structure is one of the most important properties of heterogeneous catalysts that govern their activity, selectivity, and lifetime. Notably, not only the surface reactivity but also the mass transfer of reactants and products can be tuned by modifying the pore structure. Owing to the size of most reactant molecules capable of access to the internal surfaces, the values of pore diameter that are of most interest to catalyst chemists are the orders of nanometers, with a range of approximately 0 to 50 nm. By definition, the pores in this range are classified into two groups: "micropore," with a diameter <2 nm; and "mesopore," with diameters in the range of 2 to 50 nm [1]. Since the mid-twentieth century, a large family of aluminosilicates known as zeolites have been synthesized and used as catalysts with regularly arrayed micropores [2]. On the other hand, the development of oxide frameworks with pore diameters within the mesoporous range lagged far behind, due to difficulties in their synthesis. In fact, such ordered mesoporous materials were not created successfully until 1992, when a team at the Mobil Research and Development Corporation first reported the successful synthesis of silica-based mesoporous molecular sieves (M41S) via a liquid crystal template (LCT) mechanism, where the surfactant liquid crystal phases served as the organic templates [3]. Since then, investigations to prepare organized mesoporous molecular sieves have been conducted not only for silica/aluminosilicate materials but also for oxides of Mg, Ti, Nb, Ta, and Al [4, 5]. For this, the electrostatic interaction between positively charged surfactants (S^+) and negatively charged inorganic precursors (I^-), charge-reversed interactions ($S^- I^+$), counterion-mediated interactions ($S^- M^+ I^-$ and $S^+ X^- I^+$; M^+ = metal cation, X^- = Cl$^-$, Br$^-$), and neutral interactions ($S^0 I^0$ and $N^0 I^0$; N = nonionic organics) were used to assemble series of ordered mesopores [6].

Because of their high surface areas and strong adsorption abilities, aluminas with various morphologies have been widely used as catalysts and catalyst supports, both in the laboratory and in industry [7]. It seems natural, therefore, that

Advanced Nanomaterials. Edited by Kurt E. Geckeler and Hiroyuki Nishide
Copyright © 2010 WILEY-VCH Verlag GmbH & Co. KGaA, Weinheim
ISBN: 978-3-527-31794-3

catalyst chemists would be prompted to investigate the behavior of these newly synthesized aluminas with uniform mesopores in a variety of catalytic reactions. The most likely major advantage of using mesoporous aluminas in catalytic reactions is that the regular mesoporous structures, which do not contain micropores, are free from the pore-plug problems caused by the formation of coke and other polymerized byproducts [8]. These problems, which have often been observed in catalysis by conventional aluminas, include a decrease in the surface areas and a disturbance of the diffusion of reactants and products within the porous networks. One other advantage includes the modification of surface reactivity by creating a uniform mesopore structure within the alumina. A warped bulk structure in mesoporous alumina may generate a unique surface reactivity (e.g., acidity and basicity) that is considerably different from that of normal alumina, and provides the opportunity to adjust the adsorption characteristics to optimum values by altering the mesopore sizes.

In this chapter, attention is focused on those aluminas with a regular mesoporosity, together with details of their synthesis and examples of their uses in catalytic reactions. In addition, a reference base of information up to early 2007 is included. Comparisons with conventional aluminas are frequently noted in order to emphasize the utility of mesoporous alumina as a catalytic material.

15.2
Synthesis of Mesoporous Alumina

15.2.1
Experimental Techniques

15.2.1.1 Synthesis
The initial stage of mesoporous alumina synthesis is the hydrolysis of aluminum precursors in the presence of a template (a structure-directing agent) in a solvent to afford a precipitate or gel material; this material is subsequently filtered out, washed, and dried to provide the as-made material. It should be noted that special care is required when using water for any purpose in these procedures, as this may lead to a serious structural collapse of the samples. It is recommended, therefore, that anhydrous ethanol or propanols be used in any washing processes [9]. Moreover, when aluminum alkoxides are used as precursors, care must be taken when deciding on the amount of water required for the hydrolyses, as well as how the water should be contacted with the aluminum precursors, since the very rapid hydrolyses may proceed without being influenced by the template.

The as-made products are organic/inorganic biphase composite materials in which the assembly of a hydrophobic group of templates helps to create the cylindrical inorganic walls. The template group is, on the other hand, adsorbed onto the surface of the inorganic walls by various interactions, including electrically neutral and electrostatic interactions. The most reliable way of removing the

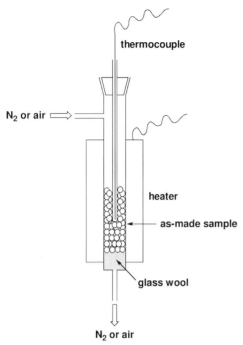

thermocouple

N₂ or air ⟹

heater

as-made sample

glass wool

N₂ or air

Figure 15.1 Apparatus for calcination of the as-made samples with organic/inorganic biphase. Quartz-made glass should be used for the tubular vessel, due to its high thermal stability.

adsorbed templates is to burn off the organic phase by calcination in air; this also promotes dehydration of the inorganic wall, transforming the framework from aluminum hydroxide to aluminum oxide (alumina). An example of the apparatus suitable for the calcination of as-made samples is shown in Figure 15.1. This flow system allows a choice of calcination atmosphere simply by changing the type of flow gas. A muffle oven can also be used if the calcination steps are to be performed under air. The required temperature for burning off the adsorbed template molecules depends on the strength of the interaction between the molecules and the inorganic walls. Thus, whilst calcination can be carried out at lower temperatures for nonionic templates, any ionic templates that are strongly adsorbed onto the inorganic wall will require higher temperatures for their removal. It should be noted that increasing the calcination temperature will enhance removal of the templates, but also result in a structural collapse by sintering. Consequently, the use of templates that bond too strongly to the inorganic surface should be avoided. Some groups have experienced problems in removing the sulfate or sulfonate surfactants, whereby substantial amounts of sulfur and organic residues remained (even at calcination temperature) that caused the mesostructure to collapse [5b, 10].

15.2.1.2 **Characterization**

The structural properties of mesoporous alumina can be evaluated by N_2 adsorption–desorption, powder X-ray diffraction (XRD), transmission electron microscopy (TEM), scanning electron microscopy (SEM), and ^{27}Al magic angle spinning nuclear magnetic resonance (MAS-NMR) spectroscopy. The N_2 adsorption–desorption method represents a powerful tool for characterizing the shape of mesoporous alumina. In Figure 15.2 the type IV isotherm is shown [11], which is typically observed for aluminas with organized mesopores.

The large uptake in the lowest partial pressure (P/P_0) region is caused by a monolayer coverage of the surface, followed by a steep increase in adsorbed volume in the P/P_0 range of 0.4–0.8, which corresponds to the filling of the mesopores. In the latter P/P_0 region, the adsorption isotherm usually does not conform to the desorption isotherm, giving rise to a hysteresis loop which may be accounted for by the presence of ink-bottle-like mesopores [12]. A further increase in P/P_0 leads to no further adsorption, which is represented by a horizontal plateau in the isotherm. In contrast, an additional steep increase is observed by the multilayer adsorption for the low-quality samples containing larger mesopores (i.e., pore diameter >8 nm) and macropores. By applying the N_2 adsorption–desorption data to the appropriate theories, it is possible to obtain the values of surface area, pore diameter, and pore volume. The Brunauer–Emmett–Teller (BET) method has been used for calculation of the surface areas [13], whilst values of pore diameter and pore volume have been calculated using the Barrett–Joyner–Halenda (BJH) or Dollimore–Heal (DH) methods [14, 15]. The Horváth–Kawazoe model, which was employed in the characterization of M41S by Beck and coworkers [3b], is not recommended because the model can give overestimated pore sizes [5a]. Čejka et al., on the other hand, attempted to calculate the surface area and mesopore volume by employing a comparison plot in which the adsorption isotherm $a(P/P_0)$ of a mesoporous alumina sample was transformed to a function $a(a_{ref})$ of the amount a_{ref} adsorbed on a nonporous reference solid (Degussa Aluminiumoxid

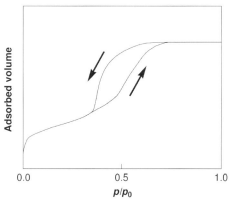

Figure 15.2 Type IV isotherm typically observed for mesoporous alumina.

C) at the same relative pressure [16]. The values obtained agreed well with those determined with the BET and BJH methods.

Powder X-Ray Diffraction   Powder XRD provides information concerning the pore and framework structure. The recommendation is to use chromium as an X-ray source because of its longer wavelength compared to the more commonly used copper or cobalt [9]. Aluminas with a regular mesoporous structure usually exhibit a single intense diffraction line at a low-angle region ($2\theta = 1$–$2°$) without further diffraction lines, indicating a hexagonal or cubic symmetry. The position of the line is shifted to a lower angle for samples with larger pore sizes, and also when the sintering of a sample is promoted by calcination at higher temperatures. It should be noted that the appearance of this single diffraction line is not sufficient evidence for the presence of organized mesopores [5n]. Other approaches, including N_2 adsorption–desorption measurements and TEM imaging, are also indispensable when discussing the quality of mesopores. The framework structure of a sample may be elucidated via the XRD pattern of the wide-angle region ($2\theta = 10$–$80°$). The detailed assignment of the XRD diffraction lines to the framework structures was described by Yanagida *et al.* [17].

Transmission Electron Microscopy (TEM)   TEM enables a direct observation of the pore size and long-range channel order. In marked contrast to the long-range hexagonal arrangement of channels observed for MCM-41 [3], the mesoporous aluminas developed to date have exhibited a regular sponge- or worm-like channel motif, without any discernible long-range order.

²⁷Al MAS NMR Spectroscopy   This procedure provides information regarding the coordination number of aluminum in mesoporous alumina. When using this method, six-, five-, and four-coordinate aluminum centers produce a peak at δ = ca. 0, 35, and 75, respectively, although in some cases the peak from the five-coordinate form is invisible due to its low intensity. It is noteworthy that the ratio of four- and six-coordinate Al estimated via the NMR spectrum of a sample may provide evidence that the framework has a γ-phase structure [5i].

15.2.2
Examples of Synthesis

An early review introducing the synthetic methods of mesoporous alumina was provided by Čejka, and included those reports made up until 2002 [9]. Čejka classified the methods into three groups – namely neutral, anionic, and cationic synthesis – according to the type of electric state of the surfactant, while others used the abbreviations, S (surfactant), I (inorganic precursors), N (nonionic organics), with their electric charges, +, −, and 0, to express the synthetic pathways of mesophases [6]. Even though the electric charge of the surfactant is known, the detailed structure of the inorganic moiety that interacts with the surfactant has, in most cases, not been elucidated in published reports that describe mesoporous alumina synthesis. For this reason, the synthetic methods are classified here as four groups, focusing only on the electric charge of the surfactant side: (i) neutral surfactant templating; (ii) anionic surfactant templating; (iii) cationic surfactant templating; and (iv) nonsurfactant templating. An overview of the various syntheses of mesoporous alumina is provided in Table 15.1.

15.2.2.1 Neutral Surfactant Templating

The synthesis of mesoporous alumina using electrically neutral surfactant templates was first developed in 1996 by Bagshaw and Pinnavaia, who used poly(ethylene oxide)s (PEOs) or PEO/poly(propylene oxide) (PPO) copolymers as surfactant templates [5a]. The former templates involve the use of Tergitol® 15-S-9 with a formula of $C_{11-15}[PEO]_9$, while the latter templates involve Pluronic® L64 with a formula of $[PEO]_{13}[PPO]_{30}[PEO]_{13}$. In the presence of the surfactant, an aluminum source of aluminum *sec*-butoxide was hydrolyzed in 2-butanol solvent to give the as-synthesized surfactant–alumina mesophase. This was subsequently washed with ethanol, dried, and calcined at 500 °C to afford the mesoporous aluminas (denoted MSU-X, where X = 1–3 depending on the type of surfactant). The XRD reflections of the MSU-X aluminas at 1–3° (of which the intensity became large after calcination) were broad, signifying a greater degree of structural disorder compared to the disordered MCM-41, HMS, and MSU-X silicas (Figure 15.3).

Subsequent TEM images of the calcined MSU-X aluminas highlighted the regular wormlike channel motif in which the diameters of pores were more or less uniform (Figure 15.4). However, no discernible long-range order was observed, which was indicative of the randomness of packing of the channel system.

Characterization by N_2 adsorption–desorption measurements showed hystereses in the isotherms of the MSU-X aluminas that had resulted from the condensation of N_2 in the mesopores, and afforded pore diameters in the range of 2.4 to 4.7 nm, as estimated by the BJH model and in agreement with values obtained from TEM images (Figure 15.5). It was found that, on the basis of the BJH pore-sizes and the d_{100} values, a larger surfactant tended to give a longer pore diameter.

The MSU-X aluminas were also characterized using ^{27}Al MAS NMR spectroscopy, which provides information concerning the coordination environment of aluminum (Figure 15.6). Three peaks appeared for both the as-synthesized and calcined samples at δ 0, 35, and 75, and these were attributed to the six-, five-, and four-coordinate aluminum centers, respectively. However, the calcination increased the peak intensities at δ 35 and 75 and decreased the intensity of the δ 0 peak through an enhancement of the dehydration and dehydroxylation.

In a succeeding study, the same group attempted to improve the thermal stability of the MSU-X aluminas by doping with rare earth ions such as Ce^{3+} and La^{3+} [5c]. In this case, the hydrolysis for the surfactant–alumina mesophase formation was performed in the presence of the corresponding rare earth nitrate. A PEO-based surfactant of Tergitol® 15-S-12 was used for synthesis of the Ce^{3+}-doped mesoporous alumina, while a PEO/PPO copolymer surfactant of Pluronic® P65 or P123 was employed for the La^{3+}-doping. The isotherms, BJH pore-sizes, XRD

Table 15.1 Overview of representative synthetic methods of mesoporous alumina.

Mesoporous alumina denotation	Aluminum source (solvent)	Template: name; formula[a]	Surface area[b] ($m^2 g^{-1}$)	Pore diameter[c] (nm)	Pore volume ($ml\,g^{-1}$)	Reference(s)
I. Neutral surfactant templating						
MSU-1	$Al[OCH(CH_3)CH_2CH_3]_3$ (2-Butanol)	Tergitol® 15-S-9; $C_{11-15}[PEO]_9$	490	3.3	0.40	[5a, 5c]
MSU-2	$Al[OCH(CH_3)CH_2CH_3]_3$ (2-Butanol)	Triton® X-114; $C_8Ph[PEO]_8$	460	3.3	0.35	[5a, 5c]
MSU-3	$Al[OCH(CH_3)CH_2CH_3]_3$ (2-Butanol)	Pluronic® L64; $[PEO]_{13}[PPO]_{30}[PEO]_{13}$	430	2.4	0.21	[5a, 5c]
MSU-γ	$[Al_{13}O_4(OH)_{24}(H_2O)_{12}]Cl_7$ (Water)	Pluronic® P84; $[PEO]_{19}[PPO]_{39}[PEO]_{19}$	299	6.4	0.73	[5i]
MSU-γ	$Al[OCH(CH_3)CH_2CH_3]_3$ (2-Butanol)	Pluronic® P84; $[PEO]_{19}[PPO]_{39}[PEO]_{19}$	306	9.0	1.15	[5i]
DL12	$Al(NO_3)_3$ (Water; pH = 7.01)	Poly(ethylene glycol) 1540; $HO-(CH_2CH_2-O)_{1540}-H$	203	6.6	0.33	[5m]
DL54	$Al_2(SO_4)_3$ (Water; pH = 10.0)	Poly(ethylene glycol) 1540; $HO-(CH_2CH_2-O)_{1540}-H$	309	6.3	Not reported	[5m]
II. Anionic surfactant templating						
C_{A383}	$Al[OCH(CH_3)CH_3]_3$ (1-Propanol)	Caproic acid; $C_5H_{11}COOH$	530	2.1	0.31	[5b]
L_{A383}	$Al[OCH(CH_3)CH_3]_3$ (1-Propanol)	Lauric acid; $C_{11}H_{23}COOH$	710	1.9	0.41	[5b]
S_{A383}	$Al[OCH(CH_3)CH_3]_3$ (1-Propanol)	Stearic acid; $C_{17}H_{35}COOH$	700	2.1	0.41	[5b]

Table 15.1 Continued.

Mesoporous alumina denotation	Aluminum source (solvent)	Template: name; formula[a]	Surface area[b] ($m^2 g^{-1}$)	Pore diameter[c] (nm)	Pore volume ($ml\,g^{-1}$)	Reference(s)
III. Caticnic surfactart templating						
ICMUV-1	Al[OCH(CH$_3$)CH$_2$CH$_3$]$_3$ (Triethanolamine)	Cetyltrimethylammonium bromide; CH$_3$(CH$_2$)$_{15}$N(CH$_3$)$_3^+$ Br$^-$	250–340	3.3–6.0	Not reported	[5d]
I, II	Al$_2$Cl(OH)$_5$ (Water)	1-Methyl-3-octylimidazolium chloride;	262–269	3.8–3.9	0.25–0.26	[5n]
IV. Nonsurfactant templatirg						
Al-TUD-1	Al[OCH(CH$_3$)$_2$]$_3$ (Ethanol and 2-propanol)	Tetra(ethylene glycol); HO(CH$_2$CH$_2$O)$_3$CH$_2$CH$_2$OH	528	4.0	0.63	[5j]
Al$_2$O$_3$-X (X = 1–3)	Al[OCH(CH$_3$)$_2$]$_3$ (Water; pH = 4.5–5.5)	D-Glucose;	337–422	3.8–5.1	0.43–0.66	[5k]
ID 7	Boehmite sol (Water)	Citric acid; HOC(COOH)(CH$_2$COOH)$_2$	320	4.1	0.44	[5l]

a Abbreviations: PEO, poly(ethylene oxide); PPO, poly(propylene oxide).
b Determined by the Brunauer–Emmett–Teller (BET) method.
c Determined by the Barrett–Joyner–Halender (BJH) method.

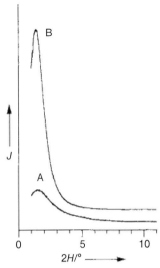

Figure 15.3 Powder X-ray diffraction patterns of the MSU-3 alumina prepared with a surfactant of Pluronic® L64. The intensity I is in arbitrary units. Spectrum A represents the as-synthesized sample after air drying at room temperature for 16 h. Spectrum B represents the sample calcined at 500 °C in air for 6 h. Reprinted with permission from Ref. [5a].

Figure 15.4 Transmission electron microscopy image of the MSU-1 alumina prepared with a surfactant of Tergitol® 15-S-9 and calcined at 500 °C. Scale bar = 60 nm. Reprinted with permission from Ref. [5a].

basal spacings, and wormhole TEM images of the rare earth-stabilized MSU-X aluminas were equivalent to those observed for the pure MSU-X aluminas, which demonstrated that the incorporation of rare earth ions had not altered the original mesopore structures. However, the surface area after calcination at 500 or 600 °C was 35% higher for the 5 mol% Ce^{3+}-stabilized MSU-1 alumina than for the pure MSU-1 alumina. In addition, the 1 mol% La^{3+}-stabilized mesostructures formed

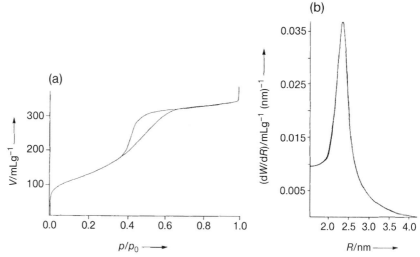

Figure 15.5 (a) N$_2$ adsorption–desorption isotherms for the MSU-3 alumina prepared with a surfactant of Pluronic® L64 and calcined at 500 °C in air for 4 h; (b) BJH pore-size distribution determined from the N$_2$ adsorption isotherm. Reprinted with permission from Ref. [5a].

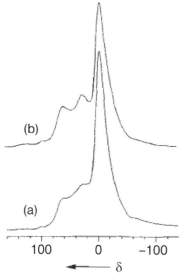

Figure 15.6 ^{27}Al MAS NMR spectra of the MSU-3 alumina prepared with a surfactant of Pluronic® L64. Spectrum A represents the as-synthesized sample after air drying at room temperature for 16 h. Spectrum B represents the sample calcined at 500 °C in air for 4 h. The chemical shifts are referenced to external [Al(H$_2$O)$_6$]$^{3+}$. Reprinted with permission from Ref. [5a].

by the Pluronic surfactants also exhibited high surface areas after calcination at 500 °C. The authors suggested that the incorporation of rare earth cations into the oxide framework took place during the assembly process of the mesostructures (but not during the calcination process), and that the site substitution of aluminum cations by rare earth cations reduces the lability of the oxide framework. These successes in increasing the thermal stability should greatly extend the potential applications of the MSU-X aluminas to catalytic reactions. González-Peña *et al.* later reported that thermally stable mesoporous aluminas could also be obtained by adding dipropylamine during formation of the surfactant–alumina mesophase [5f].

The synthesis of γ-Al$_2$O$_3$ with mesostructured forms (denoted MSU-γ) was also accomplished with the neutral surfactants by the group led by Pinnavaia [5i]. The key to synthesizing these mesostructured aluminas composed of γ-Al$_2$O$_3$ walls was the formation of a mesostructured surfactant/boehmite precursor (denoted MSU-S/B). The MSU-S/B (the structure of which was elucidated by XRD) could be formed by the hydrolysis of [Al$_{13}$O$_4$(OH)$_{24}$(H$_2$O)$_{12}$]Cl$_7$, AlCl$_3$, or Al(O-*sec*-Bu)$_3$ in the presence of a PEO/PPO copolymer surfactant such as Pluronic® P84, P123, and L64 with a formula of [PEO]$_{19}$[PPO]$_{39}$[PEO]$_{19}$, [PEO]$_{20}$[PPO]$_{69}$[PEO]$_{20}$, and [PEO]$_{13}$[PPO]$_{30}$[PEO]$_{13}$, respectively. Calcination of the MSU-S/B precursors at 550 °C afforded the MSU-γ aluminas, and their γ-Al$_2$O$_3$ walls were well-characterized by XRD, TEM, electron diffraction (ED), and ^{27}Al MAS NMR analysis. On the other hand, the XRD patterns in the small-angle region, N$_2$ adsorption–desorption isotherms, and BJH pore-size distributions of the MSU-γ aluminas were those typically observed for mesoporous materials. A TEM image of the MSU-γ alumina obtained through the hydrolysis of Al(O-*sec*-Bu)$_3$ in 2-butanol as solvent indicated that the nanoparticles were assembled into a scaffold-like array, whereas for those aluminas prepared from the Al$_{13}$ oligomer and AlCl$_3$ in aqueous media, parallel arrays were observed which gave a hierarchical, sheet-like morphology. The MSU-γ alumina from the alkoxide had thicker and longer nanoparticles and a larger pore size and pore volume compared to those from the other aluminum sauces. The pore size of the MSU-γ aluminas was also correlated with the size of surfactant, indicative of the presence of the surfactant-induced assembly process of nanoparticles during the synthesis.

Zhao *et al.* investigated the effect of the addition of poly(ethylene glycol) (PEG) on the traditional precipitation synthesis using aluminum nitrate or sulfate and ammonium carbonate as an aluminum source and a precipitator, respectively [5m]. PEG 1540 was first mixed with an aqueous solution of aluminum nitrate or sulfate, followed by the slow addition of ammonium carbonate, an adjustment of the pH of the mixture using ammonia, and aging. Subsequent filtration, washing with water, and drying yielded the as-made material, which was calcined at 550 °C in air to give the mesoporous alumina denoted DL12. A comparison of the TEM image of DL12 with that of the alumina synthesized in the absence of PEG showed that both materials exhibited a similar wormhole channel motif without any long-range order in their pore structure. The authors concluded that the PEG had not acted as a template in the synthesis process, and that the mesophase of

DL12 was formed by a self-assembly process of the inorganic precursor, which was more strongly affected by the ammonium carbonate precipitator. The main role of PEG as a surfactant in the synthesis was to create a slight enlargement of the mesopore size.

15.2.2.2 Anionic Surfactant Templating

The anionic surfactant templating mechanism involves the coordination of negatively charged surfactant molecules with aluminum sites on freshly formed aluminum hydroxides, as well as cylindrical assemblies of the surfactant alkyl groups to form the corresponding mesophases. Anionic surfactants of types $RCOO^-$, $R-XSO_3^-$ (R = long alkyl chain; X = O, C_6H_4, CH_2), and $R-OPO_3^{2-}$ are plausible candidates as structure-directing agents, although to date the successful synthesis (including complete removal of the surfactant) has only been achieved with carboxylic acid surfactants.

Vaudry and Davis, in 1996, reported that carboxylic acids with long alkyl chains could be used as templates for the alumina mesostructure assemblies [5b]. The important parameters were the type of solvent, the synthesis temperature, and the thermal history of the as-made material (i.e., temperature and atmosphere for calcination), in addition to the type of alkyl group in the templates. Branched or normal carboxylic acids from C_3 to C_{18} were found to afford mesophases, while suitable solvents were branched or normal alcohols, from C_1 to C_9. The best mesoporous alumina (denoted L_{A383}) was obtained by hydrolyzing aluminum *sec*-butoxide in 1-propanol solvent under the influence of lauric acid ($C_{11}H_{23}CO_2H$), followed by heating at 110 °C in a glass jar under static conditions. The solid was then filtered, washed with ethanol, and dried at room temperature to give the as-synthesized material, which was subsequently calcined at 430 °C to give the L_{A383} alumina. An N_2 adsorption–desorption isotherm and a BJH pore-size distribution of this alumina are shown in Figure 15.7.

The isotherm has no hysteresis loop and the distribution is centered at 1.9 nm. It was concluded, from the shape of distribution and the low-pressure hydrocarbon adsorption data (which showed that there was minimal difference between *n*-heptane uptake and neopentane uptake) that the mesoporous alumina L_{A383} possessed neither "zeolitic" micropores nor pores larger than 4.0 nm. Characterization of the L_{A383} alumina by XRD also showed a diffraction line in the small-angle region, which was attributed to the regular mesoporosity (Figure 15.8).

In addition, these figures show that the alumina is stable up to 800 °C, even under an air/water atmosphere. The shift to a higher *d*-spacing was, however, striking for the calcination in the presence of water, and indicated that sintering had taken place with increasing Al–OH groups to collapse the mesoporous structure. The authors referred to a mechanism of the alumina mesophase formation, in which very small AlO(OH) clusters with coordinated carboxylate ligands were first formed, followed by their rapid aggregation to produce the alumina mesophase (Scheme 15.1).

It should be noted that the alkyl chain of coordinated carboxylate is not in the straight-form but rather in bent-form. This is because the observed *d*-spacing of

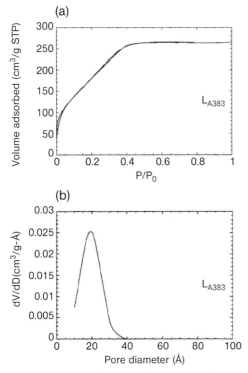

Figure 15.7 (a) N_2 adsorption–desorption isotherm and
(b) BJH pore-size distribution of the L_{A383} alumina calcined
at 430 °C for 1.5 h. Reprinted with permission from Ref. [5b].

the as-made material (2.9 nm) is much shorter than the sum of the estimated aggregate diameter, assuming a straight alkyl-chain model (3.1 nm) and the wall average width (ca. 1.0 nm). On the other hand, the diameter assuming a curved alkyl-chain model (1.9 nm) is equal to the d-spacing by adding the wall width. The minimal effect of surfactant size on pore diameter can be seen more clearly in the calcined samples (see Table 15.1, II). The factor which controls the porous properties involving pore diameter is the extent of dehydration of the inorganic walls by calcination, which should also prevent pore-size control by the choice of surfactant size. Dehydration through calcination also drastically lowered the local order and symmetry of the aluminum, increasing the amount of four- and five-fold coordinated aluminum at the expense of the six-fold type, as demonstrated by ^{27}Al MAS NMR studies (Figure 15.9; see Section 15.2.1.2 for the peak assignments).

The synthesis of mesoporous alumina using sulfate or sulfonate surfactants has been attempted by several groups, but has usually proved unsuccessful owing to the low thermal stability and difficulties of surfactant removal. Vaudry and Davis added sodium dodecylbenzenesulfonate dissolved in formamide to an aqueous solution of hydrolyzed aluminum *sec*-butoxide, followed by aging of the mixture

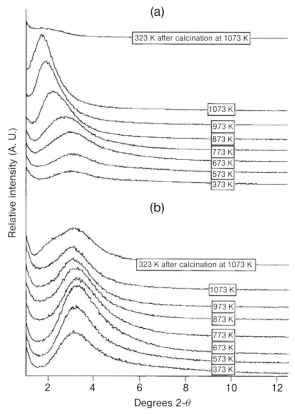

Figure 15.8 Variation of the X-ray diffraction patterns with increasing temperature under (a) an air/water atmosphere and (b) an air atmosphere. Reprinted with permission from Ref. [5b].

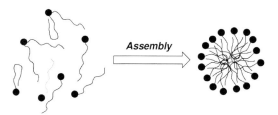

Scheme 15.1 Mechanism of the mesophase formation as proposed by Vaudry and Davis [5b]. The filled circles represent small clusters of AlO(OH), while the lines attached to the circles represent coordinated carboxylates.

at 110 °C for 2 days [5b]. The resultant solid was filtered, washed with deionized water, and dried at room temperature. The as-made material thus obtained exhibited a low thermal stability, losing the XRD line at a high *d*-spacing above a calcination temperature of 510 °C. In addition, the calcined solid still contained

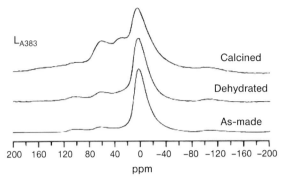

Figure 15.9 ^{27}Al MAS NMR spectra of the as-made, dehydrated, and calcined solid L$_{A383}$. Reprinted with permission from Ref. [5b].

substantial amounts of sulfur components that might have reduced the stability to high-temperature treatment. The sulfonate surfactant also could not be removed from the as-synthesized material by extraction with HCl/ethanol, HCl/water, or AcOH/water solutions.

The use of sodium dodecyl sulfate as a surfactant template was reported by Yada and coworkers [10], who noted that an aluminum-based dodecyl sulfate mesophase possessing a hexagonal structure such as MCM-41 could be synthesized by the homogeneous precipitation method, using aluminum nitrate nonahydrate and urea as an aluminum source and a pH-controlling agent, respectively. The as-made solid obtained from the mixture, of which the components were Al:surfactant:urea:water (1:2:30:60), exhibited an XRD line at a high d-spacing even after calcination at 600 °C, if it was heated up at a slow rate of 1 °C min^{-1}. However, the content of sulfur impurities in the calcined sample was not reported, and thus the purity remained doubtful. In addition, the surface area—which was in the range 93 to 365 m^2 g^{-1} depending on the calcination conditions—was much smaller than that of the mesoporous alumina synthesized with carboxylic acid surfactants.

The above two examples indicate that sulfate and sulfonate surfactants are not appropriate as structure-directing agents if the aim is to obtain pure mesoporous alumina with a high thermal stability. However, this approach may open the possibility that sulfated mesoporous alumina, which is an important material in the chemistry of strong acid catalysis, could be synthesized via a one-step process by using sulfate or sulfonate surfactants.

15.2.2.3 Cationic Surfactant Templating

Cabrera *et al.*, in 1999, first reported mesoporous alumina (denoted ICMUV-1) synthesized with a cationic surfactant template [5d]. A mixture of aluminum *sec*-butoxide and triethanolamine was added to cetyltrimethylammonium bromide (CTAB) dissolved in water, followed by aging of the resultant solution. The aged solid was filtered, washed with ethanol, dried, and calcined at 500 °C in air to give

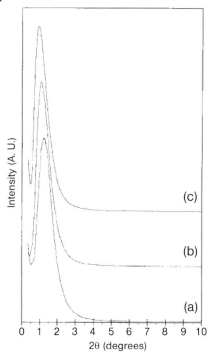

Figure 15.10 X-ray diffraction patterns of calcined ICMUV-1 aluminas synthesized with water/triethanolamine ratios of 7.4 (pattern a, $d = 6.9$ nm), 22.1 (pattern b, $d = 7.9$ nm), and 44.5 (pattern c, $d = 9.5$ nm). Reprinted with permission from Ref. [5d].

the ICMUV-1 alumina. The pore diameter and surface area of the alumina was found to be tailored by changing the Al:water:triethanolamine molar ratio in the solution from 2:111:15 to 2:195:4, while the Al:surfactant ratio was held at 2:1 for all the syntheses. Increasing the amount of water led to a monotonous increase in pore diameter, from 3.3 to 6.0 nm, whilst monotonously decreasing the surface area from 340 to 250 $m^2 g^{-1}$. The XRD patterns of some samples are shown in Figure 15.10.

Each sample exhibited an XRD line with a high d-spacing that was attributed to the organized mesopore structures, and these peaks remained unaltered with increasing temperature up to 900 °C, indicative of the high thermal stability of the mesoporous aluminas. The TEM images of ICMUV-1 aluminas showed the worm-hole-like channel motif, without any apparent order in the pore arrangement (Figure 15.11).

The increase in pore diameter with an increase in water:triethanolamine ratio can be clearly seen in the images. The ICMUV-1 aluminas were also characterized using the N_2 adsorption–desorption method (Figure 15.12). The hysteresis loop in the isotherm indicates the presence of ink-bottle-type pores, while the volume at

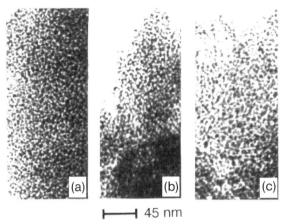

Figure 15.11 Transmission electron microscopy images of ICMUV-1 aluminas showing the wormhole-like channel motif. The aluminas were synthesized with water/triethanolamine ratios of 7.4 (a), 22.1 (b), and 44.5 (c). Reprinted with permission from Ref. [5d].

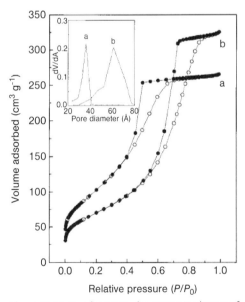

Figure 15.12 N_2 adsorption–desorption isotherms of ICMUV-1 aluminas prepared with water:triethanolamine ratios of 7.4:1 (isotherm a) and 44.5:1 (isotherm b). The corresponding BJH pore-size distributions are shown in the inset. Reprinted with permission from Ref. [5d].

the saturation vapor pressure demonstrates the absence of any macroporosity in the aluminas. It is interesting to note that the pore wall widths were approximately 3.5 nm, regardless of the water:triethanolamine ratio. Unfortunately, the contents of nitrogen and bromide impurities in the calcined solids were not reported.

The authors suggested that formation of the $(HOCH_2CH_2)_3N–Al[OCH(CH_3)CH_2CH_3]_3$ complex lowered the rate of hydrolysis, which allows the partially hydrolyzed inorganic moieties to be organized together with the surfactant molecules. In contrast, without triethanolamine, the hydrolysis of aluminum *sec*-butoxide proceeded too rapidly to be influenced by the surfactant, yielding the lamellar-type aluminum hydroxides. The fact that an increasing water:triethanolamine ratio increased the pore diameter should be attributed to a shift towards the formation of lamellar moieties.

Žilková and Čejka applied 1-methyl-3-octylimidazolium chloride to the synthesis of mesoporous alumina [5n]. To an aqueous solution of aluminum chlorohydrate and the cationic surfactant was added ammonium hydroxide solution drop-wise, followed by stirring, hydrothermal treatment, filtration of the precipitate, and washing the recovered solid with ethanol. The as-synthesized sample thus obtained was calcined at 560 °C with a temperature ramp of 1 °C min^{-1} to give the mesoporous alumina (denoted I–III, depending on the Al:surfactant molar ratio). The effect of modifying the Al:surfactant molar ratio on the structural parameters was investigated by comparing the results of N_2 adsorption–desorption of the calcined solids. The aluminas I and II prepared with a respective Al:surfactant ratio of 5.8 and 3.3 exhibited similar values of surface area, mesopore volume, and pore diameter, but increasing the ratio to 12.0 (alumina III) drastically decreased the surface area and increased the pore volume and pore diameter remarkably. The latter exhibited an N_2 adsorption–desorption isotherm that is typically observed for dispersed materials composed of small particles, and also showed a broad pore-size distribution without maximum, whereas a single XRD line at a low-angle region was observed for this alumina. Thus, the authors referred to the risk of judging the mesoporosity of materials based on XRD data alone. The aluminas I and II, on the other hand, exhibited not only the XRD lines but also the BJH pore-size distributions, with clear maxima centered at 3.8 and 3.9 nm, respectively. The XRD peaks at higher-angle regions indicated that alumina I was composed of γ-Al_2O_3. A wormhole framework structure was observed for the alumina I by TEM. The $S^+ X^- I^+$-type interaction, which has been discussed in detail elsewhere [6], was proposed for the alumina mesophase formation, where S^+, X^-, and I^+ represent the $C_{12}H_{23}N_2^+$, Cl^-, and cationic aluminum species, respectively.

15.2.2.4 Nonsurfactant Templating

Shan *et al.* succeeded in synthesizing mesoporous alumina with a high thermal stability by utilizing tetraethylene glycol as a nonsurfactant template [5j]. A controlled amount of water was added stepwise to a mixture of aluminum isopropoxide and tetraethylene glycol dissolved in ethanol and 2-propanol to produce a suspension, which was subsequently aged at room temperature and then dried in air at 60–100 °C. The solid gel thus obtained was heated at 120–190 °C in an autoclave and calcined at 600 °C to afford the mesoporous alumina (denoted Al-TUD-1). One of the most important parameters in the synthesis was the water : aluminum ratio. The wide-angle XRD patterns of the alumina prepared with an Al : H_2O molar ratio of 1 : 10–20 indicated that the bulk structure of the alumina contained crystalline phases of δ- or θ-Al_2O_3. Even though this alumina also gave an intensive diffraction line at about 1.0°, the corresponding TEM image did not show any regular mesopores in the alumina particles. On the other hand, decreasing the amount of water to an Al : H_2O molar ratio of 1 : 2 afforded the amorphous alumina with a sponge- or worm-like randomly three-dimensionally connected mesoporous network. The N_2 adsorption–desorption measurement of the latter alumina displayed the isotherm of type IV, with a hysteresis loop that indicated the presence of ink-bottle-like pores. In addition to the water content, the drying and heating conditions were also important, as they significantly affected the surface area and pore size of the final product. For instance, increasing the heating time of the solid gel resulted in a monotonous increase and decrease in pore size and surface area, respectively. Moreover, these two structural properties also could be tuned by changing the drying and heating temperature, although few examples were shown in the report and thus the tendencies remain unclear. The highlight of the Al-TUD-1 mesoporous alumina is its high stability toward high-temperature treatment. The alumina synthesized under the optimal conditions possessed a high surface area of 528 $m^2 g^{-1}$ after calcination at 600 °C, and of 414 $m^2 g^{-1}$ even after calcination at 700 °C, indicating the high potential of this material for practical uses.

Xu *et al.* synthesized mesoporous alumina in an aqueous medium using glucose as a structure-directing agent [5k]. To an aqueous solution of aluminum isopropoxide and glucose was added diluted aqueous nitric acid to adjust the pH value. After removal of the water by evaporation, the resulting solid material was calcined at 600 °C to give the mesoporous alumina (denoted Al_2O_3-X; X = 1–3). The pH value of the starting solution had a major effect on the structural property of the resultant alumina. The best alumina, Al_2O_3-2, was obtained when the solution pH was adjusted to 5.0, and showed a surface area, total pore volume, and BJH pore diameter of 422 $m^2 g^{-1}$, 0.66 $cm^3 g^{-1}$, and 5.1 nm, respectively. On the other hand, Al_2O_3-1 and 3, which were obtained from solutions at pH 4.5 and 5.5, exhibited smaller values of these parameters. The N_2 adsorption–desorption isotherm of Al_2O_3-2 was type IV, as typically observed for mesoporous materials; the isotherm contained a hysteresis loop, the shape of which was typical for ink-bottle-like pores. The most remarkable property of Al_2O_3-2 was its high thermal stability, with a high surface area of 225 $m^2 g^{-1}$ even after calcination at 800 °C. Such excellent thermal stability and simple preparation method imply a high potential for this material as a catalyst.

Zhang and coworkers reported the use of hydroxy carboxylic acids as structure-directing agents [5l]. Here, boehmite sol was first prepared by peptizing a water–boehmite suspension with nitric acid. To this sol was added a hydroxy carboxylic acid, and the mixture was then stirred at 30 °C. The resultant homogeneous mixture was dried and then calcined at 500 °C in air to give the corresponding mesoporous alumina (denoted ID 1–11). The textural properties of the mesoporous alumina depended heavily on the content of hydroxy carboxylic acid and the drying temperature before calcination. When citric acid was used as a structure-directing agent, a citric acid : Al^{3+} molar ratio in the range of 0.5 to 2.0, and a drying temperature of 100 °C, were found to be the optimal synthesis conditions. The use of excess citric acid decreased the surface area and pore volume of the mesoporous alumina, and led to a deterioration of the intrinsic crystalline structure of boehmite in the as-synthesized material. This effect was caused by the formation of larger aggregates of free citric acid molecules, as well as the enhancement of dissolution of boehmite particulates by the increased acidity of the synthetic mixture. The drying temperature, on the other hand, affected the pore structure by altering the coordination interaction between the citric acid molecules and boehmite. Raising the drying temperature increased the amount of hydroxy carboxylic acids coordinated with boehmite, but decreased the size of the aggregation of free hydroxy carboxylic acid molecules, owing to the decrease in free molecules. The size of the aggregation appeared to be crucial for the final structure of pores. Thus, the N_2 adsorption–desorption isotherm of the sample obtained by drying at 30 °C indicated the presence of larger mesopores, while that at 150 °C was analogous to the type I isotherm that is characteristic of microporous materials. Only the mesoporous alumina dried at 100 °C gave the type IV isotherm with a hysteresis loop that indicated the presence of ink-bottle-type pores. Unfortunately, the XRD patterns in the low-angle region were not reported, which makes the quality of the aluminas ambiguous.

15.3
Mesoporous Alumina in Heterogeneous Catalysis

Mesoporous alumina is superior to conventional alumina in that it possesses a larger surface area and organized mesopores that are not easily plugged by coke and other high-molecular-weight byproducts. The ideal structural properties of mesoporous alumina, however, are not always crucial factors when determining catalytic performance; the surface reactivity such as acidity and basicity is, in most cases, much more important. The reactivity of mesoporous alumina surface should depend heavily on the synthetic method, including the final calcination temperature, as is usually observed for most oxide catalysts. Unfortunately, most studies on mesoporous alumina have focused only on the synthesis and quality of mesoporous structure, and have not referred to the surface reactivity, which renders the potential for mesoporous alumina in catalysis ambiguous. At present, therefore, reliance must be placed on rather primitive trial-and-error methods to determine which mesoporous alumina is most suitable for the catalytic reaction under investigation. Nevertheless, continuous efforts are being made to utilize mesoporous alumina as a catalyst and catalyst support. In fact, the use of mesoporous alumina has in some cases led to better results than with conventional alumina, a topic which is introduced in the following sections (see Table 15.2).

Table 15.2 Survey of reactions catalyzed by mesoporous alumina and modified mesoporous alumina.

Catalyst[a]	Reaction equation	Conditions (Reactor type; solvent; reaction temperature; atmosphere)	Reference(s)
I. Base-catalyzed reactions			
A. Tishchenko reaction			
*meso*Al$_2$O$_3$; *meso*Al$_2$O$_3$/SO$_4^{2-}$		Batch; scCO$_2$ (8 MPa); 40 °C	[18]
B. Knoevenagel reaction			
*meso*Al$_2$O$_3$		Batch; scCO$_2$ (8 MPa); 40 °C	[19]
C. Double-bond migration			
Na/*meso*Al$_2$O$_3$		Batch; neat; 2 °C; *in vacuo*	[20]
Na/*meso*Al$_2$O$_3$		Batch; neat; rt; *in vacuo*	[20]

Table 15.2 Continued.

Catalyst[a]	Reaction equation	Conditions (Reactor type; solvent; reaction temperature; atmosphere)	Reference(s)
II. Epoxidation			
Au/mesoAl₂O₃	Ph⟶ **10** + TBHP ⟶ Ph—◁O **11**	Batch; benzene; 82–83 °C	[21]
M = Mn, Co mesoAl₂O₃ surface	Ph⟶ **10** + TBHP ⟶ Ph—◁O **11**	Batch; CH₃CN; 60 °C	[22]
	12 + TBHP ⟶ **13**		
III. Hydrodechlorination			
Ni/mesoAl₂O₃	**14** + H₂ ⟶ **15** + 2 HCl	Continuous flow; 300 °C	[23, 24]
Ni–Mg/mesoAl₂O₃[b]	**16** + m H₂ ⟶ hydrocarbons + n HCl	Continuous flow; 300 °C	[25]

IV. Hydrodesulfurization

MoS$_2$/*meso*Al$_2$O$_3$^{b} [thiophene, 17] + m H$_2$ ⟶ hydrocarbons + H$_2$S   Continuous flow; 280–400 °C; 1 MPa   [26]

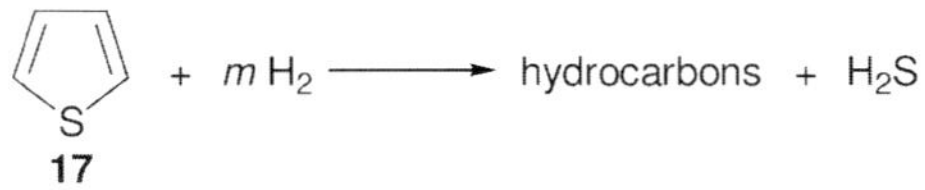

V. Olefin metathesis

A. Terminal hydrocarbon olefins

Re$_2$O$_7$/*meso*Al$_2$O$_3$ 2 [C$_6$H$_{13}$, 18] ⇌ [C$_6$H$_{13}$–C$_6$H$_{13}$, 19] + H$_2$C=CH$_2$ (20)   Batch; n-heptane; −1 °C; N$_2$; Batch; neat; 40 or 60 °C; Ar   [27–29]

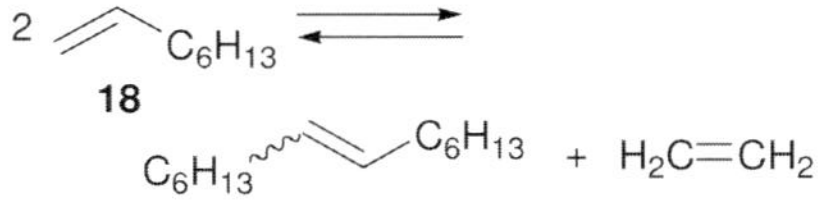

MTO/ZnCl$_2$//*meso*Al$_2$O$_3$ 2 [C$_6$H$_{13}$, 18] ⟶ [C$_6$H$_{13}$–C$_6$H$_{13}$, 19] + H$_2$C=CH$_2$ (20)   Batch; CH$_2$Cl$_2$; rt; N$_2$ [30]

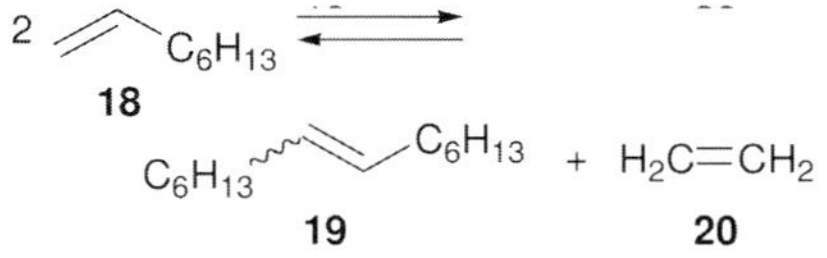

Re$_2$O$_7$/*meso*Al$_2$O$_3$ 2 [C$_8$H$_{17}$, 21] ⇌ [C$_8$H$_{17}$–C$_8$H$_{17}$, 22] + H$_2$C=CH$_2$ (20)   Batch; neat; 25, 40, or 60 °C; Ar   [28–31]

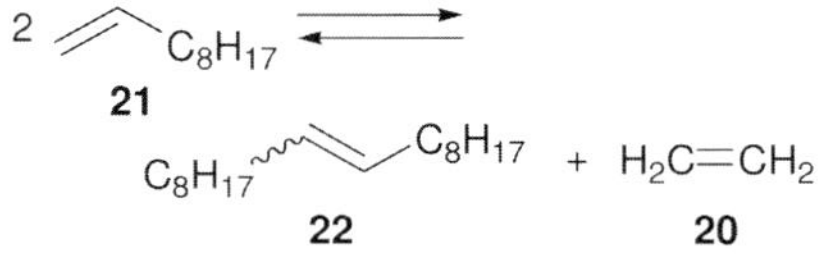

Re$_2$O$_7$/*meso*Al$_2$O$_3$ 2 [C$_{16}$H$_{33}$, 23] ⟶ [C$_{16}$H$_{33}$–C$_{16}$H$_{33}$, 24] + H$_2$C=CH$_2$ (20)   Batch; neat; 60 °C; Ar   [28]

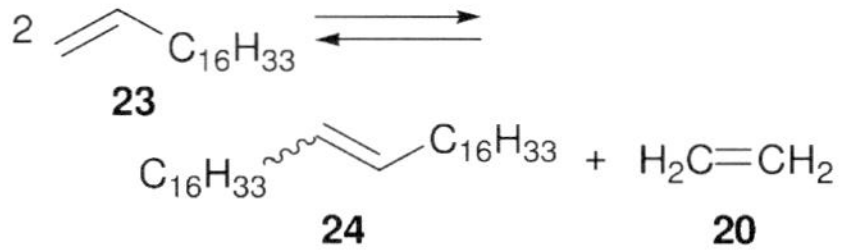

Re$_2$O$_7$/*meso*Al$_2$O$_3$ 2 [C$_4$H$_9$, 25] ⟶ [C$_4$H$_9$–C$_4$H$_9$, 26] + H$_2$C=CH$_2$ (20)   Batch; dodecane; 40 °C; N$_2$ [32]

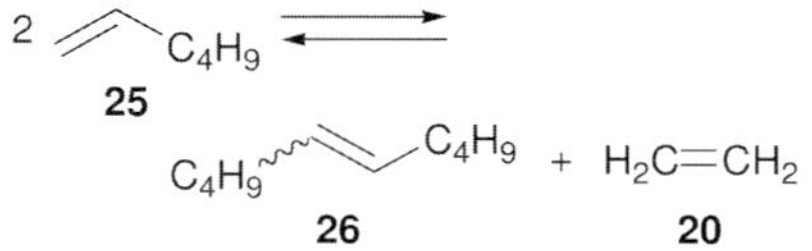

B. Internal hydrocarbon olefins

Re$_2$O$_7$/*meso*Al$_2$O$_3$ 2 [C$_6$H$_{13}$–C$_8$H$_{17}$, 27] ⇌ [C$_6$H$_{13}$–C$_6$H$_{13}$, 19] + [C$_8$H$_{17}$–C$_8$H$_{17}$, 22]   Batch; n-heptane; 50 °C; N$_2$ [27]

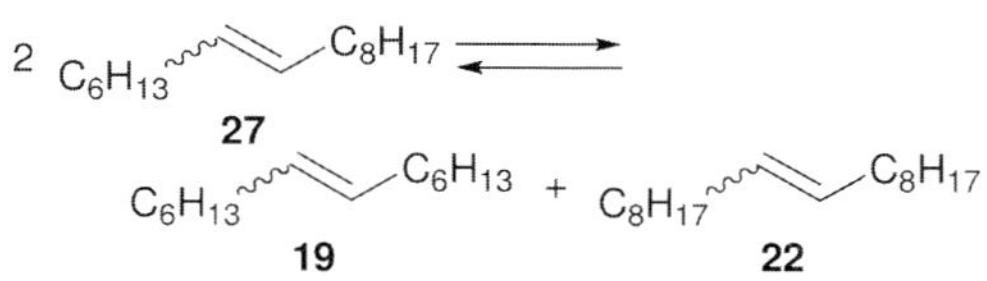

Table 15.2 Continued.

Catalyst[a]	Reaction equation	Conditions (Reactor type; solvent; reaction temperature; atmosphere)	Reference(s)
MTO/ZnCl₂//mesoAl₂O₃	$2\ C_6H_{13}\!-\!\!=\!\!-C_8H_{17}$ (**27**) ⇌ $C_6H_{13}\!-\!\!=\!\!-C_6H_{13}$ (**19**) + $C_8H_{17}\!-\!\!=\!\!-C_8H_{17}$ (**22**)	Batch; CH₂Cl₂; rt; N₂	[30]
Re₂O₇/mesoAl₂O₃	$2\ CH_3\!-\!\!=\!\!-C_2H_5$ (**28**) ⇌ $CH_3\!-\!\!=\!\!-CH_3$ (**29**) + $C_2H_5\!-\!\!=\!\!-C_2H_5$ (**30**)	Batch; neat; 25°C; Ar	[29]
Re₂O₇/mesoAl₂O₃	$2\ CH_3\!-\!\!=\!\!-C_5H_{11}$ (**31**) ⇌ $CH_3\!-\!\!=\!\!-CH_3$ (**29**) + $C_5H_{11}\!-\!\!=\!\!-C_5H_{11}$ (**32**)	Batch; neat; 40°C; Ar	[29]
C. Terminal functionalized olefins			
Re₂O₇/mesoAl₂O₃ + Me₄Sn	2 (**33**, MeO–C₆H₄–CH₂–CH=CH₂) ⇌ (**34**, MeO–C₆H₄–CH₂–CH=CH–CH₂–C₆H₄–OMe) + H₂C=CH₂ (**20**)	Batch; toluene or decane; 25°C; Ar	[31]
MTO/ZnCl₂//mesoAl₂O₃	2 (**35** $(n=8)$; **36** $(n=9)$, $CH_2=CH\!-\!(CH_2)_n\!-\!CO_2Me$) ⇌ $MeO_2C\!-\!(CH_2)_n\!-\!CH=CH\!-\!(CH_2)_n\!-\!CO_2Me$ (**37** $(n=8)$; **38** $(n=9)$) + H₂C=CH₂ (**20**)	Batch; CH₂Cl₂; rt; N₂	[30]

MTO/ZnCl₂//*meso*Al₂O₃

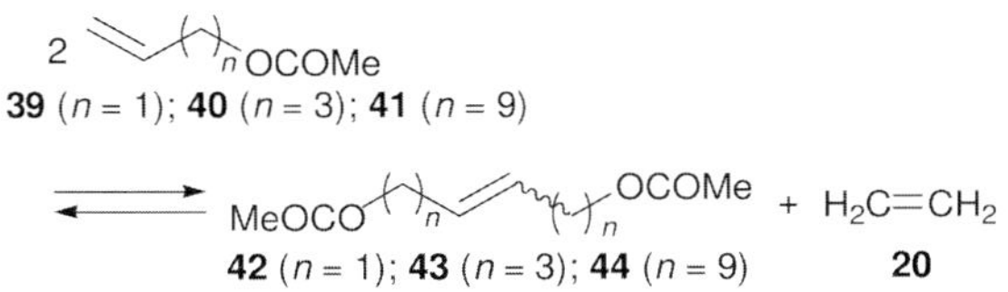

2 OCOMe

39 (n = 1); **40** (n = 3); **41** (n = 9)

MeOCO   OCOMe   +   H₂C=CH₂   **20**

42 (n = 1); **43** (n = 3); **44** (n = 9)

Batch; CH₂Cl₂; rt; N₂   [30]

MTO/ZnCl₂//*meso*Al₂O₃

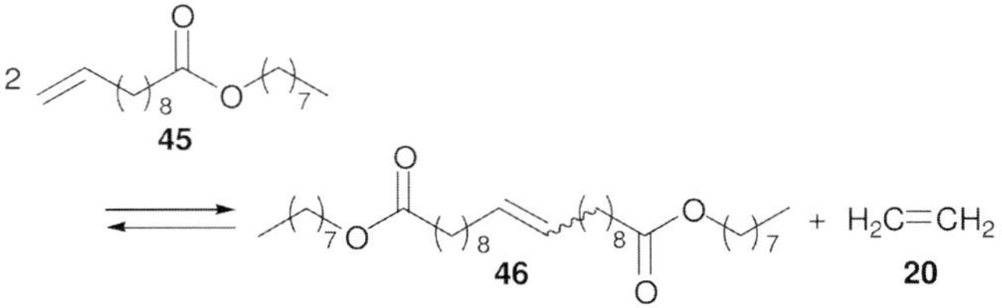

45

46   +   H₂C=CH₂   **20**

Batch; CH₂Cl₂; rt; N₂   [30]

MTO/ZnCl₂//*meso*Al₂O₃

47

48   +   H₂C=CH₂   **20**

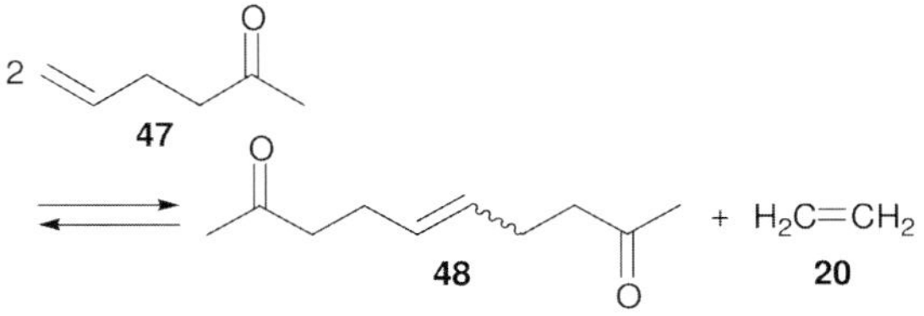

Batch; CH₂Cl₂; rt; N₂   [30]

Re₂O₇/*meso*Al₂O₃

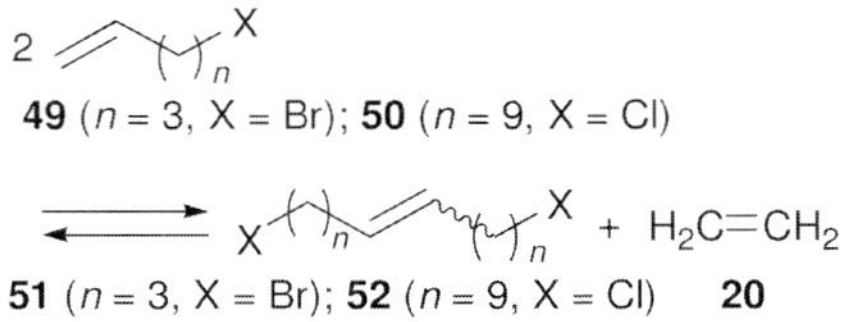

2 X

49 (n = 3, X = Br); **50** (n = 9, X = Cl)

X   X   +   H₂C=CH₂   **20**

51 (n = 3, X = Br); **52** (n = 9, X = Cl)

Batch; CH₂Cl₂; rt; N₂   [30]

D. Ring-opening metathesis polymerization (ROMP)

Re₂O₇/*meso*Al₂O₃

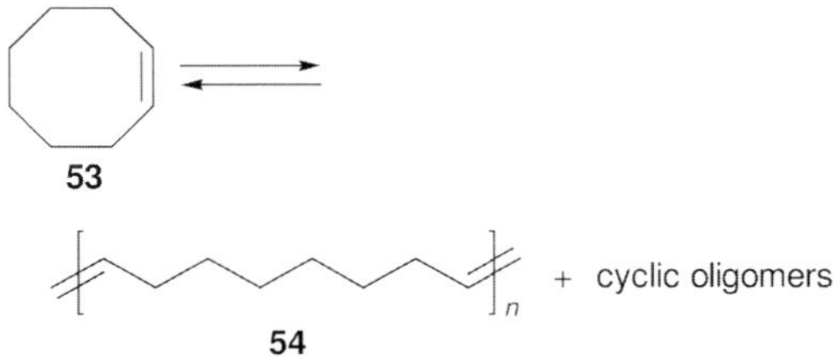

53

54   + cyclic oligomers

Batch; toluene; 40°C; Ar   [29]

Table 15.2 Continued.

Catalyst[a]	Reaction equation	Conditions (Reactor type; solvent; reaction temperature; atmosphere)	Reference(s)
E. Ring-closing metathesis (RCM)			
$Re_2O_7/mesoAl_2O_3$	**55** ⇌ **12** + $H_2C{=}CH_2$ **20**	Batch; neat or dodecane; 40°C; Ar	[29]
$Re_2O_7/mesoAl_2O_3$ + Me_4Sn	**56** ⇌ **57** (CO_2Et, CO_2Et) + $H_2C{=}CH_2$ **20**	Batch; toluene; 0–40°C; Ar	[31]
F. Polymerization of terminal dienes			
$Re_2O_7/mesoAl_2O_3$	n **58** ⇌ **59** + $n{-}1$ $H_2C{=}CH_2$ **20**	Batch; toluene; 25°C; Ar	[29]
$Re_2O_7/mesoAl_2O_3$	n **60** → **54** + $n{-}1$ $H_2C{=}CH_2$ **20**	Batch; toluene; 40°C; Ar	[29]

G. Cross metathesis

Re$_2$O$_7$/mesoAl$_2$O$_3$ Batch; neat; 40°C; Ar   [29]

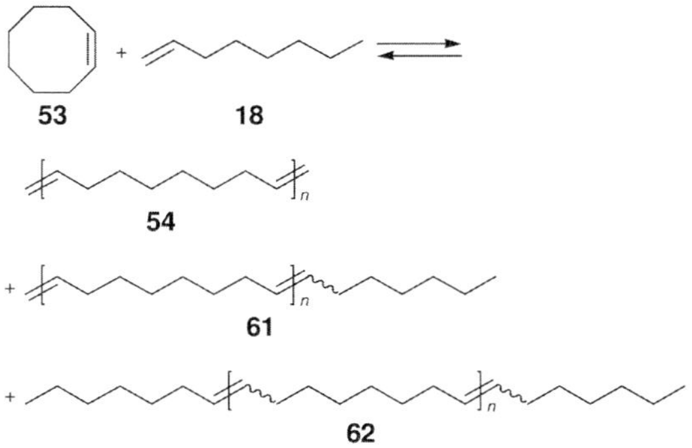

53 + **18** → **54**, **61**, **62**

Re$_2$O$_7$/mesoAl$_2$O$_3$ + Me$_4$Sn   Batch; toluene; 40°C; Ar   [31]

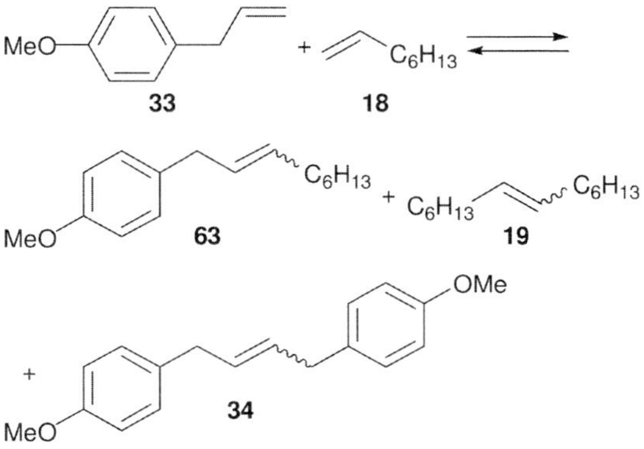

33 + **18** → **63** (C_6H_{13} + C_6H_{13}), **19**, **34**

VI. Oxidative dehydrogenation

V/mesoAl$_2$O$_3^c$
Mo–V/mesoAl$_2$O$_3^c$

$$H_3C{-}CH_3 + O_2 \longrightarrow H_2C{=}CH_2 + CO + CO_2$$
64   **20**

Continuous flow; 520–600°C; atmospheric pressure   [33, 34]

VII. Oxidative steam reforming of methanol

Pd–Zn/mesoAl$_2$O$_3^c$

$$CH_3OH + H_2O + O_2 \longrightarrow CO_2 + H_2 + CO$$
65

Continuous flow; 100–400°C: [35]

a   mesoAl$_2$O$_3$ represents mesoporous alumina.
b   Products were not reported.
c   Stoichiometries are not considered due to the complex reaction mechanism.
rt = room temperature.

15.3.1
Base-Catalyzed Reactions

Thermally activated alumina has long been recognized as a solid base catalyst [7b]. In fact, studies of the use of mesoporous alumina as a solid base catalyst were initiated by Seki and Onaka in their attempt to develop heterogeneous basic catalysis in supercritical CO_2 (scCO_2). These authors found that mesoporous alumina synthesized in a similar manner to L_{A383} (see Table 15.1, II) and activated at 500 °C under vacuum (10^{-4} Torr) exhibited a strong base catalysis for the Tishchenko reaction, even in the Lewis acidic medium of scCO_2 (Table 15.2, I, A) [18]. In addition, the activity could be increased by introducing SO_4^{2-} ions into the alumina framework. Conventional γ-Al_2O_3 and CaO, in contrast, showed quite low activities in scCO_2, even though they were very active in benzene as solvent under nitrogen atmosphere. The phthalide **2** formation over the sulfated mesoporous alumina was accelerated remarkably by adding a small amount of tetrahydrofuran (THF) as a cosolvent, whereas the addition of acetic acid (or even much less-acidic methanol) led to a fatal deactivation of the catalyst due to strong poisoning. The former result was attributed to the improvement of solubility of phthalaldehyde **1** in scCO_2, while the latter result suggested that the strong base sites on sulfated mesoporous alumina had promoted the reaction in scCO_2. The infrared (IR) spectroscopy of adsorbed pyrrole implied that the average strength of base sites on sulfated mesoporous alumina was weaker than on conventional γ-Al_2O_3. However, the poisoning by methanol indicated that it still had a small number of strong base sites that functioned even in scCO_2. It was concluded that the CO_2 molecules adsorbed onto the sulfated mesoporous alumina were less stable compared to those on conventional γ-Al_2O_3, due to the weaker surface basicity of the former, and thus could offer the active strong base sites to **1** much more easily through the surface diffusion, which in turn made a noticeable difference in activity.

In a subsequent study, the same authors reported that mesoporous alumina also exhibited a higher activity for the Knoevenagel reaction between benzaldehyde **3** and ethyl cyanoacetate **4** compared to conventional γ-Al_2O_3 in scCO_2 (Table 15.2, I, B) [19]. Although the surface area of the mesoporous alumina employed was 2.7-fold larger than that of conventional form, the numbers of active base sites on the aluminas were almost equal according to the results of temperature-programmed desorption (TPD) of CO_2. This indicated that the density of active base sites was much smaller for mesoporous alumina than for conventional γ-Al_2O_3. The authors thus suggested that the surface diffusion of CO_2 took place smoothly on mesoporous alumina owing to the large vacancy around the sites, while the CO_2 molecules adsorbed on the active base sites of conventional γ-Al_2O_3 were jam-packed, thus retarding the surface diffusion with each other. As free active base sites should appear by moving the adsorbed CO_2, it was concluded that the difference in facility of the surface diffusion of CO_2 accounted for the difference in activity between the two aluminas. In the same study, a careful examination was also made of the CO_2-TPD in the high temperature range of 527–1000 °C. The results implied that mesoporous alumina had a small number of very strong base sites, the strength of which was even higher than that of the base sites on conventional γ-Al$_2$O$_3$.

The double-bond migration of olefins is a simple, but important, process in chemical industry. Although both acid and base catalysts can promote double-bond migrations, base catalysts are usually much more selective, especially when cyclic olefins are used as the starting compounds. In addition, increasing the surface basicity increases the reaction rate, allowing the migrations to be performed even at a low reaction temperature (i.e., below zero region) by using superbase catalysts. Seki *et al.* attempted to enhance the surface basicity of mesoporous alumina by blowing an alkaline metal vapor against the activated alumina surface to improve the catalytic performance for the double-bond migrations (Table 15.2, I, C) [20]. It was revealed, via XRD and N$_2$ adsorption–desorption analysis, that the sodium addition hardly collapsed the intrinsic mesoporosity and bulk structure of mesoporous alumina. Unfortunately, the double-bond migration of 2,3-dimethyl-1-butene **6** proceeded more slowly over sodium-doped mesoporous alumina than over sodium-doped conventional γ-Al$_2$O$_3$, which should be brought about by the lower density of active strong base sites on mesoporous alumina compared to conventional alumina. Nonetheless, sodium-doped mesoporous alumina could selectively yield 2,3-dimethyl-2-butene **7** at high yield (81%), even at a low reaction temperature (2 °C), demonstrating that very strong base sites–including superbase sites–existed on the surface. The catalyst also promoted the reaction of α-pinene **8** to give β-pinene **9** in excellent selectivities of 93–96%, while the selectivity fell to 58% and 89% for parent mesoporous alumina and sodium-doped conventional γ-Al$_2$O$_3$, respectively.

15.3.2
Epoxidation

Yin and coworkers investigated the catalysis of gold nanoparticles supported on mesoporous alumina for the epoxidation of styrene **10** with *tert*-butyl hydroperoxide (Table 15.2, II) [21]. Three types of mesoporous alumina were synthesized via neutral surfactant templating (La^{3+}-doped MSU-3) [5c], cationic surfactant templating (ICMUV-1; Table 15.1, III) [5d], and an original method using chitosan as the structure-directing agent. In addition, commercially available γ-Al$_2$O$_3$ was employed for comparative purposes. Gold nanoparticles were subsequently loaded onto the aluminas via a homogeneous deposition precipitation method using HAuCl$_4$ and urea as a gold source and precipitation agent, respectively. The as-made materials were reduced under an H$_2$/He flow before being used for the reactions. Among the catalysts thus prepared, the Au loaded on ICMUV-1 exhibited the highest catalytic performance, yielding styrene oxide **11** in 84.3% conversion and 69.0% selectivity. The major byproducts were benzaldehyde and phenyl acetaldehyde, which were given in 23.0% and 3.6% selectivity, respectively. Although the other catalysts yielded **11** in similar selectivities, they were inferior to Au/ICMUV-1 in terms of conversion. In addition to the high activity, the Au/ICMUV-1 catalyst exhibited a long lifetime and could be reused at least eight times, without losing its high catalytic performance. X-ray photoelectron spectroscopy (XPS) measurements revealed that the preparation method gave Au(0) particles without any trace of higher-oxidized species, regardless of the type of support alumina. On the other hand, the CO_2-TPD profiles and isoelectric points of the alumina supports showed that the ICMUV-1 alumina had the largest number of basic sites, which was advantageous for the formation and stabilization of smaller gold particles. The TEM image, in fact, showed that the size of gold particles on Au/ICMUV-1 was uniform and smaller (<4 nm) compared to those of other catalysts. The authors thus concluded that the presence of smaller gold particles with a uniform size accounted for the highest activity of Au/ICMUV-1, because the active sites for the epoxidation were considered to be low-coordinated Au atoms existing at the corner and edge of the particles. It was also noteworthy that the pure ICMUV-1 alumina could catalyze the epoxidation, giving **11** in 50.3% conversion and 66.9% selectivity.

Chaube *et al.* immobilized Mn– and Co–salen complexes onto mesoporous alumina and applied them to the catalytic epoxidation of **10** and cyclohexene **12** with *tert*-butyl hydroperoxide (Table 15.2, II) [22]. A synthetic method similar to that for the L_{A383} (Table 15.1, II) was employed, but aluminum isopropoxide was used as an aluminum source. The mesoporous alumina thus obtained was first treated with 3-aminopropyltriethoxysilane to give the aminopropyl-modified mesoporous alumina, which was subsequently reacted with the precursor complexes, (bis-salicyl aldehyde)M (M = Mn, Co), to give the target catalysts. The IR band assignable to the C=N bond then appeared, indicating that the bis-salicyl aldehyde ligand had been converted into the salen-type ligand. The change in ligand was also demonstrated by the shift of characteristic peaks in UV-visible spectrum. A single XRD line in the low-angle region appeared, even after immobilization of the (bis-salicyl aldehyde)Co complex. In addition, the N_2 adsorption–desorption isotherm of type IV was obtained for the immobilized Co–salen complex catalyst, as well as for the parent mesoporous alumina. Moreover, thermal analysis indicated that the immobilized complexes decomposed at higher temperatures compared to the precursor complexes, which implied that the stabilities had been enhanced by the immobilizations. Thus, the entrapment of complex moieties inside the mesopores was proposed. The immobilized complexes were firmly fixed and did not dissolve into the solution during the reactions. In addition, the Mn catalyst was found to be used at least three times, without any significant loss in activity and selectivity. Unfortunately, however, the yields of the desired **11** (14–16%) and **13** (9–16%) were too low, rendering the potential of these catalysts for practical epoxide synthesis questionable.

15.3.3
Hydrodechlorination

Hydrodechlorination involves the cleavage of a C–Cl bond with H_2 under the influence of a catalyst, to yield useful hydrocarbons from hazardous chlorinated alkanes.

Kim *et al.* prepared nickel supported on mesoporous alumina and applied this to the hydrodechlorination of 1,2-dichloropropane **14** to propene **15** (Table 15.2, III) [23]. The mesoporous alumina was synthesized using stearic acid as a surfactant, using what has been termed a "post-hydrolysis" method [36]. The procedure was similar to that for the synthesis of S_{A383} alumina (Table 15.1, II), except that the hydrolyzed solution was dried directly without thermal treatment to produce the as-synthesized material that was subsequently calcined at 450 °C. Nickel was loaded onto the mesoporous alumina using two different methods. The first method involved the vapor deposition of nickel acetylacetonate (denoted Ni-VD), followed by calcination, while the second method involved impregnation using an appropriate amount of aqueous solution of nickel nitrate (denoted Ni-IMP). Characterization by XPS, temperature-programmed reduction (TPR), and UV-diffuse reflectance spectroscopy (UV-DRS) revealed that the NiO species were dominant on the Ni–VD catalyst, whereas a large amount of Ni^{2+} ions was incorporated into the nickel aluminate spinels for Ni–IMP. It was proposed that such spinels were formed through the impregnation step in which a mixed adsorption occurred between Ni^{2+} ions of nickel nitrate and Al^{3+} ions that once were dissolved into the aqueous solution from the alumina surface. Both of the catalysts promoted the hydrodechlorination of **14** in a continuous-flow fixed-bed reactor, maintaining approximately 83% selectivity for at least 10 h. However, the conversion with Ni–VD was approximately twofold higher than that with Ni–IMP. The authors concluded that the NiO species could be reduced much more smoothly into active nickel metals compared to nickel aluminate species, as demonstrated by the TPR study, and thus Ni–VD exhibited a higher activity than Ni–IMP. The larger number of active nickel sites on Ni–VD was also clearly seen in the TPD measurement of 1,2-dichloropropane. Later, the same authors reported that the activity of Ni–IMP depended heavily on the molar ratio of surfactant/$Al[OCH(CH_3)CH_2CH_3]_3$ in the preparation of the mesoporous alumina support [24]. It was observed that the amount of dissolved Al^{3+} ions during the impregnation increased as the ratio increased, leading to the formation of a larger amount of unfavorable nickel aluminate species. In fact, the Ni–IMP prepared with a lower ratio tended to exhibit a higher activity for the hydrodechlorination of **14**.

More recently, Kim *et al.* prepared nickel- and magnesium-containing mesoporous alumina as a catalyst for the hydrodechlorination of *o*-dichlorobenzene **16** (Table 15.2, III) [25]. Bimetallic nickel and magnesium stearate were used as structure-directing agents, as well as nickel and magnesium sources; these were first dissolved in 2-butanol with aluminum *sec*-butoxide, and the subsequent addition of water, drying, and calcination at 500 °C yielded the catalyst. The bimetallic stearate was obtained by dissolving magnesium stearate and nickel nitrate in 2-butanol, after which the pH was adjusted with HCl or NH_4OH. The use of NH_4OH resulted in the precipitation of a solid bimetallic stearate, which yielded the most active catalyst (denoted NiMg–NP). A TPR study clearly demonstrated the presence of homogeneously mixed Ni metals on NiMg–NP that were the active sites for the hydrodechlorination, whereas on the catalyst prepared via the HCl-treated surfactant (denoted NiMg–HS) there existed undesirable nickel aluminate species, in addition to nickel metal aggregates. Another factor that rendered NiMg–NP most active was a favorable modification of the electronic properties of Ni metals on NiMg–NP, caused by homogeneously mixed Mg. This was confirmed by H_2-TPD profiles which revealed that only NiMg–NP showed an intense desorption peak below 400 °C; this indicated that the Mg species had weakened the interaction between Ni metal surface and hydrogen. The Mg species also played an important role in extending the lifetime of the catalyst through the reaction with HCl; here, HCl was trapped on a partially formed basic MgO surface, inhibiting the conversion of active Ni metals into $NiCl_2$.

15.3.4
Hydrodesulfurization

Hydrodesulfurization is an industrially very important catalytic process for the removal of sulfur from natural gas and refined petroleum products. Such removal should lead to a reduction in SO_2 emissions from these fuels, and also extend the lifetime of the noble metal catalysts used in the subsequent reforming processes. Kaluža *et al.* employed the S_{A383} mesoporous alumina (Table 15.1, II) as a support for an active MoS_2 species, although the calcination conditions were carefully modified from those reported originally (Table 15.2, IV) [26]. The mesoporous alumina, when finally calcined at 450 °C, was found to serve as the best support. MoO_3 as a precursor was loaded onto the alumina by two different methods. The first method involved impregnation using an aqueous solution of $(NH_4)_6Mo_7O_{24}$, followed by drying at 50 °C under vacuum (denoted 30Mo/MA(CIM); 30 wt% MoO_3). The second method involved thermal spreading, where the mesoporous alumina was first ground with an appropriate amount of MoO_3 in an agate mortar and subsequently calcined at 500 °C in a stream of air (denoted 30Mo/MA(TSM); 30 wt% MoO_3). In both cases, the loaded MoO_3 was converted into active MoS_2 by treatment with an H_2S/H_2 mixture at 400 °C. The N_2 adsorption–desorption measurements revealed that the added MoO_3, despite a very high loading, did not significantly block the texture of the parent alumina support. The rate constant of 30Mo/MA(CIM), when normalized to weight for the hydrodesulfurization of thiophene 17, was almost twice as high as that of the commercial 15 wt% MoO_3/Al_2O_3 catalyst, whereas the rate constants of the two catalysts normalized to moles of Mo were almost equal. The authors concluded that the high surface area of mesoporous alumina was able to disperse a significantly higher amount of Mo than the conventional alumina. Although, the activity of 30Mo/MA(TSM) was inferior to that of 30Mo/MA(CIM), its rate constant (when normalized to weight) was still approximately 1.5-fold higher than that of the commercial catalyst, indicating that the high loading of MoO_3 could be realized also by thermal spreading. It should also be noted that 30Mo/MA(TSM) exhibited a long lifetime compared to the commercial catalyst for the hydrodesulfurization conducted at 370 °C.

15.3.5
Olefin Metathesis

Olefin metathesis involves the mutual exchange of alkylidene fragments between two olefin molecules by the action of metal catalysts. Its use in organic synthesis, particularly of polymers and pharmaceuticals with various functional groups, has been explosively expanded by the respective discoveries of Mo- and Ru-based alkylidene complex catalysts by Schrock [37a] and Grubbs [37b], which led to these authors receiving the Nobel Prize in 2005. Heterogeneous olefin metathesis catalysts such as WO_3/SiO_2, MoO_3/Al_2O_3, and Re_2O_7/Al_2O_3, on the other hand, have long been important in the petroleum industry for the manufacture of various olefins of pure hydrocarbons.

The use of mesoporous alumina in olefin metathesis was first attempted by Oikawa and Onaka (Table 15.2, V) [27]. These authors dispersed rhenium oxide, Re_2O_7, on the L_{A383} alumina (Table 15.1, II, calcined finally at 600 °C), using an impregnation method from the aqueous solution of NH_4ReO_4, with subsequent calcination at 600 °C. Despite water being used as solvent, the loading did not significantly alter the structural properties of the parent support. The catalyst (denoted $Re_2O_7/meso$-Al_2O_3) thus obtained was activated under vacuum at 500 °C; this step was found to be crucial for achieving a high conversion and selectivity. Compared to the conventional, 7 wt% Re_2O_7/γ-Al_2O_3 catalyst, 7 wt% $Re_2O_7/meso$-Al_2O_3 exhibited a higher conversion and selectivity for the metathesis of 1-octene **18** to 7-tetradecene **19** and ethylene **20** and of 7-hexadecene **27** to 7-tetradecene **19** and 9-octadecene **22**. In the case of the latter reaction, the lower conversion with a conventional catalyst was attributed to deactivation of the catalyst. The effect of changing the loading amount of Re_2O_7 was investigated for the metathesis of **18**, and showed that 7 wt% $Re_2O_7/meso$-Al_2O_3 exhibited a higher conversion and selectivity than 3.5 and 15 wt% $Re_2O_7/meso$-Al_2O_3. Characterization using X-ray adsorption spectroscopy (XAS) showed that rhenium oxide on both 7 wt% $Re_2O_7/meso$-Al_2O_3 and 7 wt% Re_2O_7/γ-Al_2O_3 were in the same oxidation state, and had the same structure. It was also revealed that the ReO_4 species on the catalysts were greatly distorted from a regular tetrahedron by their fixation onto the alumina surfaces. The authors proposed that mesoporous alumina could produce a larger amount of active Re ions in partially reduced states compared to the conventional alumina, and that the difference in the amount of Re sites caused the difference in activity. An active rhenium site model, as depicted in Figure 15.13, was suggested, in which a rhenium ion is electronically activated not only by the Re–O–Al bond derived from an acidic hydroxyl group of mesoporous alumina but also by the coordination of a Re=O moiety to an adjacent Lewis acid site of Al^{3+}.

Balcar *et al.* employed the thermal spreading method to prepare rhenium oxide supported on mesoporous alumina (denoted Re/OMA) [28]. In this method, mesoporous alumina was first ground with NH_4ReO_4 at room temperature, and then calcined at 550 °C. Before use in a reaction, the catalyst was activated at 500 °C in air. The rhenium oxide loading did not significantly affect the pore size of the parent mesoporous alumina. In addition, the decrease in surface area caused by

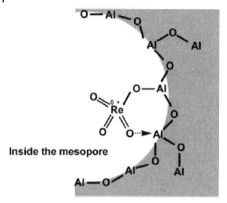

Inside the mesopore

Figure 15.13 Oikawa–Onaka model of surface rhenium oxide species on mesoporous alumina. Reprinted with permission from Ref. [27b].

the loading was much smaller for mesoporous alumina than for conventional alumina; this was due to the absence of micropores in the former case that may easily be plugged by rhenium oxide species. In the 1-decene **21** metathesis, the optimum Re loading was found to be between 9 and 15 wt%, and that the conversion depended heavily on the pore size of the catalysts. The rhenium oxide on conventional alumina (denoted Re/ALCOA) that had micropores exhibited a much lower conversion compared to Re/OMA. In addition, higher conversions were observed with Re/OMA with larger mesopores. Although the mesoporous alumina supports with different pore sizes were prepared by different methods, and thus the surface chemical properties might also be more or less different, it is reasonable to conclude that the pore size is the major factor that influences the conversion. The authors suggested that the active Re sites in larger pores may be much more easily accessible for long-chain olefin substrates, and that the larger pores were also favorable for product release from the mesoporous network.

The same group subsequently applied the Re/OMA catalysts to the metathesis of dienes and cycloalkenes to determine the effect of pore size on the product distributions [29]. One example that clearly showed the effect was the metathesis of 1,7-octadiene **55** to yield cyclohexene **12** and a small amount of 1,7,13-tetradecatriene. For this reaction, two types of mesoporous alumina-supported 9.3 wt% Re catalysts were tested; these were prepared using mesoporous aluminas with different pore diameters of 3.5 nm (denoted Re/OMA3.5) and 6.5 nm (denoted Re/OMA6.5). The yield of 1,7,13-tetradecatriene reached a maximum of approximately 5% in 25 min at 40 °C over both catalysts, but began to decrease after 25 min over Re/OMA6.5, while remaining constant for 250 min with Re/OMA3.5. This indicated that the large mesopores of Re/OMA6.5 were sufficient for the triene molecules to re-enter and react with the internal Re species to give **12**. Another example was seen in the reaction of 1,9-decadiene **60** to the corresponding dimers, trimers, and tetramers (**54**, $n = 2–4$), which was carried out in toluene solvent using the same catalysts. The formation of these compounds occurred much more rapidly with Re/OMA6.5 than with Re/OMA3.5. In addition, the final amount of products was decreased in the order of dimers > trimers > tetramers > oligomers over Re/OMA3.5, whereas the order was changed to tetramers > trimers > dimers > oligomers over Re/OMA6.5. It is also noteworthy that the final conversion was higher for Re/OMA6.5 than for Re/OMA3.5. These results demonstrated that the larger mesopore of the Re/OMA6.5 was favorable for the reaction of larger-sized molecules. The authors stressed that the use of mesoporous alumina supports with different pore sizes provided the opportunity to tune the product distributions in olefin metathesis by using a "molecular sieving" effect.

Quite recently, Balcar *et al.* reported the metathesis of oxygen-containing terminal olefins promoted by the Re/OMA catalysts and Me₄Sn as cocatalyst [31]. The main objective was, as in the former studies, to determine the effect of pore size on catalysis. The metathesis of *p*-allylanisole **33** at 25 °C in decane solvent proceeded only when Me₄Sn was present in the reaction mixture, and gave the corresponding product **34** in 100% selectivity, regardless of the pore size of the mesoporous catalysts. Conversion at the initial and final stages of the reaction, on the other hand, was increased with as the pore size of the catalyst increased. It is also noteworthy that the conventional Re/ALCOA catalyst with micropores exhibited almost no activity for the metathesis at 25 °C in toluene solvent, whereas the conversions given by the Re/OMA catalysts under the same conditions exceeded 80% in 5 h, indicating that the presence of organized mesopores (and particularly of larger mesopores) is crucial in order to achieve a high catalytic performance. This tendency was the same as observed in previous studies of nonfunctionalized 1-decene metathesis [28], and was also seen in the cross-metathesis between **33** and **18**. The ring-closing metathesis (RCM) of diethyl diallylmalonate **56** was also tested with the Re/OMA–Me₄Sn catalytic system. Under optimal conditions (i.e., a substrate : Re molar ratio of 40 : 1; reaction temperature 25 °C), a conversion of over 75% and 100% selectivity to **57** were achieved in 2 h with the Re/OMA catalysts, of which the parent supports had pore diameters of 4–7 nm. In contrast, a much lower conversion (ca. 50%) was observed for the Re/OMA prepared from the mesoporous alumina with a 3 nm pore diameter, again indicating the importance of larger mesopores also for this RCM reaction. It is of interest to note that **56** deactivated the Re/OMA catalysts more intensively than did **33**. Although the reason for this was not proposed, the present authors suggest that the strong interaction of the active Re sites with a polarized carbonyl group in **56** would inhibit the metathesis.

Aguado *et al.* investigated the metathesis of 1-hexene **25** over rhenium oxide supported on a commercially available alumina, and their original mesoporous alumina [32]. The commercial material was the MSU-type alumina synthesized with nonionic polyethylene oxides (Table 15.1, I), whereas the original alumina was prepared by the hydrolysis of aluminum isopropoxide in the presence of a cationic surfactant of CTAB at pH ≈ 1; the latter mixture was then heated at 80 °C, dried at 110 °C, and calcined at 550 °C. Rhenium oxide was loaded onto the aluminas (using an impregnation method) from an aqueous solution of $HReO_4$, followed by drying at 110 °C, calcination under a dry air flow, and cooling to room temperature under a dry N_2 flow. The decrease in surface area after rhenium oxide loading was noticeable for the commercial mesoporous alumina (MSU), with the value changing from 317 to 187 $m^2 g^{-1}$. In contrast, a much smaller decrease, from 300 to 274 $m^2 g^{-1}$, was observed for the original alumina (SGAL). The pore sizes, on the other hand, were enlarged from 5.3 to 7.4 nm and from 6.9 to 8.5 nm for MSU and SGAL, respectively. The activity order for the metathesis of **25** performed at 40 °C, $Re_2O_7/MSU > Re_2O_7/SGAL \gg Re_2O_7/\gamma\text{-}Al_2O_3$ (conventional), clearly demonstrated the superiority of mesoporous alumina as a support for rhenium oxide, where 9 wt% of Re_2O_7 was loaded onto each alumina. Selectivity for the self-metathesis product **26** exceeded 99% for all of these catalysts. It is also noteworthy that rhenium leaching was not observed by inductively coupled plasma (ICP) analysis for all of the reactions. The effect of modifying the solvent and Re_2O_7 content in $Re_2O_7/SGAL$ revealed dodecane to be the best solvent for the metathesis, and the optimum loading to be 6–10 wt%. The authors examined the surface acidity of the catalysts in detail. Subsequent ^{31}P MAS NMR spectroscopy using triethylphosphine oxide as a probe molecule indicated that the intensity of the peak at approximately 80 ppm, which is attributed to the strongest Lewis acids sites, increased in the order of $Re_2O_7/MSU > Re_2O_7/SGAL \gg Re_2O_7/\gamma\text{-}Al_2O_3$, which was also in good agreement with the activity order for the metathesis. The absence of Brönsted acid sites on the catalysts was confirmed by the diffuse reflectance infrared Fourier transform (DRIFT) spectra of adsorbed pyridine. The authors concluded that the metathesis activity was mainly governed by the surface Lewis acidity, and insisted that the importance of the interaction between the monomeric rhenium species and the Lewis acid Al^{3+} sites on mesoporous alumina would generate active rhenium sites, as had been proposed by Oikawa *et al.* (see Figure 15.13).

It seems that the major task in heterogeneous catalytic olefin metathesis chemistry is to develop functional group-tolerant, highly active catalysts. As described above, Balcar *et al.* have reported that rhenium oxide, when supported on mesoporous alumina and with Me_4Sn as cocatalyst, will promote the metathesis of olefins containing oxygen atoms [31]. Oikawa and coworkers, on the other hand, have developed methyltrioxorhenium (MTO, $MeReO_3$) doped on a $ZnCl_2$-modified mesoporous alumina as a functional group-tolerant, heterogeneous catalyst [30]. The mesoporous alumina used was the L_{A383} alumina (Table 15.1, II, calcined finally at 600 °C), while $ZnCl_2$ was doped onto the alumina using an impregnation method from an ethanolic solution, followed by drying and calcination at 400 °C. Subsequent characterization by N_2 adsorption–desorption showed that the $ZnCl_2$-modified mesoporous alumina (denoted $ZnCl_2//meso\text{-}Al_2O_3$) had regular mesopores with a relatively narrow pore-size distribution and a high surface area. The $ZnCl_2//meso\text{-}Al_2O_3$ was suspended in CH_2Cl_2, and to the suspension was added a solution of MTO in CH_2Cl_2 to yield the catalyst (denoted $MTO/ZnCl_2//meso\text{-}Al_2O_3$). Typically, 3.0 wt% MTO was loaded and an Al/Zn ratio of 16 was chosen. Olefins with ester (**35**, **36**, **39–41**, **45**), ketone (**47**), bromide (**49**), and chloride (**50**) groups were converted into the corresponding products at much higher yields with MTO/

ZnCl$_2$//*meso*-Al$_2$O$_3$ than with MTO/SiO–Al$_2$O$_3$, as developed previously by Herrmann *et al.* as a functional group-tolerant heterogeneous catalyst. When the effect of the pore size of the parent mesoporous alumina on catalytic performance was also investigated in the metathesis of methyl 10-undecenoate **35**, the results revealed that the mesoporous alumina with larger pore sizes and pore volumes would afford the corresponding product **37** at higher yields. This trend was in perfect accord with that observed by the Czech group [28, 29, 31].

The ZnCl$_2$-modified mesoporous alumina (ZnCl$_2$//*meso*-Al$_2$O$_3$) was designed in order that the mesoporous alumina should have a greater Lewis acidic character. The most interesting point here was that MTO, or a combination of MTO and ZnCl$_2$ in CH$_2$Cl$_2$, showed no catalytic ability for the olefin metathesis. However, when MTO contacted with the ZnCl$_2$//*meso*-Al$_2$O$_3$ support, the MTO/ZnCl$_2$//*meso*-Al$_2$O$_3$ catalyst promoted the metathesis efficiently because the MTO was activated by the Lewis acidic sites on ZnCl$_2$//*meso*-Al$_2$O$_3$. This was the first example of a catalytic reaction in which organometallic intermediates were activated by Lewis acidic inorganic supports. The Lewis acid character of ZnCl$_2$//*meso*-Al$_2$O$_3$ provided another feature; in the metathesis of simple olefins such as 1-octene **18** and 7-hexadecene **27**, catalyzed by MTO/ZnCl$_2$//*meso*-Al$_2$O$_3$, no olefin isomerization was observed because the Lewis acid sites had no ability to shift a double bond in the olefin. In contrast, MTO/SiO$_2$–Al$_2$O$_3$ (Herrmann's catalyst) produced a variety of olefins as SiO$_2$–Al$_2$O$_3$ was seen to be a typical Brönsted acid support and to promote olefin isomerization prior to or during the metathesis.

15.3.6
Oxidative Dehydrogenation

Oxidative dehydrogenation involves the catalytic dehydrogenation of alkanes to the corresponding alkenes in the presence of gaseous oxygen. One of the most active catalysts for this reaction is alumina-supported vanadium oxide, and this especially effective for ethane and propane. Concepción *et al.* employed mesoporous alumina as support and conducted the oxidative dehydrogenation of ethane **64** to ethylene **20** (Table 15.2, VI) [33]. The mesoporous alumina was synthesized using a method similar to that of Davis, with stearic acid as a surfactant (Table 15.1, II). Vanadium was loaded onto the alumina by impregnation from an ethanolic solution of vanadyl acetylacetonate, followed by drying with a rotary evaporator and calcination at 600 °C. The mesoporous structure was preserved up to a V-loading of 17.4 wt%. Anlayses with UV-DRS and ^{51}V wide-line and MAS NMR spectroscopy indicated that the mesoporous alumina behaved similar to conventional alumina for the V-loading, except that it permitted the incorporation of a greater amount of vanadium atoms, without forming the polymeric vanadium species "bulk-type" V$_2$O$_5$ composed of octahedral V^{5+} species. This effect can be explained in terms of the larger surface area of mesoporous alumina compared to conventional alumina. Although the specific activity per vanadium atom was lower on mesoporous alumina than on conventional alumina, the mesoporous alumina-supported vanadium catalyst could afford higher space-time yields and selectivity for the oxidative dehydrogenation of **64** to **20**, due to the larger number of active tetrahedral V^{5+} species. The mesoporous catalyst containing 9.7 wt% vanadium was best in terms of selectivity and showed a good catalytic performance at higher reaction temperatures. A V-content >9.7 wt% led to the formation of more octahedral V^{5+} species and Brönsted acid sites, which promoted an unfavorable deep oxidation of **64** and **20** to carbon oxides.

Later, the same group investigated the effect of molybdenum addition on the catalysis (Table 15.2, VI) [34]. Molybdenum (ammonium heptamolybdate) and vanadium were each loaded onto the mesoporous alumina in similar fashion, with a Mo/(Mo + V) ratio of 0.32 to 0.77. Increasing this ratio led to a decrease in the surface area, although the change was not significant (surface area range: 213–248 $m^2 g^{-1}$). Characterization by XRD and Raman spectroscopy revealed an absence of molybdovanadate species (i.e., mixed Mo–V–O species), indicative of the high metal dispersion on the catalysts due to the high surface area of mesoporous alumina. A synergetic effect between Mo and V was observed in terms of the better activity and selectivity compared to the pure Mo and V catalysts. The authors proposed two reasons for the favorable effect of Mo addition: (i) the coverage of nonselective sites of mesoporous alumina by molybdenum species, which suppressed the formation of carbon oxides and increased the selectivity to ethylene; and (ii) the formation of active molybdovanadate species, even though such species could not be observed with XRD and Raman spectroscopy. In all cases, the use of mesoporous alumina as a support led to a higher activity and selectivity than when using conventional γ-Al$_2$O$_3$, this being consistent with a former report [33].

15.3.7
Oxidative Methanol Steam Reforming

The oxidative steam reforming of methanol (OSRM) represents a promising catalytic process for the production of hydrogen from methanol, water, and oxygen. The process is a combination of the endothermic methanol steam reforming with the exothermic partial oxidation of methanol, with a zero net enthalpy. Lenarda *et al.* synthesized finely dispersed Pd–Zn on mesoporous alumina as catalyst for the OSRM reaction (Table 15.2, VII) [35]. The mesoporous alumina was synthesized with stearic acid as surfactant, using a method similar to that of Davis (Table 15.1, II, calcined finally at 500 °C). Palladium was then loaded onto the alumina by impregnation from an ethanolic solution of palladium acetate, and the resultant solid subsequently contacted with Zn(BH$_4$)$_2$ in diethyl ether solvent, followed by drying. The as-made material (denoted PZBAA) was reduced in a hydrogen-flow at 500 °C to give the mesoporous alumina-supported Pd–Zn alloy catalyst (denoted PZBAAr), which contained 1.7 wt% Pd and 12.9 wt% Zn. Both the PZBAAr and PZBAA samples showed type IV isotherms with a hysteresis loop typical of mesoporous materials. The surface area and pore volume of PZBAAr were smaller than those of the parent mesoporous alumina, but the pore-sizes were almost the same. The XRD, XPS, and IR spectra of adsorbed CO revealed that the Pd–Zn alloy was not present on PZBAA, but formed only after a high-temperature treatment with hydrogen-flow (i.e., on PZBAAr). The OSRM was carried out with a fixed-bed, continuous-flow reactor using the PZBAAr catalyst pretreated in a stream of O_2 at 400 °C, followed by H_2 at 400 °C. In order for the reaction to take place, a temperature above 250 °C was required, and the methanol conversion reached 100% at 350 °C, giving H_2 in a yield of 2.52% (defined as: (mol of H_2)/(mol of CH_3OH), theoretical value: 2.75). The reaction at 350 °C, however, also afforded 3% CO that was not the OSRM product. Both XPS and IR studies of CO adsorption implied the presence of metallic Pd on the surface, which could promote CO formation reactions such as the decomposition of methanol and the reverse water gas-shift reaction. The authors stressed that the high H_2 yield at 350 °C with the PZBAAr catalyst was unusual for such low active metal loadings.

15.4
Conclusions and Outlook

Today, mesoporous alumina is becoming a relatively "common" catalyst and catalyst support, with several synthetic methods using various types of structure-directing agents having been developed. Nonetheless, the search continues to produce mesoporous alumina with better structural qualities. Whilst mesoporous alumina itself exhibits a unique catalytic performance, as seen in the base catalysis in $scCO_2$, it serves as a much better support for metal-loading compared to conventional alumina, owing to its intrinsic high surface area. Interestingly, "molecular sieving," which is well established in zeolite catalysis chemistry, has also been observed in several olefin-metathesis reactions promoted by mesoporous, alumina-supported Re catalysts. It is expected that these fundamental studies and future investigations in this area will lead to the practical use of mesoporous alumina as both catalyst and catalyst support in large-scale production processes in the chemical industry.

References

1 Valenzuela Calahorro, C., Chaves Cano, T. and Gomez Serrano, V. (1972) IUPAC manual of symbols and terminology. Appendix 2, Part 1, Colloid and surface chemistry. *Pure Appl. Chem.*, **31**, 578.

2 (a) Breck, D.W. (1964) *J. Chem. Educ.*, **41**, 678.
(b) Csicsery, S.M. (1976) *Zeolite Chemistry and Catalysis* (ed. J.A. Rabo), ACS Monograph 171, American Chemical Society, p. 680.
(c) Weisz, P.B. (1980) *Pure Appl. Chem.*, **52**, 2091.
(d) Breck, D.W. (1984) *Zeolite Molecular Sieves*, Kreiger Publishing Company.

(e) Chen, N.Y., Garwood, W.E. and Dwyer, F.G. (1989) *Shape Selective Catalysis in Industrial Application*, Marcel Dekker.
(f) Baerlocher, C., Meier, W.M. and Olson, D.H. (2001) *Atlas of Zeolite Framework Types*, 5th edn, Elsevier, http://www.iza-structure.org/databases/.
(g) Cundy, C.S. and Cox, P.A. (2003) *Chem. Rev.*, **103**, 663.

3 (a) Kresge, C.T., Leonowicz, M.E., Roth, W.J., Vartuli, J.C. and Beck, J.S. (1992) *Nature*, **359**, 710.
(b) Beck, J.S., Vartuli, J.C., Roth, W.J., Leonowicz, M.E., Kresge, C.T., Schmitt, K.D., Chu, C.T.-W., Olson, D.H.,

Sheppard, E.W., McCullen, S.B., Higgins, J.B. and Schlenker, J.L. (1992) *J. Am. Chem. Soc.*, **114**, 10834.

4 Synthesis of mesoporous oxides other than alumina: (a) For magnesium: Takenaka, S., Sato, S., Takahashi, R. and Sodesawa, T. (2003) *Phys. Chem. Chem. Phys.*, **5**, 4968.
(b) For magnesium: Li, W.-C., Lu, A.-H., Weidenthaler, C. and Schüth, F. (2004) *Chem. Mater.*, **16**, 5676.
(c) For titanium: Antonelli, D.M. and Ying, J.Y. (1995) *Angew. Chem. Int. Ed. Engl.*, **34**, 2014.
(d) For titanium and niobium: Stone, V.F. Jr. and Davis, R.J. (1998) *Chem. Mater.*, **10**, 1468.
(e) For titanium: Yu, J.C., Wang, X. and Fu, X. (2004) *Chem. Mater.*, **16**, 1523.
(f) For niobium: Antonelli, D.M. and Ying, J.Y. (1996) *Angew. Chem. Int. Ed. Engl.*, **35**, 426.
(g) For niobium: Antonelli, D.M., Nakahira, A. and Ying, J.Y. (1996) *Inorg. Chem.*, **35**, 3126.
(h) For tantalum: Antonelli, D.M. and Ying, J.Y. (1996) *Chem. Mater.*, **8**, 874.

5 Synthesis of mesoporous alumina: (a) Bagshaw, S.A. and Pinnavaia, T.J. (1996) *Angew. Chem. Int. Ed. Engl.*, **35**, 1102.
(b) Vaudry, F., Khodabandeh, S. and Davis, M.E. (1996) *Chem. Mater.*, **8**, 1451.
(c) Zhang, W. and Pinnavaia, T.J. (1998) *Chem. Commun.*, 1185.
(d) Cabrera, S., El Haskouri, J., Alamo, J., Beltrán, A., Beltrán, D., Mendioroz, S., Dolores Marcos, M. and Amorós, P. (1999) *Adv. Mater.*, **11**, 379.
(e) Yao, N., Xiong, G., Zhang, Y., He, M. and Yang, W. (2001) *Catal. Today*, **68**, 97.
(f) González-Peña, V., Diaz, I., Márquez-Alvarez, C., Sastre, E. and Pérez-Pariente, J. (2001) *Microporous Mesoporous Mater.*, **44–45**, 203.
(g) Cruise, N., Jansson, K. and Holmberg, K. (2001) *J. Colloid Interface Sci.*, **241**, 527.
(h) Yao, N., Xiong, G., Zhang, Y., He, M. and Yang, W. (2001) *Catal. Today*, **68**, 97.
(i) Zhang, Z., Hicks, R.W., Pauly, T.R. and Pinnavaia, T.J. (2002) *J. Am. Chem. Soc.*, **124**, 1592.

(j) Shan, Z., Jansen, J.C., Zhou, W. and Maschmeyer, T. (2003) *Appl. Catal. A*, **254**, 339.
(k) Xu, B., Xiao, T., Yan, Z., Sun, X., Sloan, J., Gonzalez-Cortes, S.L., Alshahrani, F. and Green, M.L.H. (2006) *Microporous Mesoporous Mater.*, **91**, 293.
(l) Liu, Q., Wang, A., Wang, X. and Zhang, T. (2006) *Microporous Mesoporous Mater.*, **92**, 10.
(m) Zhao, R., Guo, F., Hu, Y. and Zhao, H. (2006) *Microporous Mesoporous Mater.*, **93**, 212.
(n) Žilková, N., Zukal, A. and Čejka, J. (2006) *Microporous Mesoporous Mater.*, **95**, 176.

6 (a) Huo, Q., Margolese, D.I., Ciesla, U., Feng, P., Gier, T.E., Sieger, P., Leon, R., Petroff, P.M., Schüth, F. and Stucky, G.D. (1994) *Nature*, **368**, 317.
(b) Huo, Q., Margolese, D.I., Ciesla, U., Demuth, D.G., Feng, P., Gier, T.E., Sieger, P., Firouzi, A., Chmelka, B.F., Schüth, F. and Stucky, G.D. (1994) *Chem. Mater.*, **6**, 1176.

7 (a) Ertl, G., Knözinger, H. and Weitkamp, J. (1997) *The Handbook of Heterogeneous Catalysis*, Wiley-VCH Verlag GmbH, Weinheim.
(b) Kabalka, G.W. and Pagni, R.M. (1997) *Tetrahedron*, **53**, 7999.
(c) Euzen, P., Raybaud, P., Krokidis, X., Toulhoat, H., Le Loarer, J.-L., Jolivet, J.-P. and Froidefond, C. (2002) *Handbook of Porous Materials* (eds F. Schüth, K. Sing and J. Weitkamp), Wiley-VCH Verlag GmbH, Weinheim, p. 1591.

8 Bartholomew, C.H. and Farrauto, R.J. (2006) *Fundamentals of Industrial Catalytic Process*, 2nd edn, John Wiley & Sons, Inc., Hoboken, New Jersey, p. 260.

9 Čejka, J. (2003) *Appl. Catal. A*, **254**, 327.

10 (a) Yada, M., Machida, M. and Kijima, T. (1996) *Chem. Commun.*, 769.
(b) Yada, M., Hiyoshi, H., Ohe, K., Machida, M. and Kijima, T. (1997) *Inorg. Chem.*, **36**, 5565.

11 Brunauer, S., Deming, L.S., Deming, W.E. and Teller, E. (1940) *J. Am. Chem. Soc.*, **62**, 1723.

12 McBain, J.W. (1935) *J. Am. Chem. Soc.*, **57**, 699.

13 Brunauer, S., Emmett, P.H. and Teller, E. (1938) *J. Am. Chem. Soc.*, **60**, 309.

14 Barrett, E.P., Joyner, L.G. and Halenda, P.P. (1951) *J. Am. Chem. Soc.*, **73**, 373.

15 Dollimore, D. and Heal, G.R. (1964) *J. Appl. Chem.*, **14**, 109.

16 Čejka, J., Žilkova, N., Rathouský, J. and Zukal, A. (2001) *Phys. Chem. Chem. Phys.*, **3**, 5076.

17 (a) Yamaguchi, G. and Yanagida, H. (1962) *Bull. Chem. Soc. Jpn.*, **35**, 1896.
(b) Yanagida, H. and Yamaguchi, G. (1964) *Bull. Chem. Soc. Jpn.*, **37**, 1229.
(c) Yanagida, H., Yamaguchi, H. and Kubota, J. (1965) *Bull. Chem. Soc. Jpn.*, **38**, 2194.

18 (a) Seki, T. and Onaka, M. (2005) *Chem. Lett.*, **34**, 262.
(b) Seki, T. and Onaka, M. (2006) *J. Phys. Chem. B*, **110**, 1240.
(c) Seki, T. and Onaka, M. (2006) *Catal. Surv. Asia*, **10**, 138.

19 Seki, T. and Onaka, M. (2007) *J. Mol. Catal. A*, **263**, 115.

20 Seki, T., Ikeda, S. and Onaka, M. (2006) *Microporous Mesoporous Mater.*, **96**, 121.

21 Yin, D., Qin, L., Liu, J. Li, C. and Jin, Y. (2005) *J. Mol. Catal. A*, **240**, 40.

22 Chaube, V.D., Shylesh, S. and Singh, A.P. (2005) *J. Mol. Catal. A*, **241**, 79.

23 Kim, P., Kim, Y., Kim, C., Kim, H., Park, Y., Lee, J.H., Song, I.K. and Yi, J. (2003) *Catal. Lett.*, **89**, 185.

24 Kim, P., Kim, Y., Kim, H., Song, I.K. and Yi, J. (2004) *J. Mol. Catal. A*, **219**, 87.

25 Kim, P., Kim, Y., Kim, H., Song, I.K. and Yi, J. (2005) *J. Mol. Catal. A*, **231**, 247.

26 Kaluža, L., Zdražil, M., Žilkova, N. and Čejka, J. (2002) *Catal. Commun.*, **3**, 151.

27 (a) Onaka, M. and Oikawa, T. (2002) *Chem. Lett.*, 850.
(b) Oikawa, T., Ookoshi, T., Tanaka, T., Yamamoto, T. and Onaka, M. (2004) *Microporous Mesoporous Mater.*, **74**, 93.

28 Balcar, H., Hamtil, R., Žilková, N. and Čejka, J. (2004) *Catal. Lett.*, **97**, 25.

29 Hamtil, R., Žilková, N., Balcar, H. and Čejka, J. (2006) *Appl. Catal. A*, **302**, 193.

30 Oikawa, T., Masui, Y., Tanaka, T., Chujo, Y. and Onaka, M. (2007) *J. Organomet. Chem.*, **692**, 554.

31 Balcar, H., Hamtil, R., Žilková, N., Zhang, Z., Pinnavaia, T.J. and Čejka, J. (2007) *Appl. Catal. A*, **320**, 56.

32 Aguado, J., Escola, J.M., Castro, M.C. and Paredes, B. (2005) *Appl. Catal. A*, **284**, 47.

33 Concepción, P., Navarro, M.T., Blasco, T., López Nieto, J.M., Panzacchi, B. and Rey, F. (2004) *Catal. Today*, **96**, 179.

34 Solsona, B., Dejoz, A., Garcia, T., Concepción, P., Lopez Nieto, J.M., Vázquez, M.I. and Navarro, M.T. (2006) *Catal. Today*, **117**, 228.

35 Lenarda, M., Moretti, E., Storaro, L., Patrono, P., Pinzari, F., Rodríguez-Castellón, E., Jiménez-López, A., Busca, G., Finocchio, E., Montanari, T. and Frattini, R. (2006) *Appl. Catal. A*, **312**, 220.

36 Kim, Y., Lee, B. and Yi, J. (2002) *Korean J. Chem. Eng.*, **19**, 908.

37 (a) Schrock, R.R. and Hoveyda, A.H. (2003) *Angew. Chem. Int. Ed. Engl.*, **42**, 4592.
(b) Trnka, T.M. and Grubbs, R.H. (2001) *Acc. Chem. Res.*, **34**, 18.

16
Nanoceramics for Medical Applications

Besim Ben-Nissan and Andy H. Choi

16.1
Introduction

"Nanostructured materials" refer to certain materials which have delicate structures and sizes that fall within the range of 1 to 100 nm. As a consequence of this size, an extensive development of nanotechnology has taken place in the fields of materials science and engineering during the past decade. Yet, such developments have not come as a surprise, when it is appreciated that these nanostructured materials have the ability to be adapted and integrated into biomedical devices. This is possible because most biological systems, including viruses, membranes and protein complexes, exhibit natural nanostructures.

The microstructure and properties of nanostructured materials depend in an extreme manner on the method of their synthesis method, as well as on their processing route. Consequently, it is extremely important to select the most appropriate technique when preparing nanomaterials with desired properties and property combinations. The synthesis techniques most commonly used for the production of advanced ceramics include pressing, as well as wet chemical processing techniques such as co-precipitation and sol–gel, all of which have been used to produce nanoparticles, nanocoatings, and nanostructured solid blocks and shapes.

In modern ceramics technology, *pressing* is accomplished by placing the powder into a die and applying pressure to achieve compaction. Hot pressing (HP) and hot isostatic pressing (HIP) are the most common methods used to produce bioceramics. HIP can induce the higher densities and small grain structures required by bioceramics, whereby heat and pressure are applied simultaneously and the pressure is applied from all directions via a pressurized gas such as helium or argon. In contrast, flat plates or blocks and nonuniform components are relatively easily produced using HP.

Sol–gel processing is unique in that it can be used to produce different forms, such as powders, platelets, coatings, fibers and monoliths of the same composition, merely by varying the chemistry, viscosity, and other factors of a given solution. The advantages of the sol–gel technique are numerous:

Advanced Nanomaterials. Edited by Kurt E. Geckeler and Hiroyuki Nishide
Copyright © 2010 WILEY-VCH Verlag GmbH & Co. KGaA, Weinheim
ISBN: 978-3-527-31794-3

- it is of the nanoscale
- it results in a stoichiometric, homogeneous and pure product, owing to mixing on the molecular scale
- high purity can be maintained as grinding can be avoided
- it allows reduced firing temperatures due to the small particle sizes with high surface areas
- it can be used to produce uniform, fine-grained structures
- it allows the use of different chemical routes (alkoxide or aqueous-based)
- it is easily applied to complex shapes with a range of coating techniques.

Sol–gel coatings also have the added advantages that the costs of the precursors are relatively unimportant, owing to the small amounts of materials required. Shrinkage up to a number of coatings, depending on the chemistry, is fairly uniform perpendicular to the substrate, and the coatings can dry rapidly without cracking. Shrinkage becomes an important issue, however, in monolith ceramic production.

At present, the most common materials in clinical use are those selected from a handful of well-characterized and available biocompatible ceramics, including metals, polymers, and their combinations as composites or hybrids. These unique production techniques, together with the development of new enabling technologies such as microscale, nanoscale, bioinspired fabrication (biomimetics) and surface modification methods, have the potential to drive at an unprecedented rate the design and development of new nanomaterials useful for medical applications.

The current focus is on the production of new nanoceramics that are relevant to a broad range of applications, including: implantable surface-modified medical devices for better hard- and soft-tissue attachment; increased bioactivity for tissue regeneration and engineering; cancer treatment; drug and gene delivery; treatment of bacterial and viral infections; delivery of oxygen to damaged tissues; imaging; and materials for minimally invasive surgery. A more futuristic view, which could in fact become reality within two decades, includes nanorobotics, nanobiosensors, and micronanodevices for a wide range of biomedical applications.

A *biomaterial*, by definition, is a nondrug substance that is suitable for inclusion in systems that augment or replace the function of bodily tissues or organs. A century ago, artificial devices made from materials as diverse as gold and wood were developed to a point where they could replace the various components of the human body. These materials were capable of being in contact with bodily fluids and tissues for prolonged periods of time, while eliciting little, if any, adverse reactions.

When these synthetic materials are placed within the human body, the tissues react towards the implant in a variety of ways. The mechanism of tissue interaction at a nanoscale level is dependent on the response to the implant surface, and as such three terms which describe a biomaterial, with respect to the tissues' responses, have been defined, namely *bioinert*, *bioresorbable*, and *bioactive* (Figure 16.1):

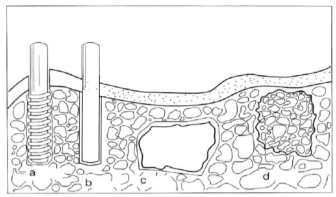

Figure 16.1 Classification of bioceramics according to their bioactivity. (a) Bioinert (alumina dental implant); (b) Bioactive, hydroxyapatite $(Ca_{10}(PO_4)_2(OH)_2)$ coating on a metallic dental implant; (c) Surface-active, bioglass or A-W glass; (d) Bioresorbable tricalcium phosphate implant $[Ca_3(PO_4)_2]$.

- *Bioinert* refers to any material that, once placed within the human body, has a minimal interaction with its surrounding tissue; examples include stainless steel, titanium, alumina, partially stabilized zirconia, and ultra-high-molecular-weight polyethylene.

- *Bioactive* refers to a material which, upon being placed within the human body, interacts with the surrounding bone and, in some cases, even soft tissue.

- *Bioresorbable* refers to a material that, upon placement within the human body, begins to dissolve or to be resorbed and slowly replaced by the advancing tissues (e.g., bone).

During the early 1970s, bioceramics were employed as implants to perform singular, biologically inert roles. The limitations of these synthetic materials as tissue substitutes were highlighted with the increasing realization that the cells and tissues of the body perform many other vital regulatory and metabolic roles. The demands of bioceramics have since changed, from maintaining an essentially physical function without eliciting a host response, to providing a more positive interaction with the host. This has been accompanied by increasing demands on medical devices that they not only improve the quality of life but also extend its duration. Most importantly, nanobioceramics–at least potentially–can be used as body interactive materials, helping the body to heal, or promoting the regeneration of tissues, thus restoring physiological functions.

The main factors in the clinical success of any biomaterial are its *biocompatibility* and *biofunctionality*, both of which are related directly to tissue/implant interface interactions. This approach is currently being explored in the development of a new generation of nanobioceramics with a widened range of medical applications. The improvement of interface bonding by nanoscale coatings, based on biomimetics, has been of worldwide interest during the past decade, and today several companies are in early commercialization stages of new-generation, nanoscale-modified implants for orthopedic, ocular, and maxillofacial surgery, as well as for hard- and soft-tissue engineering.

Biomimetic processing is based on the notion that biological systems store and process information at the molecular level, and the extension of this concept to the processing of nanocomposites for biomedical devices and tissue engineering, such as scaffolds for bone regeneration, has been brought out during the past decade [1]. Several research groups have reported the synthesis of novel bone nanocomposites of hydroxyapatite (HAp) and collagen, gelatin, or chondroitin sulfate, through a self-assembly mechanism. These self-assembled experimental bone nanocomposites have been reported to exhibit similarities to natural bone in not only their structure but also their physiological properties [2].

The term *nanocomposite* can be defined as a heterogeneous combination of two or more materials, in which at least one of those materials should be on a nanometer-scale. By using the composite approach, it is possible to manipulate the mechanical properties such as strength and modulus of the composites closer to those of natural bone, with the help of secondary substitution phases. For example, HAp–polymer composites have been shown to have an elastic modulus close to that of bone.

The fabrication of a nanocomposite can be achieved by physically mixing or introducing a new component into an existing nanosized material, which allows for property modifications of the nanostructured materials and may even offer new material functions. For example, some biopolymers and biomolecules, such as poly(lactic acid) (PLA), poly(lactic-*co*-glycolic acid) (PLGA), polyamide, collagen, silk fibrin, chitosan, and alginate have been reported to mix into nanohydroxyapatite (nanoHAp) systems.

Another form of nanocomposite which has been developed for biomedical applications is the *gel system*. For this, nanostructured materials can be entrapped in to a gel (a three-dimensional (3-D) network immersed in a fluid), such that the properties of the nanomaterials can be improved and tailored to suit the specific needs of certain biomedical devices. A *nanogel*, which is a nanosized, flexible hydrophilic polymer gel [3], is an example of a gel that can be used in drug delivery carriers. These nanogels can bind and encapsulate spontaneously (through ionic interactions) any type of negatively charged oligonucleotide drug. A key advantage of nanogels is that they allow for a high "payload" of macromolecules (up to 50 wt%), a value which normally cannot be approached with conventional nanodrug carriers [4]. Recently, a novel intracellular biosensor has been fabricated by entrapping indicator dyes into an acrylamide hydrogel [5, 6], whilst a carbon nanotube (CNT) aqueous gel has been developed as an enzyme-friendly platform for use in enzyme-based biosensors [7].

This aim of this chapter is to provide information relating to the use of bioceramics for medical applications, including tissue engineering and regeneration using scaffolds and liposomes, nanoHAp powders for medical applications, calcium phosphate nanocoatings and other surface modification techniques, simulated body fluids, and nano- and macrobioceramics for use in drug delivery and radiotherapy.

16.2
Tissue Engineering and Regeneration

16.2.1
Scaffolds

In the past, the development of bone tissue engineering has been directly related to changes in materials and nanotechnology. While the inclusion of materials requirements is standard in the design process of engineered bone substitutes, it is also critical to incorporate clinical requirements such that clinically relevant devices can be engineered.

Multiple clinical reasons exist for the development of bone tissue-engineering alternatives, including a need for better filler materials when reconstructing large orthopedic defects, and for orthopedic implants that are mechanically more suitable to their biological environment. The traditional biological methods of bone-defect management include *autografting* and *allografting*.

Bone regeneration requires four components: (i) a morphogenetic signal; (ii) responsive host cells that will respond to the signal; (iii) a suitable carrier of this signal that can deliver it to specific sites then serve as scaffolding for the growth of the responsive host cells; and (iv) a viable–and most importantly–a well-vascularized host bed [8–10]. A *scaffolding material* is used either to induce the formation of bone from the surrounding tissue, or to act as a carrier or template for implanted bone cells or other agents.

The process of *bone regeneration* is common to the repair of fractures. The incorporation of bone grafts, the skeletal homeostasis and the cascading sequence of biological events are often described as the *remodeling cycle*. Stem cells have been incorporated into a range of bioceramics and when implanted, can combine with mineralized 3-D scaffolds to form highly vascularized bone tissue. These nanoscale cultured cell–bioceramic composites can be used to treat full-thickness gaps in long bone shafts, providing an excellent integration of the ceramic scaffold with bone, and a good functional recovery (Figure 16.2).

Peroglio *et al.* [11] have produced and characterized polycaprolactone (PCL)-coated alumina scaffolds to validate the concept of polymer–ceramic composites with increased fracture resistance. Alumina scaffolds were sintered using a foam replication technique, and the polymer coating was obtained by infiltrating the scaffold with either a PCL solution or nanodispersion.

Bioactive glasses are amorphous, silica-based materials that are biocompatible, bioactive, osteoconductive and even osteoproductive. Sol–gel-derived bioactive glasses were developed by Hench *et al.* [12] to produce 3-D bioactive scaffolds with hierarchical interconnected pore morphologies similar to trabecular bone. The scaffolds consist of a pore network with macropores in excess of $500\,\mu m$

(a)   (b)

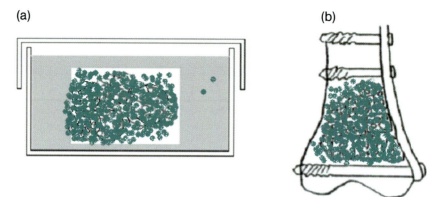

Figure 16.2 Bone regeneration therapy using marrow mesenchymal cells for bone-graft applications for increased bioactivity. (a) The marrow cell/hydroxyapatite composite scaffold is cultured in an osteogenic medium; (b) After two to three weeks of culture, the cultured bone graft/hydroxyapatite composite is inserted into a bone defect.

that are connected by pore windows with diameters in excess of 100 μm, this being the minimum pore diameter required for tissue ingrowth and vascularization in the human body. The scaffolds also have textural porosity in the mesopore range (10–20 nm). The compressive strength reported was in the range of that for trabecular bone. In conclusion, Horsch *et al.* [12] stated that, by combining these properties with those inherent to sol–gel-derived bioactive glass, these scaffolds would have high potential as scaffolds for bone tissue engineering applications.

Considerable attention has also been paid during the past two decades to bioactive composite grafts that consists of a bioactive ceramic filler in a polymeric matrix. These bioactive composite grafts are designed essentially to achieve interfacial bonding between the graft and the host tissues. HAp/collagen, HAp/polyethylene (PE), and HAp/Ti-6Al-4V are notable examples of bioactive composite grafts [13, 14].

Of particular interest is the combination of HAp with collagen as a bioactive composite, as this appears to be a natural choice for bone grafting [15]. Skeletal bones comprise mainly collagen and carbonate-substituted HAp, both of which are osteoconductive components; consequently, an implant manufactured from such components is likely to behave in similar fashion. A composite matrix, when embedded with human-like osteoblast cells, showed better osteoconductive properties compared to monolithic HAp, and produced calcification of an identical bone matrix. In addition, collagen–HAp composites proved not only to be biocompatible in both humans and animals [16], but also to behave mechanically in a superior fashion to the individual components. Moreover, the ductile properties of collagen helped to increase the poor fracture toughness of HAps.

The reconstruction of bone tissue using nanocomposite bone grafts with structure, composition, physico-chemical, biomechanical, and biological features that mimic those of natural bone, is a goal to be pursued. It is well known that natural bone consists of nanosized, plate-like crystals of HAp grown in intimate contact with an organic matrix which is rich in collagen fibers. One novel approach to fabricating nanocomposite bone grafts, using strategies found in Nature, has recently received much attention and is perceived to be beneficial over conventional methods. A variety of production methods have been employed for the formation of collagen–HAp composite gels, films, collagen-coated ceramics, ceramic-coated collagen matrices and composite scaffolds for spine and hard tissue repair [17].

Stem cells are cells from an embryo, fetus, or adult that have the ability to reproduce for long periods, and can also give rise to specialized cells that comprise the tissues and organs of the body. When implanted onto immunodeficient mice, stem cells were shown to combine with mineralized 3-D scaffolds to form a highly vascularized bone tissue. Cultured cell–bioceramic composites can be used to treat defects across the bone diaphysis, with excellent integration of the ceramic scaffold with bone, and a good functional recovery [18]. Excellent innovative studies with nanobioceramics are currently in progress, and clinical applications are becoming relatively common.

Vago and coworkers [19] have introduced a novel 3-D biomatrix obtained from the marine hydrocoral *Millepora dichotoma* as a scaffold for hard-tissue engineering. *M. dichotoma* was biofabricated under both field and laboratory conditions, and 3-D biomatrices prepared in order to convert mesenchymal stem cells (MSCs) to exemplify an osteoblastic phenotype. The effect of the biomatrices on the proliferation and differentiation of MSCs was then examined at 2, 3, 4, 7, 10, 14, 21, 28, and 42 days. The investigations included light microscopy, scanning electron microscopy (SEM) and energy dispersive spectroscopy (EDS), in addition to monitoring calcium incorporation into newly formed tissue (with Alizarin red staining), bone nodule formation (von Kossa staining), fat aggregate formation (oil red O staining), collagen type I immunofluorescence, DNA concentrations, alkaline phosphatase (ALP) activity, and osteocalcin concentrations. The MSCs seeded onto *M. dichotoma* biomatrices showed higher levels of calcium and phosphate incorporation, and higher type I collagen levels, than did control *Porites lutea* biomatrices. In addition, the ALP activity revealed that those MSCs seeded on *M. dichotoma* biomatrices were highly osteogenic compared to those on control biomatrices. The osteocalcin content of MSCs seeded on *M. dichotoma* remained constant for up to two weeks, before surpassing that of seeded *P. lutea* biomatrices after 28 days. The investigators reported that *M. dichotoma* biomatrices enhanced the differentiation of MSCs into osteoblasts, and hence showed excellent potential as bioscaffolding for hard-tissue engineering.

As emerging areas, both tissue and implant engineering are evolving to address the shortage of human tissue and organs. Feasible and productive strategies have been aimed at combining a relatively traditional approach, such as bioceramic implants, with the acquired knowledge applied to the field of cell growth and differentiation of osteogenic cells. The core of the tissue engineering and regenerative medicine is the fabrication of scaffolds, in which a given cell population is seeded, proliferated, and differentiated with the introduction of functional cell types from many different sources [20].

Nanostructured materials and their modified forms offer some attractive possibilities in the fields of tissue and implant engineering, taking advantage of the combined use of living cells and 3-D ceramic scaffolds (Figure 16.2) to deliver vital cells to the damaged site of the patient. Recently, bone-like nanostructure scaffolds have been developed using the technology of composites to imitate natural bone in bone-tissue engineering [21].

The results of recent studies have suggested that *bone marrow stromal cells* might be a potential source of osteoblasts and chondrocytes, and can be used to regenerate damaged tissues using a tissue-engineering approach. However, these strategies require the use of an appropriate scaffold architecture that can support the formation *de novo* of bone and/or cartilage tissue, as in the case of osteochondral defects. Oliveira *et al.* [14, 22] developed a novel hydroxyapatite/chitosan (HAp/CS) bilayered scaffold by combining a sintering and a freeze-drying technique, aiming to show the potential of such scaffolds to be used in tissue engineering for osteochondral defects. Subsequently, *in vitro* (Phase 1) cell culture studies were carried out to evaluate the capacity of the HAp and CS layers to separately support the growth and differentiation of goat bone marrow stromal cells (GBMCs) into osteoblasts and chondrocytes, respectively. The data showed not only that the GBMCs were able to adhere and proliferate but also that the constructs exhibited a great potential for use in tissue-engineering strategies, leading to the formation of adequate tissue substitutes for the regeneration of osteochondral defects.

The effects of surface chemistry modifications of titanium alloy (Ti-6Al-4V) with zinc, magnesium, or alkoxide-derived nanocrystalline carbonate hydroxyapatite (CHAp) on the regulation of key intracellular signaling proteins in human bone-derived cells (HBDCs) cultured on these modified Ti-6Al-4V surfaces, have been investigated in Australia by Zreiqat *et al.* [23]. The surface modification with nanocrystalline CHAp was shown to contribute to successful osteoblast function and differentiation at the skeletal tissue–device interface.

The role of *gene therapy* in aiding wound healing and treating various diseases or defects has become increasingly important in the field of tissue engineering. The use of 3-D scaffolds in gene delivery has emerged as a popular and necessary delivery vehicle for obtaining controlled gene delivery. Ko *et al.* [24] described the techniques to synthesize composite scaffolds by combining natural polymers such as agarose and alginate with calcium phosphate (CaP). *Alginate* has been used extensively in various applications such as cell encapsulation seeding, gene delivery, and antibody or growth factor entrapment and release, while *agarose* has been used as a scaffold involved in cartilage repair. The incorporation of CaP into the agarose or alginate hydrogels was performed *in situ*. Ko *et al.* concluded that, by incorporating CaP into the agarose or alginate hydrogel, they were able to synthesize a scaffold that was mechanically strong and chemically suitable for use as a gene-delivery vehicle in tissue engineering.

For many implants, a sustained and controlled release of antibacterial agents into the wound site is desirable for combating infection. A further advantage of nanostructured sol–gel-derived glasses is that silver, which is known to have antibacterial properties, can be incorporated into the glass composition. The addition of silver ions to bioactive glasses has also been investigated by Jones *et al.* [25], for the production of glasses with bactericidal properties. A bioactive glass scaffold containing 2 mol% silver was shown to release silver ions at a rate shown previously to be bactericidal in, but not cytotoxic to, bone cells.

16.2.2
Liposomes

Liposomes are the most clinically established nanometer-scale systems currently used to deliver nontoxic and antifungal drugs, genes, and vaccines; they are also being used as imaging agents. Liposomes consist of a single layer, or multiple concentric lipid bilayers, that encapsulate an aqueous compartment. The outstanding clinical profile of liposomes, compared to other delivery systems, is based on their biocompatibility, biodegradability, reduced toxicity, and capacity for size and surface manipulations [26].

Nanometer-sized particles, such as superparamagnetic iron oxides, semiconducting nanocrystals, silica nanoparticles, and calcium phosphate, each possess novel functions that include unique magnetic, optical, therapeutic, and medical properties. The encapsulation of these nanoparticles within liposomes may lead to an enhanced nanoparticle hydrophilicity, stability in plasma, and an overall improvement in their biocompatibility [26–28]. Furthermore, by utilizing the ability of liposomes to carry hydrophilic and hydrophobic moieties, combinatory therapy/imaging modalities can be achieved by incorporating therapeutics and diagnostic agents into a single liposome-delivery system [26].

Semiconductor nanocrystals, known as quantum dots (QDs) are fluorescent nanoparticles with a diameter in the range of 1 to 10 nm. QDs offer distinct spectrofluorometric advantages over traditional fluorescent organic molecules, with typical fluorescence characteristics 10 to 20-fold brighter than conventional dyes. QDs also exhibit a greater photostability, a broad excitation wavelength range, a size-tunable spectrum, and a narrow and symmetric emission spectrum. On the basis of these photophysical characteristics QDs are currently being investigated as potential imaging agents, primarily in fluorescence-based diagnostic applications [26].

The self-assembly of organized nanoscopic structures has been the subject of much interest in both colloidal and nanomaterials science. Indeed, some recent studies have shown that nanoscale liposomes can be used as a nanoscale template for the deposition of silica, so as to create a hollow silica nanoshell. These silicate materials have been used to encapsulate fluorescent dyes, enzymes, polymer particles, and liquids [27]. Moreover, the liposome–silica nanoparticle hybrid systems thus formed can be used in the design of biosensors, whereby the physical characteristics of silica can be matched with the biocompatibility and pharmaceutical and pharmacodynamic properties of liposomes [28].

Calcium phosphate-based hybrid nanoparticles have shown great promise as candidates for drug delivery and bone regeneration systems, based on the excellent biocompatibility of calcium phosphate [29]. Recently, hydroxyapatite-coated liposomes (HACLs) were successfully manufactured and filled with a model hydrophobic (lipophilic) drug, namely indomethacin (IMC) [29]. In this process, the HAp layer was precipitated onto the liposomes, the aim being to provide the HACL with two functions: (i) that the inner core liposome provides a sustained drug release; and (ii) that the outer HAp layer provides the osteoconductivity for bone cells. The liposomes were formed from 1,2-dimyristoyl-*sn*-glycero-3-phosphate (DMPA) and 1,2-dimyristoyl-*sn*-glycero-3-phosphocholine (DMPC). The results reported by Xu *et al.* [29] indicated that precipitating HAp onto the liposome reduced the release rate of IMC compared to uncoated liposomes. In fact, under the conditions used, the 5 h period required to release 70% of the IMC from the liposomes was extended to 20 h when the liposomes were coated with HAp. Perhaps, more importantly, IMC release from the uncoated liposomes occurred more rapidly at pH 7.4 than at pH 4, whereas the HAp coating reduced the release rate at pH 7.4 compared to that at pH 4. Based on these findings, Xu *et al.* suggested that this effect might open up the possibility of creating "smart" (pH-controlled) targeted drug delivery devices.

When Hang *et al.* [30] used liposome-coated HAp and tricalcium phosphate as bone implants in the mandibular bony defect of miniature swine, they found the liposome-coated materials to be biocompatible. Moreover, the clinical endpoint was enhanced compared to that in the absence of liposomes. It was hypothesized that coating hydroxyapatite and tricalcium phosphate with negatively charged liposomes might improve the nucleation process for new bone formation. In experiments conducted in miniature swine, artificial bony defects on one side were implanted with either HAp-coated or tricalcium phosphate-coated liposomes, while defects on the other side served as controls. Histology and radiography performed at three and six weeks after surgery showed the coated liposome materials to be biocompatible. At three weeks, the implant material was surrounded by dense connective tissues, whilst by six weeks, new bone formation was visible near the implanted material. Liposomes immobilized in agarose gel and implanted in the defects also showed new bony bridge formation.

16.3
Nanohydroxyapatite Powders for Medical Applications

Bone mineral is composed of nanocrystals or, more accurately, nanoplatelets which originally were described as HAp, and similar to the mineral *dahllite*. Today, it is agreed that bone apatite may be better described as CHAp, and approximated by the formula $(Ca,Mg,Na)_{10}(PO_4CO_3)_6(OH)_2$. The composition of commercial CHAp is similar to that of bone mineral apatite. Bone pore sizes range from 1 to 100 nm in normal cortical bone, and from 200 to 400 nm in trabecular bone tissue, with the pores being interconnected.

Orthopedic implants used mainly for joint replacement and fracture fixation include metallic (cobalt chromium or titanium alloys) implants, screws, plates and nails, and their various permutations and combinations. The most important parameters for these implants are that they have the necessary wear resistance, allow for an adequate attachment to bone, and display the required strength, ductility, and elasticity. At a bone-implant–load-carrying interface, the greater the implant material stiffness, the greater load it can carry, compared to the surrounding tissues. This imbalance in load, which is known as *stress shielding*, can cause the bone tissue to be resorbed. An implant that is too rigid may also increase the likelihood of bone fracture, as the bone becomes osteoporotic (thinned) due to the excessive protection generated by the stress-shielding effect of the implant.

Macro-textured implants nanocoated with calcium phosphate and possessing the appropriate bioactivity characteristics, bonding ability, and design, may be the answer to this serious problem. A range of new nanomaterial production companies are in the process of applying this new technology in orthopedic, cardiovascular, and dental implants.

Nanotechnology has opened up novel techniques for the production of bone-like synthetic nanopowders and coatings of HAp. Indeed, the availability of HAp nanoparticles has opened up new opportunities for the design of superior biocompatible coatings for implants, and the development of high-strength orthopedic and dental nanocomposites.

Although, bone-like HAp nanopowders and nanoplatelets (Figure 16.3) can be synthesized by a range of production methods, one very promising approach is to synthesize these materials via a sol–gel solution. The results of earlier studies have shown that, while biphasic sol–gel HAp products are easily synthesized, monophasic HAp powders and coatings are more difficult to produce. Many companies have successfully synthesized HAp nanoparticles with diameters in the range of 15 to 20 nm, and with HAp coatings 70 nm thick (Nanocoatings Pty. Ltd., Australia). The nanoparticles and nanoplatelets of HAp provide excellent bioactivity for integration into bone, which arises from their very high surface areas [31, 32].

Figure 16.3 Nanocrystalline carbonate apatite platelets formed by a sol–gel process.

Several new production and surface-modification techniques, including some sol–gel techniques, chemical vapor deposition (CVD), and plasma spray, have resulted in bonds with an excellent adhesive strength between the HAp and the substrate material, while others may be poor (e.g., electrophoretic deposition, various solution dip-coating systems, thermal spray). Currently, several companies and research groups are producing nanocomposites (e.g., NanoCoatings Pty Ltd, Australia; Mitsubishi Materials; ApaTech Ltd; Dentsply International; BioMet Int.; Wyeth BioPhamia; and Medtronic Sofamor Danek) that incorporate the macroparticles and nanoparticles of HAp and organic and biogenic materials (e.g., polyethylene, synthetic peptides and collagen, growth factors). Such a combination provides mechanical strength that is not achievable by using nanoparticles alone. In addition, for some of these materials an enhanced bioactivity and mechanical properties have been reported in orthopedic and dental applications, such as bone cements and dental fillings [33].

Saiz and coworkers [34] have focused on the sintering of porous HAp scaffolds fabricated using two techniques based on manipulation of the HAp slurries, namely *infiltration of the polymer foams* and *robocasting*:

- The first method involves the infiltration of a polymer sponge with a ceramic slurry until the inner polymer walls are completely coated by the ceramic powders. Subsequently, the sample is fired to remove the polymer and form a ceramic skeleton that is strengthened by sintering at high temperature.

- The second technique involves the use of computer-driven rapid prototyping techniques to produce porous ceramic with anisotropic microstructures. This so-called "robocasting" is a simple technique used to produce porous ceramic parts with complex shapes. In robocasting, a ceramic ink is extruded through a thin nozzle to build a part layer-by-layer, following a computer design. Sintering in air at temperatures ranging between 1100 °C and 1200 °C yields dense materials with narrow, grain-sized distributions.

It has been stated that both techniques can be used to fabricate scaffolds with an adequate pore size capable of promoting bone ingrowth.

Khalyfa and coworkers [35] have developed a powder mixture comprising tetracalcium phosphate (TTCP) as the reactive component and β-tricalcium phosphate (β-TCP) or calcium sulfate as a biodegradable filler, which can be printed with an aqueous citric acid solution. Two post-processing procedures – a sintering and a polymer infiltration process – were established to substantially improve the mechanical properties of the printed devices. Specimens of different shapes and sizes have been printed to study the usability of the developed powder-binder systems in the 3-D printing process; in this way a 3-D scaffold with a thoroughly open channel system was produced. The printing of a human cranial segment, together with all its filigree structures, was successfully achieved using both powder-binder systems, based on computed tomography scanning data of a human cranium. In all cases, the printed objects were strong enough to be handled manually, without damaging the integrity of the devices. Preliminary examinations on relevant application properties, including *in vitro* cytocompatibility testing, indicated that the new powder-binder system represented an efficient approach to the creation of patient-specific ceramic bone substitutes and scaffolds for bone-tissue engineering.

Ordered tubular structures with open porosity were created by de Sousa and Evans [36] by using the microextrusion freeforming of a tubular latticework. The extrudate was a suspension of fine HAp powder in isopropyl alcohol with a polyvinyl butyral binder. The extruder consisted of a stepper-driven syringe fitted with a miniature tube extrusion die. In this way, tubular lattice scaffolds and microsprings were successfully prepared from crowded ceramic suspensions of HAp. The lattices were then sintered at 1250 °C to produce a ceramic that had potential as a bone scaffold and could accommodate growth promoters in a slow release form.

16.4
Nanocoatings and Surface Modifications

16.4.1
Calcium Phosphate Coatings

When considering an ideal material to replace and mimic bone, synthetic calcium phosphates are an obvious choice, as they can replicate the structure and composition of HAp, a bone mineral. However, despite having a similar composition and chemistry to that of human bone, the mechanical properties of calcium phosphate are far from being close to those of human bone, which limits their use for load-bearing applications.

Today's solutions of materials for bone replacement are still far from ideal, with metallic implants remaining the first choice for load-bearing applications. As all metallic orthopedic and dental implants are bioinert and do not bond chemically to bone, the only means of fixation is by mechanical interlocks, whereby the implant must be manufactured in such a way that it possesses suitable surface roughness by micro- and macro-texturing. By increasing the surface roughness, the surface area is increased and this in turn increases the area of fixation. Other current methods for fixing implants firmly in place are the use of screws or bone cement, which are both used in dental and orthopedic implants.

Most published information on HAp is classified under calcium phosphate, to which HAp belongs. Therefore, the chemical properties will be viewed from the standpoint that HAp is calcium phosphate, although it will have different solubility and reactivities from other phosphates within the physiological environments.

Calcium phosphates are characterized by particular solubilities, such as when bonding to the surrounding tissues, and their ability to degrade and be replaced by advancing bone growth. As the calcium phosphate or HAp comes into contact with body fluid, its surface ions can be exchanged with those of the aqueous solution; alternatively, various ions and molecules, such as collagen and proteins, can be adsorbed onto the surface [31, 32].

The goal of calcium phosphate as a bioactive coating is to achieve a rapid biological attachment to bone. *Biological fixation* is defined as the process by which prosthetic components become firmly bonded to the host bone by bone in-growth, without the use of adhesive or mechanical fixation.

Coatings offer the possibility of modifying the surface properties of surgical-grade materials to achieve improvements in performance, reliability, and biocompatibility. More recently, techniques such as physical vapor deposition, thermal and electron beam evaporation, plasma metalorganic chemical vapor deposition (MOCVD), electrochemical vapor deposition, thermal or diffusion conversion and sol–gel processing have been used to produce both macro- and nanocoatings.

HAp-coated implants have demonstrated extensive bone apposition in animal models. The development of good implant–bone interfacial strength is thought to be a result of the biological interactions of released calcium and phosphate ions. Quality HAp-coated implants heal faster and attach more completely to the bone. The long-term performance of a calcium-phosphate-coated implant depends on coating properties such as thickness, porosity, phases and crystallinity, implant surface roughness, and overall design.

In addition to the effects of surface topography and chemistry, thin depositions of HAp or calcium phosphate crystals on implants were found to accelerate early bone formation and increase the strength of the bond between implant and bone. A histologic and histomorphometric evaluation of the implant–bone interface was carried out by Orsini and coworkers [37] to determine the effects of a novel surface treatment created by discrete crystalline deposition of nanometer-sized calcium phosphate particles added to the dual acid-etched surface of dental implants placed in the human posterior maxilla. The bone–implant contact evaluations indicated that an increase in osteoconduction along the calcium phosphate-treated surface occurs during the first two months after implant placement. These authors also stated that their results suggested that the nanometric deposition of calcium phosphate crystals could be clinically advantageous for shortening the implant healing period, providing earlier fixation, and minimizing micromotion, thus allowing earlier loading and restoration of function for implants placed in areas with low-density bone.

During the past 20 years, four general industrial coating methods have been adapted for the production of bioactive coatings:

- *Spray coating* was developed by Ducheyne and colleagues, who used relatively thick (100 μm to 2 mm) calcium phosphate coatings for bone in-growth [38].
- *Thick bioglass coating* was initiated and developed by Hench and colleagues for surface bioactivity [39].
- The "self-assembly" of *biomimetics* by precipitation in a simulated body fluid (SBF) solution [40].
- *Nanocoatings*, using a range of methods including dipping in sol–gel HAp solutions to produce strong and bioactive coatings [31].

The sol–gel method in particular represents an attractive and versatile method, as it can be used to produce ceramic coatings from solutions by chemical means. It is relatively easy to perform, and complex shapes can be coated. It has also been shown that the nanocrystalline grain structure results in improved mechanical properties. Sol–gel processing is also unique in that it can be used to produce different forms, such as powders, platelets, coatings, fibers, and monoliths of the same composition, merely by varying the chemistry, viscosity, and a number of factors of a given solution.

The advantages of the sol–gel technique are numerous: it is of the nanoscale; it results in a stoichiometric, homogeneous and pure product, owing to mixing on the molecular scale; high purity can be maintained as grinding can be avoided; it allows reduced firing temperatures due to its small particle sizes with high surface areas; it has the ability to produce uniform fine-grained structures; it allows the use of different chemical routes; and it is easily applied to complex shapes with a range of coating techniques, including dip, spin, and spray deposition (Figure 16.4). Furthermore, sol–gel coatings have the added advantages that the costs of precursors are relatively unimportant, owing to the small amounts of materials required [31].

The lower processing temperature has another advantage, namely that it avoids the phase transition (~1156 K) observed in titanium-based alloys used for biomedical devices.

Currently, companies are producing HAp nanoparticles with diameters in the range of 15 to 20 nm. Various sol–gel routes have been used for the production of synthetic HAp. A number of studies have been carried out on a range of precursors to produce pure nanocrystalline apatites for medical applications. A coating thickness in the range of 70–90 nm has been reported by some investigators [31, 32].

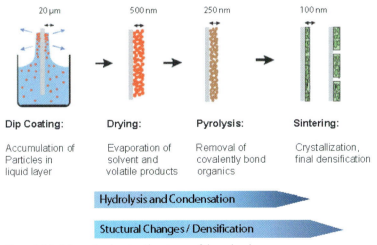

Dip Coating:
Accumulation of
Particles in
liquid layer

Drying:
Evaporation of
solvent and
volatile products

Pyrolysis:
Removal of
covalently bond
organics

Sintering:
Crystallization,
final densification

Hydrolysis and Condensation

Stuctural Changes / Densification

Figure 16.4 Schematic showing the stages of the sol–gel dipping process and densification stages of nanocrystalline coating process.

Ti metal forms a nanoscale sodium titanate hydrogel layer on its surface, when it is soaked in $5\,M$ NaOH solution at $60\,°C$ for $24\,h$. In Japan, a large number of patients have been reported to receive artificial total hip joints of Ti alloy, modified with titanium beads subjected to NaOH treatments.

16.4.2
Sol–Gel Nanohydroxyapatite and Nanocoated Coralline Apatite

Current bone graft materials are mainly produced from coralline HAp. Due to the nature of the conversion process, commercial coralline HAp has retained coral or $CaCO_3$ and the structure possesses nanopores within the inter-pore trabeculae, resulting in high dissolution rates. Under certain conditions, these features reduce durability and strength, respectively, and are not utilized where high structural strength is required. To overcome these limitations, a new double-stage conversion technique was developed by Ben-Nissan and coworkers [23, 31, 32].

The current technique involves a two-stage application route whereby, in the first stage, a complete conversion of coral to pure HAp is achieved. In the second stage, a sol–gel-derived HAp nanocoating is applied directly to cover the meso- and nanopores within the intra-pore material, while maintaining the large pores for appropriate bone growth. The process is shown in Figure 16.5.

The compression and biaxial strengths, fracture toughness, and Young's modulus were each improved as a result of this unique double treatment. Application of the treatment method is expected to result in an enhanced bioactivity due to the nanograin size – and hence large surface area – that increases the reactivity of the nanocoating. It is anticipated that this new material could be applied to load-bearing bone-graft applications where high-strength requirements are pertinent.

(a)   (b)   (c)

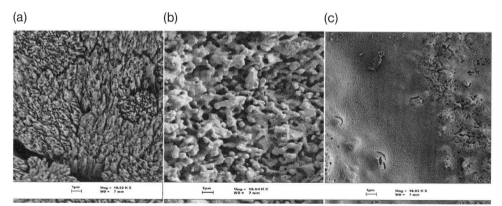

Figure 16.5 Stages of nanocrystalline hydroxyapatite-coated coralline apatite formation. (a) An enlarged micrograph of a coral skeleton spine area that has very sharp meso- and nanopore platelet regions; (b) Surface morphological changes of the coral after conversion to hydroxyapatite with the hydrothermal method; (c) Stage 2 covering of the mesopores and nanopores with a nanohydroxyapatite coating.

(a) (b)

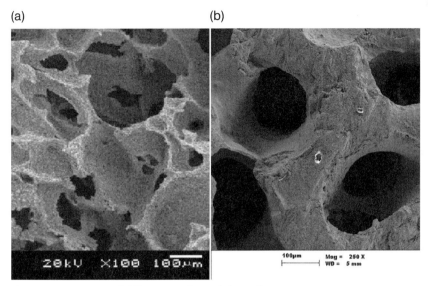

Figure 16.6 (a) Structural differences and morphology of (a) a synthetic tricalcium phosphate and (b) a natural Australian coral skeleton, showing pore size, distribution, and interconnectivity.

For these studies, the coral was obtained from the Australian Great Barrier Reef and contained micropores of 100 to 300 μm size (Figure 16.6). The coral was shaped in the form of a block, and treated with boiling water and 5% NaClO solution. A hydrothermal conversion was carried out in a Parr reactor (Parr Instrument Company, USA) with a Teflon liner at 250 °C and 3.8 MPa pressure with excess $(NH_4)_2HPO_4$. A total conversion to HAp was achieved in this way.

Nanocrystalline ceramic coatings were produced using the sol–gel process. For this, the precursor solution was formed using the previously reported method. Coatings were formed using these solutions, followed by subsequent heat treatments. Mechanical testing involved a standard, four-point bend test according to ASTM C1161, to measure the flexural strength and flexural modulus of the natural coral. Comparative compression and biaxial strength tests were also carried out. Fracture surfaces were then viewed using SEM, which was performed on a LEO-Supra55VP instrument. Samples were analyzed using X-ray diffraction (XRD; Siemens D-5000, Karlsruhe, Germany), with scans being carried out from 20.0 to 60.0 in 0.020 steps at a step time of 2.0 s. A combined thermogravimetric analysis (TGA)/differential thermal analysis (DTA) was performed using a TA Instruments SDT 2960, at a heating rate of 10 °C min^{-1}.

Characterization studies of the natural and converted corals using XRD, SEM, DTA/TGA, nuclear magnetic resonance (NMR) and Raman spectroscopy have been reported previously.

These results showed a large increase in all mechanical properties, specifically the compression strength, due to hydrothermal conversion and nanocoating methods. The bioactivity was enhanced through the nanocrystalline formation, due to the HAp nanocoating.

16.4.3
Surface Modifications

The surfaces of nanostructured materials can be modified and functionalized with different reagents, using a variety of physical, chemical, and/or biological methods. An enhanced solubility or stability of nanosized materials in aqueous media, as well as new material functions and properties, can be achieved via the surface modification of nanostructured materials.

A variety of physical, chemical, and biological surface modifications to increase bioactivity or mechanical properties have been proposed and investigated by many groups. Molecular coating, surface entrapment, and physical treating with plasma, ozone, or ultraviolet (UV) light have emerged as the leading strategies for surface modifications of nanostructured materials using physical methods. Through physical modifications, a range of functional molecules and entities, varying charges or active chemical groups can be introduced onto the surfaces of nanostructured materials, leading to functionalization and activation of the surfaces of materials.

Functional molecules may also be linked to the surfaces of nanostructured materials via certain chemical reactions. Compared to certain physical methods, a chemical modifications can be used not only to activate the surfaces of nanostructured materials to a greater extent, but also to offer stronger interactions between the linking molecules and the material surfaces through stable chemical bonds. A number of different chemical reagents and methods can be used for the surface modification of nanosized materials.

The biological modification of nanoparticle surfaces is often necessary for nanoparticle functionality. By employing chemical or physical methods, biospecific molecules and devices can be incorporated into the nanoparticles, thereby offering biospecific sites for the further immobilization of ligands specific to these molecules. The immobilizations of specific ligands can be performed through biologically specific reactions, for example antibody–antigen and receptor–ligand [21].

Different biomedical devices and applications require different properties and functions of materials. Therefore, methods to modify nanostructured materials in order to meet the needs of various biomedical systems will vary. A brief summary of the basic methods and technologies used to modify nanostructured materials for biomedical devices is presented below.

Today, highly porous scaffolds with an open structure represent the best candidates for cancellous bone substitution. In addition to natural and ceramic materials, many polymers have been proposed for medical applications. Each of these presents different biological and mechanical properties, allowing a choice of the correct polymer for the correct application. However, polymers usually present low elastic modulus values, creep resistance and chemical constituents, compared to bone, and this is the major reason that limits their clinical use for hard- and soft-tissue substitution.

One example was that reported by Peroglio and coworkers [11], who produced PCL-coated alumina scaffolds that were then characterized to validate the concept of polymer–ceramic composites with an increased fracture resistance. The alumina scaffolds were processed using a classical foam replication technique, and then sintered to produce an open-porous structure with ~70% porosity and a mean pore size of 150 μm. The polymer coating was obtained by infiltrating the scaffold with either a PCL solution or PCL nanodispersion. An emulsion–diffusion technique, using a nonionic surfactant, was used for the latter process. Subsequently, after infiltration with PCL, and irrespective of which quantity or infiltration technique was used, no change was seen in the Young's modulus. However, this was to be expected as the elastic modulus of PCL is negligible compared to that of alumina. Nonetheless, the addition of PCL completely altered the mechanical behavior of the scaffold during a four-point bending test, with a 10–20 vol% addition of PCL to the alumina scaffold leading to seven- to 13-fold increases in the apparent fracture energy. A further examination of the material, using SEM, indicated that the toughening was the result of the polymer fibrils bridging the cracks.

In recent years, nanoparticle systems have attracted increasing attention for use as potential drug-delivery systems. Despite the advantages of nanoparticles – such as their small size, which allows them to penetrate small capillaries and be taken up by cells – a number of problems, including a relatively short blood circulation time, have limited their clinical application. The efficiency and targeting ability of a nanoparticle drug-delivery system are often hampered by the rapid recognition of the carrier system by the body. As the main concern for nanoparticle drug carriers is a long circulation time in the blood, numerous approaches to the design and engineering of long-circulating-time carriers have been investigated. Among these, the surface modification of nanoparticles with a range of nonionic surfactant or polymeric macromolecules has proved to be the most successful for maintaining nanoparticle presence in the blood for prolonged periods [41]. Suitable and effective modifications of the nanoparticles are also required, however, to overcome a number of technical problems and possible issues of toxicity.

16.5
Simulated Body Fluids

Artificial materials implanted into bone defects might be encapsulated from time to time by a fibrous tissue, leading to their isolation from the surrounding bone. Any improvement in bone-implant bonding to the tissues requires the correct implant surface morphology and chemistry to generate a mechanical interlock and good surface activity. Interactions between the bone and the implant will be controlled with appropriate biological interactions.

During the past three decades, many investigators have proposed that the essential requirement for an artificial material to bond to living bone is the formation of a bone-like apatite on its surface when implanted in the living body. In 1991, Kokubo *et al.* proposed that *in vivo* apatite formation on the surfaces of many biomedical materials could be reproduced in a SBF with ion concentrations almost equal to those of human blood plasma. In essence, this means that the *in vivo* bone bioactivity of a material can be predicted from the apatite formation on its surface in SBF (Figure 16.7) [42].

Hydroxyapatite layers can be easily produced on various organic and inorganic substrates when submerged in SBF and indeed, in 1989, Kokubo and Takadama [42] showed that, after immersion in SBF, a wide range of biomaterial surfaces initiated very fine crystallites of carbonate ion-containing apatite. Subsequently, many reports have described the ability of osteoblasts to proliferate and differentiate on this apatite layer. Based on these findings, other SBFs have been produced in order to provide insight into the reactivity of the inorganic component of blood plasma, and to predict the bioactivity of implants and bone scaffolds, as well as other novel biomaterials. SBF has also been used to prepare bioactive composites by forming HAp on various types of substrate.

The SBF solutions have been shown to induce apatitic calcium phosphate formation on any metal, ceramic, or polymer soaked in them. The SBF solutions, which closely resemble Hank's balanced salt solution (HBSS) [39], are prepared with the aim of simulating the ion concentrations present in human plasma. Hence, the solutions are prepared with relatively low calcium and phosphate ion concentrations (i.e., 2.5 and 1.0 mM, respectively), while the pH is adjusted to a physiological value of 7.4 by using organic buffers (e.g., Tris or HEPES).

Typically, a SBF solution will have ionic concentrations of 142.0 mM Na^+, 5.0 mM K^+, 1.5 mM Mg^{2+}, 2.5 mM Ca^{2+}, 147.8 mM Cl^-, 4.2 mM HCO_3^-, 1.0 mM HPO_4^{2-}, and 0.5 mM SO_4^{2-}, with a pH of 7.4, all of these values being almost equal to those of human blood plasma at 36.5 °C. The SBF is usually prepared by dissolving reagent-grade NaCl, $NaHCO_3$, KCl, $K_2HPO_4 \cdot 3H_2O$, $MgCl_2 \cdot 6H_2O$, $CaCl_2$ and Na_2SO_4 in distilled water and buffering at pH 7.4 with Tris(hydroxymethyl)aminomethane ($(CH_2OH)_3CNH_3$) and 1.0 M hydrochloric acid at 36.5 °C.

Since their ionic composition is more or less similar to that of human blood plasma, the HBSS or SBF formulations have only limited power with regards to the precipitation of apatitic calcium phosphates. As a direct consequence, the nucleation and precipitation of calcium phosphates from HBSS or SBF solutions is rather slow. The time taken to achieve total surface coverage of a $10 \times 10 \times 1$ mm titanium or titanium alloy substrate immersed in a 1.5× or 2× SBF solution is typically two to three weeks, with frequent (every 36–48 h) replenishment of the solution [43].

Among the metallic oxide gels prepared using a sol–gel method, those consisting of SiO_2, TiO_2, ZrO_2, and Ta_2O_5 were found to have apatite formation on their surfaces in SBF, as shown in Figure 16.7. These results indicated that the Si–OH, Ti–OH, Zr–OH, and Ta–OH groups on the surfaces of these gels were effective in inducing apatite formation on their surfaces within the body environment (Figure 16.8).

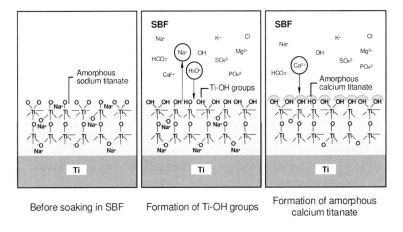

Before soaking in SBF   Formation of Ti-OH groups   Formation of amorphous calcium titanate

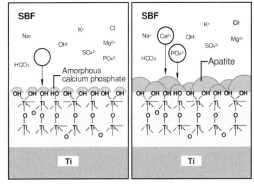

Figure 16.7 Treatment of a titanium alloy surface with NaOH to produce a bioactive surface on which hydroxyapatite particles nucleate and grow quite readily within the SBF solution [40].

A variety of studies have been conducted using SBF solutions to deposit apatite on both two-dimensional (2-D) and 3-D scaffolds. For example, Wu and coworkers [41] developed a novel bioactive, degradable and cytocompatible bredigite ($Ca_7MgSi_4O_{16}$) scaffold with a biomimetic apatite layer (BTAp) for bone-tissue engineering. For this, porous bredigite scaffolds were first prepared using the polymer sponge method. A BTAp was then applied to the scaffolds by soaking them in SBF (pH 7.4) at 37 °C for 10 days, with a solution volume-to-scaffold mass ratio of $200\,ml\,g^{-1}$. After soaking, the scaffolds were dried at 120 °C for one day, such that bredigite scaffolds with BTAps were obtained. The porosity and *in vitro* degradability of the BTAp scaffolds were investigated. Likewise, the osteoblast-like cell morphology, proliferation and differentiation on BTAp scaffolds were evaluated and compared with those of β-tricalcium phosphate (β-TCP) scaffolds. The results showed that the bredigite scaffolds possessed a highly porous structure with a large pore size (300–500 μm). This biomimetic process mimics biomineralization and leads to the formation of a bone-like apatite layer on the scaffold

Figure 16.8 Apatite formation on silica, titania, zirconia, and tantalum oxide gel-covered substrates within the SBF solution [40].

surface. The obtained BTAp scaffolds also possessed a high porosity (90%) and pore interconnectivity. When compared to β-TCP scaffolds, the cells on BTAp scaffolds showed a higher proliferation rate and differentiation level.

Cromme and coworkers [44] investigated the activation of regenerated cellulose 2-D model thin films and 3-D fabric templates with calcium hydroxide, $Ca(OH)_2$. For this, the Langmuir–Blodgett (LB) film technique was applied to manufacture model thin films using a trimethylsilyl derivative of cellulose (TMS-cellulose). Regenerated cellulose films were obtained by treating the TMS-cellulose LB-films with hydrochloric acid vapors. For the 3-D templates, regenerated cellulose fabrics were used, and the templates activated with a $Ca(OH)_2$-suspension and subsequently exposed to 1.5× SBF to induce the *in situ* formation of calcium phosphate phases. The calcium phosphates were identified using Fourier transform infrared (FTIR) and Raman spectroscopy as highly carbonated apatites (CA) lacking hydroxyl ions. Such 3-D fabric templates of regenerated cellulose covered with a biomimetic coating of apatite might be of particular interest for novel scaffold architectures in bone repair and tissue engineering.

Lim and coworkers [45] noted that bone-like apatite could be more efficiently coated onto the scaffold surface by using polymer/ceramic composite scaffolds rather than polymer scaffolds, and by using an accelerated biomimetic process to enhance the osteogenic potential of the scaffold. The creation of a bone-like, apatite-coated polymer scaffold was achieved by incubating the scaffolds in SBF. Apatite growth on the porous poly(D,L-lactic-*co*-glycolic acid)/nanohydroxyapatite (PLGA/HAp) composite scaffolds was significantly faster than on the porous PLGA scaffolds. In addition, the distribution of coated apatite was more uniform on the PLGA/HAp scaffolds than on the PLGA scaffolds. After a five-day incubation period, the mass of apatite coated onto the PLGA/HAp scaffolds incubated in 5× SBF was 2.3-fold higher than on the PLGA/HAp scaffolds incubated in 1× SBF. Furthermore, when the scaffolds were incubated in 5× SBF for five days, the mass of apatite coated onto the PLGA/HAp scaffolds was 4.5-fold higher than on the PLGA scaffolds. These results indicated that the SBF-initiated apatite coating could be accelerated by using a polymer/ceramic composite scaffold and concentrated SBF. It was reported that, when seeded with osteoblasts, the apatite-coated PLGA/HAp scaffolds exhibited significantly higher cell growth, alkaline phosphatase (ALP) activity, and mineralization *in vitro* compared to the PLGA scaffolds coated only with HAp. In conclusion, the biomimetic apatite coating could be accelerated by both introducing nucleation sites into polymer scaffolds and using concentrated SBF. When seeded with osteoblasts, scaffolds with accelerated apatite coating significantly enhanced cell growth, ALP activity, and mineralization *in vitro*.

The SBF method was used by Kolos *et al.* [46] to fabricate calcium phosphate fibers for biomedical applications. A natural cotton substrate was first pretreated with phosphorylation and a $Ca(OH)_2$-saturated solution, and then soaked in SBF of two different concentrations, namely 1.5× and 5.0× the ion concentration of blood plasma. The cotton was then burned out by sintering the ceramic coating at 950 °C, 1050 °C, 1150 °C, and 1250 °C, such that hollow calcium phosphate fibers approximately 25 μm in diameter and with a 1 μm wall thickness were successfully manufactured. However, the 5.0× SBF produced a thicker and more crystalline coat of greater uniformity. Kolos *et al.* further reported that osteoblastic cells were able to cover the entire surface of the cotton fibers; more surprisingly, the cell coverage seemed to be independent of the surface roughness and the fibers' Ca : P ratio.

Although of micron size rather than nanosize, a bioactive CHAp layer on cellulose fabrics was developed by Hoffman *et al.* [47]. Nonwoven cellulose (regenerated, oxidized) fabrics were coated with CHAp using a procedure based on the SBF method. For this, SBF with a high degree of supersaturation (5× SBF) was applied to accelerate the biomimetic formation of bone-like apatite on the cellulose fabrics. After creating calcium phosphate nuclei on the cellulose fibers in an initial 5× SBF with high Mg^{2+} and HCO_3^- concentrations, the cellulose fabrics were additionally soaked in a second 5× SBF which was optimized with regards to accelerated crystal growth by reduced Mg^{2+} and HCO_3^- concentrations. The carbonated apatite layer thickness was increased from $6\,\mu m$ after a 4h soaking in the latter solution, to $20\,\mu m$ after 48h. The amount of CO_3^{2-} substituting PO_4^{3-} in the HAp lattice of the precipitates could be varied by changing the soaking time.

16.6
Nano- and Macrobioceramics for Drug Delivery and Radiotherapy

16.6.1
Nanobioceramics for Drug Delivery

For drug delivery, the primary aim is to target drugs to specific sites within the body, and to release them in a controllable fashion. However, for many current delivery systems the guest molecules are often released upon dispersion of the carrier/drug composites in water. This type of premature release is particularly undesirable and problematic when the guest molecule (e.g., an anti-tumor drug) is cytotoxic and might potentially harm healthy cells and tissues before being delivered to the affected sites [48].

In the case of ceramics, the critical pore and grain size may be varied from a few nanometers up to microns in order to control the ease of delivery and dispersion of a material to the targeted area. A variety of nanoceramic drug-delivery systems are currently undergoing clinical evaluation. In addition to reducing toxicity to nondiseased cells, these systems have the potential to increase drug efficiency, which translates to significant cost savings for the expensive drug treatments that currently are being engineered. On the basis of their physical size, nano drug-delivery systems also have the extraordinary characteristic of being able to target and control drug release with very high precision.

Mesoporous silica nanoparticle (MSN) materials, such as mobile composition of materials (MCM)-41/48, can be synthesized by utilizing surfactants as structure-directing templates to generate a range of mesoporous structures with high surface areas ($>900\,m^2\,g^{-1}$), tunable pore sizes in the range of 2 to 20nm, and uniform pore morphologies. Recent breakthroughs in terms of morphology control and the surface functionalization of MSN materials have resulted in a range of new materials that can be used as stimuli-responsive, controlled-release delivery-carriers for many biotechnological and biomedical applications. As with some other conventional drug-delivery agents with high loading capacities (e.g., polymer and liposomes), MSNs can encapsulate large quantities of drugs with various sizes, shapes, and functionalities [46]. In contrast to many biodegradable polymeric delivery systems, in which the loading of drug molecules requires organic solvents, the molecules of interest can be encapsulated inside the porous framework of the MSN by capping the openings of the mesoporous channels covalently with size-defined "caps," which physically block the drugs from leaching out. Drug molecules loaded into the pores may then be released by the introduction of "uncapping triggers," with the rate of release being controlled by the concentration of the trigger molecules. Prior to uncapping, the capped MSN system exhibits a negligible release of drug molecules. This "zero-release" feature of a capped MSN delivery system, along with an ability to tune the rate of release by varying stimulant concentrations, are important prerequisites for developing delivery systems with many site-specific applications, such as highly toxic anti-tumor drugs, hormones, and neurotransmitters to certain cells types and tissues [48].

An MCM-41-type MSN-based, controlled-release delivery system has been synthesized and characterized using surface-derivatized cadmium sulfide (CdS) nanocrystals as chemically removable caps to encapsulate several drug molecules and neurotransmitters inside the organically functionalized MSN mesoporous framework. Lai and coworkers [49] studied the stimuli-responsive release profiles of vancomycin- and adenosine triphosphate (ATP)-loaded MSN delivery systems by using disulfide bond-reducing molecules, such as dithiothreitol (DTT) and mercaptoethanol (ME), as release triggers. The biocompatibility and delivery efficiency of the MSN system with neuroglial cells (astrocytes) *in vitro* was demonstrated. In contrast to many current delivery systems, the molecules of interest were encapsulated inside the porous framework of the MSN by capping the openings of the mesoporous channels with size-defined CdS nanoparticles, so as to physically prevent the drugs/neurotransmitters from leaching out.

Porous aluminosilicate ceramics were investigated by Byrne and Deasy [50] for their potential to act as extended-release drug-delivery systems. The aluminosilicate pellets were obtained either commercially, produced by extrusion-spheroization, or by cryopelletization. It was reported that each product had a highly interconnected porous microstructure, with the porosity and pore-size distribution being product-dependent. Drugs were loaded into the pellets using a vacuum impregnation technique, with the concentration of the drug loading solution and pellet porosity influencing the loading obtained. Each product provided an extended release of the incorporated drug, with the rate-determining step of release being the diffusion of the drug from the porous pellet interior into the bulk dissolution medium. Byrne and Deasy [50] showed that this rate was influenced by the pellet size, its porosity, pore-size distribution and porous microstructure, and by electrostatic interactions between the pellet surfaces and the drug. The solubility of the drug in the dissolution medium and its molecular weight also influenced the release rate. It was concluded that porous aluminosilicate pellets represent a particularly versatile class of extended-release drug-delivery system, as the drug is incorporated into the pellets after their production.

A new TiO_2 nanostructured bioceramic device was synthesized by López and coworkers [51], using a sol–gel process in order to control the pore-size distribution and particle size. The objective was to obtain a constant drug release rate for anti-epileptic drugs directly into the central nervous system (CNS). This method of drug delivery, using small reservoirs, is very important in pharmaceutical applications, as it offers advantages such as the elimination of secondary effects, a long duration of pharmacological activity, and protection of the drug against enzymatic degradation or pH variations. Among the best-developed and most studied materials, Titania has been shown to be an excellent candidate because of the possibility to manipulate both the structure and the number of OH groups. Point defects were generated in the Titania network in order to obtain the desired interaction between a highly polar drug and the Titania device. The device contained an anti-convulsant drug, valproic acid (VPA), which could be released directly into the temporal lobe of the brain, at a constant rate. During the initial stage of the synthesis, VPA was added until a completely homogeneous solution was created; this solution was then maintained under constant stirring until a gel was formed. Under these conditions, the titanium dioxide underwent nucleation, whilst at the same time it was restructuring around the VPA in such a way that the chemical and polar properties of the anticonvulsant were preserved. The reaction was seen to take place under well-controlled conditions of pressure and temperature, which were purposely kept low. During the gelation period, both the electronic and molecular properties of the material were preserved. Finally, a dry gel was obtained, in which the anticonvulsant was occluded within the pore structure. The release of VPA into the temporal lobe of the brain, and its effect on epileptic rats, was observed by using the Kindling method. Several important conclusions were drawn from these studies, notably that a Titania device charged with an anticon-vulsant drug could be successfully implanted in the rat's temporal lobe. Pore-blocking with higher concentrations of VPA led to a fall in the initial rate of drug release, while insertion of the device caused a drastic reduction in the animal's epileptic activity.

16.6.2
Microbioceramics for Drug Delivery

The particulate forms of ceramic materials have found application in both medical and non-medical fields. Particles in the form of microspheres are especially applicable when treating tumors located in organs that are supplied by a single afferent arterial blood supply. More traditionally, microspheres and nanostructurally modified ceramics have been used in the targeted drug delivery of chemotherapeutic and radiotherapeutic agents.

In recent years, the preparation of surface-modified hollow microspheres has attracted considerable attention because of their unusual properties, notably their large specific surface area due to the nanolayer-modified surfaces, their low density and their encapsulation properties. Consequently, these materials should be very useful for novel applications such as drug- and protein-delivery systems. In certain applications, the efficacy of microparticulate materials can be greatly improved if they can act simultaneously as carriers for biologically active molecules. In this sense, porous and surface-modified materials have an advantage, as they present an additional surface area that greatly influences the loading capacity and release rates.

In dental clinical applications, such as the treatment of severe periodontitis, where massive alveolar bone loss occurs, bone defect filling and intensive systemic long-term antibiotics administration are often required. Nanohydroxyapatite microspheres intended for use as injectable bone filling material or as enzyme-delivery matrices may also be used as antibiotic-releasing materials. Ferraz and coworkers [52] developed a novel injectable drug-delivery system with a drug-releasing capability for treating periodontitis. Their aim was to use HAp, loaded with an antibiotic, as a local drug-delivery system; consequently amoxicillin (with and without clavulanic acid) and erythromycin – both of which are used orally and parenterally to treat periodontitis – were tested. The results showed that the micro-spheres exhibited chemical compositions, porosities, and surface areas which provided adequate conditions for the sustained-release profile of the drugs inves-tigated. Bactericidal tests indicated that the released antibiotics had an inhibitory effect on the bacteria present, while osteoblastic cells were seen to proliferate well on the surface of the microspheres, with cell growth being enhanced by the pres-ence of antibiotics. Ferraz *et al.* concluded that these microspheres might also be considered as an alternative carrier system for antibiotics and to enhance bone regeneration while treating periodontitis.

16.6.3
Microbioceramics for Radiotherapy

The partial surgical removal of an organ afflicted with cancer generally results in a total or partial loss of organ function, that may or may not be recovered postop-eratively. It would, therefore, be desirable to develop a cancer treatment that could destroy only cancerous cells, such that the normal healthy cells could regenerate after treatment and organ function be maintained. Although *radiotherapy* shows such potential, the irradiation is generally applied from an external source, often causing the cancer to receive an insufficient dose, especially if it is deep-seated. Moreover, the irradiation may cause severe damage to healthy tissues. In order to overcome this problem, microsphere-assisted radiation therapy is currently under development for a range of clinical applications, and has indeed shown a degree of success.

An *in situ* microsphere-assisted radiation method has been clinically applied using $17Y_2O_3$–$19Al_2O_3$–$64SiO_2$ (mol%) (YAS) glass microspheres that have been prepared using a conventional melt–quench method. Although the yttrium-89 ([89]Y) in this glass is a nonradioactive isotope, neutron bombardment will activate [89]Y to form the β-emitter ^{90}Y, which has a half-life of 64.1 h. When radioactive glass microspheres 20–30 μm in diameter are injected into a target organ (e.g., a liver tumor), they become trapped inside the small blood vessels in the tumor, blocking the nutritional supply to the tumor, and delivering a large, localized dose of short-range, highly ionizing β-rays. As the β-rays have a short penetration range of approximately 2.5 mm in living tissue, only minimal radiation damage will occur among the neighboring healthy tissues. These microspheres show a high chemical durability and, after their administration, the radioactive ^{90}Y remains essentially within the microspheres so that it does not affect any neighboring healthy tissues. The radioactivity of ^{90}Y decays to a negligible level within 21 days after its initial preparation; hence, the microspheres become inactive soon after the cancer treat-ment. These microspheres have been used clinically to treat liver cancer in Canada, the USA, and China. Likewise, radioactive yttrium-containing resin microspheres 30–35 μm in diameter have also been used clinically to treat liver cancer in Australia, China, New Zealand, and Singapore [53].

Kawashita and coworkers [53] have successfully prepared hollow Y_2O_3 microspheres of 20–30 μm diameter. For this, an aqueous carboxymethylcellulose sodium salt solution containing urease was atomized into an aqueous yttrium nitrate solution containing urea, using a spray gun located 2 m above the yttrium nitrate solution. The resultant solid materials were heat-treated at 1300 °C. An investigation of the structure and chemical durability of these microspheres showed the outer surfaces to be smooth and dense, and the inner parts to have a honeycomb structure. In SBFs at pH 6 and 7, the hollow Y_2O_3 microspheres showed a high chemical durability.

It has been accepted that cancer cells generally perish at a temperature of approximately 43 °C because their oxygen supply via the blood vessels becomes insufficient; this is in contrast to normal cells, which do not suffer damage at higher temperatures of approximately 48 °C. A tumors is also more easily heated than the surrounding normal tissues, as its blood vessels and nervous system are poorly developed. It follows, therefore, that *hyperthermia* might represent a very effective treatment for cancer, with few adverse side effects. A variety of techniques has been employed to heat the tumors, including hot water, IR radiation, ultrasound, and microwaves but, unfortunately, deep-seated tumors cannot be heated either effectively or locally using these methods. Ferrimagnetic microspheres 20–30 μm in diameter (and more recently nanospheres) have been shown to be valuable as *thermoseeds* for inducing hyperthermia in cancers, especially in those tumors located deep inside the body. In this situation, the spheres are implanted through the blood vessels and become entrapped in the capillary bed of the tumor. The application of an alternating magnetic field close to the tumor then causes the spheres to generate heat by their hysteresis loss; this results in the tumor being heated locally and strongly to a point where the cancerous cells are killed.

Kawashita *et al.* [54] prepared Fe_3O_4 microspheres by melting powders in a high-frequency induction thermal plasma, and also by precipitation from an aqueous solution. The preparation in a high-frequency induction thermal plasma involved pure Fe_3O_4 powders being completely melted in a plasma flame of argon gas, which was produced by high-frequency induction heating. The solidified products were sieved using a nylon mesh in order to obtain particles of 20–30 μm in size. These particles were then heat-treated to 600 °C and allowed to cool under a reduced pressure of 5.1×10^3 Pa. Fe_3O_4 powders were added into 600 ml of a 1 wt% aqueous solution of hydrofluoric acid (HF) for the preparation of microspheres in aqueous solution. Silica glass microspheres of 12.4 μm average diameter were soaked in Fe–HF solution at 30 °C for 24 days, with vigorous stirring. The products were then heat-treated at 600 °C for 1 h, and allowed to cool in a CO_2–H_2 gas atmosphere. The microspheres prepared in a high-frequency induction thermal plasma were shown to be composed mainly of Fe_3O_4, accompanied by a small amount of FeO. The surfaces of the microspheres were smooth before heat treatment, but became slightly rough after treatment. This increase in surface roughness might be attributed to the formation of α-Fe_2O_3 and/or the crystal growth of Fe_3O_4. The saturation magnetization of the microspheres was $92\,emu\,g^{-1}$. The heat generation was as low as $10\,W\,g^{-1}$, under $300\,Oe$ and $100\,kHz$. Microspheres, 20–30 μm in diameter, composed of small crystals of Fe_3O_4 and $50\,nm$ in size, were prepared by precipitation from an aqueous solution and subsequent heat treatment. Notably, the β-FeOOH was seen to precipitate on the surface of silica glass microspheres, while its layer thickness increased with the increasing soaking period. The deposited β-FeOOH was transformed into Fe_3O_4 by heat treatment above $400\,°C$ in a CO_2–H_2 atmosphere. The silica glass microspheres soaked in Fe–HF solution showed ferrimagnetism, with a saturation magnetization of $53\,emu\,g^{-1}$, a coercive force of $156\,Oe$ and a heat generation of $41\,W\,g^{-1}$, under $300\,Oe$ and $100\,kHz$, under the same magnetic field, which was fourfold higher than for pure Fe_3O_4 microspheres prepared in a high-frequency induction thermal plasma.

16.7
Nanotoxicology and Nanodiagnostics

Today, nanomaterials and nanotechnologies represent a rapidly expanding discipline, with major potential to improve the lifestyle of humankind, notably by offering new solutions to many engineering and biomedical problems.

One major concern has arisen, however, regarding possible risks related to the impact of synthetic nanoparticulate matter on human health. However, at present insufficient data are available to make any accurate predictions concerning the long-term consequences of using these materials. Nevertheless, unintentional nanopollution is amply present in the current environment, and the use of nanomaterials for medical applications, as described here, is expected to increase as their benefits are recognized. Yet, it is somehow pertinent to mention that, there is no reason why unintentional pollution should differ from the intentional pollution, as far as its impact on health is concerned.

Gatti et al. [55], in their European project "Nanopathology" (QOL-2002-147), succeeded in developing a novel diagnostics tool which showed the presence of microsize and nanosized inorganic particles (including nanoceramics in pathological tissues) by using an environmental scanning electron microscope equipped with an X-ray microprobe of an energy-dispersive system. This study, which included approximately 700 tissues affected by cryptogenic pathologies, showed that in many cases where phlogosis (the inflammation of external parts) was dominant, then microsized and nanosized inorganic particles were present, and their elemental chemistry had been clearly identified and reported.

Among the pathologies investigated by Gatti et al. were cryptogenic granulomatosis and different forms of cancer, including lymphomas, carcinomas of the lung, colon and liver, and leukemia. Deep-vein thrombosis was also investigated, to demonstrate the possible thrombogenic activity of environmental pollutants. In most cases, an inflammatory component was evident (e.g., in granulomas) when comparatively large particles (usually >10 μm) were identified. Particles the size of an erythrocyte or smaller were commonly found in thrombi, while particles down to 10 nm in size were identified in the cell nuclei. From a chemistry standpoint, Gatti *et al.* [55] identified a very wide variety of elemental compositions. As result of their investigations, the group advised that the use of biodegradable nanoceramics such as calcium phosphates should be seriously considered for biomedical applications, due to their non-nanotoxicity. However, they suggested that extreme care be taken with all other synthetic nanobiomaterial medical applications, until a better understanding and a more comprehensive combat strategy is available regarding these materials.

References

1 Chang, M.C., Ko, C.C. and Douglas, W.H. (2003) *Biomaterials*, **24**, 2853.

2 Zhang, W., Liao, S.S. and Cui, F.Z. (2003) *Chem. Mater.*, **15**, 3221.

3 Vinogradov, S.V., Bronich, T.K. and Kabanov, A.V. (2002) *Adv. Drug Delivery Rev.*, **54**, 135.

4 Vinogradov, S.V., Batrakova, E.V. and Kabanov, A.V. (2004) *Bioconj. Chem.*, **15**, 50.

5 Clark, H.A., Hoyer, M., Philbert, M.A. and Kopelman, R. (1999) *Anal. Chem.*, **71**, 4831.

6 Park, E.J., Brasuel, M., Behrend, C., Philbert, M.A. and Kopelman, R. (2003) *Anal. Chem.*, **75**, 3784.

7 Gavalas, V.G., Law, S.A., Ball, J.C., Andrews, R. and Bachas, L.G. (2004) *Anal. Biochem.*, **329**, 247.

8 Croteau, S., Rauch, F., Silvestri, A. and Hamdy, R.C. (1999) *Orthopaedics*, **22**, 686–695.

9 Harakas, N.K. (1984) *Clin. Orthop. Rel. Res.*, **188**, 239–251.

10 Zhang, N., Nichols, H.L., Taylor, S. and Wen, X. (2007) *Mater. Sci. Eng.*, **27**, 599–606.

11 Peroglio, M., Gremillard, L., Chevalier, J., Chazeau, L., Gauthier, C. and Hamaide, T. (2007) *J. Eur. Ceram. Soc.*, **27**, 2679–2685.

12 Jones, J.R., Ehrenfried, L.M. and Hench, L.I. (2006) *Biomaterials*, **27**, 964–973.

13 Murugan, R. and Ramakrishna, S. (2005) *Compos. Sci. Technol.*, **65**, 2385–2406.

14 Oliveira, J.M., Rodrigues, M.T., Silva, S.S., Malafaya, P.B., Gomes, M.E., Viegas, C.A., Dias, I.R., Azevedo, J.T., Mano, J.F. and Reis, R.L. (2006) *Biomaterials*, **27**, 6123–6137.

15 TenHuisen, K.S., Martin, R.I., Klimkiewicz, M. and Brown, P.W. (1995) *J. Biomed. Mater. Res.*, **29**, 803–810.

16 Scabbia, A. and Trombelli, L. (2004) *J. Clin. Periodontol.*, **31**, 348–355.

17 Wahl, D.A. and Czernuszka, J.T. (2006) *Eur. Cell. Mater.*, **11**, 43–56.

18 Yoshikawa, T., Ohmura, T., Sen, Y., Iida, J., Takakura, Y., Nokana, I. and Ichijama, K. (2003) in *Bioceramics 15* (eds B. Ben-Nissan, D. Sher and W. Walsh), Trans Tech Publications, Uetikon-Zurich, pp. 383–386.

19 Abramovitch-Gottlib, L., Geresh, S. and Vago, R. (2006) *Tissue Eng.*, **12**, 729–739.

20 Stock, U.A. and Vacanti, J.P. (2001) *Annu. Rev. Med.*, **52**, 443–451.

21 Xu, T., Zhang, N., Nichols, H.L., Shi, D. and Wen, X. (2007) *Mater. Sci. Eng. C*, **27**, 579–594.

22 Oliveira, A.L., Gomes, M.E., Malafaya, P.B. and Reis, R.L. (2003) in *Bioceramics 15* (eds B. Ben-Nissan, D. Sher and W. Walsh), Trans Tech Publications, Uetikon-Zurich, p. 101.

23 Zreiqat, H., Valenzuela, S.M., Ben-Nissan, B., Roest, R., Knabe, C., Radlanski, R.J., Renz, H. and Evans, P.J. (2005) *Biomaterials*, **26**, 7579–7586.

24 Ko, H.F., Sfeir, C. and Kumta, P.N. (2007) *Mater. Sci. Eng. C*, **27**, 479–483.

25 Jones, J.R., Ehrenfried, L.M., Saravanapavan, P. and Hench, L.L. (2006) *J. Mater. Sci.: Mater. Med.*, **17**, 989–996.

26 Al-Jamal, W.T. and Kostarelos, K. (2007) *Nanomedicine*, **2**, 85–98.

27 Schmidt, H.T. and Ostafin, A.E. (2002) *Adv. Mater.*, **14**, 532–535.

28 Moura, S.P. and Carmona-Ribeiro, A.M. (2006) *Cell. Biochem. Biophys.*, **44**, 446–452.

29 Xu, Q., Tanaka, Y. and Czernuszka, J.T. (2007) *Biomaterials*, **28**, 2687–2694.

30 Huang, J., Liu, K., Cheng, C., Ho, K., Wu, Y., Wang, C., Cheng, Y., Ko, W. and Liu, C. (1997) *Kaohsiung J. Med. Sci.*, **13**, 213.

31 Ben-Nissan, B. and Choi, A.H. (2006) *Nanomedicine*, **1**, 311–319.

32 Choi, A.H. and Ben-Nissan, B. (2007) *Nanomedicine*, **2**, 51–61.

33 Oonishi, H., Clarke, I.C., Good, V., Amino, H., Ueno, M., Masuda, S., Oomamiuda, K., Ishimaru, H., Yamamoto, M. and Tsuji, E. (2003) in *Bioceramics 15* (eds B. Ben-Nissan, D. Sher and W. Walsh), Trans Tech Publications, Uetikon-Zurich, p. 735.

34 Saiz, E., Gremillard, L., Menendez, G., Miranda, P., Gryn, K. and Tomsia, A.P. (2007) *Mater. Sci. Eng. C*, **27**, 546–550.

35 Khalyfa, A., Vogt, S., Weisser, J., Grimm, G., Rechtenbach, A., Meyer, W. and Schnabelrauch, M. (2007) *J. Mater. Sci.: Mater. Med.*, **18**, 909–916.

36 Gomes de Sousa, F.C. and Evans, J.R.G. (2005) *Adv. Appl. Ceram.*, **104**, 30–34.

37 Orsini, G., Piattelli, M., Scarano, A., Petrone, G., Kenealy, J., Piattelli, A. and Caputi, S. (2007) *J. Periodontol.*, **78**, 209–218.

38 Ducheyne, P., Beight, J., Cuckler, J., Evans, B. and Radin, S. (1990) *Biomaterials*, **11**, 244–254.

39 Hench, L.L. and West, J.K. (1990) *Chem. Rev.*, **90**, 33–72.

40 Kukobo, T., Kim, H.M., Kawashita, M. and Nakamura, T. (2000) *J. Aust. Ceram. Soc.*, **36**, 37–46.

41 Wu, C., Chang, J., Zhai, W. and Ni, S. (2007) *J. Mater. Sci.: Mater. Med.*, **18**, 857–864.

42 Kukobo, T. and Takadama, H. (2006) *Biomaterials*, **27**, 2907–2915.

43 Tas, A.C. and Bhaduri, S.B. (2004) *J. Mater. Res.*, **19**, 2742–2749.

44 Cromme, P., Zollfrank, C., Müller, L., Müller, F.A. and Greil, P. (2007) *Mater. Sci. Eng. C*, **27**, 1–7.

45 Kim, S.S., Park, M.S., Gwak, S.J., Choi, C.Y. and Kim, B.S. (2006) *Tissue Eng.*, **12**, 2997–3006.

46 Kolos, E.C., Ruys, A.J., Rohanizadeh, R., Nuir, M.M. and Roger, G. (2006) *J. Mater. Sci.: Mater. Med.*, **17**, 1179–1189.

47 Hofmann, I., Mülloer, L., Greil, P. and Müller, F.A. (2006) *Surf. Coat. Technol.*, **201**, 2392–2398.

48 Giri, S., Trewyn, B.G. and Lin, V.S.Y. (2007) *Nanomedicine*, **2**, 99–111.

49 Lai, C.Y., Trewyn, B.G., Jeftinija, D.M., Jeftinija, K., Xu, S., Jeftinija, S. and Lin, V.S.Y. (2003) *J. Am. Chem. Soc.*, **125**, 4451–4459.

50 Byrne, R.S. and Deasy, P.B. (2005) *J. Microencapsul.*, **22**, 423–437.

51 López, T., Ortiz, E., Quintana, P. and González, R.D. (2007) *Colloids Surf. A: Physicochem. Eng. Aspects*, **300**, 3–10.

52 Ferraz, M.P., Mateus, A.Y., Sousa, J.C. and Monteiro, F.J. (2007) *J. Biomed. Mater. Res. A*, **81**, 994–1004.

53 Kawashita, M., Takayama, Y., Kokubo, T., Takaoka, G.H., Araki, N. and Hiraoka, M. (2006) *J. Am. Ceram. Soc.*, **89**, 1347–1351.

54 Kawashita, M., Tanaka, M., Kokubo, T., Inoue, Y., Yao, T., Hamada, S. and Shinjo, T. (2005) *Biomaterials*, **26**, 2231–2238.

55 Gatti, A. and Montanari, S. (2005) in *Handbook of Nanostructured Biomaterials and Their Application in Nanobiotechnology* (ed. H.S. Nalwa), American Scientific Publishers, pp. 347–369.

17
Self-healing of Surface Cracks in Structural Ceramics

Wataru Nakao, Koji Takahashi, and Kotoji Ando

17.1
Introduction

Self-healing is the most valuable phenomenon to overcome the integrity decreases that are caused by the damages in service. Thus, self-healing should occur automatically as soon as the damages occur, and the healed zone should have high integrity as it was before damaging. When proposing self-healing materials, one must know the nature of the damage and the service, conditions of the materials. In the case of the structural ceramics, the severest damage is surface cracks, which is possible to be introduced by crash, fatigue, thermal shock, and corrosion during their service time. Over the past 30 years, ceramics have become the key materials for structural use at high temperature due to their enhanced quality and good processability. Structural ceramics are also expected to be applied in the corrosion environments such as air, because of its chemical stability. Thus, self-healing of surface cracks in the structural ceramics is an important issue to ensure the structural integrity of ceramic components.

In this chapter the mechanism and effects of self-healing of surface cracks in structural ceramics are introduced. Apart from this, the fracture manner of ceramics is also discussed. This will help the readers to understand the self-healing phenomena in ceramics and its benefits. The history of crack healing is also included in the text. Furthermore, new methodology to ensure the structural integrity using crack-healing effect and advanced ceramics having self–crack-healing ability are mentioned. Finally, it is assayed the self-healing ceramics can be applied to blades and nozzles of high-temperature gas turbine.

17.2
Fracture Manner of Ceramics

Ceramics tend to have brittle fracture that usually occurs in a rapid and catastrophic manner. Brittle fracture is usually caused by the stress concentration at

Advanced Nanomaterials. Edited by Kurt E. Geckeler and Hiroyuki Nishide
Copyright © 2010 WILEY-VCH Verlag GmbH & Co. KGaA, Weinheim
ISBN: 978-3-527-31794-3

the tip of the flaws. For brittle fracture under pure mode, I, loading, under which crack is subjected to opening, the failure criteria is that the stress intensity factor, K_I, is equal to the fracture toughness, K_{IC}. The stress intensity factor is an indicator of the magnitude of stress near a crack tip or the amplitude of the elastic field. The value of K_I can be obtained from the liner elastic mechanics as-follows:

$$K_I = \sigma \cdot Y \sqrt{\pi \cdot a} \tag{17.1}$$

where a is the crack length, σ is the tensile stress applying to crack perpendicularly, and Y is a dimensionless parameter that is determined from the crack and loading geometries. Thus the stress at fracture is given by

$$\sigma_c = \frac{1}{Y} \cdot \frac{K_{IC}}{\sqrt{\pi \cdot a}} \tag{17.2}$$

From the fracture criterion, one can understand that the fracture strength of ceramic components is not intrinsic strength but is fracture toughness and crack geometry.

In general, ceramic components contain flaws at which the stress concentration causes brittle fracture before their end-use. These flaws have particular figurations and sizes and are introduced mainly during manufacturing. Figure 17.1 shows a schematic diagram of the flaw populations that could exist in ceramics. In this example, the severest flaws are surface cracks, perhaps resulting from machining. The next severest flaws are voids and pores introduced by sintering. Voids and pores would become the main fracture of origin, if the surface cracks are much smaller or removed, such as healed. The variation in the fracture sources leads to the large strength, distribution in ceramics.

There are few scenarios that can generate or introduce surface cracks during service. One possibility is by contact events, for example, impact, erosion, corrosion, and wear. Contact events may cause high stresses to the vicinity of the contact site, leading to crack formation. Sudden changes in temperature can also lead to stresses, known as *thermal stress* or *thermal shock*. The introduced surface cracks would be more severe than the pores and voids. As a result, these crackings cause

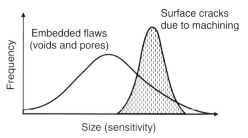

Figure 17.1 Populations of the strength-controlling flaws.

large strength decrease to the component. If these scenarios occur again, it is possible that the component fractures catastrophically.

As mentioned above, ceramics usually fracture when $K_I = K_{IC}$ but, in studies of fracture [1–3], it is sometimes found that crack growth can occur at lower value of K_I The mechanisms have been analyzed to describe the slow crack growth behavior, including chemical reaction kinetics and interfacial diffusion. At the stressed crack tip, it is found that environmental species, for example, moisture, reacts and breaks the bonds at the crack tip resulting in stress corrosion cracking. The kinetics depends on the concentration of the environmental species. Actually it was found that the crack growth velocity in toluene is less than that in air [2]. This implies that the presence of moisture enhances the behavior. Furthermore, at high temperature, it is also found that the localized creep damage can give rise to the slow crack growth.

17.3
History

In 1966, the study on the strengthening behavior of ceramics by heat treatment was reported by Heuer and Roberts [4]. Then, Lange and Gupta [5] reported the strengthening of ZnO and MgO by heat treatment, and used the term "crack healing" for the first time in 1970. Now, we can find more than 200 reports on the strengthening effects by heat treatment for cracking ceramics. The crack-healing mechanisms in these reports can be roughly categorized into

1. re-sintering
2. relaxation of tensile residual stress at the indentation site
3. cracks bonding by oxidation.

Re-sintering [5–11], that is, diffusive crack-healing process, is an older crack-healing concept and commences with a degradation of the primary crack. This regression generates regular arrays of cylindrical voids in the immediate crack tip vicinity. Also, some studies [9–11] on the model and the kinetics of diffusive crack healing in single crystalline and polycrystalline ceramics have been proposed. However, as this crack healing requires the high crack-healing temperature, grain growth might also be generated. In some cases, the strength decreases than it before heat-treatment, although large strength recovery due to the crack healing is attained. The relaxation of the tensile residual stress at the indentation site leads to strength recovery. However, this phenomenon does not heal cracks. The crack bonding by oxidation has been first reported by Lange [12]. He investigated the strength recovery of the cracked polycrystalline silicon carbide (SiC) by heat treatment in air at 1.673 K, and reported that the average bending strength of the specimens heat-treated for 110 h was 10% higher than that of the unheat-treated specimens. The same phenomenon in polycrystalline silicon nitride (Si_3N_4) was reported by Easier et al. [13]. The heat treatment temperature required for this crack-healing mechanism by oxidation is less than that required for the re-sintering crack healing. The other important aspect is that the cracks healed by this mechanism are filled with the formed oxides. A further mechanism and method for crack-healing has been proposed. As an example, Chu *et al.* [14] proposed the crack-healing method using penetrating glasses. They succeeded to repair cracks in alumina (Al_2O_3) and have found that the repaired part becomes even stronger than the base alumina.

As mentioned above, many other investigators have shown their interest on the crack healings of ceramics. In the field of ceramic nanocomposites, there are many reports also available on crack healing. The original impulsion for research in crack healings of ceramic nanocomposites originates from the works of Niihara and coworkers [15–17]. They observed that the strength of the alumina containing 5 vol% of submicrometer-sized SiC particles can be enhanced by annealing at 1573 K for 2 h in Argon. Since the original report, various mechanisms have been proposed to explain this phenomenon. Nowadays, this mechanism is confirmed to be driven by the oxidation of the dispersed SiC particles. Thompson *et al.* [18] observed that the partial healing of indentation cracks occurred when 5 vol% 0.15 μm SiC particles reinforced alumina were annealed at 1573 K for 2 h. Chou *et al.* [19] have also investigated the crack length and the lending strength of alumina/5 vol% 0.2 μm SiC particles nanocomposite after annealing at 1573 K for 2 h in Argon or air, concluding that the crack healing occurs by the oxidation of SiC particles. The similar conclusion was also derived by Wu *et al.* [20]. However, Chou *et al.* [19] noted that a uniform reaction layer was not formed between the crack walls, because the lower SiC content (only 5 vol%) results in a small quantity of the formed oxide.

Ando and coworkers observed that the similar crack healing in mullite ($3Al_2O_3 2SiO_2$) [21–23], Si_3N_4 [24–27] and alumina [28–31] based composites containing more than 15 vol% SiC particles can recover the cracked strength completely. They found that the healed zone is mechanically stronger than the base material and proposed the following requirements to obtain a strong healed zone:

1. Mechanically strong products (compared to the base material) should be formed by the crack-healing reaction.
2. The volume between crack walls should be completely filled with the products formed by the crack-healing reaction.
3. The bond between the product and crack wall should be strong enough.

Crack-healings reports can be classified into three generations. as shown in Table 17.1.

First generation, that is, the crack healing driven by the re-sintering is only to recover the cracked strength. Second generation, that is, the crack healing driven by oxidation of less than 10 vol% SiC can be triggered by damage and occur under service conditions, but the strength recovery is inadequate. Third generation, that is, the crack healing proposed by Ando *et al.* can be attained with all the requirements. Consequently, the third generation crack healing is confirmed to be "true" self-healing for structural ceramics.

Table 17.1 Categorization of self-healing ceramics.

Types	Mechanism	Triggered by damage	Valid under service condition	Strong healed part
First generation	Re-sintering	No	No	Yes
Second generation	Oxidation of SiC (<5 vol% SiC)	Yes	Yes	No
Third generation	Oxidation of SiC (>10 vol% SiC)	Yes	Yes	Yes

17.4
Mechanism

To keep the structural integrity of ceramic, an efficient self-healing should occur. This is possible if healing the surface cracks obeys the following conditions:

1. healing must be triggered by cracking;

2. healing must occur at high temperature [as structural ceramics are expected to typically operate at high temperature (~1273 K) in air] in the corrosion atmosphere, such as air;

3. Strength of the healed zone must be superior to the base material.

Self-crack healing driven by the oxidation of silicon carbide (SiC) can be qualitatively understood to satisfy the requirements 1 and 2. Figure 17.2 shows the schematic of the crack healing in the ceramics containing SiC particles heated at high temperature in the presence of air Cracking allows the SiC particles located on the crack walls to react with oxygen in the atmosphere resulting in healing. The details of the valid conditions are discussed later (Section 17.6). Subsequently, the crack is completely healed as oxidation progresses. As mentioned earlier, if the three important conditions of achieving strong healed zone is satisfied, then fracture initiation changes from the surface crack to the other flaws such as embedded flaw. This behavior is well demonstrated in Figures 17.3a and b [29].

The following equation showing oxidation of SiC also supports the above findings.

$$SiC + 3/2\, O_2 = SiO_2 + CO \tag{17.3}$$

There exist two important features in the above mentioned process. One is the increase in the volume of the condensed phase and the other is the generation of the huge exothermic heat. Because the mole number of silicon is held constant during the oxidation, the volume increase is found to be 80.1%. As the oxidation

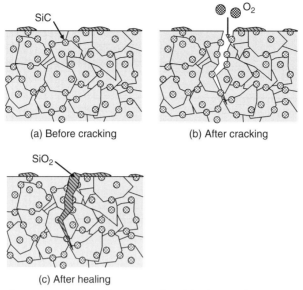

(a) Before cracking   (b) After cracking

(c) After healing

Figure 17.2 Schematic illustration of crack-healing mechanism.

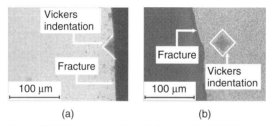

(a)   (b)

Figure 17.3 Fracture initiation of alumina/15 vol% 0.27 μm SiC particles composite: (a) as-cracked and (b) crack healed at 1573 K for 1 h in the presence of air.

progresses, the crack walls are covered with the formed oxide. Finally, the space between the crack walls is completely filled with the formed oxide. For the complete infilling of the space between crack walls, it is necessary to contain more than 10 vol% SiC (Section 17.5). Another important parameter for attaining the complete infilling is the size of crack.

From Figure 17.4 [31], one can find the critical crack size that can be completely crack healed. As an example, the critical crack size of alumina/30 vol% SiC particles composite is 300 μm. This value is the surface length of a semi-elliptical crack with an aspect ratio (crack depth/half of surface length) of 0.9 introduced by indentation method. Below this value, the crack-healed specimens exhibit the same strength, because the space between crack walls is completely filled with the formed oxide and because the fracture initiates from an embedded flaw. When

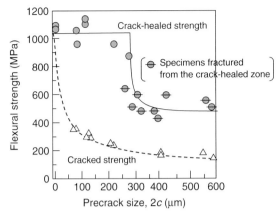

Figure 17.4 Flexural strength of the crack-healed alumina/30 vol% 0.27 μm SiC particles composite as a function of surface length of a semi-elliptical crack.

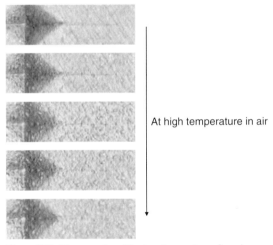

At high temperature in air

Figure 17.5 Photographs of *in situ* observation of crack healing, in which alumina/15 vol% 0.27 μm SiC particles composite containing cracks with an indentation is heat-treated at 1573 K in the presence of air.

the value is above the critical crack size, the space between the cracks walls is too large to be filled with the formed oxide.

Alternatively, the reaction heat makes the formed oxide and the base material to react or to once melt. The second low enthalpy change of the reaction can be evaluated to be −945 kJ using the thermochemical data [32] of the pure substance. This phenomenon might lead to strong bonding between the reaction products and crack walls. The crack-healing mechanism is clearly demonstrated by *in situ* observation, as shown in Figure 17.5.

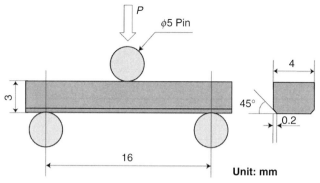

Figure 17.6 Dimensions of three-point bending and the bar specimens.

The phenomenon was observed while alumina/15 vol% 0.27 μm SiC particles composite containing an indentation crack was heat-treated at 1573 K in the presence of air. The features of the crack healing behavior are as follows: (i) the reaction products like sweats appear from the cracks and surface as the reaction progresses; (ii) cracks are perfectly covered and filled with the reaction products; (iii) the reaction products form with bubbling; and (iv) there are no changes in the indentation figuration. From the observation, it is noted that the high temperature makes the reaction products, as well as base material melting, and the bubble, including carbon monoxide (CO) gas, forming strong crack-healed zone.

Furthermore, to estimate the strength of the healed zone in detail, it is necessary to take account of the following issues, that is:

1. Effective volume should be so small that most fracture initiates at the crack-healed zone.
2. The strain energy at failure should be so low that fracture initiation is identified easily.

Ando and coworkers adapted a three-point bending method as shown in Figure 17.6 for fracture tests.

The span of the geometry is 14 mm less than that of Japan Industry Standard (JIS) [33]. The crack-healed specimens had higher strength than the smooth mirror-polished specimens (Section 17.6). Polishing was carried out according to JIS standard [33].

17.5
Composition and Structure

17.5.1
Composition

Most important factor to decide the self–crack-healing ability is the volume fraction of SiC. As mentioned in Section 17.4, it is necessary for achieving strong

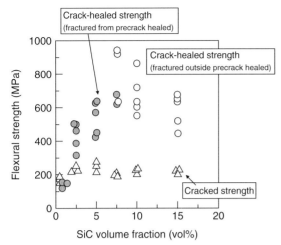

Figure 17.7 Crack healed and cracked strength of alumina-SiC composites as a function of SiC volume fraction.

crack-healed zone that the volume between crack walls is completely filled with the products formed by the crack-healing reaction. Therefore, there is lower limit of the SiC volume fraction to endow with adequate self–cracking-healing ability. Figure 17.7 shows the cracked and crack-healed strengths of alumina containing various volume fractions of SiC particles, which has mean particle size of 0.27 μm. As a result, the crack-healed strength varies with SiC volume fraction and shows a maximum at SiC volume fraction of 7.5, as shown in Figure 17.7. From the strength dispersion, the crack-healing ability cannot be estimated by the strength recovery behavior alone. However, the fracture surface observations can reveal whether the crack is completely healed. There are two kinds of fracture mode. One is the fracture initiated from the crack-healed zone, as shown in Figure 17.8a. This means the formed oxide is not enough to heal the precrack completely. This fracture mode is observed in some specimens of the alumina mixed with 7.5 vol% SiC particles and all specimens that contain less than 5 vol% SiC particles. The other is the fracture initiated outside the crack-healed zone. as shown in Figure 17.8b. This means that the formed oxide is enough to heal the cracks. The enough quantity of the oxide is formed by the oxidation of more than 10 vol% SiC particles. From the results, the ceramics have to contain at least 10 vol% SiC to produce strong crack-healed zone.

17.5.2
SiC Figuration

The SiC figuration also affects the crack-healing ability. Especially, SiC whisker that has high aspect ratio causes intrinsic change to the micromechanism of the crack healing.

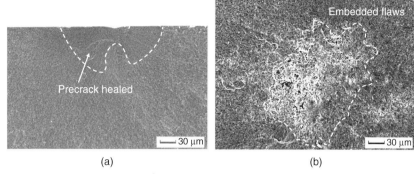

Figure 17.8 Fracture initiations of (a) alumina/7.5 vol% SiC particles composite in which fracture initiates from the precrack healed and (b) alumina/10 vol% SiC particles composite in which fracture initiates from an embedded flaws.

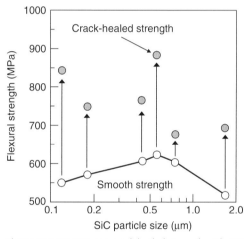

Figure 17.9 Variation in crack-healed strength with SiC particle size in mullite containing 20 vol% SiC particles composite.

Sato *et al.* [34] investigated the relation between SiC particle size and the crack-healed strength in the case of mullite ($3Al_2O_3 2SiO_2$)/20 vol% SiC composite. From the results shown in Figure 17.9, they concluded that the crack healing at 1573 K for 1 h in the presence of air causes 100–300 MPa strength enhancement to all specimens, which shows a maximum with SiC particle size of 0.56 µm.

Ceramics containing SiC whiskers also show the self–crack-healing ability, but there are some differences between the crack-healing behaviors driven by oxidations of the SiC whiskers and that of SiC particles. This difference arises from the geometric relation between the SiC whiskers and the crack wall. The SiC whiskers

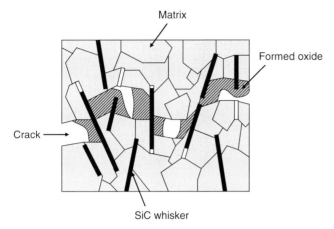

Figure 17.10 Schematic illustration of crack-healing mechanism by SiC whiskers.

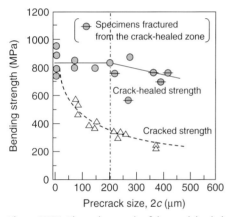

Figure 17.11 Flexural strength of the crack-healed alumina/30 vol% SiC whiskers (diameter = 0.8–1.0 μm, length = 100 μm) composite as a function of surface length of a semi-elliptical crack.

stick out at the crack wall and bridge between the crack walls as illustrated in Figure 17.10.

Owing to this geometry, partial bondings between the crack walls can be formed despite the small amount of the oxide formation. The partial bondings [31, 35] was observed in the crack-healed zone of alumina/20 vol% SiC whiskers (diameter = 0.8–1.0 μm, length = 30–100 μm). The partial bondings enhance the strength recovery of the crack healing and at primary stage on large cracks, as shown in Figure 17.11 [31] (cf. Figure 17.4). Both in the primary stage and on large crack, the amount of the formed oxide is too less to completely fill the crack. In this situation, a large strength recovery could not be attained without the partial bonding. Therefore, composites with SiC whiskers do not only improve the fracture roughness but also have the advantage on crack-healing ability. However, the reliability of the crack-healed zone comprised by the partial bonding is inferior to that of the crack-healed zone completely filled with the formed oxide. Therefore, composites containing SiC whiskers as well as SiC particles show the excellent self–crack-healing ability.

17.5.3
Matrix

Since the self–crack-healing ability can only be seen in case of SiC composites, there is no restriction in selecting the matrix. Ando and coworkers succeeded to endow silicon nitride [24–27], alumina [28–31, 35], and mullite [21–23] with the self–crack-healing ability. Also monolithic SiC [36, 37] has excellent self–crack-healing ability.

These composites can be prepared from commercially available powders using ball mill mixing and hot pressing techniques. Sintering additives does not show any influence on the crack-healing ability. Moreover, a further improvement in mechanical properties can be obtained by employing the optimized sintering conditions. For example, the entrapped SiC particles [28] presented in the alumina matrix grains, when alumina containing 15 vol% SiC particles composite is hot pressed at 1873 K for 4 h. The entrapped SiC particles can inhibit the glide deformation of the alumina grains above 1273 K and this increases the temperature limit for bending strength, as shown in Figure 17.12 [28].

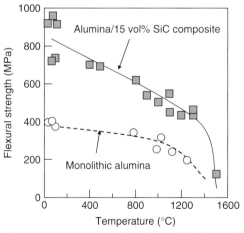

Figure 17.12 Temperature dependence of the flexural strength of alumina containing 15 vol% SiC particles, which are entrapped in the alumina grins, compared with that of monolithic sintered alumina.

17.6
Valid Conditions

17.6.1
Atmosphere

The annealing atmosphere has an outstanding influence on the extent of the crack healing and the resultant strength recovery, as shown in Figure 17.13 [29].

From the figure, it can be clearly seen that the presence of oxygen causes the self-healing phenomenon, as the crack healing is driven by the oxidation of SiC. However, the threshold oxygen partial pressure can be expected to be quite low. Therefore, the self-crack healing must be valid in the atmosphere, except in deoxidized conditions, for example, the atmosphere containing hydrogen. Also embedded flaws cannot be healed, because the SiC particles present in the embedded flaws cannot react with oxygen.

Annealings in vacuum, argon (Ar), and nitrogen (N_2) result in a slight strength recovery. Wu *et al.* [20] discussed this phenomenon to be the release of the tensile residual stress at the indentation site. Furthermore, Fang *et al.* [38] used a satellite indentation technique to show that, after 2 h at 1573 K, the degree of the annealing-induced relaxation in the stress intensity factor of the residual stress at the indentation site was ~26% for alumina/5 vol% SiC nanocomposite. Using this result, one can predict that annealing in atmosphere without oxygen leads to 10% strength recovery.

Recently, two large efforts [39, 40] made to know the quantitative influence of oxygen partial pressure on the self-healing behavior. From the results, the crack-healing rate is significantly decreased with decreasing oxygen partial pressure, as shown in Figure 17.14, in which the bending strengths of the alumina/20 vol%

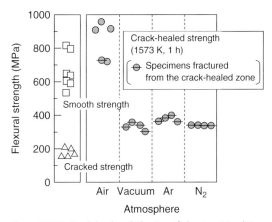

Figure 17.13 Crack-healing behavior of alumina/15 vol% 0.27 μm SiC particles composite under several atmospheres.

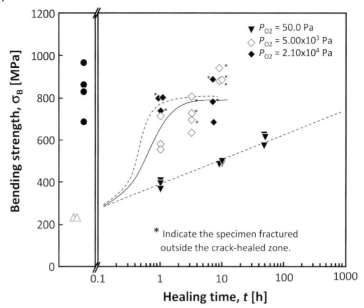

Figure 17.14 Bending strength of the specimens crack-healed at 1573 K under several oxygen partial pressures as a function of the healing time, with the bending strength of healed smooth specimen and the as-cracked specimen.

SiC particles composite crack-healed under several oxygen pressure atmospheres. Moreover, the detail of the kinetics will be discussed later (Section 17.10).

17.6.2
Temperature

The ceramic components are usually operated at high temperatures. The self-healing relies on the oxidation of SiC, thereby leading to the self-crack healing. Thus, it is important to know the valid temperature range for self-crack healing.

As the crack healing is induced by chemical reaction, the strength recovery rate decreases exponentially with decreasing temperature. For example, Figure 17.15 [28] shows the relationship between the crack-healing temperature and the strength recovery for alumina/15 vol% 0.27 μm SiC parades composite.

In order to completely heal a semi-elliptical crack of 100 μm in surface length within 1 h, heating above 1573 K is required. In the similar way, heating above 1473 and 1273 K is needed in order to completely heal the surface crack within 10 and 300 h, respectively. The relation between the crack-healing temperature and the strength recovery rate follows Arrhenius' equation.

Figure 17.16 [29] shows the Arrhenius plots on the crack healing of several ceramics having the self–crack-healing ability, in which the crack-healing rate is defined as the inverse of the time when complete strength recovery is attained at

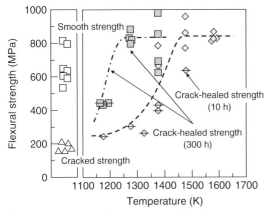

Figure 17.15 Relationship between crack-healing temperature and strength recovery for alumina/15 vol% 0.27 μm SiC particles composite. (Centered line symbols indicate specimens fractured from the crack-healed zone.)

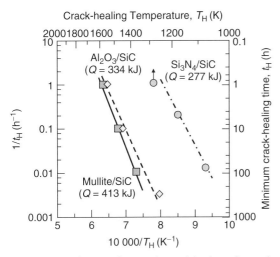

Figure 17.16 Arrhenius' plots on the crack healing of several ceramics having the self-crack healing ability.

elevated temperatures. Apparent activation energies of the crack healing can be evaluated from Figure 17.16. The question is why the activation energy of $Si_3N_4/$ SiC composite differs from those of alumina/SiC and mullite/SiC composites. The reason could be the crack healing of $Si_3N_4/$SiC composite is driven by the oxidation of SiC as well as Si_3N_4. Using these values, one can estimate the time for which a semi-elliptical crack of 100 μm in surface length can be completely healed at several temperatures.

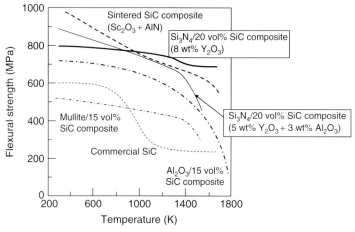

Figure 17.17 Temperature dependences of the flexural strength of several typical crack-healed ceramics.

The refractoriness of the crack-healed zone restricts the determination of the upper limit of the valid temperature range of self-crack healing. The temperature dependence of the flexural strength of the several typical ceramics crack healed [22, 25, 28, 36, 37] is shown in Figure 17.17. Except the dependence of $Si_3N_4/20\,wt\%$ SiC particles composite containing $8\,wt\%$ Y_2O_3 as sintering additives, all dependences of the crack-healed specimens have the temperature at which the strength decreases abruptly, and this has been determined as the temperature limit for strength. The temperature limit is affected by the features of the oxide formed by the self-crack healing. The commercial sintered SiC [36] was found to have considerably low temperature limit of $873\,K$ because the formed oxide is in glassy phase. Modifying the sintered additives to Sc_2O_3 and AlN, Lee *et al.* [37] succeeded in improving the temperature limit of the crack-healed zone significantly. The similar behavior was observed in the $Si_3N_4/20\,wt\%$ SiC particles composites [25]. When the sintered additive is $5\,wt\%$ Y_2O_3 and $3\,wt\%$ Al_2O_3, the formed oxide and grain boundary are in glassy phase. Alternatively $Si_3N_4/20\,wt\%$ SiC particles composite containing $8\,wt\%$ Y_2O_3 as sintering additives forms the crystalline oxide, such as $Y_2Si_2O_7$ by crack healing. The difference gives rise to the difference in the temperature limit. Both alumina containing $15\,vol\%$ SiC particles composite [28] and mullite containing $15\,vol\%$ SiC particles composite [22] form the crystalline phase because the formed oxide reacts with the matrix and forms mullite. These temperature limits are summarized in Table 17.2.

The temperature range at which the self-crack healing is valid is limited by the crack-healing rate and the high-temperature mechanical properties. Assuming that fracture by the second damage allows complete healing of surface cracks introduced by the first damage in $100\,h$, one can evaluate the valid temperature range of the self-crack healing, as listed in Table 17.3.

Table 17.2 Temperature limit of several ceramics.

Materials	Temperature limit (K)
Si_3N_4/20 vol% SiC particles composite (8 wt% Y_2O_3)	1573
Si_3N_4/20 vol% SiC particles composite (5 wt% Y_2O_3 + 5 wt% Al_2O_3)	1473
Alumina/15 vol% SiC particles composite	1573
Mullite/15 vol% SiC particles composite	1473
SiC sintered with Sc_2O_3 and A/N	1673
Commercial SiC sintered	873

Table 17.3 Valid temperature region of self-crack healing for several ceramics.

Materials	Valid temperature range (K)
Si_3N_4/20 vol% SiC particles composite (8 wt% Y_2O_3)	1073–1573
Alumina/15 vol% SiC particles composite	1173–1573
Mullite/15 vol% SiC particles composite	1273–1473
SiC sintered with Sc_2O_3 and A/N	1473–1673

17.6.3
Stress

Stress applied to the components is also one of the most important factors to decide the valid condition of the self-crack healing. Structural components generally suffer various kinds of stresses. The applied stress is possible to cause the slow crack growth. If the applied stress exceeds the critical value, it would rise to catastrophic fracture. Therefore, it is important to know the threshold stress that could be safely applied to the cracks during self-crack healing.

Ando et al. [41] reported for the first time that the surface crack in the mullite containing 15 vol% SiC particles composite can be healed, although the tensile stress is applied to the cracks. Their results revealed that the crack healing occurs, although the precrack is grown by the applied stress, and the specimens crack healed under stress had the same strength as the specimens crack healed under no stress and at same temperature. Furthermore, Ando et al. [42] reported that surface cracks in the mullite containing 15 vol% SiC particles composite can be healed even though dynamic stress such as cyclic stress, opens and closes the crack.

Nakao et al. [43] investigated the threshold stresses during self-crack healing for several oxide ceramics. For example, the threshold static stress during crack healing for a semi-elliptical surface crack (surface length = 100 μm) in alumina containing 15 vol% SiC particles has been determined to be 150 MPa, as shown in Figure 17.18 [43].

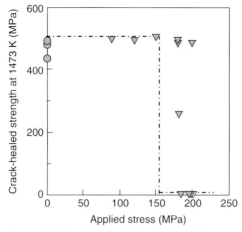

Figure 17.18 Crack-healing behavior at 1473 K under static stress for alumina/15 vol% SiC particles composite.

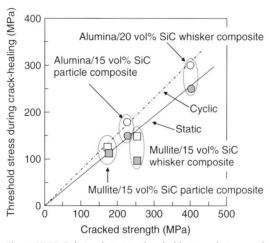

Figure 17.19 Relation between threshold stress during crack healing and the corresponding cracked strength.

Figure 17.18 demonstrates that the tensile static stress of 180 MPa is possible to fracture the specimen during crack healing, whereas the stress less than 150 MPa never fractures the specimens during crack-healing.

Figure 17.19 shows the determined threshold stress as a function of the flexural strength of the specimen containing the same surface crack for several cracks in the oxide ceramics–SiC composite [43–46]. Except the threshold stresses of mullite containing 15 vol% SiC whiskers composite, all data satisfies the proportional relation, although the crack healings were subjected to different conditions. The crack healing ability for mullite containing 15 vol% SiC whiskers composite has been found to be so low that the crack healing part was weaker than the other parts as only partly welding occurs, not satisfying the proportional relation. The proportional constants for the relations between the threshold static and cyclic stresses between the cracked strength have been found to be 64 and 76%, respectively. The threshold stress imposes an upper limit to the crack growth rate, thereby limiting the crack length to less than the critical crack length before the crack healing starts. This implied that the crack growth behavior of all specimens is time dependent rather than cyclic dependent at high temperature. Therefore, applying static stress could be confirmed to be the easiest condition for fracture during the crack healing under stress, and the threshold stresses of every condition during the crack healing have been found to be the threshold static stresses. The stress intensity factors at the tip of the precrack during the crack-healing treatment, K_{HS}, were estimated. Since a tensional residual stress was introduced during precracking by using an indentation method, it is necessary to consider the stress intensity factor of the residual stress, K_R, as expressed by the following equation:

$$K_{HS} = K_{ap} + K_R \tag{17.4}$$

where K_R can be evaluated by using the relation proposed by Kim *et al.* [47] and $K_R = 0.35 \times K_{IC}$. Also, by interpolating the threshold static stress during the crack healing and the geometry for the precrack into Newman–Raju equation [48], one can obtain K_{ap}. From the evaluation, it was found that ceramic components having the adequate crack healing ability can be crack healed under the stress intensity factor below 56% fracture toughness.

17.7
Crack-healing Effect

17.7.1
Crack-healing Effects on Fracture Probability

The crack-healing can simplify the complexity in the flaws associated with fracture, because surface cracks that are severest flaws in ceramic are completely healed. As a result, a fracture probability can be easily described after the crack-healing. Furthermore the crack-healing has a large contribution to decrease the fracture probability.

Fracture probability is one of the most important parameters for structural components. If the fracture probability is too high, one needs to either change the design or substitute high strength materials. The fracture probability can be obtained from the failure statistics. As indicated in Section 17.2, ceramics contain many flaws that can vary in size and figuration, causing the wide strength distribution. Thus the empirical approach needs to describe the strength distribution of a structural ceramic. Once the strength of a material is fitted to the distribution, the fracture probability can be predicted for any applied stress. A common empirical approach to describe the strength distribution of a structural ceramic is the Weibull approach. The two-parameter Weibull function, which is given by

$$F(\sigma) = 1 - \exp\left\{-\left(\frac{\sigma}{\beta}\right)^m\right\}$$

(17.5)

can express the strength distribution of structural ceramics well, where $F(\sigma)$ is the fracture probability at the tensile stress of σ, m the Weibull modulus, and β the scale parameter. The Weibull modulus describes the width of the strength distribution. High Weibull modulus implies that the strength has a low variability. Values of m for ceramics are in the range of 5–20. The scale parameter describes the stress when $F(\sigma) = 63.2\%$. To analyze the strength distribution, Equation (17.5) is usually expanded as follows:

$$\ln\ln\left(\frac{1}{1-F(\sigma)}\right) = m\ln\sigma - m\ln\beta$$

(17.6)

Thus, a plot of the left-hand side of Equation (17.6) has a linear relation versus the natural logarithm of the strength. In such a procedure tire, a fracture probability is needed for each test specimen. This is usually estimated using

$$F(\sigma) = \frac{i - 0.3}{n + 0.4}$$

(17.7)

The strength data of n specimens are organized from weakest to strongest and given a rank i with $i = 1$ being the weakest specimen. Equation (17.7) is well known as *median rank method*.

As an example, the Weibull plot of the crack-healed alumina containing 20 vol% silicon carbide (SiC) particles composite is shown in Figure 17.20. The healed crack is a semi-elliptical surface crack having a surface length of 100 μm and a depth of 45 μm. In comparison, those of the as-cracked specimen and the smooth specimen having a miror finish surface are shown in Figure 17.20. Assuming that the data obey the two-parameter Weibull function, one can apply a least-squares fitting. From the obtained line profiles, the values of m and β can be obtained for the crack-healed specimens the as-cracked specimens and the smooth specimens.

The crack-healing causes slight increase to the value of m compared to the smooth specimens. Furthermore, the strength distribution of the crack-healed specimen is in a good agreement with the two-parameter Weibull function, although that of the smooth specimen differs from the function significantly. The flaw population in ceramics leads to this behavior. The Weibull modulus m of the crack-healed specimen was smaller than that of the as-cracked specimen. All the as-cracked specimens fractured from a crack introduced by the Vickers indentation, while fractures of most crack-healed specimens occurred outside of

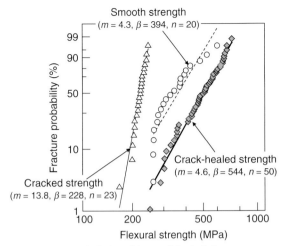

Figure 17.20 Weibull plot of the crack-healed alumina containing 20 vol% silicon carbide (SiC) particles composite.

the crack-healed zone, as shown in Figure 17.3b, because cracks were completely healed. Since embedded flaws as the fracture initiation of the crack-healed specimens have different sizes, the fracture stresses exhibit a large scatter.

This specimen was tested at room temperature and exhibited the fracture stress of 526 MPa. To improve the reliability and the quality of the structural ceramics, it is therefore necessary to remove the specimens with large embedded flaws by proof test, even if surface cracks were completely healed. The scale parameter of the crack-healed specimens has a higher value than that of the as-cracked specimens as well as that of the smooth specimens, because cracks introduced by machinings which existed even in the smooth specimens, were also completely healed.

To show the considerable merit of the crack healing, the fracture probabilities of three specimens (smooth specimens, as-cracked specimens, and crack-healed specimens) for the proof test stress of 435 MPa were compared. The fracture probabilities of the smooth specimens, the as-cracked specimens, and the crack-healed specimens were 80, 100, and 30%, respectively. Therefore, it can be concluded that the crack healing drastically increases the survival probability by the proof test, and thus increases the working stress of structural ceramics.

17.7.2
Fatigue Strength

The effect of the self-crack healing on the fatigue strength is greater than that on the monotonic strength. The fatigue degradation of ceramics progresses by the stress corrosion cracking at the tip of cracks as mentioned in section 17.2. Therefore the presence of surface cracks affects strongly the fatigue strength. Figure 17.21 [50] shows the dynamic fatigue results of the crack-healed mullite containing

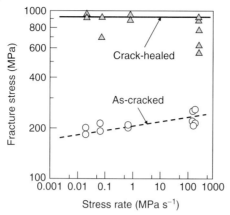

Figure 17.21 Dynamic fatigue results of the crack-healed mullite containing 15 vol% SiC whiskers and 10 vol% SiC particles composite with that of the composite having a semi-elliptical crack of 100 μm in surface length.

15 vol% SiC whiskers and 10 vol% SiC particles composite with that of the composite having a semi-elliptical crack of 100 μm in surface length. From logarithmic plots of the dynamic strength versus stressing rate, the effect of the crack healing on the fatigue behavior can be clearly understood. The positive slope implies that the slow crack growth occurs. As the data on the specimens containing the surface crack shows the positive slope, the slow crack growth has been included in the fatigue behavior. On the other hand, the data on the crack-healed specimen is almost constant over the whole stressing rate. Therefore, the crack healing makes the fatigue sensitivity decrease significantly. Actually, the fracture initiator in the crack-healed mullite containing 15 vol% SiC whiskers and 10 vol% SiC particles composite under the every stressing rate is an embedded flaw, which did not grow under the applied stress.

In high temperature fatigue, there is another interesting phenomenon, in order that the self-crack healing occurs at the same time as fatigue damage. For example, Figure 17.22 [50] shows a logarithmic plot of life time in terms of the applied stress for the crack-healed mullite containing 15 vol% SiC whiskers and 10 vol% SiC particles composite at 1273 K. In general, that is, the slow crack growth is included, the life time increases as the applied stress decreases. However, all the crack-healed test specimens survived up to the finish time of 100 h obeying the JIS standard [51] under static stresses of 50 MPa less than the lower limit of the monotonic strength at the same temperature. Alternatively, the specimens fractured at less than 100 s under stresses corresponding to the lower limit of the flexural strength. This fracture is not fatigue, but rather rapid fracture. Therefore, it is confirmed that the crack-healed composite is not degraded by the static fatigue at 1273 K. The behavior would result from the fact that self-crack healing occurs rapidly compared with the fatigue damage.

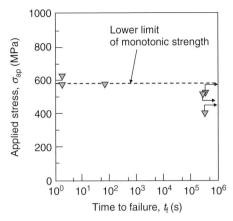

Figure 17.22 Logarithmic plot of life time in terms of the applied stress for the crack-healed mullite containing 15 vol% SiC whiskers and 10 vol% SiC particles composite at 1273 K.

17.7.3
Crack-healing Effects on Machining Efficiency

An important alternative aspect is that the self-crack healing is a most valuable surface treatment. Applying the crack-healing process into the manufacturing for ceramic components can reduce the manufacturing cost. Machining processes included in the manufacturing reduce the reliability of the components [52] because it causes many cracks to the surface of the component. To remove the nonacceptable cracks, final machining processes, such as polishing and lapping, are generally required. Although these processes leave behind many minute cracks, these are expensive processes. It is, therefore, anticipated that substituting the creak-healing process for the final machining processes leads to economical manufacturings for ceramic components secured with high reliability.

Figure 17.23 [53] demonstrates that the nonacceptable cracks introduced by a heavy machining can be completely crack healed. The machining cracks were introduced at the bar specimens surface of alumina/20 vol% SiC whiskers composite by a ball-drill grinding. The ball-drill grinding was performed along the direction perpendicular to the long side of the specimens, as shown in shown in Figure 17.6, which consequently fabricated a semi-circular groove whose depth and curvature were 0.5 and 2 mm, respectively. As a result, the machined specimens contained many machining cracks perpendicular to the tensile stress at the bottom of the semi-circular groove. The horizontal variable is a cut depth by one pass, which is an indicator of the machining efficiency. For example, 14 to 40 cycles are needed to fabricate the semi-circular groove by the grinding with the cut depth by one pass of 15 and 5 μm, respectively. Alternatively, the vertical axis indicates the local fracture stress of the machined specimens and the machined specimen healed. From the load as the specimens fractured, P_F, the section

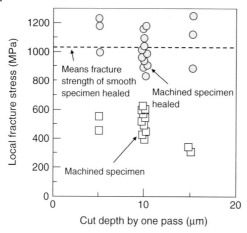

Figure 17.23 Effect of depths of cut by one pass on the local fracture stress at room temperature of the machined specimens healed.

modulus, Z, and the stress concentration factor, α, the local fracture stress at the bottom of the semi-circular groove, σ_{LF}, was evaluated as follows:

$$\sigma_{LF} = \frac{\alpha P_F l}{4Z} \tag{17.8}$$

where l is a span length. Under this geometry, the value of Z and α were 4.2 and 1.4 [54], respectively. The local fracture stress of the machined specimens decreased with increasing the cutting depth. This behavior implies that the cut depth by one pass also means the degree of the machining heaviness. Throughout the range of the cut depth by one pass a complete strength recovery was found to be attained by the crack-healing treatment for 10 h at 1673 K, because these average strengths were almost equal to that of the complete crack-healed specimens [35]. The crack healing is possible for relatively large cracks initiated by a heavy machining for cutting depths up to 15 μm. However, the heavy machining for cutting depths above 15 μm makes the diamond grain to drop out of the ball drill significantly, reducing the machining efficiency. Therefore, with a simple operation of heating, one can ensure the reliability over ceramic components machined by the limiting conditions of the grinding tool (ball-drill). To not cause the outstanding strength to decrease by fabricating the semi-circular groove, it is necessary to perform not only on machining with the cut depth by one pass of less than 5 μm but also on lapping at the bottom. Thus, a high machining efficiency can be attained by the use of the crack-healing process.

It is important to note the difference in the optimized condition between the crack healing for indented cracks and the machining cracks. The optimized crack healing condition for the indentation cracks in alumina/20 vol% SiC whiskers composite was found to be 1573 K for 1 h [53]. However, the crack healing above

Table 17.4 Weibull modulus (m) and shape parameter (β) of ball-grind alumna composites containing 20 vol% of SiC whiskers.

Sample type	Weibull modulus, m (MPa)	Shape parameter (β)
Machined specimen	6.69	549
Machined specimen healed at 1573 K for 1 h	6.70	796
Machined specimen healed at 1673 K for 10 h	11.5	1026
Healed (1573 K for 1 h) smooth specimen	8.15	1075

1673 K for more than 10 h is required to attain the complete strength recovery for the machined alumina/20 vol% SiC whiskers composite. This behavior can be clearly understood by the statistical analysis using the two-parameter Weibull function given by Equation (17.5). Table 17.4 shows the Weibull modulus and the shape parameter evaluated from the analyses. The values of m and β of the machined specimen healed at 1673 K for 10 h were 11.5 and 1026 MPa, respectively. The values were almost equal to those of complete crack-healed specimen. From this statistical and χ^2 analysis, it was found that the cracks introduced by machining were completely healed by the crack healing process at 1673 K for 10 h. On the other hand, the machined specimen healed at 1573 K for 1 h had lower values of m and β than those of the smooth specimen healed; thus, it can be concluded that this crack-healing condition is inadequate. Two approaches are confirmed to be reasonable to explain this difference: (i) the difference in the state of the subsurface residual stress associated with the different crack geometries and (ii) the oxidation of SiC by the heat generation during the machining. The machining crack was closed by the action of the compressive residual stress resulting in a redaction in the supply of oxygen to the crack walls. Moreover, before the crack-healing treatment, if SiC particles were already covered with a thin oxidation layer, by the grinding heat, this would lead to the decrease in the oxidation activity of the SiC particles.

Furthermore, it was found that the various cracks initiated by machining into various machined figurations could be healed by crack-healing treatment at 1673 K for 10 h, as shown in Figure 17.24 [53].

17.8
New Structural Integrity Method

17.8.1
Outline

A combination of the crack healing and the proof testing ultimately guarantees the structural integrity of the ceramic components. As mentioned above, by

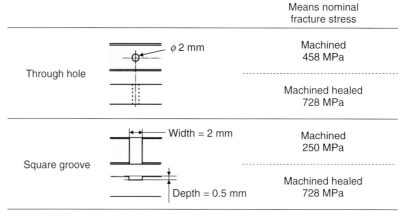

Figure 17.24 Effect of crack-healing condition on strength recoveries of various machined specimens.

eliminating not only the cracks introduced during manufacturing but also the cracks introduced during service, the crack healing can ensure perfectly against risk of the fracture initiated from surface flaws. The proof testing, in which components are over-stressed print to use, can determine the maximum critical size of embedded flaws associated with fracture. Embedded flaws, such as voids, may be present in a material as a result of the processing, and could not be generated during service. Therefore, the minimum fracture stress caused by the embedded flaws [55, 56] or probabilistic fatigue S–N curves [57–59] has been estimated from the proof testing stress on the basis of the linear fracture mechanics.

However, two important points must be taken account of while using the proof testing. First is to ensure that engineering ceramics show the nonlinear fracture behavior [60, 61]. The other is that the evaluated minimum fracture stress is valid only at the proof testing temperature. Therefore, if ceramic components are used at high temperature, the proof testing must be conducted at the operating temperature. Ando *et al.* [62] proposed a theory to evaluate the temperature dependence of the guaranteed (minimum) fracture stress of a proof tested sample based on nonlinear fracture mechanics. This approach allows the ceramic components proof tested at room temperature to operate at the arbitrary temperatures.

17.8.2
Theory

On the basis of the process zone size failure criterion, the minimum fracture stress at high temperature can be guaranteed from the proof testing stress at room temperature. The criterion proposed by Ando *et al.* [62] has been obtained by the size of process zone, which is the plastic deformation region at the crack tip, well expressing the nonlinear fracture behavior of ceramics. From the criterion, the process zone size at fracture D_C, is given by the following equation:

$$D_C = \frac{\pi}{8}\left(\frac{K_{IC}}{\sigma_0}\right)^2 = a_c\left\{\sec\left(\frac{\pi\sigma_C}{2\sigma_0}\right) - 1\right\} \tag{17.9}$$

where σ_C and σ_0 are the fracture stress at the fracture caused by the flaws having a_c and the fracture strength of the plain specimen (intrinsic bending strength), respectively. K_{IC} plane strain fracture toughness and a_c, equivalent crack length, which is given by the equation as

$$K_C = \sigma_C\sqrt{\pi a_e} \tag{17.10}$$

By expanding Equation 17.9 around a_c, one can write the equivalent crack size of the flaws associated with fracture as shown in the following equation:

$$a_e = \frac{\pi}{8}\left(\frac{K_{IC}}{\sigma_0}\right)^2\left\{\sec\left(\frac{\pi\sigma_C}{2\sigma_0}\right) - 1\right\}^{-1} \tag{17.11}$$

Since K_{IC} and σ_0 is the function of temperature, the fracture strength associated with a_e needs to be also expressed as a function of temperature:

$$\sigma_C^T = \frac{2\sigma_0^T}{\pi}\arccos\left\{\frac{\pi}{8}\frac{1}{a_e}\left(\frac{K_{IC}^T}{\sigma_0^T}\right)^2 + 1\right\}^{-1} \tag{17.12}$$

where the superscript T is the value at elevated temperature.

The maximum size of the flaw. a_e^P, that is able to present in the sample proof tested under σ_p, at room temperature can be given by

$$a_e^P = \frac{\pi}{8}\left(\frac{K_{IC}^R}{\sigma_0^R}\right)^2\left\{\sec\left(\frac{\pi\sigma_p}{2\sigma_0^R}\right) - 1\right\}^{-1} \tag{17.13}$$

where the superscript R is the value at room temperature. Assuming that the sizes of the residual embedded flaws do not vary with change in the temperature, one can determine the minimum fracture stress guaranteed of the proof tested sample, σ_G, as the fracture strength associated with a_e^P at arbitrary temperatures. Thus the value of σ_G can be expressed as

$$\sigma_G = \frac{2\sigma_0^T}{\pi}\arccos\left\{\left(\frac{K_{IC}^T}{K_{IC}^R}\right)^2\left(\frac{\sigma_0^R}{\sigma_0^T}\right)^2\left\{\sec\left(\frac{\pi\sigma_p}{2\sigma_0^R}\right) - 1\right\} + 1\right\}^{-1} \tag{17.14}$$

and can be evaluated from the data on the temperature dependences of K_{IC} and σ_0.

17.8.3
Temperature Dependence of the Minimum Fracture Stress Guaranteed

Using the above theory, one can estimate the minimum fracture stresses at elevated temperatures for the sample proof tested at room temperature. Ono *et al.* [49] (evaluated the temperature dependence of σ_G for alumina/20 vol% SiC particles composite. Moreover, by comparing the evaluated σ_G with the measured fracture stress of the proof tested sample at elevated temperature, the validity of this estimation was given by them. Their results and discussion are presented in the following text.

Before discussing the temperature dependence of σ_G, temperature dependences of the plane strain fracture toughness, K_{IC}, and the intrinsic bending strength, σ_0, are noticed. Ono *et al.* [49] investigated these temperature dependences for alumina/20 vol% SiC particles composite. The σ_0, which is determined as the average fracture stress of 5% of the highest strengths of the crack-healed specimens at the temperatures, has large temperature dependence and the tendency is almost linear and negative up to 1373 K. Moreover, the K_{IC} is almost constant against temperature. The features affect the correlativeness between the fracture stresses as a function of the equivalent crack length, a_e, at room temperature and at high temperature. The schematic is shown in Figure 17.25.

The fracture stress associated with small flaw, that is, a_e is low, is equal to the σ_0, and varies considerably as temperature varies. On the other hand, the fracture stress associated with large flaw, that is, a_e, is high, is determined by linear fracture mechanics as expressed by Equation (17.10) and changes scarcely with temperature change. Therefore, the negative temperature gradient of the σ_G increases with increasing the proof testing stress, σ_p^R, as shown in Figure 17.26. Since high σ_p^R qualifies a_e^P to be a low value, the negative temperature gradient of the σ_G becomes considerably high. Alternatively, since low σ_p^R allows the large flaws to present

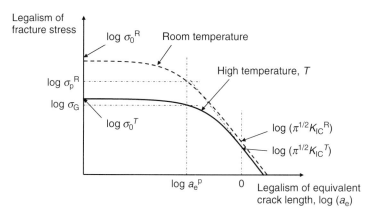

Figure 17.25 Schematic illustration of proof-test theory and the effect of equivalent crack on fracture strength at room temperature and at high temperature.

in the proof tested specimen, the σ_G is almost constant as a function of temperature.

The values of the evaluated σ_G have good agreements with the measured minimum fracture stress of the proof tested specimens, σ_F^{min}. Figure 17.27 shows the data on the fracture stress of the crack-healed and the proof-tested specimens as a function of temperature with the evaluated σ_G for the crack-healed alumina/20 vol% SiC particles composite when $\sigma_p = 435$ MPa. Except the data at 1373 K, all specimens have higher strength than the σ_G at all the temperature.

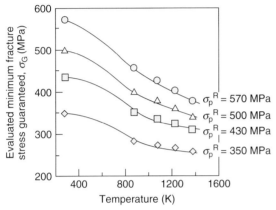

Figure 17.26 Temperature dependence of minimum fracture stress guaranteed of the proof tested under several proof testing stress for alumina containing 20 vol% SiC particles composite.

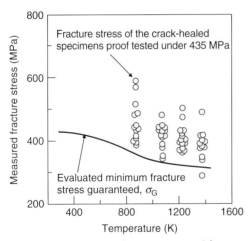

Figure 17.27 Comparison between measured fracture stress and the evaluated minimum fracture stress guaranteed for the crack-heated alumina/20 vol% SiC particles composite proof tested under 435 MPa.

Also, the minimum values of the experimental fracture stress are almost equal to the σ_G at all temperatures. At 1373 K, the σ_F^{min} is 6.8% less than σ_G, but the value exists in the dispersion evaluated from the K_{IC} and the σ_0 that have large scatters. Moreover, in the case of $\sigma_p = 530$ MPa, the evaluated σ_G have good agreements with the σ_F^{min} as well as with all the proof-tested specimens fractured under the tensile stress more than σ_G. Therefore, the results well demonstrate the validity of the guaranteed method.

Moreover, Ono *et al.* [49] and Ando *et al.* [62, 63] reported that the guaranteed theory can be applied to different conditions and the different materials. The obtained results can be seen in Figure 17.28, where the measured σ_F^{min} is plotted as a function of the evaluated σ_G. *N* in the figure denotes the number of samples used to obtain σ_F^{min}. Four open diamonds indicate the data on the ceramic coil spring made of silicon nitride. Also, a open square differs from the closed circles in the crack healing condition, that is, the open square employs 1373 K for 50 h and the closed circle employs 1573 K for 1 h. However, all σ_F^{min} shows good agreement with σ_G. Therefore, using Equation (17.14) one can estimate the σ_G at higher temperatures of every material having the crack healing ability and every crack healing condition.

On the other point of view, it is interesting whether this estimation is reversible for temperatures, that is, the stress evaluated in Equation (17.14) can guarantee the minimum fracture stress at room temperature of the specimen proof tested at high temperature. For example, the σ_G at room temperature of the alumina/20 vol% SiC particles composite proof tested under 335 MPa ($=\sigma_p$) at 1073 K is evaluated to be 435 MPa. Alternatively, the experimental minimum fracture stress was 410 MPa. The σ_G existed in the dispersion obtained from K_{IC} and σ_0, which have large scatters.

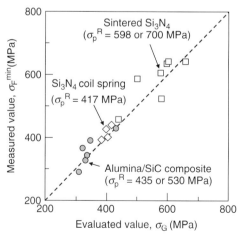

Figure 17.28 Comparison between minimum fracture stresses guaranteed and measured minimum fracture stress.

17.9
Advanced Self-crack Healing Ceramics

17.9.1
Multicomposite

Ceramic composite containing both SiC whiskers and SiC particles, called SiC *multicomposite*, enhance fracture strength and toughness as well as endow the ceramics with the good self-crack healing ability. As mentioned in Section 17.5, the reinforcement by SiC whiskers can not only improve fracture toughness but can also generate the self-crack healing ability. However, it is difficult to disperse a large amount of SiC whiskers uniformly, and so the aggregated SiC whiskers decreases the fracture strength. SiC multicomposits containing both whiskers and particles improves the self-crack healing ability endowed by SiC whiskers alone without adversely affecting the composite strength. Therefore, ceramic–SiC multicomposites exhibit high strength, high fracture toughness, and excellent self–crack-healing ability.

Especially, SiC multicomposites perform better than the mullite-based composites. Mullite and mullite-based composites have been expected to be advanced ceramic spring because they exhibit the same low level elastic constant as metal and excellent oxidation resistance. However, mullite has remarkably low fracture toughness. Therefore, it is necessary to endow mullite with self–crack-healing ability for actualizing mullite-based ceramic springs. The mechanical properties of mullite/SiC composites [21, 46] and multicomposites [64] were investigated as shown in Table 17.5.

The fracture strength increases with SiC content increasing up to 20 vol%, above which it remains almost constant. The fracture toughness increased with an increase in SiC whiskers content. Clearly, it is confirmed that crack bridging and pulling out due to SiC whiskers lead to increase in fracture toughness. All the

Table 17.5 Mullite/SiC composites having self-crack healing ability.

Sample descriptions	Content (vol%)		
	Mullite	SiC particle (diameter = 0.27 µm)	SiC whisker (diameter = 0.8–1.0 µm, length =30–100 µm)
MSI5P	85	15	0
MS15W	85	0	15
MS20W	80	0	20
MS25W	75	0	25
MS15W5P	80	5	15
MS15W10P	75	10	15

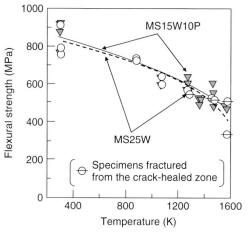

Figure 17.29 Temperature dependence of the crack-healed strength of mullite/25 vol% SiC whiskers composite (MS25W) and mullite/15 vol% SiC whiskers/10 vol% SiC particles composite (MS15W10P), in which the center lined symbols indicate specimens fractured from the precrack healed.

mullite/SiC composites, which are listed in Table 17.5, can exhibit a large strength recover by the crack healing. However, mullite/15 vol% SiC whiskers composite (MS15W) [46] cannot attain the complete strength recovery despite the optimized crack healing condition. Complete strength recovery of the crack-healed specimens is defined as the strength of the strength of the specimens, whose fracture initiation is embedded flaw. This implies the complete elimination of surface cracks that can be attained and the strength of the crack healed part is superior to that of the base materials. The crack healed part in mullite/25 vol% SiC whiskers composite (MS25W) holds higher strength than the base materials below 1273 K, as shown in Figure 17.29 [64]. On the other hand, the crack healed part in mullite/15 vol% SiC whiskers/10 vol% SiC particles composite (MS15W10P) holds higher strength than the base materials at the whole of the experimental temperature region, as shown in Figure 17.29.

Figure 17.30 [64] shows the maximum shear strains of the mullite/SiC multi-composites as a function of SiC content. The maximum shear strain corresponds to the deformation ability as spring. The value of the maximum shear strain showed a maximum at a SiC content of 20 vol%, above which it slightly decreased because Young's modulus increased with an increase in SiC content, but the fracture strengths were almost constant above SiC content of 20 vol%. MS15W10P has the best potential as a material for the ceramic springs used at high temperatures, because it has a shear deformation ability that was almost two times greater than monolithic mullite as well as an adequate crack healing ability.

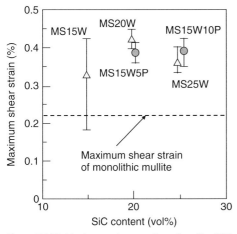

Figure 17.30 Maximum shear strains of mullite/SiC composites as a function of SiC content.

17.9.2
SiC Nanoparticle Composites

Nanometer-sized SiC fine particles enhance the self-crack healing rate because it gives a large increment in the reactive area and makes the surface of SiC particles active. This effect give a large benefit to the self-crack healing at relatively low temperatures at which the self-crack healing is completed in more than 100 h.

Reaction synthesis is a promising process for directly fabricating nanocomposites which are difficult to obtain by the normal sintering of nanometer-sized starting powder compacts. Reaction synthesis to fabricate alumina–SiC nanocomposites [65–71] were reported. Using the reaction synthesis (17.15)

$$3(3Al_2O_3 2SiO_2) + 8Al + 6C = 13Al_2O_3 + 6SiC \qquad (17.15)$$

Zhang *et al.* [71] succeeded in fabricating alumina nanometer-sized SiC particles nanocomposite, in which the formed SiC particles are mainly entrapped inside the alumina grains. Employing the similar process to prepare alumina–SiC nanocomposite, Nakao *et al.* [72] investigated the effect of nanometer-sized SiC particle on the crack-healing behavior, as shown in Figure 17.31.

The result demonstrates that the nanometer-sized SiC with particle size of 20 nm can significantly increase the self-crack healing rate compared to the commercial 270 nm SiC particles. Furthermore, the nano-SiC particles can attain the complete strength recovery within 10 h at 1023 K, which is a 250 K lower temperature compared to the commercial SiC particles. Although nanometer-sized SiC particles makes the crack-healing reaction activated at lower temperatures, it gives same level of refractoriness as the alumina containing commercial SiC particles

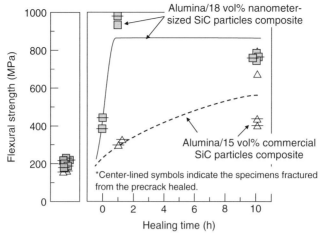

Figure 17.31 Crack-healing behavior at 1373 K on alumina containing 18 vol% nanometer-sized SiC particles composite and alumina containing 15 vol% commercial SiC particles composite, in which center lined symbols indicate the specimen fractured from the precrack healed.

composite. Therefore, it is noted that the use of SiC nanoparticles is a most valuable route to enhance the valid temperature region of the self crack healing.

17.10
Availability to Structural Components of the High Temperature Gas Turbine

The structural component of the high temperature gas turbine is one of the most attractive applications for the self healing ceramics. The application requires high temperature durability and high mechanical reliability because of its service condition consisting of high temperature and oxidizing atmosphere, which is valid condition of the self-healing driven by the SiC oxidation. Actually the turbine nozzle and blade of the 1500 °C-class gas turbine are exposed in high temperatures ranging from 1273 to 1773 K and oxygen partial pressures ranging from 8000 to 10 000 Pa. To discuss the availability of the self-healing driven by SiC oxidation to the turbine blades and nozzles requires knowing the self healing kinetics as a function of temperature and oxygen partial pressure, because the rate is sensitive to temperature and oxygen partial pressure changes.

Osada *et al.* [39] proposed the kinetics model in which the rate of self-crack healing, v_H, is expressed as a function of both the healing temperature, T_H, and oxygen partial pressure, P_{O2}. Two assumptions of the kinetic model are that the reaction order, n, is independent on temperature and that obeying Arrhenius' law, the reaction constant, k, depends on only temperature. By using the kinetic model, k is expressed as a following function of T_H and P_{O2};

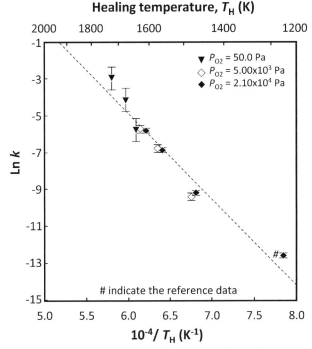

Figure 17.32 Temperature dependence on natural logarithm of the rate constant, k, calculated from various levels of P_{O2}.

$$k = \frac{v_H}{P_{O2}^n} = A \exp\left(-\frac{Q_H}{RT_H}\right) \qquad (17.16)$$

where A is constant, Q_H activation energy, and R gas constant. Validity of the model was assayed from the experimental data on alumina/15 vol% SiC composite, as shown in Figure 17.32. The vertical axis indicates the logarithm of $1/t_H \cdot P_{O2}^n$, corresponding to the logarithm of k, since where t_H is the minimum time until complete healing, and the value of v_H is determined to be the inverse of t_H. As all the plots can be fitted by only one straight line, it is confirmed that the proposed model can be available in the temperature region from 1273 to 1473 K. Furthermore, the minimum time for complete healing was obtained as;

$$t_H = 1.44 \times 10^{-6} \cdot P_{O2}^{-0.835} \exp\left(\frac{4.65 \times 10^4}{T_H}\right) \qquad (17.17)$$

Also it is important to know the threshold stress during self-crack healing under low oxygen partial pressure. Figure 17.33 show the bending strength of Si_3N_4/SiC composite crack-healed at 1473 K for 5 h in P_{O2} of 500 Pa under several tensile stress. The tensile stress is applied by three point bending and applied to one face containing an indentation crack of 0.1 mm in surface length. The specimens

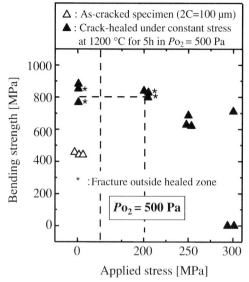

Figure 17.33 Bending strength of crack-healed specimen as a function of applied stress during crack-healing in $P_{O2} = 500$ Pa.

applied stress of 200 MPa and 250 MPa survived during crack-healing process. For the applied stress of 200 MPa, the specimens fractured outside crack-healed zone, whereas the applied stress of 250 MPa the specimens fractured from crack-healed zone. The average values of bending strength of the specimens crack-healed under 200 MPa are comparable to those of the specimen healed without stress. Therefore, the threshold tensile stress for crack-healing at 1473 K in $P_{O2} = 500$ Pa is determined to be 200 MPa. In the same method, for the specimens crack-healed in $P_{O2} = 5000$ Pa and 21 000 Pa (air), the threshold tensile stresses for crack-healing were determined to be 200 MPa. From these results, the crack-healing can be achieved in low oxygen partial pressure of $P_{O2} = 500$ Pa and can elongate the lifetime and the reliability.

References

1 Williams, L.S. (1956) *Transactions of the British Ceramic Society*, **55** (**5**), 287–312.

2 Evans, A.G. (1972) *Journal of Materials Science*, **7** (**10**), 1131–1146.

3 Dwivedi, P.J. and Green, D.J. (1995) *Journal of the American Ceramic Society*, **78** (**8**), 2122–2128.

4 Heuer, A.H. and Roberts, J.P. (1966) *Proceedings of the British Ceramic Society*, **6**, 17–27.

5 Lange, F.F. and Gupta, T.K. (1970) *Journal of the American Ceramic Society*, **53** (**1**), 54–55.

6 Davies, L.M. (1966) *Proceedings of the British Ceramic Society*, **6**, 29–53.

7 Lange, F.F. and Radford, K.C. (1970) *Journal of the American Ceramic Society*, **53** (**7**), 420–421.

8 Roberts, J.T.A. and Wrona, B.J. (1973) *Journal of the American Ceramic Society*, **56** (**6**), 297–299.

9 Bandyopadhyay, G. and Roberts, J.T.A. (1976) *Journal of the American Ceramic Society*, **59 (9–10)**, 415–419.

10 Gupta, T.K. (1976) *Journal of the American Ceramic Society*, **59 (9–10)**, 448–449.

11 Evans, A.G. and Charles, E.A. (1977) *Acta Metallurgica*, **25**, 919–927.

12 Lange, F.F. (1970) *Journal of the American Ceramic Society*, **53 (5)**, 290.

13 Easler, T.E., Bradt, R.C. and Tressler, R.E. (1982) *Journal of the American Ceramic Society*, **65 (6)**, 317–320.

14 Chu, M.C., Cho, S.J., Yoon, K.J. and Park, H.M. (2005) *Journal of the American Ceramic Society*, **88 (2)**, 491–493.

15 Niihara, K. and Nakahira, A. (1998) "Strengthening of oxide ceramics by SiC and Si_3N_4 dispersions", *Proceeding of the Third International Symposium on Ceramic Materials and Components for Engines*, American Ceramics Society, Westerville, pp. 919–926.

16 Niihara, K. (1991) *Journal of the Ceramic Society of Japan*, **9 (10)**, 974–982.

17 Niihara, K., Nakahira, A. and Sekino, T. (1993) *Materials Research Society Symposium Proceedings*, **286**, 405–412.

18 Thompson, A.M., Chan, H.M. and Harmer, M.P. (1995) *Journal of the American Ceramic Society*, **78 (3)**, 567–571.

19 Chou, I.A., Chan, H.M. and Harmer, M.P. (1998) *Journal of the American Ceramic Society*, **81 (5)**, 1203–1208.

20 Wu, H.Z., Lawrence, C.W., Roberts, S.G. and Derby, B. (1998) *Acta Materialia*, **46 (11)**, 3839–3848.

21 Chu, M.C., Sato, S., Kobayashi, Y. and Ando, K. (1995) *Fatigue and Fracture of Engineering Materials and Structures*, **18 (9)**, 1019–1029.

22 Ando, K., Tsuji, K., Hirasawa, T., Kobayashi, Y., Chu, M.C. and Sato, S. (1999) *Journal of the Society of Materials Science, Japan*, **48 (5)**, 489–494.

23 Ando, K., Tsuji, K., Ariga, M. and Sato, S. (1999) *Journal of the Society of Materials Science, Japan*, **48 (10)**, 1173–1178.

24 Ando, K., Ikeda, T., Sato, S., Yao, F. and Kobayasi, Y. (1998) *Fatigue and Fracture of Engineering Materials and Structures*, **21**, 119–122.

25 Ando, K., Chu, M.C., Kobayashi, Y., Yao, F. and Sato, S. (1991) *The Japan Society of Mechanical Engineering, International Journal Series A*, **65A**, 1132–1139.

26 Ando, K., Chu, M.C., Yao, F. and Sato, S. (1999) *Fatigue and Fracture of Engineering Materials and Structures*, **22**, 897–903.

27 Yao, F., Ando, K., Chu, M.C. and Sato, S. (2001) *Journal of the European Ceramic Society*, **21**, 991–997.

28 Ando, K., Kim, B.S., Chu, M.C., Saito, S. and Takahashi, K. (2004) *Fatigue and Fracture of Engineering Materials and Structures*, **27**, 533–541.

29 Kim, B.S., Ando, K., Chu, M.C. and Saito, S. (2003) *Journal of the Society of Materials Science, Japan*, **52 (6)**, 667–673.

30 Ando, K., Kim, B.S., Kodama, S., Ryu, S.H., Takahashi, K. and Saito, S. (2003) *Journal of the Society of Materials Science, Japan*, **52 (11)**, 1464–1470.

31 Nakao, W., Osada, T., Yamane, K., Takahashi, K. and Ando, K. (2005) *Journal of the Japan Institute of Metals*, **69 (8)**, 663–666.

32 Chase M.W. Jr. (ed.) (1998) *NIST-JANAF Thermochemical Tables*, 4th edn, American Chemistry Society and American Institute of Physics for the National Institute of Standards and Technology.

33 Japan Industrial Standard R1601 (1993) *Testing Method for Flexural Strength of High Performance Ceramics*, Japan Standard Association, Tokyo.

34 Sato, S., Chu, M.C., Kobayashi, Y. and Ando, K. (1995) *Journal of the Japan Society of Mechanical Engineers*, **61**, 1023–1030.

35 Takahashi, K., Yokouchi, M., Lee, S.K. and Ando, K. (2003) *Journal of the American Ceramic Society*, **86 (12)**, 2143–2147.

36 Lee, S.K., Ishida, W., Lee, S.Y., Nam, K.W. and Ando, K. (2005) *Journal of the European Ceramic Society*, **25 (5)**, 569–576.

37 Lee, S.K., Ando, K. and Kim, Y.W. (2005) *Journal of the American Ceramic Society*, **88 (12)**, 3478–3482.

38 Fang, J., Chan, H.M. and Harmer, M.P. (1995) *Materials Science and Engineering A*, **195**, 163–167.

39 Osada, T., Nakao, W., Takahashi, K., Ando, K. and Saito, S., (2009) *Journal of the American Ceramics Society*, **92 (4)**, 864–869.

40 Jung, Y.S., Nakao, W., Takahashi, K., Ando, K. and Saito, S., (2008) *Journal of the Society of the Materials Science Japan*, **57** (**11**), 1132–1137.

41 Ando, K., Furusawa, K., Chu, M.C., Hanagata, T., Tuji, K. and Sato, S. (2001) *Journal of the American Ceramic Society*, **84** (**9**), 2073–2078.

42 Ando, K., Furusawa, K., Takahashi, K., Chu, M.C. and Sato, S. (2002) *Journal of the Ceramic Society of Japan*, **110** (**8**), 741–747.

43 Nakao, W., Takahashi, K. and Ando, K. (2006) *Materials Letters*, **61**, 2711–2713.

44 Ando, K., Yokouchi, M., Lee, S.K., Takahashi, K., Nakao, W. and Suenaga, H. (2004) *Journal of the Society of Materials Science, Japan*, **53** (**6**), 599–606.

45 Nakao, W., Ono, M., Lee, S.K., Takahashi, K. and Ando, K. (2005) *Journal of the European Ceramic Society*, **25** (**16**), 3649–3655.

46 Ono, M., Ishida, W., Nakao, W., Ando, K., Mori, S. and Yokouchi, M. (2004) *Journal of the Society of Materials Science, Japan*, **54** (**2**), 207–214.

47 Kim, B.A., Meguro, S., Ando, K. and Ogura, N. (1990) *Journal of the High Pressure Institute of Japan*, **28**, 218–223.

48 Newman, J.C. and Raju, I.S. (1981) *Engineering Fracture Mechanics*, **15**, 185–192.

49 Ono, M., Nakao, W., Takahashi, K., Nakatani, M. and Ando, K. (2007) *Fatigue and Fracture of Engineering Materials and Structures*, **30** (**7**), 599–667.

50 Nakao, W., Nakamura, J., Yokouchi, M., Takahashi, K. and Ando, K. (2006) *Transactions of JSSE*, **51**, 20–26.

51 Japan Industrial Standard R1632 (1998) *Test Method for Static Bending Fatigue of Fine Ceramics*, Japan Standard Association, Tokyo.

52 Kanematsu, W., Yamauchi, Y., Ohji, T., Ito, S. and Kubo, K. (1992) *Journal of the Ceramic Society of Japan*, **100** (**6**), 775–779.

53 Osada, T., Nakao, W., Takahashi, K., Ando, K. and Saito, S. (2007) *Journal of the Ceramic Society of Japan*, **115** (**4**), 278–284.

54 Nishida, M. (1967) *Stress Concentration*, Morikita Publishing, Tokyo, pp. 572–574.

55 Ritter J.E. Jr., Oates, P.B., Fuller E.R. Jr. and Wiederhorn, S.M. (1980) *Journal of Materials Science*, **15**, 2275–2281.

56 Ritter J.E. Jr., Oates, P.B., Fuller, E.R. Jr. and Wiederhorn, S.M. (1980) *Journal of Materials Science*, **15**, 2282–2295.

57 Hoshide, T., Sato, T. and Inoue, T. (1990) *Journal of the Japan Society of Mechanical Engineers A*, **56**, 212–218.

58 Hoshide, T., Sato, T., Ohara, T. and Inoue, T. (1990) *Journal of the Japan Society of Mechanical Engineers A*, **56**, 220–223.

59 Ando, K., Sato, S., Sone, S. and Kobayashi, Y. "Probabilistic study on fatigue life of proof tested ceramics spring", in *Fracture From Defects, Proceedings of ECF-12* (eds M.W. Brown, E.R. de los Rios and K.J. Miller), EMAS Publishing, London, 1998, pp. 569–574.

60 Ando, K., Kim, B.A., Iwasa, M. and Ogura, N. (1992) *Fatigue and Fracture of Engineering Materials and Structures*, **15**, 139–149.

61 Ando, K., Iwasa, M., Kim, B.A., Chu, M.C. and Sato, S. (1993) *Fatigue and Fracture of Engineering Materials and Structures*, **16**, 995–1006.

62 Ando, K., Shirai, Y., Nakatani, M., Kobayashi, Y. and Sato, S. (2002) *Journal of the European Ceramic Society*, **22**, 121–128.

63 Ando, K., Takahashi, K., Murase, H. and Sato, S. (2003) *Journal of the High Pressure Institute of Japan*, **41**, 316–326.

64 Nakao, W., Mori, S., Nakamura, J., Yokouchi, M., Takahashi, K. and Ando, K. (2006) *Journal of the American Ceramic Society*, **89** (**4**), 1352–1357.

65 Chaklader, A.C.D., Gupta, S.D., Lin, E.C.Y. and Gutowski, B. (1992) *Journal of the American Ceramic Society*, **75** (**8**), 2283–2285.

66 Borsa, C.E., Spiandorello, F.M. and Kiminami, R.H.G.A. (1999) *Materials Science Forum*, **299–300**, 57–62.

67 Amroune, A., Fantozzi, G., Dubois, J., Deloume, J.P., Durand, B. and Halimi, R. (2000) *Materials Science and Engineering A*, **290**, 11–15.

68 Amroune, A. and Fantozzi, G. (2001) *Journal of Materials Research*, **16**, 1609–1613.

69 Lee, J.H., An, C.Y., Won, C.W., Cho, S.S. and Chun, B.S. (2000) *Materials Research Bulletin*, **35**, 945–954.

70 Pathank, L.C., Bandyopadhyay, D., Srikanth, S., Das, S.K. and Ramachandrarao, P. (2001) *Journal of the American Ceramic Society*, **84** (**5**), 915–920.

71 Zhang, G.J., Yang, J.F., Ando, M. and Ohji, T. (2004) *Journal of the American Ceramic Society*, **87** (**2**), 299–301.

72 Nakao, W., Tsutagawa, Y. and Ando, K. *Journal of Intelligent Material Systems and Structures*, **19**, 407–410.

18
Ecological Toxicology of Engineered Carbon Nanoparticles

Aaron P. Roberts and Ryan R. Otter

18.1
Introduction

Advancements in the science of engineered nanoparticles (materials with at least one dimension <100 nm) have created a great deal of promise for their application in a wide variety of fields. Today, the "nanosized" properties of a range of materials are being harnessed for use not only in industrial applications, but also in biomedical applications [1–3] and personal care products [4], among others. A great deal of concern has been expressed, however, regarding the potential safety of these materials to biological systems. Reviews in *Science* [5] as well as a United States National Research Council report [6] have noted a general lack of information on the potential human and environmental health effects of exposure to these compounds.

In particular, relatively few resources have been devoted to examining the potential ecological fate and effects of engineered nanoparticles, and there is a general lack of conclusive data regarding the fate and potential effects of these materials in natural ecosystems. Despite safeguards and environmental protection legislation, industrial and personal products are routinely deposited in the environment, including freshwater and marine ecosystems [7]. For example, Kolpin *et al.* [7] reported that in a study of 139 streams in the United States, 80% contained measurable levels of organic wastewater contaminants including caffeine, insecticides, cholesterol, antimicrobial agents, and fire retardants. With over 300 commercial products containing some type of nanotechnology currently available on the market [8], and growth expected to exceed US$1 trillion by 2015 [5], it is likely that inputs of nanomaterials into the environment will increase over time [9].

In order to estimate the potential risk posed by engineered nanoparticles to ecological systems, several objectives must be met:

- **Realistic estimates of environmental discharge must be made:** Current research on the ecological effects of engineered nanoparticles is carried out without an estimate of what *environmentally relevant* concentrations of nanoparticles might be. Accurate environmental discharge estimates are necessary for ecotoxicologists

Advanced Nanomaterials. Edited by Kurt E. Geckeler and Hiroyuki Nishide
Copyright © 2010 WILEY-VCH Verlag GmbH & Co. KGaA, Weinheim
ISBN: 978-3-527-31794-3

to refine test methods and ascertain mechanisms of toxicity which are most likely to occur in ecological settings.

- **The behavior of engineered nanoparticles in soils and water must be understood:** In order for toxic effects to occur, a material must be bioavailable and taken up by a biological organism. Thus, the environmental compartments that engineered nanomaterials move into, and how they behave within those compartments, must be understood in order to estimate biological–nanoparticle interaction.

- **The mechanisms by which engineered nanoparticles exert effects on ecologically important species must be understood:** Current research on the toxicity of engineered nanoparticles has focused on mammalian systems as a model for potential human health impacts. While biochemical, molecular, and tissue-level outcomes described by these studies are likely relevant for ecologically important species, other – more subtle – effects such as shifts in energetics or effects on organism behavior may also be significant and unexplored by the biomedical community.

In this chapter we will discuss some of these issues and how they relate to the potential ecological impacts of engineered carbon nanoparticles following their release into aquatic ecosystems.

18.2
Fate and Exposure

18.2.1
General

In order to assess the risk posed by engineered carbon nanomaterials to ecological systems, both the exposure scenarios as well as potential effects must be characterized. The behavior of engineered carbon nanoparticles in aqueous systems, their bioavailability to aquatic biota, and their potential to alter the transport and availability of co-occurring chemical contaminants, will be examined in the following sections.

18.2.2
Stability in Aquatic Systems

The fate and behavior of engineered carbon nanoparticles in aquatic ecosystems depend on the specific properties of the individual material. Most carbon-derived nanoparticles are largely insoluble in water, and tend to aggregate into larger macroparticles which cannot maintain their place in the water column [10]. Thus, these materials, similar to many other organic chemical contaminants, are likely to partition into sediments following their discharge into aqueous environments (Figure 18.1). In the sediments, these materials will be available to benthic

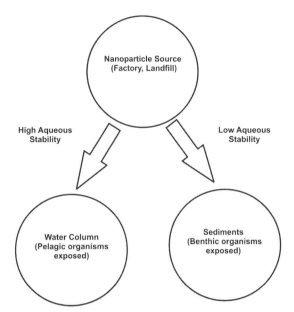

Figure 18.1 The potential fate of engineered carbon nanomaterials in aqueous systems.

organisms and, possibly, to pelagic organisms through resuspension or food chain transport.

Potentially more relevant to aquatic ecotoxicology are engineered carbon nanoparticles which have been functionalized for technological applications that require increased stability in aqueous environments. Engineered carbon nanoparticles can be stabilized in aqueous environments using surfactants, organic solvents, and biomacromolecular coatings such as lipids and proteins [10–13]. However, little is known about how relevant these solubilization strategies are for natural systems, or how they influence the behavior of engineered carbon nanoparticles in freshwater and marine environments.

Perhaps more relevant for ecological systems is the discovery that some nanomaterials, including multiwalled carbon nanotubes (CNTs), can be stabilized in aqueous environments using natural organic matter (Figure 18.2) [10]. The latter is a complex mixture of organic molecules of varying size and chemical properties that originates from degraded plant and animal material, and is found at varying concentrations in virtually all aquatic ecosystems.

Natural organic matter contains both lipophilic and hydrophilic components which are able to interact with engineered carbon nanoparticles and thus increase their solubility and stability in aqueous environments. It is important to note that all of these methods (coatings, solvents, organic matter) often result in only stable *suspensions* and not true solutions. Thus, the suspended nanoparticles will come out of suspension over time. For example, Roberts *et al.* [12] found that lipid-coated single-walled CNTs, while far more stable in suspension than uncoated tubes, still

Figure 18.2 Flasks containing: (a) organic-free water; (b) 1% sodium dodceylsulfate solution; (c) 100 mg natural organic matter (NOM) solution with 500 mg l^{-1} multi-walled nanotubes (MWNTs); and (e) river water containing NOM and 500 mg l^{-1} MWNTs. The 100 mg carbon per liter (C/L) SR-NOM solution and river water without MWNT addition are shown in flasks (d) and (f). Reproduced from Ref. [10].

precipitated approximately 20% out of suspension in moderately hard freshwater over a 48 h period. Regardless, increased stability in aqueous environments will increase the period for which aquatic organisms are exposed to engineered carbon nanoparticles.

18.2.3
Bioavailability and Uptake

The term *bioavailable* refers to those chemical contaminants which can interact with, and be taken up by, an organism into its tissues. Bioavailable contaminants may enter an organism through a number of pathways, but the most common route is by uptake across the skin, the ingestion of contaminated food and water, or via respiratory processes. Not all contaminants in the environment are bioavailable. For example, the bioavailability of heavy metals such as cadmium and copper are reduced in aqueous environments with high water hardness [14, 15]. The bioavailability of metals in sediments can be reduced by the presence of acid-volatile sulfides [16]. Organic contaminants such as polycyclic aromatic hydrocarbons can be sequestered by dissolved organic carbon and natural organic matter, thus preventing their uptake by aquatic biota [17]. This reduction in bioavailability coincides with a reduction in toxic effects due to a decrease in the internal dose of the toxicant.

The potential bioavailability of engineered carbon nanoparticles in aquatic ecosystems is largely unknown, with few studies having been conducted to measure the uptake of engineered carbon nanoparticles by aquatic biota. Potential routes of exposure include uptake across the gills and skin, as well as ingestion [9]. Roberts *et al.* [12] demonstrated that grazing freshwater zooplankton (*Daphnia magna*) were able to ingest lipid-coated single-walled CNTs through their normal feeding behavior. The organisms were able to digest the lipid coating from the tubes and excrete the uncoated, insoluble nanotubes. Similar results were obtained

(a)   (b)

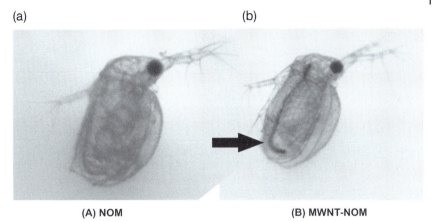

(A) NOM   **(B) MWNT-NOM**

Figure 18.3 Micrographs of the zooplankton *Ceriodaphnia dubia*, following exposure to (a) natural organic matter and (b) natural organic matter and multiwalled carbon nanotubes (MWCNTs). The dark line visible inside the organism in panel (b) is the intestinal tract (arrow), containing large amounts of MWCNTs.

(a)   (b)

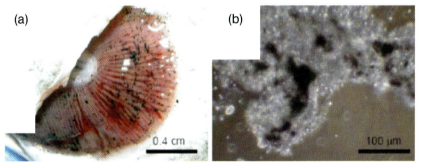

Figure 18.4 (a) Single-walled carbon nanotubes adhered to the mucous secretions around the gills of rainbow trout; (b) Phase-contrast micrograph of a mucus smear, showing nanotubes associated with the mucous proteins. Reproduced from Ref. [19].

with another smaller species (*Ceriodaphnia dubia*) using natural organic matter-stabilized multiwalled CNTs (Figure 18.3).

The dietary uptake of single-walled CNTs has also been reported in estuarine copepods (*Amphiascus tenuiremis*), another zooplankton species [18]. Also reported here was the presence of compact aggregations of nanotubes in the fecal pellets of the copepods. Fewer studies exist on the uptake of nanotubes by fish; in rainbow trout, nanotubes ingested from the water column were shown to adhere to the mucous secretions of the gill tissues (Figure 18.4) [19].

However, it is unknown what fraction (if any) of the nanotubes crossed the epithelium and were taken up into the soft tissues of the organisms in any of these studies, although this is largely due to a lack of effective methods for measuring nanotubes at the low concentrations found in biological tissues. The limit of detection for analytical methods requires samples to be concentrated to such a high degree that there may be great uncertainty in the accuracy of their determination.

Fullerenes and some fullerene-derivatives are known to be lipophilic [20]. Fullerenes are easily extracted from tissues using organic solvents, and their concentration determined using spectrophotometry. In one study, Oberdorster *et al.* [21] documented an increasing uptake of fullerenes by zooplankton from the water column over a 96 h period [21]. It has been suggested that adding functional groups to both fullerenes and nanotubes decreases their toxicity *in vitro* [22–24], though this may be due to alterations in the hydrophobicity and thus uptake of the material by cells [24].

18.2.4
Tissue Distribution

Once internalized by the organism, chemical contaminants are distributed to the tissues via the circulatory system. Again, there is a paucity of data regarding the tissue distribution of carbon nanoparticles in aquatic organisms. Oberdorster *et al.* [21] found that exposure to uncoated fullerenes resulted in an increased oxidative stress in the brains of largemouth bass, leading to the possibility that the materials were able to cross the blood–brain barrier [11]. These results were supported by another study, in which the distribution of fluorescent, nanosized polystyrene spheres in the tissues of the medaka, an aquarium fish commonly used in aquatic toxicological studies, was investigated [25]. Kashiwada [25] found that fluorescing polystyrene nanoparticles accumulated within the gut and gills of the fish, and were also detected in the brain, testis, liver, and blood. However, the most intense fluorescence was observed in the gill tissues, indicating that uptake across the gill epithelium may be a major route of exposure. A previous study had shown that rats dosed orally with fullerenes excreted most of the material via the feces, whereas rats injected intravenously with fullerenes retained the materials for over one week, with the majority being distributed to the liver [26]. These findings support the hypothesis that, in fish, uptake across the gills may be a more important route of exposure than dietary ingestion.

18.2.5
Food Web

Several chemical compounds found in the environment – and especially those which are not easily metabolized, such as mercury and polychlorinated biphenyls – have been shown to move through aquatic food chains [27–29]. These materials are taken up by organisms at low trophic levels such as bacteria, algae, and zooplankton and, because they are not easily metabolized and excreted, are passed

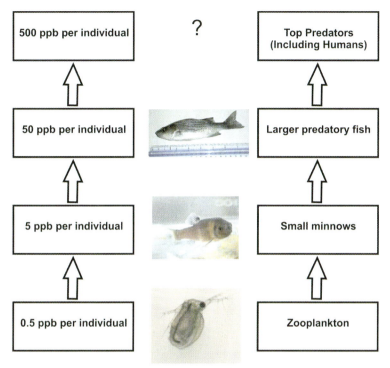

| 500 ppb per individual | ? | Top Predators (Including Humans) |

| 50 ppb per individual | | Larger predatory fish |

| 5 ppb per individual | | Small minnows |

| 0.5 ppb per individual | | Zooplankton |

Figure 18.5 Biomagnification through an aquatic food chain. Contaminants may be found at relatively low concentrations at lower trophic levels (zooplankton). Concentrations will increase as predatory organisms consume large numbers of prey items, but do not have the ability to metabolize or excrete the chemical contaminant.

on to higher trophic levels when those organisms are subsequently ingested by predators. This results in increasing concentrations of chemical within organisms at higher levels of the food chain, a process known as *biomagnification* (Figure 18.5).

The biomagnification of these materials can pose significant risks to the health of organisms located at higher trophic levels which might otherwise not be exposed, including humans [30–34]. For example, mercury consumption warnings in seafood are a result of biomagnification of methyl mercury through the food web.

Studies conducted by Templeton *et al.* [12] and Roberts *et al.* [18] have indicated that zooplankton in aquatic systems can ingest nanoparticles from their environment through normal feeding behavior. Bacteria, which form the bulk of aquatic biofilms, have also been shown to take up nanosized particles. Both, biofilms and zooplankton occupy lower trophic levels in aquatic ecosystems and provide an important food base for a variety of organisms, including fish. Considering the lipophilicity of some nanoparticles, and the lack of evidence of true metabolism, the potential for biomagnification clearly exists.

18.2.6
Effects on the Uptake of Other Contaminants

Due to the relative ease with which some engineered carbon nanoparticles are taken up by cells, there is also concern regarding the potential for these materials to facilitate the transport and uptake of other, adsorbed chemical contaminants into cells. Polycyclic aromatic hydrocarbons (PAHs) are a ubiquitous contaminant, and are discharged into the environment as a result of the incomplete combustion of fossil fuels. They are generally considered to be carcinogenic and to have a range of toxic effects [35–37]. Yang *et al.* [38] showed that PAHs readily adsorb to fullerenes and nanotubes [38], while Moore *et al.* [9] also found that coexposure to a model PAH (anthracene) and a nanosized polyester sucrose particle increased both the uptake and cellular toxicity of anthracene in mussels. Contaminants which are readily adsorbed onto carbon nanoparticles, or are closely associated with them as a result of manufacturing processes (such as transition metals), may be delivered more rapidly to cells in the presence of nanosized particles.

18.3
Effects

18.3.1
General

The limited data relating to the toxicological effects of engineered carbon nanoparticles has derived from research groups that have focused largely on mammalian models (both *in vitro* and *in vivo*). Such studies have been carried out primarily to determine whether these compounds cause, or have the potential to cause, human health effects following occupational exposure. Although there are exceptions, the bulk of the data suggests that engineered carbon nanoparticles are toxic to some degree, and points to oxidative stress, inflammatory reactions, and immunological effects as the key features of carbon nanoparticle toxicity [39, 40].

Relatively few studies, however, have been carried out regarding nanotoxicological effects on ecologically important aquatic species, such as fish and zooplankton (Table 18.1). The aim of this section is to summarize and review the ecotoxicological effects of engineered carbon nanoparticles, focusing primarily on aquatic eukaryotic systems (e.g., fish and zooplankton), as has the bulk of reports made to date.

18.3.2
Oxidative Stress and Nanoparticles

Aerobic organisms use oxygen to oxidize (burn) carbon- and hydrogen-rich substrates (foods) to obtain the chemical and heat energy that is essential for life. However, when molecules are oxidized with oxygen, the oxygen molecule itself

Table 18.1 Summary of the ecotoxicological literature on nanoparticles.

Reference	Species	Nonmaterial tested	Preparation method	Concentrations tested	Endpoints examined
[11]	*Micropterus salmoides*	nC_{60}	THF	0.5–1.0 ppm	Lipid peroxidation, protein oxidation, glutathione
[21]	*Daphnia magna* *Pimephales promelas* *Hyalella azteca* *Oryzias latipes*	nC_{60}	Water-stirred	0.5–30 ppm	Mortality, CYP P450 isozymes, PMP70
[41]	*Daphnia magna*	nC_{60}	Sonication THF	0.2–880 ppm	Mortality, behavior
[42]	*Daphnia magna* *Pimephales promelas*	nC_{60}	Water-stirred THF	0.5 ppm	Lipid peroxidation, CYP2K1, CYP2M1
[42]	*Stylonychia mytilus*	MWCNT	AEDP	0.1–200 μg ml^{-1}	Cytotoxicity cell morphology, structural alterations
[43]	*Danio rerio*	nC_{60} nC_{70} $nC_{60}(OH)_{24}$	DMSO DMSO DMSO	100–500 ppb 100–500 ppb 500–5000 ppb	Development effects, apoptosis

Table 18.1 Continued.

Reference	Species	Nonmaterial tested	Preparation method	Concentrations tested	Endpoints examined
[44]	*Bacillus subtilis* *Escherichia coli*	nC_{60} $nC_{60}(OH)_{22-24}$	THF	$0.04-4\,mg\,l^{-1}$ $5\,mg\,l^{-1}$	Growth
[45]	*Danio rerio*	nC_{60}	Stirring and sonication THF	Dilution of stock solution	Mortality, behavior, global gene expression
[12]	*Daphnia magna*	SWCNT-lipid coated	LPC	$2.5-20\,mg\,l^{-1}$	Mortality
[42]	*Danio rerio*	nC_{60} $nC_{60}(OH)_{16-18}$	Acetone/benzene/ THF water	$1.5\,mg\,l^{-1}$ $50\,mg\,l^{-1}$	Mortality, developmental effects, hatching rate, heartbeat, pericardial edema
[19]	*Oncorhynchius mykiss*	SWCNT	SDS, sonication	$0.1-0.5\,mg\,l^{-1}$	Behavior, gill ventilation rates, plasma ions, histopathology, Na^+K^+-ATPase, oxidative stress
[46]	*Daphnia magna*	nC_{60} $nC_{20}HxC_{20}Hx$	THF	260 ppb	Behavior

THF, tetrahydrofuran; AEDP, 2-Amino-ethyene-l,1-bis-phosphonic acid; LPC, lysophophatidylcholine; SDS, sodium dodecylsulfate.

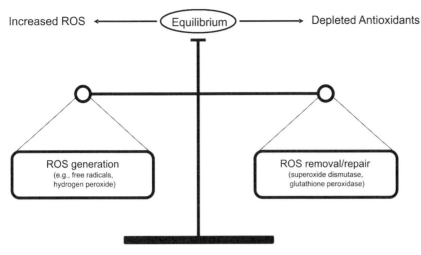

Arrows indicate a shift in balance

Figure 18.6 Oxidative stress occurs naturally, but biological systems maintain an equilibrium by counteracting the deleterious effects of reactive oxygen species (ROS) with antioxidants. Increased oxidative stress, a common mode of action for a number of chemical toxicants, can occur through either a depletion of a biological system's antioxidants or an increase in the number of ROS.

becomes reduced and forms intermediates called reactive oxygen species (ROS); these include free radicals and hydrogen peroxide. *Oxidative stress* is the term used to describe the level of oxidative damage in a cell or tissue caused by ROS, the latter being a class of molecules that are derived from the metabolism of oxygen and which exist inherently in all aerobic organisms. In fact, ROS are produced constantly during normal aerobic metabolism, but are normally safely removed by a variety of biological antioxidants, such as superoxide dismutase (SOD), catalase, or glutathione (GSH) peroxidase. The amount of oxidative stress in a cell or tissue can be thought of as a balance between the overall generation and the overall removal/repair of ROS by antioxidants (Figure 18.6). Therefore, oxidative stress within a cell or tissue can result from two factors: (i) an increase in oxidant generation; and/or (ii) a decrease in antioxidant protection.

Research has shown that xenobiotics such as metals and hydrocarbons can shift the oxidative stress balance, thus causing an overproduction of ROS or the inhibition of antioxidant systems [47–49]. This shift in balance leads to excessive molecular damage and tissue injury, an example being lipid peroxidation. The latter process is defined as the free-radical oxidation of polyunsaturated fatty acids in biological systems, and can be used as an end-point to measure oxidative stress. The most common biomarkers used to measure oxidative stress include thiobarbituric acid reactive substances (TBARS), which measures the presence of lipid peroxides and the reduction of GSH, an antioxidant which occurs naturally within cells and is known to scavenge ROS [49–52].

The majority of mammalian-based reports point to oxidative stress as the main mode of action of engineered carbon nanoparticles (e.g., Ref. [40]), and so the assumption is typically made that this is the main mode of action in all species. With the exception of one report [11], the present authors are unaware of any other *ecotoxicological* literature providing evidence that exposure to engineered carbon nanoparticles results in oxidative stress. These examinations provide insight into the possibility that carbon-engineered nanoparticles may affect aquatic organisms in a different way from mammalian systems. For example, Roberts *et al.* [12] have suggested an energetics mechanism and interference with normal feeding behavior in zooplankton.

One major issue which concerns all toxicological studies is to ensure that any observed effects can be attributed to the compound of interest, and not other confounding factors. This is of particular concern when using organic solvents to either solubilize or stabilize a material in aqueous media. Some carbon nanoparticles, including C_{60}, have a very low solubility in water; consequently, organic solvents have been used (notably tetrahydrofuran;THF) to increase the solubility of engineered carbon nanomaterials. In the first report addressing the question of whether exposure to engineered carbon nanoparticles would result in ecotoxicological effects, Oberdorster [11] used THF to create nC_{60} solutions for exposure to largemouth bass. These preliminary study results indicated that nC_{60} caused significant lipid peroxidation in the brains of the exposed fish, since which time the preparation of nC_{60} using THF has become common practice. However, data obtained by Henry *et al.* [45] indicated that, at least in zebrafish, observable toxic effects could be attributed to THF and its breakdown products, as opposed to fullerenes [45]. Although the exact mode of action of THF-prepared nanoparticles is unknown, it has been suggested that THF might cross the cross the blood–brain barrier, unlike carbon nanoparticles [53], thus explaining the findings of Oberdorster [11].

It is also important to note that not all mammalian-based data have pointed to oxidative stress as the toxic mode of action of carbon nanoparticles. In fact, it has been suggested that nC_{60} may function in an opposite manner, and act as a free-radical scavenger [53]. An *in vitro* experiment in rats investigated the antioxidant effect of nC_{60} by pretreating rats with nC_{60}, followed by an injection of carbon tetrachloride (CCl_4), a chemical which is known to cause oxidative stress in the liver [53]. The study results showed that nC_{60} protected the liver against oxidative stress, which led the research team to conclude that the mechanism behind the protective role of nC_{60} could be attributed to the ability of nC_{60} to scavenge large numbers of free radicals.

18.3.3
Effects on Specific Tissues

18.3.3.1 Brain

The brain is arguably the most complex organ of all organisms, and may be incredibly sensitive to insult, with any disruption or interference with correct brain functions possibly leading to a loss of fitness and/or death. Although the effects of engineered carbon nanoparticles on the brain are poorly understood, behavioral changes in multiple aquatic species exposed to engineered carbon nanoparticles have been observed and have provided evidence that these particles might impair normal brain function [19, 46]. For example, when Smith *et al.* [19] exposed rainbow trout (*Oncorhynchus mykiss*) to single-walled CNTs, they observed aggressive behavior in the exposed fish that led to severe fin nipping. Although no biochemical changes were observed in the brains of these fish, a histological analysis identified what appeared to be aneurysms (swelling of the blood vessels), which indicated that the blood supply – and hence the oxygen supply – of the brain may have been compromised. Behavioral effects were also observed in *Daphnia magna* after exposure to nC_{60} [41, 46]. In these studies, both juvenile and adult *D. magna*, when exposed to nC_{60}, showed an abnormal behavioral response to the exposure that resulted in sporadic swimming and disorientation.

Following initial studies conducted with mammalian cell lines, the investigations into the toxic mode of action focused on oxidative stress [5]. As noted above, Oberdorster [11] had suggested that nC_{60} could be selectively transported to the brain, where it would cause toxic effects; this idea was based on the significant increase in lipid peroxidation that had been shown in the brains of largemouth bass exposed to nC_{60}. Unfortunately, these results, and their subsequent interpretation, may have been compromised by the use of an organic solvent in the experiment.

18.3.3.2 Gills

In fish, the gills function in similar fashion to the lungs in mammals, and are vital not only as the main site for gas exchange but also as an important component of ionoregulation. Carbon-engineered nanoparticles, based on their size and chemical properties, have the potential to cause harm in both a physical (e.g., abrasions) and chemical (e.g., disruption of gas exchange) manner. A study conducted by Smith *et al.* [19] documented increases in both ventilation rate and mucus secretion, as well as an enlargement of mucocytes on the gills of trout exposed to single-walled CNTs. This led the authors to conclude that the CNTs had clearly caused respiratory distress. Interestingly, the trout had responded to the exposure by increasing a natural defense mechanisms, namely an increased mucus secretion. This reaction is common in fish exposed to aqueous pollutants [54], and is a short-term defense mechanism designed to prevent the toxicant from reaching the sensitive gill epithelium [19].

18.3.3.3 Liver

In higher-level organisms, such as fish (and humans), the liver performs the enormous duty of maintaining the metabolic homeostasis of the body. Its responsibilities include protein synthesis, nutrient homeostasis, and the filtration of particulates. The liver is also the main location for the detoxification of toxicants and is, therefore, susceptible to chemical exposure and toxicity. Results obtained from studies in mammals have shown that intravenously administered fullerenes can be retained in the body for up to one week, with the majority (>70%) lodging in the liver [26].

Whilst the potential effects of these compounds on correct liver function are relatively unknown, Smith *et al.* in 2007, observed apoptosis (programmed cell death) in the liver of trout following their exposure to single-walled CNTs. Although the exact toxic mode of action was unknown, the location of the injured liver cells (close to the blood vessels) raised concern that the nanoparticles were being delivered systemically via the blood supply [19].

18.3.3.4 Gut

In theory, the gut would be expected to serve as the main site of uptake for nanoparticles associated with dietary exposure and drinking water. As noted earlier in the chapter, carbon-engineered nanoparticles are potentially bioavailable and can enter an organism via a number of pathways, including the ingestion of food and water. Although, observations have been made of carbon nanoparticle aggregates in the gut lumen of fish [19] and zooplankton [12], the bioavailability of these aggregates and their potential to cross the epithelial lining is largely unknown, as discussed previously.

18.3.4
Developmental Effects

Organisms are typically the most sensitive to insult (chemical or physical) during their early development and in the early life stages. Chemical compounds may disrupt cellular signaling, alter apoptosis patterns, or damage DNA, thus causing organisms to develop abnormally (*teratogenicity*). Teratogens can result in obvious effects such as physical deformities, or in more subtle changes such as behavioral abnormalities. These changes early in life can have profound effects on the ecological fitness of individuals, and result in deleterious population-level outcomes.

A host of investigations have utilized zebrafish embryos when investigating the developmental effects of engineered carbon nanoparticles [42, 43]. For example, Usenko *et al.* [43] observed delayed development, morphological malformations (e.g., of the body axis, eye, snout, jaw, otic vesicle, notochord, heart, brain, somite, and fins), pericardial edema (Figure 18.7), yolk sac edema (Figure 18.7), and behavioral abnormalities (e.g., hyperactivity, hypoactivity, paralysis) in fish exposed to nC_{60}. Of particular interest in this study was the constant abnormal development of the fin regions in all exposed organisms, which was indicative of cell signaling perturbations during early development (Figure 18.7). In another study, Zhu *et al.* [42] found that hatching rates in nC_{60}-exposed zebrafish embryos showed notable developmental delay and toxicity, slower heart rates, and increased pericardial edema.

The developmental effects such as those observed by Usenko *et al.* [43] and Zhu *et al.* [42] may have far-reaching implications since, in nature, individuals that experience abnormal development rarely survive to reproduce. Therefore, compounds found to cause developmental effects have the potential to affect entire populations and communities, and thus deserve special attention.

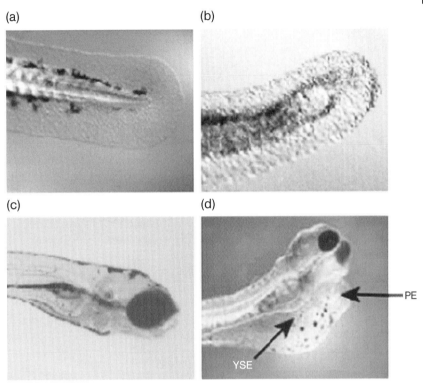

Figure 18.7 (a,b) Representative images of the caudal fins for (a) control and (b) 200 ppb nC_{60}-exposed animals; (c,d) Representative images of (c) 1% dimethylsulfoxide-exposed control larval head and (d) 200 ppb nC_{60}-exposed larval head. The arrows indicate the pericardial edema (PE) and yolk sac edema (YSE). Reproduced from Ref. [43].

18.4
Summary

The "nanotechnology revolution" has spurred interest and concern regarding the safety of engineered nanoparticles. In particular, attention has been paid to the potential effects that such materials might have on humans, either through the use of consumer products containing "nanotechnology," or through occupational exposure during the manufacturing process. Unfortunately, less attention has been paid to the potential impacts that these materials might have on ecological systems, despite the increased manufacture of engineered carbon nanomaterials being likely to result in their deposition into aquatic ecosystems, either from spills or as a result of the waste stream, in a fashion similar to other chemicals and consumer products [7, 9]. As noted in this chapter, there is a paucity of data regarding the behavior and fate of engineered carbon nanoparticles in aquatic ecosystems, their potential bioavailability to aquatic organisms, and the effects that they might have on those organisms.

The field of ecotoxicology generally operates within the framework of ecological risk assessment in order to determine the risk posed by a particular chemical to the natural environment. This framework relies on accurate exposure assessment – How much of the material is likely to be found in the environment and is the material bioavailable? – as well as accurate effects assessment – At what exposure concentrations does the material result in harm? Although preliminary investigations have indicated that fullerenes and nanotubes can be taken up by aquatic organisms, and result in deleterious effects under certain conditions, the realism of these exposure scenarios, and particularly of the exposure concentrations, is unknown. Are these exposure concentrations environmentally relevant? Are we likely to see these concentrations of fullerenes and nanotubes being discharged into aquatic systems? The effects that site-specific parameters such as pH, hardness, and temperature (which all have major impacts on the bioavailability of other contaminants) have on the bioavailability of engineered carbon nanomaterials are unknown. Until questions regarding these bioavailability and exposure scenarios are answered, it will remain difficult to ascertain the risk and potential effects that engineered carbon nanomaterials might have on aquatic ecosystems.

References

1 Daroczi, B., Kari, G., McAleer, M.F., Wolf, J.C., Rodeck, U. and Dicker, A.P. (2006) In vivo radioprotection by the fullerene nanoparticle DF-1 as assessed in a zebrafish model. *Clin. Cancer Res.*, **12**, 7086–7091.

2 Kohli, P. and Martin, C.R. (2005) Template-synthesized nanotubes for biotechnology and biomedical applications. *J. Drug Deliv. Sci. Technol.*, **15**, 49–57.

3 Mroz, P., Pawlak, A., Satti, M., Lee, H., Wharton, T., Gali, H., Sarna, T. and Hamblin, M.R. (2007) Functionalized fullerenes mediate photodynamic killing of cancer cells: Type I versus Type II photochemical mechanism. *Free Radic. Biol. Med.*, **43**, 711–719.

4 Nohynek, G.J., Lademann, J., Ribaud, C. and Roberts, M.S. (2007) Grey goo on the skin? Nanotechnology, cosmetic and sunscreen safety. *Crit. Rev. Toxicol.*, **37**, 251–277.

5 Nel, A., Xia, T., Madler, L. and Li, N. (2006) Toxic potential of materials at the nanolevel. *Science*, **311**, 622–627.

6 Council, N.R. (2006) *A Matter of Size: Triennial Review of the National Nanotechnology Initiative*, National Academies Press, p. 130.

7 Kolpin, D.W., Furlong, E.T., Meyer, M.T., Thurman, E.M., Zaugg, S.D., Barber, L.B. and Buxton, H.T. (2002) Pharmaceuticals, hormones, and other organic wastewater contaminants in US streams, 1999–2000: a national reconnaissance. *Environ. Sci. Technol.*, **36**, 1202–1211.

8 Maynard, A.D., Aitken, R.J., Butz, T., Colvin, V., Donaldson, K., Oberdorster, G., Philbert, M.A., Ryan, J., Seaton, A., Stone, V., Tinkle, S.S., Tran, L., Walker, N.J. and Warheit, D.B. (2006) Safe handling of nanotechnology. *Nature*, **444**, 267–269.

9 Moore, M.N., Lowe, D.M., Soverchia, C., Haigh, S.D. and Hales, S.G. (1997) Uptake of a non-calorific, edible sucrose polyester oil and olive oil by marine mussels and their influence on uptake and effects of anthracene. *Aquat. Toxicol.*, **39**, 307–320.

10 Hyung, H., Fortner, J.D., Hughes, J.B. and Kim, J.H. (2007) Natural organic matter stabilizes carbon nanotubes in the aqueous phase. *Environ. Sci. Technol.*, **41**, 179–184.

11 Oberdorster, E. (2004) Manufactured nanomaterials (Fullerenes, C-60) induce oxidative stress in the brain of juvenile largemouth bass. *Environ. Health. Perspect.*, **112**, 1058–1062.

12 Roberts, A.P., Seda, B., Mount, A.S., Lin, S.J., Ke, P.C., Qiao, R. and Klaine, S.J. (2007) In vivo biomodification of a lipid coated carbon nanotube by *Daphnia magna*. *Environ. Sci. Technol.*, **41**, 3025–3029.

13 Wu, Y., Hudson, J.S., Lu, Q., Moore, J.M., Mount, A.S., Rao, A.M., Alexov, E. and Ke, P.C. (2006) Coating single-walled carbon nanotubes with phospholipids. *J. Phys. Chem. B*, **110**, 2475–2478.

14 Brinkman, S.F. and Hansen, D.L. (2007) Toxicity of cadmium to early life stages of brown trout (*Salmo trutta*) at multiple water hardnesses. *Environ. Toxicol. Chem.*, **26**, 1666–1671.

15 Van Genderen, E., Gensemer, R., Smith, C., Santore, R. and Ryan, A. (2007) Evaluation of the Biotic Ligand Model relative to other site-specific criteria derivation methods for copper in surface waters with elevated hardness. *Aquat. Toxicol.*, **84**, 279–291.

16 Ogendi, G.M., Brumbaugh, W.G., Hannigan, R.E. and Farris, J.L. (2007) Effects of acid-volatile sulfide on metal bioavailability and toxicity to midge (*Chironomus tentans*) larvae in black shale sediments. *Environ. Toxicol. Chem.*, **26**, 325–334.

17 Weinstein, J.E. and Oris, J.T. (1999) Humic acids reduce the bioaccumulation and photoinduced toxicity of fluoranthene fish. *Environ. Toxicol. Chem.*, **18**, 2087–2094.

18 Templeton, R.C., Ferguson, P.L., Washburn, K.M., Scrivens, W.A. and Chandler, G.T. (2006) Life-cycle effects of single-walled carbon nanotubes (SWNTs) on an estuarine meiobenthic copepod. *Environ. Sci. Technol.*, **40**, 7387–7393.

19 Smith, C.J., Shaw, B.J. and Handy, R.D. (2007) Toxicity of single walled carbon nanotubes to rainbow trout (*Oncorhynchus mykiss*): respiratory toxicity, organ pathologies, and other physiological effects. *Aquat. Toxicol.*, **82**, 94–109.

20 Braun, M. and Hirsch, A. (2000) Fullerene derivatives in bilayer membranes: an overview. *Carbon*, **38**, 1565–1572.

21 Oberdorster, E., Zhu, S.Q., Blickley, T.M., McClellan-Green, P. and Haasch, M.L. (2006) Ecotoxicology of carbon-based engineered nanoparticles: effects of fullerene (C-60) on aquatic organisms. *Carbon*, **44**, 1112–1120.

22 Cagle, D.W., Kennel, S.J., Mirzadeh, S., Alford, J.M. and Wilson, L.J. (1999) In vivo studies of fullerene-based materials using endohedral metallofullerene radiotracers. *Proc. Natl Acad. Sci. USA*, **96**, 5182–5187.

23 Chen, X., Tam, U.C., Czlapinski, J.L., Lee, G.S., Rabuka, D., Zettl, A. and Bertozzi, C.R. (2006) Interfacing carbon nanotubes with living cells. *J. Am. Chem. Soc.*, **128**, 6292–6293.

24 Qiao, R., Roberts, A.P., Mount, A.S., Klaine, S.J. and Ke, P.C. (2007) Translocation of C60 and its derivatives across a lipid bilayer. *Nano Lett.*, **7**, 614–619.

25 Kashiwada, S. (2006) Distribution of nanoparticles in the see-through medaka (*Oryzias latipes*). *Environ. Health Perspect.*, **114**, 1697–1702.

26 Yamago, S., Tokuyama, H., Nakamura, E., Kikuchi, K., Kananishi, S., Sueki, K., Nakahara, H., Enomoto, S. and Ambe, F. (1995) In-vivo biological behavior of a water-miscible fullerene–C-14 labeling, absorption, distribution, excretion and acute toxicity. *Chem. Biol.*, **2**, 385–389.

27 Cleckner, L.B., Garrison, P.J., Hurley, J.P., Olson, M.L. and Krabbenhoft, D.P. (1998) Trophic transfer of methyl mercury in the northern Florida Everglades. *Biogeochemistry*, **40**, 347–361.

28 Drevnick, P.E., Horgan, M.J., Oris, J.T. and Kynard, B.E. (2006) Ontogenetic dynamics of mercury accumulation in Northwest Atlantic sea lamprey (*Petromyzon marinus*). *Can. J. Fish. Aquat. Sci.*, **63**, 1058–1066.

29 Johnson-Restrepo, B., Kannan, K., Addink, R. and Adams, D.H. (2005) Polybrominated diphenyl ethers and polychlorinated biphenyls in a marine foodweb of coastal Florida. *Environ. Sci. Technol.*, **39**, 8243–8250.

30 Donato, F., Magoni, M., Bergonzi, R., Scarcella, C., Indelicato, A., Carasi, S. and Apostoli, P. (2006) Exposure to polychlorinated biphenyls in residents near a chemical factory in Italy: The food chain as main source of contamination. *Chemosphere*, **64**, 1562–1572.

31 Drevnick, P.E. and Sandheinrich, M.B. (2003) Effects of dietary methylmercury on reproductive endocrinology of fathead minnows. *Environ. Sci. Technol.*, **37**, 4390–4396.

32 Drevnick, P.E., Sandheinrich, M.B. and Oris, J.T. (2006) Increased ovarian follicular apoptosis in fathead minnows (*Pimephales promelas*) exposed to dietary methylmercury. *Aquat. Toxicol.*, **79**, 49–54.

33 Langer, P., Kocan, A., Tajtakova, M., Petrik, J., Chovancova, J., Drobna, B., Jursa, S., Radikova, Z., Koska, J., Ksinantova, L., Huckova, M., Imrich, R., Wimmerova, S., Gasperikova, D., Shishiba, Y., Trnovec, T., Sebokova, E. and Klimes, I. (2007) Fish from industrially polluted freshwater as the main source of organochlorinated pollutants and increased frequency of thyroid disorders and dysglycemia. *Chemosphere*, **67**, S379–S385.

34 Tillitt, D.E., Ankley, G.T., Giesy, J.P., Ludwig, J.P., Kurita-Matsuba, H., Weseloh, D.V., Ross, P.S., Bishop, C.A., Sileo, L. *et al.* (1992) Polychlorinated biphenyl residues and egg mortality in double-crested cormorants from the Great Lakes. *Environ. Toxicol. Chem.*, **11**, 1281–1288.

35 Hoffmann, J.L. and Oris, J.T. (2006) Altered gene expression: a mechanism for reproductive toxicity in zebrafish exposed to benzo[a]pyrene. *Aquat. Toxicol.*, **78**, 332–3340.

36 Neff, J.M., (1979) *Polycyclic Aromatic Hydrocarbons in the Aquatic Environment*, Applied Science, London.

37 Oris, J.T. and Giesy, J.P. (1987) The photo-induced toxicity of polycyclic aromatic hydrocarbons to larvae of the fathead minnow *Pimephales promelas*. *Chemosphere*, **16**, 1395–1404.

38 Yang, K., Zhu, L.Z. and Xing, B.S. (2006) Adsorption of polycyclic aromatic hydrocarbons by carbon nanomaterials. *Environ. Sci. Technol.*, **40**, 1855–1861.

39 Lam, C.W., James, J.T., McCluskey, R. and Hunter, R.L. (2004) Pulmonary toxicity of single-wall carbon nanotubes in mice 7 and 90 days after intratracheal instillation. *Toxicol. Sci.*, **77**, 126–134.

40 Shvedova, A.A., Castranova, V., Kisin, E.R., Schwegler-Berry, D., Murray, A.R., Gandelsman, V.Z., Maynard, A. and Baron, P. (2003) Exposure to carbon nanotube material: assessment of nanotube cytotoxicity using human keratinocyte cells. *J. Toxicol. Environ. Health A*, **66**, 1909–1926.

41 Lovern, S.B. and Klaper, R. (2006) *Daphnia magna* mortality when exposed to titanium dioxide and fullerene (C-60) nanoparticles. *Environ. Toxicol. Chem.*, **25**, 1132–1137.

42 Zhu, X.S., Zhu, L., Li, Y., Duan, Z.H., Chen, W. and Alvarez, P.J.J. (2007) Developmental toxicity in zebrafish (*Danio rerio*) embryos after exposure to manufactured nanomaterials: Buckminsterfullerene aggregates (nC(60)) and fullerol. *Environ. Toxicol. Chem.*, **26**, 976–979.

43 Usenko, C.Y., Harper, S.L. and Tanguay, R.L. (2007) In vivo evaluation of carbon fullerene toxicity using embryonic zebrafish. *Carbon*, **45**, 1891–1898.

44 Fortner, J.D., Lyon, D.Y., Sayes, C.M., Boyd, A.M., Falkner, J.C., Hotze, E.M., Alemany, L.B., Tao, Y.J., Guo, W., Ausman, K.D., Colvin, V.L. and Hughes, J.B. (2005) C60 in water: nanocrystal formation and microbial response. *Environ. Sci. Technol.*, **39**, 4307–4316.

45 Henry, T.B., Menn, F.M., Fleming, J.T., Wilgus, J., Compton, R.N. and Sayler, G.S. (2007) Attributing effects of aqueous C-60 nano-aggregates to tetrahydrofuran decomposition products in larval zebrafish by assessment of gene expression. *Environ. Health Perspect.*, **115**, 1059–1065.

46 Lovern, S.B., Strickler, J.R. and Klaper, R. (2007) Behavioral and physiological changes in *Daphnia magna* when exposed to nanoparticle suspensions (titanium dioxide, nano-C-60, and C(60)HxC(70) Hx). *Environ. Sci. Technol.*, **41**, 4465–70.

47 Choi, J. and Oris, J.T. (2000) Anthracene photoinduced toxicity to PLHC-1 cell line (*Poeciliopsis lucida*) and the role of lipid

peroxidation in toxicity. *Environ. Toxicol. Chem.*, **19**, 2699–2706.

48 Choi, J. and Oris, J.T. (2000) Evidence of oxidative stress in bluegill sunfish (*Lepomis macrochirus*) liver microsomes simultaneously exposed to solar ultraviolet radiation and anthracene. *Environ. Toxicol. Chem.*, **19**, 1795–1799.

49 Roberts, A.P. and Oris, J.T. (2004) Multiple biomarker response in rainbow trout during exposure to hexavalent chromium. *Comparative Biochemistry and Physiology C: Toxicology and Pharmacology*, **138**, 221–228.

50 Lehmann, D.W., Levine, J.F. and Law, J.M. (2007) Polychlorinated biphenyl exposure causes gonadal atrophy and oxidative stress in *Corbicula fluminea* clams. *Toxicol. Pathol.*, **35**, 356–365.

51 Possamai, F.P., Fortunato, J.J., Feier, G., Agostinho, F.R., Quevedo, J., Wilhelm, D. and Dal-Pizzol, F. (2007) Oxidative stress after acute and sub-chronic malathion intoxication in Wistar rats. *Environ. Toxicol. Pharmacol.*, **23**, 198–204.

52 Verlecar, X.N., Jena, K.B. and Chainy, G.B.N. (2007) Biochemical markers of oxidative stress in *Perna viridis* exposed to mercury and temperature. *Chem. Biol. Interact.*, **167**, 219–226.

53 Gharbi, N., Pressac, M., Hadchouel, M., Szwarc, H., Wilson, S.R. and Moussa, F. (2005) [60]Fullerene is a powerful antioxidant in vivo with no acute or subacute toxicity. *Nano Lett.*, **5**, 2578–2585.

54 Handy, R.D., Sims, D.W., Giles, A., Campbell, H.A. and Musonda, M.M. (1999) Metabolic trade-off between locomotion and detoxification for maintenance of blood chemistry and growth parameters by rainbow trout (*Oncorhynchus mykiss*) during chronic dietary exposure to copper. *Aquat. Toxicol.*, **47**, 23–41.

19
Carbon Nanotubes as Adsorbents for the Removal of Surface Water Contaminants

Jose E. Herrera and Jing Cheng

19.1
Introduction

The level and complexity of water contamination worldwide has reached unprecedented levels in such a way that the complexity of this problem seems intractable. While there is a wide variety of different contaminant species, the US environmental protection agency divides most of the common contaminants found in surface water into six categories [1]:

- Inorganic compounds: these include toxic heavy metal ions (e.g., cadmium, chromium, lead, copper, mercury), oxoanions (arsenates, chromates) and nonmetal anions such as fluoride.

- Organic compounds: benzene, 1,2-dichloroethene (1,2-DCE), dioxins, polyaromatic hydrocarbons (PAHs), tetrachloroethene (PCE), trichloroethene (TCE), polychlorinated biphenyls (PCBs), etc.

- Disinfection byproducts: trihalomethanes and haloacetic acids.

- Disinfectants: chlorine (as Cl_2), chloramines (as Cl_2) and chlorine dioxide.

- Radionuclide: atoms with unstable potentially radioactive nuclei, such as phosphorus-32, promethium-147 or americium-243.

- Microorganisms: these include bacteria, fungi and archaea, but not viruses nor prions.

Different methods employed for the removal of pollutants from water include chemical precipitation, membrane filtration, ion exchange, and adsorption. Many types of adsorbent, such as silica, alumina, activated carbon, rare earth oxides, amorphous iron hydroxide, polymers, ion-exchange fibers, and lanthanum-based compounds, have been developed as adsorbents for the removal of pollutants from water [2–7]. Recently, carbon nanotubes (CNTs) have been considered as materials for the removal of trace pollutants from water. A number of contributions in this field have been devoted to the experimental and theoretical study of contaminant

Advanced Nanomaterials. Edited by Kurt E. Geckeler and Hiroyuki Nishide
Copyright © 2010 WILEY-VCH Verlag GmbH & Co. KGaA, Weinheim
ISBN: 978-3-527-31794-3

adsorption by CNTs. One of the earliest studies, conducted by Long *et al.* [8], reported that CNTs had a significantly higher dioxin removal efficiency than activated carbon. One year later Li *et al.* [9] suggested that CNTs had a high adsorption capacity for the removal of lead from water. The same group [10] also showed CNTs to be excellent fluoride adsorbents, with a removal capability superior to that of activated carbon. Peng *et al.* [11] indicated that CNTs were efficient adsorbents for the removal of 1, 2-dichlorobenzene (1, 2-DCB) from water, and could be used over a wide pH range. Lu and coworkers [12] noted that CNTs also displayed significant high trihalomethane removal efficiencies.

Together, the results of these studies have indicated that CNTs possess an enormous potential for applications in environmental remediation. Consequently, we present in this chapter a comprehensive overview of recent progress in laboratory studies of CNTs for contaminant removal from water. Particular aspects concentrate on the preparation, characterization and adsorption results of the nanotube materials, while specific examples of experimental studies with CNTs for the adsorption of common environmental contaminants in water are also discussed.

19.2
Structure and Synthesis of Carbon Nanotubes

Carbon nanotubes can be formally depicted as a graphene honeycomb rolled either into a seamless single-walled cylinder or into several concentric cylinders. The former structure is termed single-walled nanotube (SWNT) (Figure 19.1), and the latter multi-walled nanotube (MWNT) (Figure 19.2). The diameter of a typical

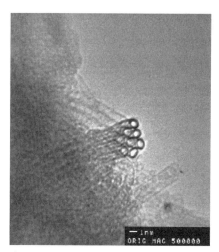

Figure 19.1 Transmission electronic microscopy image of a single-walled carbon nanotube rope. Reproduced from F.L. Darkrima, P. Malbrunot, G.P. Tartaglia, *Int. J. Hydrogen Energy* (2002), **27**, 193–202.

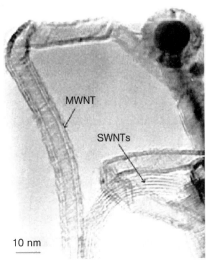

Figure 19.2 Multiwalled carbon nanotube (MWNT) as seen by transmission electron microscopy. The different walls constituting the material appear in a longitudinal view. A micrograph of a SWNT is included for comparison. Reproduced from B. Bai, A.-L. Hamon, A. Marrauda, B. Jouffreya, V. Zymlab, *Chem. Phys. Lett.* (2002), **365**, 184.

MWNT ranges from a few to a few tens of nanometers, while their length is of the order of one micron. In the case of SWNTs, the diameter is in the order of one nanometer, but the length can reach several micrometers, or more.

Each SWNT structure is fully described by two integers (n, m) which specify the number of unit vectors $\vec{a_1}$ and $\vec{a_2}$ in the graphene structure that constitute the chiral vector $\vec{V} = n\vec{a_1} + m\vec{a_2}$. The graphene structure is rolled-up in such a way that the chiral vector $\vec{V}$ forms the nanotube circumference. These indices determine the nanotube diameter, and also the orientation of the carbon hexagons with respect to the nanotube axis; this orientation is the termed the "chirality" of the nanotube.

Single-walled CNTs have unique chemical, electronic and mechanical properties, combined with a very light weight. Depending on their chirality and diameter, the nanotubes may be either electrically metallic or semiconductor. At the same time, they have shown evidence for high stiffness (Young's modulus), a very high resilience, and an ability to reversibly buckle and collapse. These properties have led to SWNTs becoming promising candidates in the fabrication of strong fibers with a light weight and high electrical conductivity. Nanotubes can also be functionalized with different chemical moieties, and this greatly broadened the scope of their applications in fields ranging from conductive coatings to molecule-specific nanosensing.

The synthesis of CNT materials can be divided into high-temperature routes (laser ablation, arc discharge) [13–18] and medium-temperature processes, based on either catalytic decomposition or a carbon-containing molecule (a saturated or unsaturated hydrocarbon or carbon monoxide) [19, 20]. For the high-temperature routes, pure graphite or a mixture of graphite and a metallic catalyst are vaporized at very high temperatures (2000–4000 °C), and the nanotubes are formed during the cooling process at lower temperatures [21]. The catalytic decomposition of a carbon-containing molecule requires a relatively low temperature, and has the potential for high-yield productions.

Iijima [13] was the first to report the presence of multi-walled CNTs in the soot of an arc-discharge chamber. The tubes were produced using an electric arc-discharge evaporation method similar to that used for the synthesis of fullerene. In this method, an electric arc discharge is generated between two graphite electrodes under an inert atmosphere of helium or argon. (Figure 19.3a) [22]. The first successful production of MWNTs at the gram level was developed in 1992 by Ebbesen and Ajayan [23]. To synthesize SWNTs, a metal catalyst is required [24], and the first report of the synthesis of substantial amounts of these materials, by Bethune and coworkers, appeared in 1993 [25].

In 1996, Smalley and coworkers produced high yields (>70%) of SWNTs by utilizing the laser-ablation (vaporization) of graphite rods doped with small amounts of Ni and Co catalysts (Figure 19.3b) [26]. In this process, a piece of graphite is vaporized by laser irradiation under an inert atmosphere. The resultant soot containing the nanotubes is deposited on the walls of a quartz tube when the reaction mixture has cooled. Although this process is known to produce CNTs of the highest quality and highest purity, a purification step is generally required to eliminate any carbonaceous impurities from the as-produced material. This purification is normally achieved by carbon gasification.

Since 1960, carbon filaments and fibers have also been produced by the thermal decomposition of carbon-containing molecules (this process is also known as chemical vapor decomposition; CVD) in the presence of a catalyst [27, 28]. A similar approach was used by Yacaman [29] in 1993, and later by Nagy and collaborators [30, 31], to grow MWNTs. Following these initial reports, the CVD technique has been subsequently improved and optimized. In general, these processes involve the catalyst-assisted decomposition of hydrocarbons (usually ethylene or acetylene) in a tube reactor at temperatures between 550–1000 °C. The growth of CNTs occurs on the catalyst surface as the gas molecule decomposes (Figure 19.3c). This process has several advantages over the arc discharge and laser ablation techniques. In fact, its amenable nature has led to its use in the large-scale synthesis of aligned CNTs [32, 33].

Since all of these synthesis methods generate, in addition to the nanotubes, relatively large amounts of impurities, a number of techniques have been proposed to improve the quality of the materials through post-synthesis purification. These methods range from concentrated acid treatments, dispersion by sonication in a surfactant and filtration, to high-temperature heating under neutral or lightly oxidizing conditions [34, 35]. A detailed account of the different methods for CNT synthesis and purification is beyond the scope of this chapter, and these topics will be addressed only in the context of the synthesis or chemical treatment affecting the adsorption properties of the materials.

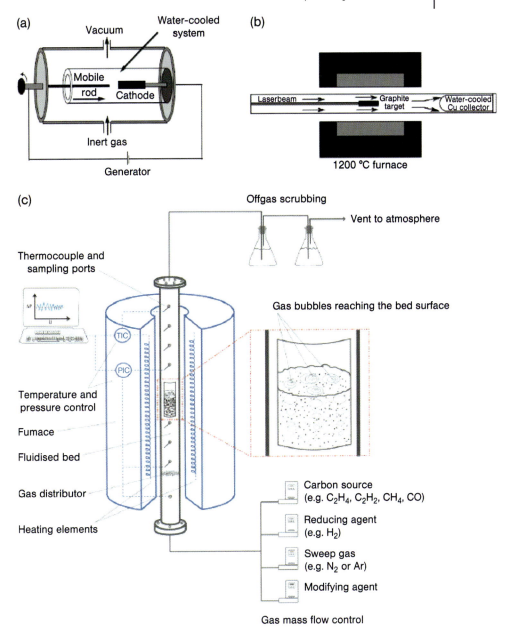

Figure 19.3 Schematic representations of various processes used to produce CNTs. (a) Electric-arc method apparatus. Reproduced from Ref. [22]; (b) Schematic representation of an oven laser-vaporization apparatus. Reproduced from Ref. [26]; (c) A chemical vapor deposition set-up in which the catalytic bed is fluidized. Reproduced from C.H. See, A.T. Harris, *Ind. Eng. Chem. Res.* (2007), **46**, 997–1012.

19.3
Properties of Carbon Nanotubes

As described above, CNTs have unique structures, with remarkable properties that can be grouped as electronic, mechanical, thermal, and optical. Most of the physico-mechanical properties of CNTs are dependent on the sp^2 bond network present on their structure [36] and their diameter, length, and chirality. As these properties have been discussed extensively elsewhere [37–39], some of the most important results in this area will be presented briefly, with emphasis placed on those properties that affect the adsorption capacity of these materials.

19.3.1
Mechanical, Thermal, Electrical, and Optical Properties of Carbon Nanotubes

The mechanical properties of a solid must ultimately depend on the strength of its interatomic bonds. Both, experimental and theoretical, results have predicted that CNTs have the highest Young's modulus of all different types of nanostructures, with similar tubular forms such as BN, BC_3, BC_2N, C_3N_4, CN, and so on. Furthermore, due to the high in-plane tensile strength of graphite, both single and multiwalled CNTs, are expected to have large bending constants. The results of experimental and theoretical studies have indeed indicated that CNTs can be very flexible, able to elongate, twist, flatten, or bend into circles, before fracturing [37].

The thermal and electrical properties of CNTs include conductivity. The specific heat and thermal conductivity are determined primarily by the nanotube's electronic and phononic structures [38], with theoretical and experimental results demonstrating superior electrical properties for these materials. Carbon nanotubes have electric current-carrying capacities which are 1000-fold higher than that of copper wires [40]. In fact, theoretical calculations based on the tight-binding model approximation within the zone folding scheme show that one-third of the possible SWNT structures are metallic, while two thirds are semi-conducting (Figure 19.4) [37, 41].

19.3.2
Adsorption-Related Properties of Carbon Nanotubes

Early studies investigating the adsorption of nitrogen onto both MWNTs [42, 43] and SWNTs [44] have highlighted the porous nature of these materials. Indeed, due to their uniformity in size and surface properties, CNTs are considered as ideal model sorbent systems to study the effect of nanopore size and surface morphology on sorption and transport properties.

The surface area of a CNT has a very broad range, depending on the nanotube number of walls, the diameter, length, wall defects and, in the case of SWNTs, the number of nanotubes in a nanotube bundle [45]. An isolated SWNT with an open end (this may be achieved through oxidation treatment) has a theoretical surface area equal to that of a single, flat graphene sheet of $2700 \, m^2 g^{-1}$ [46]; however, reported experimental values indicate lower adsorption capacities [47]. In the case

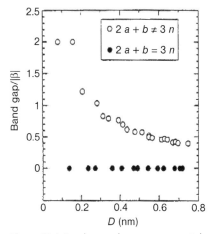

Figure 19.4 Band-gap values versus nanotube diameters defining SWNTs as metallic (●) or semi-conducting (○). Reproduced from Ref. [41].

Table 19.1 Adsorption properties and type of adsorption sites in SWNTs and MWNTs. Reproduced from Ref. [49].

Type of nanotube	Porosity ($cm^3 g^{-1}$)	Surface area ($m^2 g^{-1}$)	Adsorption site	Surface area per site ($m^2 g^{-1}$)
SWNT bundle	Microporous Vmicro: 0.15–0.3	400–900	Surface groove Pore Interstitial	483 783 45
MWNT	Mesoporous	200–400	Surface pore; aggregated pores	–

V_{micro} = micropore volume.

of SWNTs, the diameters of the tubes and number of tubes in the bundle have the strongest effects on the nanotube surface area. In the case of MWNTs, chemical treatments are reported to be useful for promoting microporosity. Surface areas as high as $1050 m^2 g^{-1}$ have been reported for MWNT subjected to alkaline treatment [48]. A two-step activation treatment (acid + CO_2 activation) has been also reported to increase the specific surface area of MWNT materials. It has been proposed that these treatments open the ends of the nanotube structure, enabling adsorption onto the nanotube inner openings [49]. Some representative results of the surface area and pore volume of SWNTs and MWNTs are listed in Table 19.1.

An important issue to address when considering adsorption onto nanotubes is to identify the adsorption sites. For instance, the adsorption of gases into a SWNT bundle can occur inside the nanotubes (pore), in the interstitial triangular channels between the tubes, on the outer surface of the bundle, or in the grooves formed at the contacts between adjacent tubes outside of the bundle (Figure 19.5).

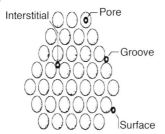

Figure 19.5 Sketch of the cross-sectional view of a SWNT bundle, illustrating the four different adsorption sites. Reproduced from B. Bhushan, *Springer Handbook of Nanotechnology*, 2nd revision, Springer, Berlin, New York (2007).

In addition, the surface functionalization of CNTs by chemical methods has been found to be a powerful tool for improving the adsorption capacity. Selective adsorption can be achieved through a controlled modification of the nanotube's physical and chemical properties, such as surface area, hydrophilicity, and permeability. For instance, Vermisoglou and Georgakilas [50] have studied the sorption properties of pristine and chemically modified [functionalized with oleylamine and poly(sodium 4-styrene sulfonate)] nanotubes using adsorbates with different polarities. Based on their measurements, these authors concluded that the sorption behavior of the CNTs was greatly modified by chemical treatment. In fact, a chemical modification that increased the hydrophilicity of the nanotube walls enhanced the adsorption selectivity for water over *n*-hexane. Chemical modification of the nanotube wall was verified using infrared (IR) spectroscopy (Figure 19.6). When comparing the IR spectra of pristine SWNTs with those of chemically treated CNTs, new peaks corresponding to aliphatic chains were observed in the case of hydrophobic nanotubes, whereas peaks corresponding to more polar bonds were observed in the case of hydrophilic materials.

Yu *et al.* [51] have investigated the adsorptive performance on modified MWNTs by using mechanical ball milling. For these materials, the adsorptive performance for aniline in aqueous solution indicated that the adsorptive capacity of milled, short open-ended MWNTs increased from $15\,mg\,g^{-1}$ to $36\,mg\,g^{-1}$ compared to the unmilled MWNTs. The measurements of pore size distribution proved that the inner pore diameter of 3 nm remained constant after milling, but the aggregated pore diameter had decreased.

19.4
Carbon Nanotubes as Adsorbents

Carbon nanotubes have superior capabilities for the adsorption of a wide range of toxic substances. The earliest reports of CNT use related to the removal of organic pollutants, notably dioxins, from water [8], though later they were reported also as

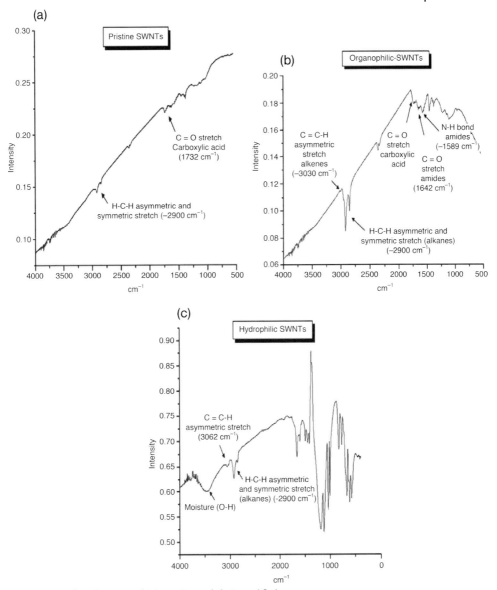

Figure 19.6 Infrared spectra of (a) pristine and (b,c) modified CNTs, illustrating the characteristic peaks for hydrophilic (b) and organophilic (c) samples. Reproduced from Ref. [51].

having an exceptional ability to adsorb inorganic contaminants, such as fluoride [52]. In both cases, the CNTs displayed a superior performance compared to "traditional" adsorbents such as activated carbon. These pioneering studies opened a new field of CNT applications, with many subsequent reports noting CNTs to be excellent adsorbents for the removal of other contaminants. For example, CNTs were shown to adsorb up to 30 mg of a trihalomethane molecule per gram from a 20 mg l^{-1} solution [11]. Other reports indicated that SWNTs could act as "molecular sponges" for small organic molecules such as CCl$_4$ [53]. A similar case was demonstrated for inorganic contaminants, with CNTs again showing superior performance; measurements of the adsorption capacity of a MWNT material showed that it could adsorb 13.5-fold more fluoride than a typical high-surface-area alumina adsorbent (see below). These early results led to the suggestion that CNTs might indeed serve as effective adsorbents for removing polluting agents from water. Consequently during the past few years some extensive laboratory studies have established the role of CNTs as effective adsorbents for common contaminants from water, including a wide variety of organic compounds and inorganic ions. Although a large number of studies have been conducted into the use of CNTs as adsorbents of gas contaminants [54–56], this chapter will focus on the removal of contaminants from surface water (the process being complementary to the role of adsorbates in the gas phase). Consequently, below is presented a discussion of recent results demonstrating the huge potential of CNTs for the removal of contaminants from surface water.

19.4.1
Adsorption of Heavy Metal Ions

The effects of heavy metals such as lead, copper, zinc, nickel, and chromium on human health have been widely studied, based on findings that the ingestion of some of these species can cause accumulative poisoning, cancer, and nervous system damage [57]. A variety of technologies exist for the removal of heavy metals, including filtration, surface complexation, chemical precipitation, ion exchange, adsorption, electrode deposition, and membrane processing [58]. Among these procedures, adsorption is considered one of the most attractive processes for heavy metal removal from solution, as the adsorbents are generally easier to handle and provide a greater operating flexibility [59]. Recent increasingly stringent standards for the quality of drinking water have also catalyzed a growing effort in the development of new, highly efficient adsorbents. The high surface area and chemical stability of CNTs offer exciting possibilities for a new generation of adsorbents; hence, some recently acquired data relating to the adsorption of a wide variety of water contaminants are reviewed below. Here, emphasis is placed on the CNT processing method, the adsorption capacities observed, and the prospects for their successful application.

19.4.1.1 Adsorption of Lead (II)

Li and collaborators [9] were the first to report experimental data on lead adsorption from water using CNTs. Lead is ubiquitous in the environment, and its ingestion is extremely hazardous; the consumption of drinking water containing high levels of lead causes serious disorders, including anemia, kidney disease, and mental retardation [60]. Li and colleagues have shown CNTs to show exceptional adsorption capacities and a high adsorption efficiency of Pb(II) removal from

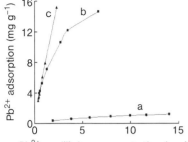

Figure 19.7 Isotherms for Pb(II) adsorption by acid-refluxed CNTs at different pH values. Curve a, pH = 3.0; curve b, pH = 5.0; curve c, pH = 7.0. Reproduced from Ref. [9].

water. The CNTs used in these investigations were pretreated with nitric acid in order to increase the adsorption capacities and to remove most of the catalyst particles within the raw material. The amount of Pb(II) adsorbed onto the CNTs was determined by the difference between the initial Pb(II) concentration and the equilibrium Pb(II) concentration of the solution. The authors also monitored the effect of varying the pH of the solution on lead adsorption.

The acid treatment was seen to have a major impact on the nanotubes' adsorption capacities. For example, whereas pristine as-grown CNTs had an adsorption capacity for Pb(II) of $1\,mg\,g^{-1}$ at pH 5, the capacity was increased remarkably (to $15.6\,mg\,g^{-1}$) when the CNTs were refluxed with concentrated nitric acid, again at pH 5. It appears that acid oxidation of the CNTs leads to the introduction of many functional groups, such as hydroxyl, carboxyl, and carbonyl onto the CNT surface [61], which in turn leads to improved adsorption capacity. Li also reported that the removal of Pb(II) from water by acid-refluxed CNTs was highly dependent on the solution pH, as this affects the surface charge of the adsorbents and the degree of ionization and speciation of the adsorbates. The data in Figure 19.7 show that the Pb(II) adsorption capacity of the CNTs was increased as the pH value was increased from 3.0 to 7.0. It was proposed that, at low pH values, the adsorption of Pb(II) was very weak due to the competition of H^+ with Pb(II) species for the adsorption sites (Figure 19.7, curve a). It was also proposed that, at pH 5, the adsorption capability had increased due to role of functional groups present on the nanotube surface (Figure 19.7, curve b), and was further increased at pH 7 (Figure 19.7, curve c). This might be the result of a combined effect of adsorption and a change in the speciation of the lead ions. The results of other experiments have indicated that, at pH 5 and room temperature, the amount of Pb(II) adsorbed onto the acid-refluxed CNTs increased rapidly during the first 8 min ($16.4\,mg\,g^{-1}$ adsorbent, 81.6% removal), with equilibrium being reached after 40 min ($17.5\,mg\,g^{-1}$, 87.8% removal).

More recently, the same group [62] reported the adsorption thermodynamics and kinetics of Pb(II) adsorption on CNTs, by evaluating various thermodynamic parameters and employing a pseudo second-order kinetic model to describe the

Table 19.2 Summary of results for lead adsorption in terms of nanotube surface area, pore specific volume, pore size distribution, and particle size distribution. Adapted from Ref. [63].

	S_{BET} (m²g⁻¹)	V_p (cm³g⁻¹)	D_p (cm³g⁻¹)	S_p (µm)
Sample 1	47	0.18	3.4	30 and 570
Sample 2	62	0.26	2.4 and 3.2	23 and 450
Sample 3	154	0.58	3.6	8 and 55
Sample 4	145	0.54	3.6	19 and 70

S_{BET} (m² g⁻¹) = BET surface area; V_p (cm³g⁻¹) = pore specific volume; D_p (cm³g⁻¹) = mean pore diameter: S_p (µm) = particle size.

adsorption processes. Based on their results, the authors concluded that the adsorption of Pb(II) onto CNTs was endothermic; they also suggested that the CNT material not only possessed a larger adsorption capacity but also showed a good desorption rate. Such benefits could result in a significant reduction in the overall costs for adsorbent recycling. Desorption studies also revealed that Pb(II) could be easily removed from the CNTs by altering the pH values of the solution using both HCl and HNO_3.

The results of another series of studies has further underlined the influence of the morphologies of CNTs on the removal of lead, in terms of specific surface area, particle size distribution, and type of functional group introduced to the CNT wall [63]. The specific surface area and pore volume of CNTs exposed to four different types of oxidation treatments were compared, and the pore and particle size distributions of the CNTs evaluated. The results (as summarized in Table 19.2) indicated that the CNTs with the highest surface area, smallest particle size and with a relative larger number of functional groups attached to their walls, showed a maximum adsorption capability of 82.6 mg g⁻¹ from a solution with an initial lead concentration of 10 mg l⁻¹. Under the same conditions, the adsorption capacities of samples with a lower number of functional groups attached, and with larger wall defects, achieved adsorption capacities of only approximately 10 mg g⁻¹.

19.4.1.2 Adsorption of Chromium (VI)

Chromium (VI) is one of the most toxic metals found in various industrial wastewaters. Public health considerations of chromium are mostly related to Cr(VI) compounds that are strong irritants due to their high solubility and diffusivity in tissue; certain Cr(VI) compounds have also been shown to be carcinogenic and mutagenic. The toxic effects of Cr(VI) ions in humans include liver damage, internal hemorrhages, respiratory disorders, dermatitis, skin ulceration, and chromosome aberrations [64].

Di and collaborators [65] have investigated the suitability of CNTs for Cr(VI) (as dichromate oxoanion) adsorption, and compared their behavior to that of activated

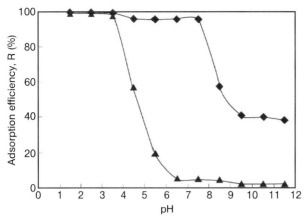

Figure 19.8 Adsorption of Cr(VI) ions onto CNTs (◆) and activated carbon (▲) as a function of pH. Cr(VI) ion concentration = 5 mg g^{-1} CNT; carbon nanotube concentration = 1 g l^{-1}; contact time = 12 h. Reproduced from Ref. [65].

carbon. In these studies, following pretreatment of both as-prepared CNTs and activated carbon with nitric acid and hydrofluoric acid, the CNTs were shown to have superior adsorption capabilities and efficiencies for the removal of Cr(VI) ions from water over the pH range 4.0 to 7.5 (see Figure 19.8). Activated carbon was seen to adsorb chromate ions very rapidly at low pH values, but the adsorption capacity declined very sharply at pH 3.5. In contrast, the adsorption efficiency of the CNTs was maintained at over 90% over a wide pH range, although the Cr(VI) ion adsorption capacity of CNTs fell sharply at pH 8. The authors attributed this behavior to the influence of the nanotube's zeta potential. Subsequently, it was shown experimentally that the isoelectric point of the CNT material was 7.7; at higher pH values the surfaces of the CNTs became negatively charged and thus inaccessible to the chromate anions. The authors also proposed competition of the hydroxyl ions for the few adsorption sites available at these pH values. The data in Figure 19.8 indicate that the highest Cr(VI) ion adsorption capacity was observed over the pH range 4.0 to 7.5. At pH ≤3, the CNTs and activated carbon showed a similar adsorption capacity, but at a higher pH the capacity of the CNTs was greater. The maximum CNT adsorption capacity (20.56 mg g^{-1}) occurred at pH 7.5, when the Cr(VI) concentration was 33.28 mg l^{-1}. The kinetic curves showed the adsorption rate of Cr(VI) ions to be relatively high over the first 20 min, reaching an adsorption capacity of 15 mg g^{-1}.

A few months later, improved values for chromium adsorption were reported in a new study [66]. Here, a composite of aligned CNTs supported in ceria nano-particles (CeO$_2$/CNTs) was used as the adsorbent, the material being prepared by the chemical reaction of CeCl$_3$ with NaOH in the presence of a CNT suspension, followed by heat treatment. Scanning electron microscopy (SEM) images showed

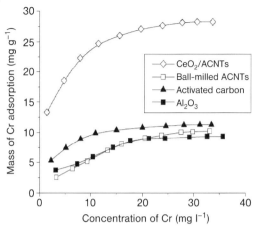

Figure 19.9 Adsorption isotherms of Cr(VI) on CeO$_2$/CNT materials compared with activated carbon, Al$_2$O$_3$ and ball-milled CNTs (at pH 5.0 and 25 °C). Reproduced from Ref. [66].

that the CNT alignment was uniform, with lengths reaching 200 μm and diameters ranging from 20 to 80 nm. Transmission electron microscopy (TEM) analysis indicated a homogeneous distribution on the ceria particles in the CNT network. Overall, the study results indicated that highest capacity for Cr(VI) adsorption occurred at pH values ranging from 3.0 to 7.4, with values of 30.3 mg Cr(VI) g^{-1} nanotube being observed at pH 7.0. The authors also compared the Cr(VI) adsorption isotherm obtained on the CeO$_2$/CNT material with those obtained on activated carbon and γ-Al$_2$O$_3$. As shown in Figure 19.9, the adsorption capacity of the CeO$_2$/CNT material was 1.5-fold higher than that observed for activated carbon, and twofold larger than for Al$_2$O$_3$.

The authors proposed that the small size of the CeO$_2$ particles, and their uniform distribution on the surface of the aligned nanotubes, contributed to the observed high Cr(VI) adsorption. They also suggested that nanotube wall defects, produced by the CVD synthesis process, could offer active sites for Cr(VI) adsorption on the outer surfaces of the aligned nanotube array. The inner cavities and the opened ends present in the inter-aligned nanotube space might also have contributed to the effective adsorption of Cr(VI) ions.

19.4.1.3 Adsorption of Cadmium (II)

Cadmium (II) represents a very high risk to human health due to its extremely high toxicity, even in very small quantities. Drinking water with a cadmium content in excess of permitted levels (0.005 mg l^{-1}) can cause nausea, salivation, diarrhea, muscular cramps, renal degradation, lung insufficiency, bone lesions, cancer, and hypertension [67]. Li *et al.* [68] have analyzed the suitability of CNT materials for Cd(II) adsorption and its removal from water; the same group evaluated the efficacy of several chemical treatments using three different oxidizing agents, namely H$_2$O$_2$, KMnO$_4$, and HNO$_3$. The chemically treated CNTs were shown to have a larger

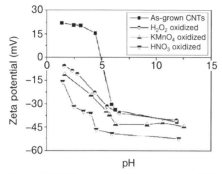

Figure 19.10 Zeta potential curves versus pH for pristine and oxidized CNTs. Reproduced from Ref. [68].

adsorption capacity for Cd(II) than did the pristine, as-grown CNTs. Subsequently, the authors measured the physico-chemical properties of oxidized CNTs and evaluated their Cd(II) adsorption capacity. The specific surface area and pore-size distributions of the as-grown and oxidized CNTs were measured using nitrogen adsorption, with the BET (Brunauer–Emmett–Teller) method. The functional groups on oxidized CNTs were assessed quantitatively using Boehm's titration method [69], and the zeta potentials of the as-grown and oxidized CNTs were also evaluated. Based on these results, it was proposed that the Cd(II) adsorption capacities for the three types of oxidized CNT were increased due to functional groups having been introduced by oxidation, compared to the as-grown, pristine CNTs. The observed Cd(II) adsorption capacity of the as-grown CNTs reached only $1.1\,mg\,g^{-1}$, compared to values of 2.6, 5.1, and $11.0\,mg\,g^{-1}$ for nanotubes treated with H_2O_2, HNO_3 and $KMnO_4$, respectively. The authors linked these results to the increase in surface area observed following each chemical treatment. The data obtained regarding the particle size distribution and suspensibility of these materials indicated that oxidation with H_2O_2 and $KMnO_4$ only partially broke up the nanotubes, whilst oxidation with HNO_3 cut short completely most of the CNTs.

The observed dependence of the zeta potential of the as-grown and oxidized CNTs on pH is shown graphically in Figure 19.10. At the same pH value, the zeta potential for the three types of oxidized CNTs followed the order $H_2O_2 < KMnO_4 < HNO_3$, and suggests that the amounts of acid-functional groups increase following the same order. The adsorption isotherms of Cd(II) also indicated that the functional groups introduced by oxidation improved the ion-exchange capabilities of the CNTs and thus led to corresponding increases in the Cd(II) adsorption capacities. A removal efficiency close to 100% at a CNT dosage of $0.08\,g\,100\,ml^{-1}$ was observed for the $KMnO_4$-oxidized CNTs, which suggested that this treatment represented an effective means of improving the Cd(II) adsorption capacity.

19.4.1.4 Adsorption of Copper (II)

Despite being one of the most widespread environmental contaminants, copper is essential to human life and health, yet is potentially toxic in larger quantities. In humans, the ingestion of relative large quantities of copper salts may cause

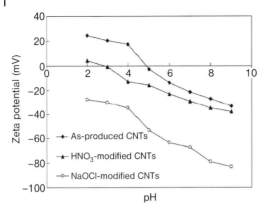

Figure 19.11 Zeta potential curves versus pH for pristine and oxidized CNTs. Reproduced from Ref. [70].

severe abdominal pain, vomiting, diarrhea, hemolysis, hepatic necrosis, hematuria, proteinuria, hypotension, tachycardia, convulsions, coma, and death. The major sources of copper in industrial effluents are metal cleaning and electroplating.

Wu [70] has evaluated the Cu(II) adsorption efficiency of pristine and chemically modified CNTs, the latter being functionalized using HNO_3 and NaOCl. This chemical treatment caused increases in both the pore volume and average pore size of the CNTs, while the value of the isoelectric point was shown to decrease. A comparison of the infrared spectra of the as-produced and modified CNTs indicated that several functional groups had been generated on the surface of modified CNTs. The zeta potential values of the as-produced and modified CNTs, as shown in Figure 19.11, indicated that the surface of the as-produced and HNO_3-modified CNTs was positively charged in solution, whereas the zeta potentials of NaOCl-modified CNTs were all negative. These negatively charged surfaces would electrostatically favor the adsorption of Cu(II), as was observed by Wu and coworkers. Indeed, the Cu(II) adsorption capacity followed the order NaOCl-modified > HNO_3-modified > as-produced CNTs. These findings suggest that modifying the surface of the CNTs may not only provide a more negatively charged and hydrophilic surface but also generate a variety of functional groups, markedly promoting the adsorption of Cu(II) onto the modified CNTs. The maximum adsorption capacities observed in this study at different temperatures are summarized in Table 19.3.

19.4.1.5 Adsorption of Zinc (II)

Whilst zinc (II) is essential for human health, large amounts can be harmful. The consequences of a relatively large intake of Zn(II) include lethargy, light-headedness, ataxia, oropharyngeal cancer, gastric burns, epigastric tenderness, pharyngeal edema, hematemesis, and melena [71]. The suitability of CNTs (both MWNT and SWNT) to adsorb Zn(II) from water was studied by Lu *et al.* [72], whose data showed the adsorption capacity of CNTs to be greatly improved following a specific

Table 19.3 Summary of copper adsorption capacities (q_m) of CNTs at different temperatures in terms of the adsorbed mass of Cu(II) in solution (mg) per mass of nanotube (g). Reproduced from Ref. [70].

Temperature (°C)	Adsorption capacity (q_m) (mg g^{-1})		
	As-produced CNTs	HNO$_3$-modified CNTs	NaOCl-modified CNTs
7	6.39	12.46	44.64
17	7.87	13.10	45.87
27	8.25	13.87	47.39
37	9.34	15.11	49.02
47	10.17	16.04	51.81

chemical treatment that renders the CNTs more hydrophilic, and thus more effective in adsorbing Zn(II).

Lu and coworkers showed that the adsorption capacity of Zn(II) onto CNTs increased in line with a rising pH of the solution over the range 1 to 8, was maximal at pH 8 to 11, and then decreased at pH 12. A comparative study on the adsorption of Zn(II) between CNTs and commercially available, powdered activated carbon, was also conducted. The maximum adsorption capacities for Zn(II) observed were 43.66, 32.68, and 13.04 mg g^{-1} with SWNTs, MWNTs, and activated carbon, respectively. The short contact time required to reach equilibrium, as well as the high adsorption capacity, suggests that both SWNTs and MWNTs possess a high potential for the removal of Zn(II) from water. The same group also suggested that the higher adsorption capacity observed for SWNTs over MWNTs might be due to the higher surface area observed for SWNTs (423 m^2 g^{-1} compared to 297 m^2 g^{-1} observed for MWNTs), and also to the larger proportion of defects present on the MWNTs (as observed using Raman spectroscopy).

However, an interesting result was observed for the case of activated carbon. Although the surface areas of purified SWNTs and MWNTs were much lower than that of activated carbon, the adsorption capacities of Zn(II) onto purified SWNTs and MWNTs were much higher than was observed for activated carbon. This superior adsorption capacity was attributed to a larger number of hydrophilic groups present in the CNT walls.

Another study which focused on the adsorption kinetics and equilibrium of Zn(II) adsorbed onto CNTs nanotubes has also been reported [73]. This thermodynamic analysis revealed that the sorption of Zn(II) onto CNTs was endothermic and spontaneous, and that the Zn(II) ions could easily be removed from the surface sites of SWNTs and MWNTs by the action of a 0.1 M nitric acid solution. Moreover, the original adsorption capacity was maintained after 10 cycles of this sorption–desorption process. Such data suggest that both types of CNT material could be reused through many cycles of water treatment and regeneration.

19.4.1.6 **Adsorption of Nickel (II)**

Nickel is a toxic metal ion that is present in wastewaters. More than 40% of all nickel produced is used in steel factories, in nickel batteries, and in the production of some alloys, which leads to an increase in the Ni(II) burden on the ecosystem and a deterioration in water quality. If ingested, Ni (II) is harmful, and may cause vomiting, chest pain, and a shortness of breath [74].

Lu *et al.* [75] have analyzed the effect of CNT mass, agitation speed, initial Ni(II) concentration, and solution ionic strength on the Ni(II) adsorption capacity of CNTs. The effects of agitation speed and solution ionic strength on Ni(II) sorption by oxidized CNTs are shown in Figures 19.12 and 19.13. The adsorption capacity of both SWNTs and MWNTs was shown to increase as the agitation speed increased, but to decrease when the ionic strength of the solution increased. The same group also reported that SWNTs showed a better performance for Ni(II) adsorption than did MWNTs. A similar conclusion was reached by Chen *et al.* [58], when studying the adsorption of Ni(II) onto oxidized MWNTs as a function of contact time, pH, ionic strength, MWNT amount, and temperature. The results showed that Ni(II) adsorption onto MWNTs was heavily dependent on the pH and, to a lesser extent, the ionic strength. Kinetic data showed the adsorption process to achieve equilibrium within 40 min, and that the process followed a pseudo second-order rate equation. The adsorption data fitted the Langmuir model and, together with thermodynamic data, indicated the spontaneous and endothermic nature of the process. The results of a desorption study showed that Ni(II) adsorbed onto oxidized MWNTs could easily be desorbed at pH <2. The authors proposed that ion exchange might be the predominant mechanism for Ni(II) adsorption on oxidized MWNTs.

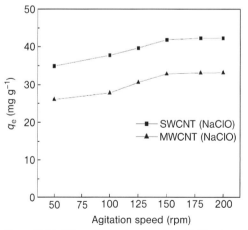

Figure 19.12 Effect of agitation speed on Ni(II) adsorption by oxidized CNTs. Reproduced from Ref. [75].

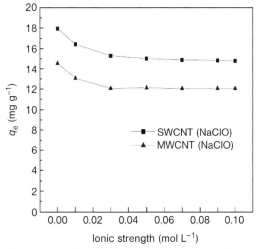

Figure 19.13 Effect of solution ionic strength on Ni(II) adsorption by oxidized CNTs. Reproduced from Ref. [75].

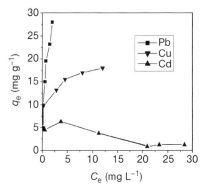

Figure 19.14 Competitive adsorption data for three ions of Pb(II), Cu(II) and Cd(II) onto CNTs at room temperature and pH 5.0. Reproduced from Ref. [76].

19.4.1.7 Competitive Adsorption of Heavy Metals Ions

Whilst most reports on heavy metal adsorption using CNTs has focused on a single metal ion, Li and colleagues [76] were the first to conduct a study on the *competitive adsorption* of Pb(II), Cu(II) and Cd(II) onto HNO$_3$-treated MWNTs. These studies showed the affinity order of these metal ions towards CNTs to vary in the order: Pb(II) > Cu(II) > Cd(II) (Figure 19.14). The Langmuir adsorption model represented the experimental data for Pb(II) and Cu(II) well, but did not provide a good fit for the adsorption data of Cd(II). It was also observed that, at a low pH, the adsorption percentages were negligible, but that at pH values between 1.8 and 6.0 the proportions of Pb(II) and Cu(II) increased sharply, almost attaining values of 100%, while only a small increase was noted for Cd(II).

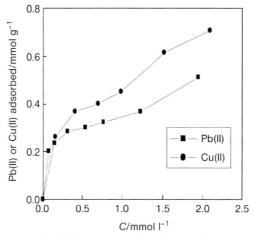

Figure 19.15 Adsorption data for Pb(II) and Cu(II) onto magnetic CNTs composites (pH = 5.0, T = 20 °C). Reproduced from Ref. [78].

Another study, conducted by Hsieh *et al.* [77], focused on the competitive adsorption of Pb(II), Cu(II) and Cd(II) by CNTs grown on microsized Al_2O_3 particles. The authors noted that the adsorption behavior of these metal ions on CNTs on Al_2O_3 particles followed Langmuir's adsorption model, with observed adsorption capacities on the CNTs for Pb(II), Cu(II) and Cd(II) of 32, 18, and 8 mg g^{-1}, respectively. These results confirmed that CNTs supported on Al_2O_3 particles showed potential for the removal of soluble heavy metals from aqueous solutions.

Peng and collaborators [78] conducted a similar study to determine the adsorption capacity of Pb(II) and Cu(II) on CNT–iron oxide composites. These ferric composites offer the advantage of a continuous contaminant adsorption from a liquid effluent whilst, when adsorption is complete, the adsorbent can be separated from the liquid phase simply by using a magnet. Both, X-ray diffraction (XRD) and SEM studies indicated the presence of an entangled network of CNTs with attached iron oxide nanoclusters. The adsorption isotherms obtained for Pb(II) and Cu(II) adsorbed onto these magnetic composites are shown in Figure 19.15. The maximum adsorption capacities for Pb(II) and Cu(II) in the concentration range studied were 105.67 and 45.12 mg g^{-1}. After adsorption, a magnetic separation process was carried out using a permanent magnet, providing a 98% recovery of the Pb(II) and Cu(II) ion mass adsorbed.

19.4.2
Adsorption of Other Inorganic Elements

In addition to heavy metal ions, other common inorganic pollutants in drinking water include ionic forms of fluoride, arsenate, and americium-243(III). Although fluoride is added via drinking water, it is often present in surface waters due to natural erosion or to discharges from fertilizers and aluminum factories. The ingestion of fluoride can cause bone disease and mottled teeth in children. Arsenate is occasionally present due to erosion of natural deposits, or it may be released into surface waters via the runoffs from orchards or from glass and electronics production wastes. Arsenate ingestion causes skin damage and circulatory problems; it is also a well-known carcinogen [2]. Americium-243(III) contributes significantly to the radiotoxicity of nuclear waste, and may be released into the environment during nuclear waste storage, processing, or disposal. Exposure to small traces of americium-243(III) increases the risk of cancer.

19.4.2.1 Adsorption of Fluoride

The acceptable fluoride concentration in drinking water is generally in the range of 0.5 to 1.5 mg l^{-1} [79]. Higher concentrations will affect the metabolism of calcium and phosphorus in the human body, and lead to dental and bone fluorosis [80]. Many methods have been adopted to remove excess fluoride from drinking water, the most common approach being adsorption with activated alumina, which has a good adsorption capacity and selectivity for fluoride. Unfortunately, the optimum capacity for fluoride removal in alumina occurs only at pH values below 6.0, which strongly limits the practical applications of this material [81].

Li and collaborators [52] have reported that amorphous Al_2O_3 supported on CNTs represents a major candidate for fluoride adsorption from water. These authors used CNTs as a supports for Al_2O_3, and showed the composite to have a high potential for removing fluoride from drinking water. Based on their adsorption isotherms (Figures 19.16 and 19.17), Li and coworkers found that Al_2O_3/CNT composites showed a high fluoride adsorption capacity over a pH range from 5.0 to 9.0. They also found the adsorption capacity for the Al_2O_3/CNT composite to be about 13.5-fold higher than that of activated carbon, fourfold higher than for γ-alumina, and also higher than that of the commercial polymeric resin, IRA-410. This broad range of pH values and high adsorption capacities observed for the Al_2O_3/CNT composite makes this material very attractive for fluoride removal from water.

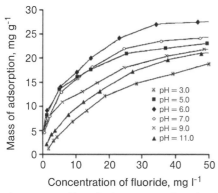

Figure 19.16 Effect of solution pH on fluoride adsorption onto an Al_2O_3/CNT composite. Reproduced from Ref. [52].

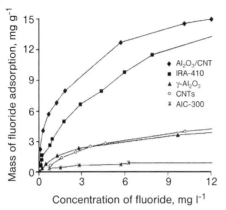

Figure 19.17 Adsorption isotherms of fluoride on activated carbon (AlC-300), CNT, γ-Al$_2$O$_3$, a commercial resin (IRA-410) and Al$_2$O$_3$/CNT composite. The data for activated carbon and IRA-410 were fitted with a Langmuir isotherm, while the data for CNT and Al$_2$O$_3$/CNTs were fitted with a Freundlich isotherm. Reproduced from Ref. [52].

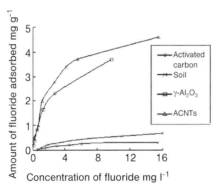

Figure 19.18 Adsorption isotherm of fluoride on CNTs (at pH 7 and 25 °C) compared with γ-Al$_2$O$_3$, soil, and activated carbon. Reproduced from Ref. [10].

A year later, the same group presented improved results using an array of aligned MWNTs [10]. For this, the authors studied the kinetics of the fluoride adsorption process, and the effect of pH on fluoride adsorption capacity. The kinetic data indicated that the fluoride adsorption rate was rapid during the first 60 min, and quickly reached an adsorption capacity of 3.0 mg g^{-1}, with equilibrium achieved after 180 min. A mild dependence of the adsorption capacity on the pH of the solution was also observed, with the highest adsorption capacity being observed at pH 7, and reaching 4.5 mg g^{-1} at fluoride concentrations of 15 mg l^{-1}. The adsorption isotherms obtained for this material, compared to those obtained in γ- Al$_2$O$_3$, activated carbon, and a soil, are shown in Figure 19.18. At low fluoride initial concentrations ($<1\,mg\,l^{-1}$), the nanotube material and alumina had the same adsorption capacities, but at higher fluoride concentrations the fluoride adsorption capacity of the CNTs was higher. These results again indicated the potential of CNTs in fluoride removal.

19.4.2.2 Adsorption of Arsenic

Arsenic is required as a micronutrient for the human body, yet it may be carcinogenic if consumed constantly. Thus, it is of great importance to remove arsenic from water before it is used for drinking. Two main forms of arsenic are encountered in natural water, namely trivalent (As(III), arsenite) and the higher-oxidized form, pentavalent (As(V), arsenate). Whilst either of these species can be found in natural waters, As(III) is more common in groundwater, and As(V) is more common in surface water [82].

When a novel ceria–CNT composite was proposed as an alternative for the removal of arsenate from water [83], the results indicated that arsenate adsorption on these materials was pH-dependent. The presence of Ca(II) and Mg(II) also significantly enhanced the adsorption capacity. These very promising results suggest that these materials might represent a promising adsorbent for drinking water purification. The effect of pH on the adsorption of As(V) onto CeO_2/CNT is shown in Figure 19.19.

These results indicate that the adsorption of As(V) is pH-dependent, and that the pristine composite has a higher adsorption capacity than the chemically modified nanotubes. The authors proposed that the dependence of adsorption on pH was due to variations in the surface charge on the nanotube composites. This was corroborated by zeta potential measurements. Moreover, the authors also measured the influence of Ca(II) and Mg(II) on the adsorption capability of the CeO_2/CNT material. From the results shown in Figure 19.20 it is clear that both Ca(II) and Mg(II) significantly enhance the adsorption capacity of the nanotube composite. In fact, an increase from 0 to $10\,mg\,l^{-1}$ in the concentration of Ca(II) and Mg(II) resulted in an almost one order of magnitude increase in the amount of As(V)

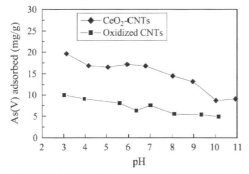

Figure 19.19 Effect of pH on As(V) adsorption by chemically treated CNTs (■) and CeO_2/CNT composite (◆). Reproduced from Ref. [83].

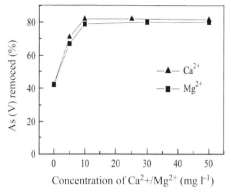

Figure 19.20 Effect of Ca(II) and Mg(II) on As(V) adsorption (initial concentration of As(V) = 20 mg l^{-1}). Reproduced from Ref. [83].

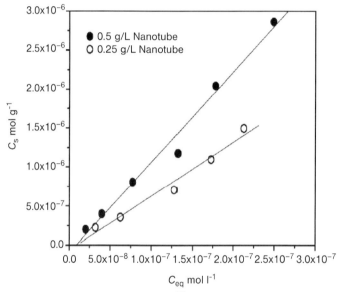

Figure 19.21 Adsorption isotherms for ^{243}Am(III) onto MWNTs in polyethylene tubes using Milli-Q water in the presence of 0.1 M NaClO. The contact time was 4 days. Reproduced from Ref. [85].

adsorbed. The authors explained this observed increase in adsorption capacity by proposing the formation of a ternary surface complex between calcium or magnesium, arsenate, and the ceria surface [84].

19.4.2.3 Adsorption of Americium-243 (III)
Wang and collaborators [85] examined the use of MWNTs as adsorbents for the radionuclide americium-243 (III). The data obtained (see Figure 19.21) indicated

that the sorption isotherms were linear, and confirmed that the sorption of Am(III) onto MWNTs did not reach saturation under these conditions (up to $3 \times 10^{-6} M$ of Am(III)). A dependence of ^{243}Am(III) sorption by MWNTs on pH and ionic strength was also reported. Approximately 80% of ^{243}Am(III) was adsorbed by the nanotubes at pH 5, and this increased to approximately 90% at pH 10. The authors explained this behavior in terms of surface complexation.

19.4.3
Adsorption of Organic Compounds

Although, as detailed above, CNTs have major potential for the removal of inorganic substances from water, the number of studies conducted on the adsorption of organic pollutants onto CNTs is relatively small. A report published by the U.S. Environmental Agency in 2002 [86] stated that the most common organic contaminants found in drinking water ranged from disinfection byproducts (e.g., trihalomethanes) to other organic chemicals, including carbon tetrachloride, benzene, polychlorinated benzenes, dioxins, and pesticides. The similar chemical characteristics of most of these organic pollutants give rise to similar acute health effects at high doses, with the primary target organ generally being the central nervous system. Sustained or very high exposures to these organic chemicals can be fatal. An overview of the results obtained on the removal of some of these chemicals from water, using CNTs as adsorbents, is provided in the following section.

19.4.3.1 Adsorption of Dioxins
Dioxins and related compounds (e.g., polychlorinated dibenzofurans and biphenyls) are highly toxic pollutants. The toxicity of dioxins varies with the number of chlorine atoms; for example, 2,3,7,8-tetrachlorodibenzo-*p*-dioxin (TCDD) is a known human carcinogen. In addition to cancer, dioxins also adversely affect the immune and endocrine systems, as well as normal fetal development [87]. Dioxins are mainly generated from the combustion of organic compounds in waste incinerators. Due to the extreme toxicity and chemical inertness of dioxins, it is necessary to improve any current technologies based on activated carbon adsorption.

Long *et al.* [8] have reported that CNTs can attract and trap a large amount of dioxins in a more efficient manner than the activated carbon or other adsorbents currently used. Long's group attributed this observation to the stronger interaction forces that exist between dioxin molecules and the curved hydrophobic surface of the nanotubes. Another series of investigations, based on dioxin temperature-programmed desorption [8, 88], reported that the interaction of dioxins with CNTs is much stronger than that with activated carbon. The desorption temperatures, desorption activation energy and Langmuir constants obtained for dioxin desorption on CNTs, activated carbon and γ-Al_2O_3 are listed in Table 19.4. These results clearly indicate that the interaction of dioxins with CNTs is significantly higher than with the other adsorbents.

A density functional theory (DFT) study conducted by Kang *et al.* [89] focused on the molecular interactions between dioxin and metal-doped (Li, Na, Fe) SWNTs. The calculations indicated that doping large-diameter SWNTs with calcium atoms

Table 19.4 Position of the maximum in the dioxin temperature programmed desorption profiles of carbon nanotubes, activated carbon and alumina at different heating rates. Data for calculated values of activation energies for desorption and Langmuir constants are also presented. Reproduced from Refs [8] and [88].

Sorbent	Peak desorption temp. (°C) at different heating rates				Desorption activation energy (kJ mol⁻¹)	Langmuir constant B at 25 °C (l atm⁻¹)
	2 °C min⁻¹	5 °C min⁻¹	10 °C min⁻¹	20 °C min⁻¹		
Carbon nanotubes	588	609	620	634	315	2.7×10^{52}
ZX-4 carbon (Mitsubishi)	481	517	543	?	119	1.3×10^{18}
γ-Al$_2$O$_3$	306	353	394	?	47.9	4.5×10^{5}

can introduce a strong cooperative binding of the carbon π system with the dioxin. A band structure analysis suggested that the charge transfer model could explain these observations. Iron atoms, which are commonly used as a catalyst for CNT synthesis, can also significantly facilitate dioxin binding. In a sense, this suggests that the adsorption of small molecules, particularly those with delocalized π electrons, can be significantly enhanced by the presence of metallic catalyst remnants following the accomplishment of SWNT synthesis.

More recently, a theoretical *ab initio* calculation of 2,3,7,8-tetrachlorinated dibenzo-*p*-dioxin interaction with pristine, "defective" as well as B-, N-, and Si-doped SWNTs, was reported by Fagan *et al.* [90]. Their results predicted that the interaction between SWNTs and dioxins would depend on the geometric configuration of the approaching dioxin (Figure 19.22), the dopant metal, and the number of defects in the nanotube wall. The results (see Table 19.5) suggested that the adsorption of dioxin was more effective for the case of defective nanotubes. This theoretical prediction shows promise from a practical point of view, as structural defects on the tube walls are naturally formed during nanotube growth and purification processes. Moreover, the results also indicated that doping with B, N, and Si does not improve the absorption capacities of dioxins.

19.4.3.2 Adsorption of 1,2-Dichlorobenzene

Chlorobenzenes such as monochlorobenzene, dichlorobenzene (DCB), and trichlorobenzene, which are present in some surface and ground-waters, have been identified as priority pollutants by the US Environmental Protection Agency [91]. Among chlorobenzenes, DCB is one of the most chemically stable, and its degradation in soil and aquatic environments is extremely limited [92]. The widespread use of chlorinated aromatic chemicals over several decades has resulted in contamination of the environment and human exposure to DCB. The methods

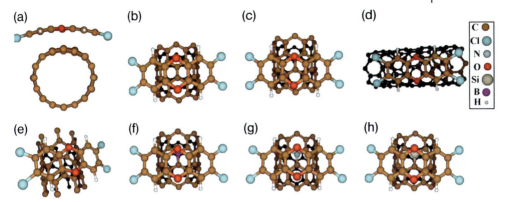

Figure 19.22 Schematic views of the most stable configurations of different SWNT/dioxin adsorption geometries. Views (b) to (d) show configurations for pristine SWNT interacting with a chlorinated dioxin in different configurations. View (a) depicts a lateral view of the SWNT/dioxin configuration shown in (b). In view (e) the SWNT has a defect in the form of a vacancy. Views (f) to (h) show N-, B-, and Si-doped SWNTs interacting with dioxin, respectively. Reproduced from Ref. [90].

Table 19.5 Calculated values for the binding energy for the SWNT/dioxin configurations shown in Figure 19.22. Reproduced from Ref. [90].

System[a]	Binding energy (eV)
SWNT/dioxin (Figure 19.1b)	−0.10
SWNT/dioxin (Figure 19.1c)	−0.35
SWNT/dioxin (Figure 19.1d)	−0.77
SWNT/vac_dioxin (Figure 19.1e)	−1.21
SWNT/B_dioxin (Figure 19.1f)	−0.43
SWNT/N_dioxin (Figure 19.1g)	−0.45
SWNT/Si_dioxin (Figure 19.1h)	−0.30

a Figure numbers relate to Ref. [90].

employed to remove DCB from water are either destructive oxidation or adsorption. Peng *et al.* [11] have reported the use of as-grown CNTs and graphitized CNTs as adsorbents to remove 1,2-DCB from water. The as-grown CNTs were prepared by catalytic pyrolysis of a propylene–hydrogen mixture at 750 °C using Ni particles as catalysts. The graphitized CNTs were prepared by treating the as-grown CNTs in a nitrogen atmosphere at 2200 °C for 2 h. The maximum amounts of DCB adsorbed by the as-grown and graphitized CNTs were 30.8 and 28.7 mg g^{-1}, respectively. The short time required to achieve equilibrium suggested that the CNTs had very high adsorption efficiencies, with the removal rate remaining almost constant over a pH range from 3 to 10. The adsorption isotherms obtained are

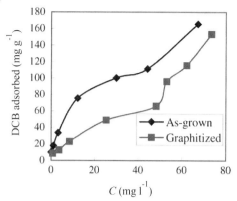

Figure 19.23 Isotherms for DCB adsorption onto as-grown and graphitized CNTs at 25 °C (pH = 5.5, t = 24 h). Reproduced from Ref. [11].

reproduced in Figure 19.23. Taken together, the results indicated that the as-grown CNTs were superior to the graphitized material in terms of DCB adsorption.

Recently, *ab initio* theoretical calculations of 1,2-DCB adsorption onto metallic SWNTs were reported by Fagan *et al.* [93]. The values for the binding energies obtained in terms of interaction between the DCB and SWNTs, indicated that a decrease in nanotube diameter and a consequent enhancement in tube curvature favored a more effective adsorption of DCB on the tube surface. Charge distribution calculations indicated that the adsorption of DCB occurred through π–π stacking; the calculations also suggested that the DCB–SWNT interaction was larger for metallic nanotubes than for semi-conducting nanotubes.

19.4.3.3 Adsorption of Trihalomethanes

Trihalomethanes (THMs) are generated during the disinfection of drinking water with chlorine [94], and are recognized as potentially hazardous and carcinogenic substances [95]. More stringent requirements for the removal of THMs from drinking water in recent years have led to the development of innovative, cost-effective alternatives to remove these byproducts.

In 2004, Lu *et al.* [12] reported the use of CNT materials to remove THMs from water, where the nanotube samples were fabricated by the catalytic decomposition of a CH_4/H mixture at 700 °C using Ni particles as catalyst. Based on FTIR spectra the authors noted that, after a mild acid treatment, the CNT material became more hydrophilic and effective for the adsorption of the relatively polar THMs molecules. It was also suggested that the diffusion mechanisms controlled the adsorption of THMs onto the CNTs. In fact, the smallest molecule, $CHCl_3$, was seen to adsorb preferentially onto the nanotubes, followed by $CHBrCl_2$, $CHBr_2Cl$, and $CHBr_3$. The adsorption isotherms obtained in these studies are shown in Figure 19.24. The observed adsorption capacity for $CHCl_3$ was highest, followed by $CHBrCl_2$, $CHBr_2Cl$ and $CHBr_3$. The maximum adsorbed amounts of $CHCl_3$,

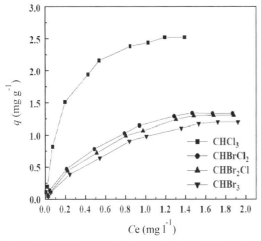

Figure 19.24 Adsorption isotherms for four trihalomethanes onto CNTs. Reproduced from Ref. [12].

$CHBrCl_2$, $CHBr_2Cl$ and $CHBr_3$ were 2.41, 1.23, 1.08 and 0.92 $mg\,g^{-1}$, respectively, from an initial concentration of 1 $mg\,l^{-1}$ of each THM.

Another study conducted by the same group [96] included a comparison of the adsorption of THMs onto powdered activated carbon. In contrast to data obtained with CNTs, for activated carbon the largest molecule, $CHBr_3$, was seen to be preferentially adsorbed, followed by $CHBr_2Cl$, $CHBrCl_2$ and $CHCl_3$. With an initial concentration of 1 $mg\,l^{-1}$, the amounts of $CHCl_3$, $CHBr_2Cl$, $CHBrCl_2$ and $CHBr_3$ adsorbed were 1.2, 1.68, 2.19, and 2.75 $mg\,g^{-1}$ CNTs, respectively. The adsorption of THMs onto CNTs also occurred more rapidly compared to activated carbon. Although the surface area of the nanotube material (295 $m^2\,g^{-1}$) was much less than that for activated carbon (900 $m^2\,g^{-1}$), the adsorption capacity of $CHCl_3$ onto CNTs was approximately twice that for activated carbon.

The thermodynamic and kinetic parameters of the adsorption processes of THMs on MWNTs was also reported [96]. Here, the amount of THMs adsorbed onto CNTs was seen to decrease with a rise in temperature, with the highest adsorption capacities observed at 5 °C and 15 °C. Under the same conditions, the purified CNT material displayed two- to threefold greater adsorption capacities for $CHCl_3$ than did activated carbon. A thermodynamic analysis revealed the adsorption of THMs onto CNTs to be exothermic in nature.

19.4.3.4 Adsorption of Polyaromatic Compounds

A recent study [97] investigating the adsorption of polycyclic aromatic hydrocarbons (PAHs) (e.g., naphthalene, phenanthrene, pyrene) from water onto six different carbon nanomaterials, including fullerenes, SWNTs, and MWNTs, has been reported. The results showed that the adsorption capacities of the nanomaterials for the different polycyclic aromatic molecules was related to the PAH

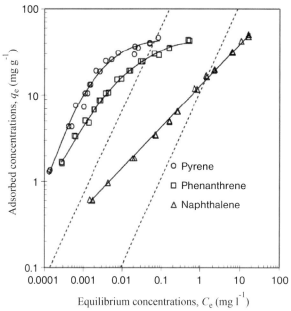

Figure 19.25 Adsorption isotherms of pyrene, phenanthrene, and naphthalene onto MWNTs. Reproduced from Ref. [97].

molecular size (Figure 19.25), which prompted the authors to suggest that, due to their size, some of these aromatic molecules could not access the nanomaterial pores. Among the different carbon nanomaterials, adsorption appeared to relate well with the adsorbent surface area, micropore, and mesopore volume. Except for the SWNTs, a linear relationship was observed between the adsorbed capacity and the carbon nanomaterial.

19.5
Summary of the Results, and Conclusions

Among the above-described data – some of which are summarized in Table 19.6 – it appears that, except for those used by Peng *et al.* [11], all CNTs employed in these experiments have been purified by pretreating with acid, before adsorption measurements were made. Such pretreatment had a major impact on the adsorption capacity of the CNTs in terms of contaminant removal. An analysis of the aforementioned results indicated that the acid treatment not only removed the amorphous carbon that coats the CNT surface but also increased the specific nanotube surface area and pore volume. However, it seems that the most important factor responsible for the improved adsorption capacities seen with oxidized CNTs is the incorporation of surface functional groups (hydroxyl, carboxyl and carbonyl) generated by their treatment with oxidizing acids.

Table 19.6 Summary of selected cited results.

Reference(s)	Type of adsorbent used	Adsorbate	Pretreatment of adsorbent	Parameters affecting adsorption efficiency evaluated
[9]	Acid-refluxed CNTs	Pb(II)	Acid treatment	pH of solution
[62]	Acid-refluxed CNTs	Pb(II)	Acid treatment	Adsorption/desorption study
[63]	Oxidized CNTs	Pb(II)	Acid treatment	CNT morphology
[65]	Acid-refluxed CNTs	Cr(VI)	Acid treatment	pH of solution
[66]	CeO$_2$/CNT composites	Cr(VI)	Acid treatment	pH of solution
[68]	Pristine and oxidized CNTs	Cd(II)	Oxidation by H$_2$O$_2$ HNO$_3$ and KMnO$_4$	CNT surface area CNT pore and particle size distribution CNT zeta potential Adsorbate/adsorbent ratio
[70]	Pristine and oxidized CNTs	Cu(II)	Oxidation by HNO$_3$ and NaClO	pH of solution CNT zeta potential

Table 19.6 Continued.

Reference(s)	Type of adsorbent used	Adsorbate	Pretreatment of adsorbent	Parameters affecting adsorption efficiency evaluated
[72, 73] [75]	Purified CNTs	Zn(II) Ni(II)	Acid treatment	Effect of consecutive adsorption/desorption cycles
[58] [76] [77]	Oxidized CNTs Oxidized MWNTs Acid-treated CNTs	Ni(II) Pb(II), Cu(II) and Cd(II)	Acid treatment	Agitation speed pH of solution
[78] [52] [10]	Al₂O₃/CNT composite FeOx/CNT magnetic composite	Pb(II), Cu(II) and Cd(II) Pb(II),	Acid treatment	Ionic strength of solution
[83]	Al₂O₃/CNT composite Acid-treated CNTs CeO₂/CNT composite	Cu(II) fluoride fluoride Arsenic	Acid treatment	Adsorbate/adsorbent ratio Contact time; Desorption capacity; pH of solution; Ionic strength of solution; Adsorbate/adsorbent ratio Competitive adsorption study; Effect of consecutive adsorption/desorption cycles; pH of solution; pH of solution; Influence of Ca(II) and Mg(II) as coadsorbates
[85] [8] [11]	MWNTs	243 Am(III) dioxin 1, 2-DCB	Acid treatment	pH of solution
[90] [12] [96]	MWNTs As-grown and graphitized CNTs Metallic SWNTs	1, 2-DCB THMs THMs PAHs	Acid treatment	Contact time; Structure and electronic properties of CNTs; pH of solution
[97]	CNTs MWNTs CNTs and fullerenes		Acid treatment	Theoretical study; pH of solution; Temperature; Adsorbed volume capacity Molecular size of adsorbate

The observed nanotube adsorption capacity is also dependent on pH, temperature, and contact time, whilst the effects of ionic strength, agitation speed, adsorbate initial concentration and CNT dosage have also been discussed. Zeta potential measurements have indicated that chemically treated CNTs have a larger number of negatively charged surface sites than do raw nanotube materials. The results of comparative studies have shown that CNTs exhibit much higher adsorption capabilities than other commonly used adsorbents. In a word, the excellent adsorption performance observed for CNTs indicates their high potential for water remediation applications. A number of fundamental issues remain, however, which limit the application of CNT materials for the removal of water contaminants, including their solubility in aqueous solution, their ultimate environmental fate, and the cost-effectiveness of a nanotube-based technology. Clearly, further field-scale studies must be conducted to determine not only the effectiveness of CNTs but also their potential, the target being to raise the limits currently faced by this technology.

References

1 USEPA US Environmental Protection agency (NPL) (2007) Superfund National Priorities List, http://www.epa.gov/superfund/sites/npl/index.htm (accessed 7 January, 2008).

2 Raichur, A.M. and Panvekar, V. (2002) *Sep. Sci. Technol.*, **37**, 1095.

3 Wasay, S.A., Tokunaga, S. and Park, S.W. (1996) *Sep. Sci. Technol.*, **31**, 1501.

4 Pierce, M.L. and Moore, C.B. (1982) *Water Res.*, **16**, 1247.

5 Suzuki, T.M., Tanco, M.L. and Tanaka, D.A.P. (2001) *Sep. Sci. Technol.*, **36**, 103.

6 Liu, R., Guo, J. and Tang, H. (2002) *J. Colloid Interface Sci.*, **248**, 268.

7 Tokunaga, S., Wasay, S.A. and Park, S.W. (1997) *Water Sci. Technol.*, **35**, 71.

8 Long, R.Q. and Yang, R.T. (2001) *J. Am. Chem. Soc.*, **123**, 2058.

9 Li, Y.H., Wang, S., Wei, J., Zhang, X., Xu, C., Luan, Z., Wu, D. and Wei, B. (2002) *Chem. Phys. Lett.*, **357**, 263.

10 Li, Y., Wang, S., Zhang, X., Wei, J., Xu, C., Luan, Z. and Wu, D. (2003) *Mater. Res. Bull.*, **38**, 469.

11 Peng, X., Li, L., Di, Z., Wang, H., Tian, B. and Jia, Z. (2003) *Chem. Phys. Lett.*, **376**, 154.

12 Lu, C., Chung, Y.L. and Chang, K.F. (2005) *Water Res.*, **39**, 1183.

13 Iijima, S. (1991) *Nature*, **354**, 56.

14 Dillon, A.C., Jones, K.M., Bekkedahl, T.A., Kiang, C.H., Bethune, D.S. and Heben, M.J. (1997) *Nature*, **386**, 377.

15 Maddox, M.W. and Gubbins, K.E. (1995) *Langmuir*, **11**, 2988.

16 Maddox, M.W., Sowers, S.L. and Gubbins, K.E. (1996) *Adsorption*, **2**, 23.

17 Darkrim, F.L. and Levesque, D.J. (1998) *Chem. Phys.*, **109**, 4981.

18 Darkrim, F.L. and Levesque, D. (2000) *J. Chem. Phys.*, **104**, 6773.

19 Yan, H., Li, Q., Zhang, J. and Liu, Z. (2004) CVD Synthesis of SWNTs, in *The Chemistry of Nanostructured Materials* (ed. P. Yang), World Scientific Pub, pp. 101–26.

20 Colbert, D.T., Zhang, J., McClure, S.M., Nikolaev, P., Chen, Z., Hafner, J.H., Owens, D.W., Kotula, P.G., Carter, C.B., Weaver, J.H., Rinzler, A.G. and Smalley, R.E. (1994) *Science*, **266**, 1218.

21 Journet, C. and Bernier, P. (1998) *Appl. Phys. A*, **67**, 1.

22 Liu, Y. and Gao, L. (2005) *Carbon*, **1**, 47.

23 Ebbesen, T.W. and Ajayan, P.M. (1992) *Nature*, **358**, 220.

24 Iijima, S. and Ichihashi, T. (1993) *Nature*, **363**, 603.

25 Bethune, D.S., Kiang, C.H., de Vries, M.S., Gorman, G., Savoy, R. and Vazquez, J. (1993) *Nature*, **363**, 605.

26 Thess, A., Lee, R., Nikolaev, P., Dai, H., Petit, P., Robert, J., Xu, C., Lee, Y.H., Kim, S.G., Rinzler, A.G., Colbert, D.T., Scuseria, G.E., Tomane'k, D., Fischer, J.E. and Smalley, R.E. (1996) *Science*, **273**, 483.

27 Baker, R.T.K. (1989) *Carbon*, **27**, 315.

28 Tibbetts, G.G. (1989) *Carbon*, **27**, 745.

29 Yacaman, M.J., Yoshida, M.M., Rendon, L. and Santiesteban, J.G. (1993) *Appl. Phys. Lett.*, **62**, 202.

30 Ivanov, V., Nagy, J.B., Lambin, P., Lucas, A., Zhang, X.B., Zhang, X.F., Bernaerts, D., Van Tendeloo, G., Amelinckx, S. and Van Landuyt, J. (1994) *Chem. Phys. Lett.*, **223**, 329.

31 Amelinckx, S., Zhang, X.B., Bernaerts, D., Zhang, Z.F., Ivanov, V. and Nagy, J.B. (1994) *Science*, **265**, 635.

32 Li, W.Z., Xie, S.S., Qian, L.X., Chang, B.H., Zou, B.S., Zhou, W.Y., Zhao, R.A. and Wang, G. (1996) *Science*, **274**, 1701.

33 Zhang, L., Li, Z., Tan, Y., Lolli, G., Sakulchaicharoen, N., Requejo, F.G., Mun, B.S. and Resasco, D.E. (2006) *Chem. Mater.*, **18**, 5624.

34 Rinzler, A.G., Liu, J., Dai, H., Nikolaev, P., Human, C.B., Rodriguez-Macias, F.J., Boul, P.J., Lu, A.H., Heymann, D., Colbert, D.T., Lee, R.S., Fisher, J.E., Rao, A.M., Eklund, P.C. and Smalley, R.E. (1998) *Appl. Phys. A: Solids Surf.*, **67**, 29.

35 Vaccarini, L., Goze, C., Aznar, R., Micholet, V., Journet, C., Bernier, P., Metenier, K., Beghin, F., Gavillet, J. and Loiseau, A.C. (1999) *Sci. Paris*, **327 (IIb)**, 935.

36 Popov, V. (2004) *Mater. Sci. Eng. R.*, **43** (3), 61.

37 Dresselhaus, M.S., Dresslhaus, G. and Eklund, P.C. (1996) *Science of Fullerenes and Carbon Nanotubes*, Academic Press, San Diego, USA.

38 Ando, T. (2003) Physics of carbon nanotubes, in *Advances in Solid State Physics*, Vol. 43, Springer, Berlin/Heidelberg, p. 301.

39 Dresselhaus, M., Dresselhaus, G. and Saito, R. (1995) *Carbon*, **33** (7), 883.

40 Collins, P.G. and Avouris, P. (2000) *Sci. Am.*, **283**, 62.

41 Hamada, N., Sawada, S. and Oshiyama, A. (1992) *Phys. Rev. Lett.*, **68**, 1579.

42 Yang, Q.H., Hou, P.X., Bai, S., Wang, M.Z. and Cheng, H.M. (2001) *Chem. Phys. Lett.*, **345**, 18.

43 Inoue, S., Ichikuni, N., Suzuki, T., Uematsu, T. and Kaneko, K. (1998) *J. Chem. Phys.*, **102**, 4689.

44 Eswaramoorthy, M., Sen, R. and Rao, C.N.R. (2005) *Chem. Phys. Lett.*, **304**, 207.

45 Peigney, A., Laurent, C., Flahaut, E., Bacsa, R.R. and Rousset, A. (2001) *Carbon*, **39**, 507.

46 Kuznetsova, A., Mawhinney, D.B., Naumenko, V., Yates, J.T. and Smalley, R.E. (2000) *Chem. Phys. Lett.*, **321**, 292.

47 Frackowiak, E., Delpeux, S., Jurewicz, K., Szostak, K., Cazorla-Amorou, D. and Beguin, F. (2002) *Chem. Phys. Lett.*, **336**, 35.

48 Raymundo, E.P., Azais, P., Cacciaguerra, T., Amorou, D.C., Solano, A.L. and Beguin, F. (2005) *Carbon*, **43**, 786.

49 Delpeux, S., Szostak, K., Frackowiak, E. and Beguin, F. (2005) *Chem. Phys. Lett.*, **404**, 374.

50 Vermisoglou, E.C., Georgakilas, V., Kouvelos, E., Pilatos, G., Viras, K., Romanos, G. and Kanellopoulos, N.K. (2007) *Microporous Mesoporous Mater.*, **99**, 98.

51 Yu, S., Ai, Z., Yin, Y., Yu, D., Cui, C., Zhang, X. and Hong, J. (2007) *Mater. Chem. Phys.*, **101**, 30.

52 Li, Y.H., Wang, S., Cao, A., Zhao, D., Zhang, X., Xu, C., Luan, A., Ruan, D., Liang, J., Wu, D. and Wei, B. (2001) *Chem. Phys. Lett.*, **350**, 412.

53 Kongdratyuk, P. and Yates, J.T. (2004) *Chem. Phys. Lett.*, **383**, 314.

54 Anastasios, I., Skoulidas, D., Sholl, J. and Karl, J. (2006) *J. Chem. Phys.*, **124**, 47.

55 Byl, O., Kondratyuk, P. and Yates, J.T. (2003) *J. Phys. Chem. B*, **107**, 4277.

56 Díaz, E., Ordóñez, S. and Vega, A. (2007) *J. Colloid Interface Sci.*, **305**, 7.

57 Friberg, L., Nordberg, G.F. and Vouk, B. (1979) *Handbook on the Toxicology of Metals*, Elsevier, Amsterdam.

58 Chen, C. and Wang, X. (2006) *Ind. Eng. Chem. Res.*, **45**, 9144.

59 Dabrowski, A. (2001) *J. Colloid Interface Sci.*, **93**, 135.

60 Calderon, J., Navarro, M.E., Jimenez-Capdeville, M.E., Santos-Diaz, M.A., Golden, A., Rodriguez-Leyva, I.,

Borja-Aburato, V. and Diaz-Brriga, F. (2001) *Environ. Res.*, **85**, 69.

61 Denizli, A., Buyuktuncel, E., Tuncel, A., Bektas, S. and Genc, O. (2000) *Environ. Technol.*, **21**, 609.

62 Li, Y.H., Di, Z., Ding, J., Wu, D., Luan, Z. and Zhu, Y. (2005) *Water Res.*, **39**, 605.

63 Li, Y., Zhu, Y., Zhao, Y., Wu, D. and Luan, Z. (2006) *Diamond Relat. Mater.*, **15**, 90–4.

64 Katz, S.A. and Salem, H. (2006) *J. Appl. Toxicol.*, **13**, 217.

65 Di, Z.C., Li, Y., Luan, Z. and Liang, J. (2004) *Adsorpt. Sci. Technol.*, **22**, 6.

66 Di, Z.C., Ding, J., Peng, X.-J., Li, Y.-H., Luan, Z.-K. and Liang, J. (2006) *Chemosphere*, **62**, 861.

67 Mohan, D. and Singh, K.P. (2002) *Carbon*, **36**, 2304.

68 Li, Y., Wang, S., Luan, Z., Ding, J., Xu, C. and Wu, D. (2003) *Carbon*, **41**, 1057.

69 Toles, C.A., Marshall, W.E. and Johns, M.M. (1999) *Carbon*, **37**, 1207.

70 Wu, C.H. (2007) *J. Colloid Interface Sci.*, **311**, 338.

71 Rainbow, P.S., Hopkin, S.P. and Crane, M. (2001) *Forecasting the Environmental Fate and Effects of Chemicals*, John Wiley & Sons, Inc., New York.

72 Lu, C. and Chiu, H. (2006) *Chem. Eng. Sci.*, **61**, 1138.

73 Lu, C., Chiu, H. and Liu, C. (2006) *Ind. Eng. Chem. Res.*, **45**, 2850.

74 Kadirvelu, K., Senthilkumar, P., Thamaraiselvi, K. and Subburam, V. (2002) *Bioresour. Technol.*, **81**, 87.

75 Lu, C. and Liu, C. (2006) *Chem. Technol.*, **81**, 1932.

76 Li, Y., Ding, J., Luan, Z., Di, Z., Zhu, Y. and Xu, C. (2003) *Carbon*, **41**, 2787.

77 Hsieh, S.H., Horng, J.-J. and Tsai, C.-K. (2006) *J. Mater. Res.*, **5**, 1269.

78 Peng, X., Luan, Z., Di, Z., Zhang, Z. and Zhu, C. (2005) *Carbon*, **43**, 855.

79 Hichour, M., Persin, F., Molenat, J., Sandeaux, J. and Gavach, C. (1999) *Desalination*, **122**, 53.

80 Hichour, M., Persin, F., Sandeaux, J. and Gavach, C. (2000) *Sep. Purif. Technol.*, **18**, 1.

81 Bishop, P.L. and Sansoucy, G. (1978) *J. Am. Water Works Assoc.*, **70**, 554.

82 Buchanan, W.D. (1962) *Toxicity of Arsenic Compounds*, Amsterdam, New York.

83 Peng, X., Luan, Z., Ding, J., Di, Z., Li, Y. and Tian, B. (2005) *Mater. Lett.*, **59**, 399.

84 Zhang, Z., Liu, L., Zhao, H. *et al.* (1996) *J. Colloid Interface Sci.*, **182**, 158.

85 Wang, X., Chen, C., Hu, W., Ding, P., Xu, D. and Zhou, X. (2005) *Environ. Sci. Technol.*, **39**, 2856.

86 National Primary Drinking Water Regulations, U.S. Environmental Protection Agency 816-F-03-016, June 2003. http://www.epa.gov/safewater/contaminants/index.html#listmcl.

87 Hileman, B. (2000) *Chem. Eng. News*, **29**, 13.

88 Yang, R.T., Long, R.Q., Padin, J., Takahashi, A. and Takahashi, T. (1999) *Ind. Eng. Chem. Res.*, **38**, 2726.

89 Kang, H.S. (2005) *J. Am. Chem. Soc.*, **127**, 9839.

90 Fagan, S.B., Santos, E.J., Filho, A.G., Filho, J.M. and Fazzio, A. (2007) *Chem. Phys. Lett.*, **437**, 79.

91 Roberge, F., Gravel, M.J., Deschenes, L., Guy, C. and Samson, R. (2001) *Water Sci. Technol.*, **44**, 287.

92 Hill, G.A., Tomusiak, M.E., Quail, B. and Cleave, K.M.V. (1991) *Environ. Prog.*, **10**, 147.

93 Fagan, S.B., Gira, E.C., Mendesfilho, O.J. and Souzafilho, A.G. (2006) *Int. J. Quantum Chem.*, **106**, 2558.

94 Rook, J.J. (1974) *Water Treatment Exam.*, **23**, 234.

95 Bull, R.J., Brinbaum, L.S., Cantor, K.P., Rose, J.B., Butterworth, B.E., Pegram, R. and Tuomisto, J. (1995) *J. Fundam. Appl. Toxicol.*, **28**, 155.

96 Lu, C., Chung, Y. and Chang, K. (2006) *J. Hazard. Mater.*, **138**, 304.

97 Yang, K., Zhu, L. and Xing, B. (2006) *Environ. Sci. Technol.*, **40**, 1855.

20
Molecular Imprinting with Nanomaterials

Kevin Flavin and Marina Resmini

20.1
Introduction

Molecular recognition plays a key role in the natural world, from DNA transcription to immune responses and enzyme substrate recognition. Important biological processes depend on the capability of small molecules to recognize their targets in a selective and efficient manner. During the past century, a great deal of research was performed involving molecular recognition, not only to further our comprehension of its role in biological processes but also to translate any findings to practical applications. The design of artificial receptors, capable of selective molecular recognition, has been a goal of research groups in different disciplines for use in a variety of processes such as catalysis, sensing, biological assaying, and the separation of complex chemical mixtures.

One interesting approach for the development of artificial receptors is molecular imprinting, and its application to polymeric materials. In this chapter we will discuss the concept of molecular imprinting, and highlight recent advances in the synthesis of molecularly imprinted polymers (MIPs) on the nanoscale. Included will be methodologies for the preparation of nanoparticles, with a view to a more rational and commercially viable receptor design, and imprinting methodologies which utilize various other nanomaterials, such as nanowires, quantum dots, fullerene, and dendrimers. The applications of these imprinted nanomaterials will be discussed, highlighting those properties which are advantageous to each application, together with details of the changes that will be needed in future in order to maintain the practical potential of molecular imprinting.

20.1.1
Molecular Imprinting: The Concept

Molecular imprinting is a process where functional and crosslinking monomers are copolymerized in the presence of the target analyte (the imprint molecule), which acts as a molecular template [1]. The functional monomers initially form a

Advanced Nanomaterials. Edited by Kurt E. Geckeler and Hiroyuki Nishide
Copyright © 2010 WILEY-VCH Verlag GmbH & Co. KGaA, Weinheim
ISBN: 978-3-527-31794-3

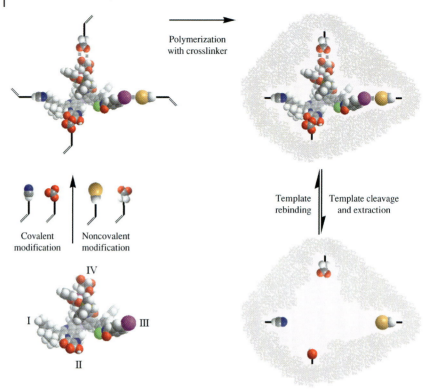

Figure 20.1 Schematic representation of the imprinting process showing some of the interactions used in creating affinity in the binding site for the template. I, reversible covalent interaction; II, semicovalent method; III, electrostatic interaction with an oppositely charged monomer; IV, noncovalent H-bonding.

complex with the imprint molecule and, following polymerization, their functional groups are held in position by the highly crosslinked polymeric structure. Upon removal of the imprint molecule, a cavity or recognition site is created which is complementary to the original template used. A molecular memory is therefore created in the polymeric matrix, which is now capable of recognizing and rebinding the analyte with a very high specificity. The interaction between the template and the functional monomers can be either covalent, noncovalent, or combinations of both, and the choice is largely dependent on the chemical structure of both the template and monomers, and the intended application (Figure 20.1).

The evolution of molecular imprinting, as a technique, will be briefly discussed in the following sections, together with descriptions of some of the chemical strategies utilized for molecular imprinting, in terms of imprint molecule–monomer interactions.

20.1.1.1 **History of Molecular Imprinting**

Towards the beginning of the last century, much effort was devoted to the development of new materials and techniques for applications in chromatography [2]. Among many of those scientists active in this field was the soviet chemist M.V. Polyakov, who reported unexpected absorption properties in silica particles prepared by a novel synthetic procedure [3]. Silica prepared, in the presence of benzene, toluene or xylene, was observed to selectively readsorb the additive present during the preparation [4, 5]. This appears to be the first point in the literature at which selective adsorption to polymeric materials had been observed and explained in terms of a template effect [6].

During the following decade, Pauling developed a theory of the structure and process of formation of antibodies that laid the foundations for the development of molecular imprinting [7]. Dickey, in 1949, reported the preparation of silica gel, having specific affinities for their predetermined substrates, using the same mechanistic approach described by Pauling [8]. Silica gels were prepared in the presence of four different dyes, namely methyl-, ethyl-, n-propyl- and n-butyl orange (Figure 20.2). The dye was subsequently removed and in rebinding experiments the presence of any of these "pattern molecules" would bind the pattern molecule in preference to the other three dyes. Several research groups pursued the preparation of specific adsorbents utilizing Dickey's method [9, 10]. After a steady flow of publications for 15 years, interest in imprinted silica materials experienced a decline as a result of limitations in the stability of these preparations and the reproducibility of the results obtained [2].

The decline in research into imprinting with silica coincided with the publication of a report by Wulff and Sarhan in 1972, on the first example of molecular imprinting in synthetic organic polymers [11]. This report has been accredited with stemming the current interest in imprinted materials as artificial antibodies. By utilizing a covalent imprinting approach, Wulff and Sarhan succeeded in preparing polymers with chiral cavities for the separation of racemic mixtures. Research with imprinted materials accelerated when Mosbach *et al.* introduced a simpler synthetic approach, which utilized noncovalent interactions of the "template" species with the functional monomers [12]. At present, research groups worldwide are focusing attention on the preparation of novel imprinted polymers and new analytical and synthetic applications of molecularly imprinted materials. In the following sections, we describe the covalent and noncovalent imprinting approaches initially reported by Wulff and Mosbach, respectively, which are

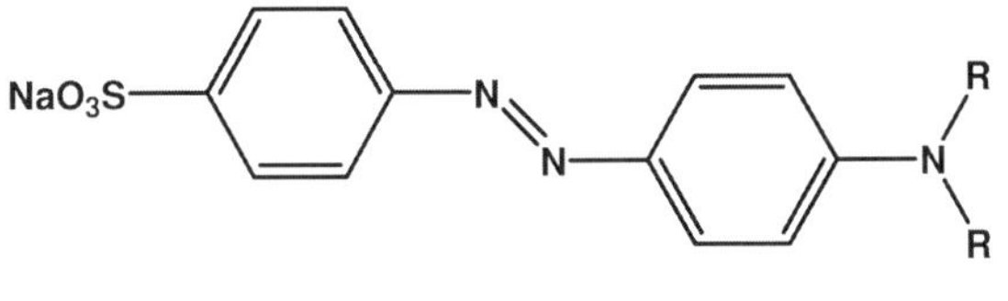

R = methyl, ethyl, n-propyl, n-butyl

Figure 20.2 Chemical structures of the alkyl orange dyes.

widely accredited to be the foundation for the field of molecular imprinting, as it is known today.

20.1.1.2 Covalent Imprinting

In the covalent imprinting strategy, the key step is the formation of a reversible covalent bond between the template and the functional monomer. After polymerization the template is removed, leaving the binding site with the functional group available for further covalent rebinding. Wulff utilized the reversible formation of boronate ester-linkages very successfully, initially for the imprinting of glyceric acid [11], and later for a number of carbohydrate derivatives [13, 14]. In one example, Wulff described the imprinting of a sugar (phenyl-α-D-mannopyranoside) which was coupled with a vinyl-derivatized phenylboronic acid [13]. The template monomer, illustrated in Figure 20.3, was polymerized by free radical initiation in the presence of a porogenic solvent and a large amount of crosslinking agent. The template was subsequently removed through hydrolysis with water or methanol, resulting in a template specific cavity.

The major advantage of the covalent approach is that the high-energy interactions will lead to high association constants, and therefore increase template recognition and reduce nonspecific binding between the template and polymer. Although, covalent strategies have been used successfully with boronate esters [13], Schiff's bases (imines) [15], ketals and acetals [16], only certain compounds such as alcohols, aldehydes, ketones, amines and carboxylic acids can be imprinted using this approach. Moreover, the synthetic difficulties associated with the preparation of a suitable polymerizable unit limit the applicability of this approach.

20.1.1.3 Noncovalent Imprinting

Currently, the most widely used technique to generate molecularly imprinted binding sites is represented by the noncovalent approach developed by Mosbach

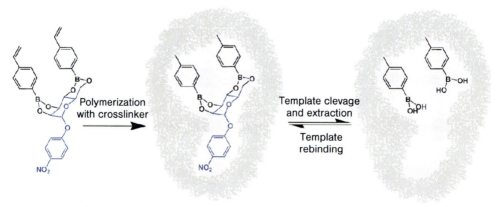

Figure 20.3 Schematic representation of a covalent imprinting approach, using a sugar coupled with a vinyl-derivatized phenylboronic acid [13].

[17]. This is based on a noncovalent self-assembly of the template with functional monomers prior to polymerization with a crosslinking monomer and subsequent template extraction, followed by rebinding via noncovalent interactions, such as H-bonding, ion pairing, and dipole–dipole interactions. The most widely used functional monomer for noncovalent imprinting is methacrylic acid (MAA), which was initially proposed by Mosbach *et al.* [18]; however, many other functional monomers, such as 4-vinylpyridine [19] and acrylamide [20] are commercially available. A number of custom-designed functional monomers, such as *N,N′*-diethyl(4-vinylphenyl)amidine [21], for use in catalysis, and *trans*-4-[*p*-(*N,N*-Dimethylamino)styryl]-*N*-vinylbenzylpyridinium chloride [22], used for fluorescent sensing, have also been reported.

Figure 20.4 illustrates the well-known noncovalent imprinting strategy of theophylline using MAA as the functional monomer and ethylene glycol dimethacrylate as the crosslinker [23].

The monomer–template association complex can be obtained *in situ* simply by adding the components to the reaction mixture. After polymerization, the template is removed by extracting the polymer with acidified methanol. The guest binding

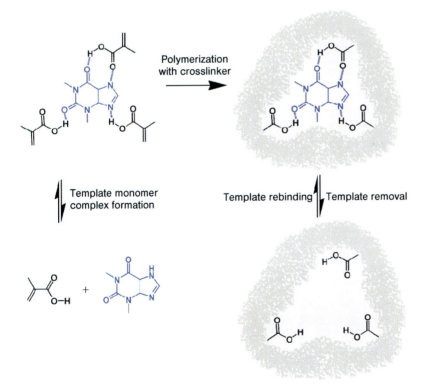

Figure 20.4 Noncovalent imprinting of theophylline, using methacrylic acid as the functional monomer and ethylene glycol dimethacrylate as the crosslinker [23].

by the polymer occurs through the corresponding noncovalent interactions. The carboxylic acid function of the methyl methacrylate forms ionic interactions with the amino groups and hydrogen bonds with polar functions. Dipole–dipole and hydrophobic interactions are also thought to contribute.

The advantages of the noncovalent strategy are that the synthesis of covalent monomer–template conjugates is unnecessary, and that the imprint molecule is easily removed from the polymer under very mild conditions, as it is only weakly bound by noncovalent interactions. Also, guest binding and guest release via noncovalent interactions result in faster binding kinetics than occur in the covalent approach. The main disadvantages are that the imprinting process may not be very efficient, as the monomer–template adduct is labile and not strictly stoichiometric. In order to overcome these limitations, the polymerization conditions must be carefully chosen so as to maximize noncovalent adduct formation in the mixtures.

20.1.1.4 Alternative Molecular Imprinting Approaches

The two most extensively used strategies for molecular imprinting have been the noncovalent and covalent approaches, respectively. However, a number of variations of these methods have been investigated in recent years, with the intention of overcoming the disadvantages inherent in each approach. One such strategy employs the stoichiometric noncovalent approach [24], in which the complex between the functional monomer and template is strong enough to ensure that the equilibrium lies well in favor of complex formation; this can be achieved when the association constant is $\geq 10^3 M^{-1}$ [25]. Semicovalent imprinting is another frequently used strategy, which attempts to merge the advantages of the covalent and the noncovalent approach. In this method, the template is bound covalently to the functional monomer, either directly [26] or via a spacer [27]; however, rebinding occurs via noncovalent interactions only.

20.1.2
Towards Imprinting with Nanomaterials

The final application of a target-imprinted polymer plays a key role in determining the type of polymeric matrix to be used. Since the first report of an example of molecular imprinting, "bulk" polymerization has been the most frequently used method for the production of MIPs. However, in this method the "bulk" material must be ground mechanically, which results in a large size distribution of irregularly shaped particles and sieved; this is not only time-consuming but can also result in a substantial loss of material. Considerable advances have been made over the past decade, in the basic methodologies available for the production of receptor sites, and in the optimization of conditions for imprinting and subsequent rebinding [28]. In order to avoid the disadvantages associated with "bulk" polymerization, alternative polymeric materials have been used, for example, imprinted microbeads, films, and membranes [28, 29]. In the case of beads that are designed for chromatographic purposes, particle diameters of 25–38 μm are suitable for efficient separation with reasonable back-pressures [30]. However, when considering MIPs intended for catalysis, facilitated synthesis, sensing and assaying, a progression from the micro- to the nanoscale has been a major aim of many research groups. Aside from the obvious advantages that miniaturization presents, good accessibility to as many recognition sites as possible and shorter diffusion times to facilitate faster equilibrium between release and reuptake of the template in the cavity, are central to requirements [31]. Ultimately, improvements in the efficiency of synthetic receptors, with respect to response times and reaction rates, may depend on how closely the actual biological systems can be mimicked. To do so would require significant developments of MIPs on the nanoscale.

20.2
Molecular Imprinting in Nanoparticles

A variety of approaches for preparing imprinted particles in the nanometer size range have proved successful. In this section we will review the various approaches, and highlight recent developments and important applications.

20.2.1
Emulsion Polymerization

The first reported synthesis of imprinted nanoparticles was over a decade ago by Takagi and coworkers [32], who successfully prepared a Cu^{2+}-selective resin using a novel template polymerization technique. Here, the target molecule (hydrophilic) and the host–monomer (amphiphilic) approach and arrange themselves at the interface of an oil-in-water emulsion. The oil phase, containing the monomer and crosslinker, was subsequently polymerized, the result being a resin with particle sizes ranging from 400 to 500 nm. Although standard emulsion polymerization has proved successful for molecular imprinting in nanoparticles [32–34], core–shell emulsion polymerization is probably a more useful approach, as it enables a much better control over particle size, polydispersity, and the location of recognition sites.

20.2.1.1 Core–Shell Emulsion Polymerization
Core–shell emulsion polymerization consists of two steps that begin with the preparation of a core latex in the absence of an emulsifier. The core, which may be prepared from a variety of materials [35–37], is then mixed with another monomer or mixture of monomers that produces the shell. The core–shell particles possess two types of property, one being endowed by the core (spherical monodisperse nanoparticles) and the other by the shell (recognition layer).

Pérez *et al.* reported the synthesis of core–shell particles, in which a sacrificial spacer method was used, for the imprinting of cholesterol [38]. Initially, 30–40 nm MMA or styrene core particles were prepared, which were subsequently used in a second-stage polymerization with ethyleneglycol dimethacrylate (EGDMA), in

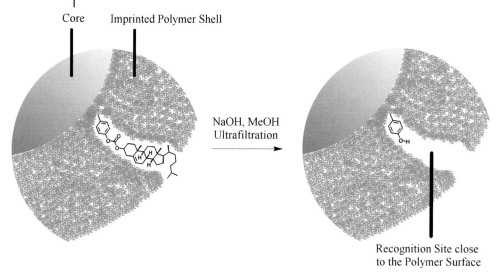

Figure 20.5 Schematic diagram of core–shell nanoparticles with cholesterol-imprinted shells Adapted from Ref. [38].

the presence of a polymerizable cholesterol derivative (Figure 20.5). Following template cleavage (carbonate ester hydrolysis) and removal, the recognition properties of the imprinted core–shell polymers were evaluated.

On comparison of the imprinted and nonimprinted polymers, a considerably greater binding of cholesterol to the imprinted core–shells was evident. Similar imprinted nanoparticles were also prepared with a magnetic core, which demonstrated similar cholesterol rebinding properties to those particles prepared with nonmagnetic cores; however, in the presence of a magnetic field, a more efficient particle sedimentation – and thus removal from solution – was demonstrated. Subsequently, Pérez *et al.* reported the preparation of imprinted nanoparticles characterized by a poly(divinylbenzene) shell over a crosslinked poly(styrene) core [39]. In this case, the second-stage polymerization was performed in the presence of polymerizable surfactants (pyridinium 12-(4-vinylbenzyloxycarbonyl)dodecanesulfate (PS) and pyridinium 12-(cholesteryloxycarbonyloxy)dodecanesulfate (TS), that also acted as a template. Figure 20.6 illustrates the noncovalent imprinting approach used, where the surfactant-like template aligns itself at the polymer–water interface with the template region in the monomer phase. Following polymerization of the monomer phase, template removal creates hydrophobic recognition sites at the surface of the beads. As the polymerized surfactant tails are also hydrolyzed, a polar surface is exposed which differentiates it from the imprinted regions.

Pérez and Mayes subsequently reported the successful imprinting of propranolol by the more classical noncovalent approach in core–shell nanoparticles [40]. Although, the binding capacity was lower than that usually measured for bulk polymers imprinted with propranolol, this investigation proved that noncovalent imprinting in core–shell particles was possible, despite the fact that they were

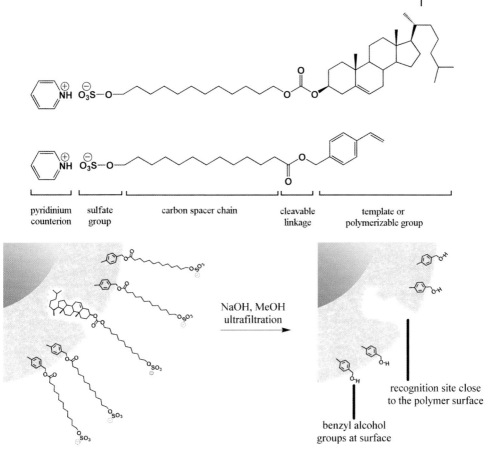

| pyridinium counterion | sulfate group | carbon spacer chain | cleavable linkage | template or polymerizable group |

Figure 20.6 Schematic diagram of core–shell nanoparticles with cholesterol-imprinted shells accompanied by structures of the template surfactant and the polymerizable surfactant. Adapted from Ref. [38].

prepared in the presence of an aqueous continuous phase. The same group later reported the synthesis of similar propranolol-imprinted nanoparticles, where a fluorescent monomer was successfully incorporated into the particle core [41]. The presence of the imprinted shell had no effect on the fluorescence of the core; neither did the fluorescent monomer in the core affect the imprinting in the shell. This type of particle would be expected to find applications in assay technology.

Carter and Rimmer also used core–shell emulsion polymerization for the non-covalent imprinting of caffeine and propranolol [42–44]. The application of these nanoparticles is for the selective extraction of caffeine/propranolol from mixtures with a structural analogue theophylline/atenolol (Figure 20.7) [44]. Particles were prepared with a styrene/divinylbenzene core, and ranged in diameter from 180 to 214 nm. The shell was prepared with a mixture of EGDMA and binding monomer

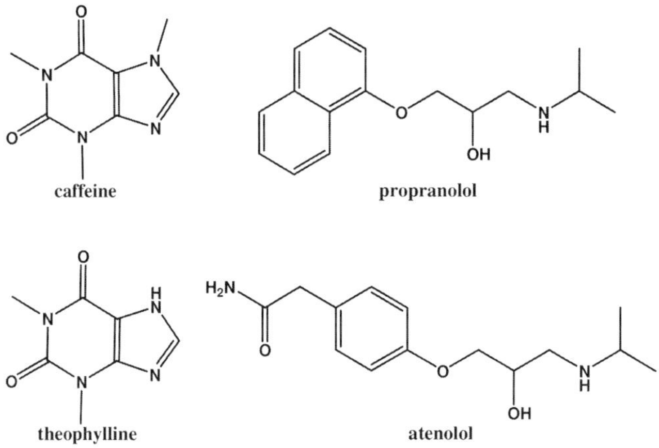

Figure 20.7 Chemical structures of template molecules (caffeine and propranolol) and their structural analogues (theophylline and atenolol).

(oleyl phenyl hydrogen phosphate) in the presence of template, and displayed thickness ranging from 2 to 20 nm.

20.2.1.2 Mini-Emulsion Polymerization

A *mini-emulsion* is a type of emulsion, where the monomer droplet size is reduced to the range of 50 to 300 nm diameter by the application of sheer forces (ultrasonication or high-pressure homogenization) [45]. In order to stabilize the homogeneous dispersed phase, a costabilizer is added to prevent diffusion processes from occurring in the continuous phase. This in turn inhibits the occurrence of Ostwald ripening during the polymerization process. The main difference between emulsion and mini-emulsion polymerization is the solubility of the initiator. In mini-emulsion polymerization the initiator is only soluble in the dispersed phase, as opposed to emulsion polymerization where the initiator is soluble in the continuous phase [46]. Nucleation, as a consequence, occurs within the dispersed nano-droplets creating small "nanoreactors," with the monomer and template already present at the start of the polymerization.

Tovar and coworkers have mainly used this method to prepare imprinted nanoparticles [47–50]. Initially, they produced nanoparticles (yield of 98 ± 2%) from varying ratios of EGDMA and MAA, imprinted with enantiomers of boc-phenylalanine anilide [47]. Although dynamic light scattering (DLS) indicated a particle diameter of 200 ± 20 nm, transmission electron microscopy (TEM) verified a much larger polydispersity with particles ranging from 50 to 300 nm. During recognition studies, the quantity of L-boc-phenylalanine anilide that rebound was fourfold greater in the case of an L-imprinted MIP than in the corresponding NIP, and 10-fold greater than the binding of the D-enantiomer in the L-imprinted nanoparticles. This imprinting system was subsequently used to demonstrate the use of microcalorimetry, to monitor the heat of binding during rebinding experiments with the nanoparticles, and to demonstrate the enthalpic basis of chiral recognition in molecularly imprinted polymers [48].

The imprinted nanoparticles produced by Tovar and coworkers were later used for the separation of enantiomers. Using the mini-emulsion approach, the imprinted nanoparticles were coated onto the surface of a polyamide membrane for enantiomeric separation [49]. The dense particle layer on the surface of the membrane resulted in a large imprinted surface area and a low flow rate, which was advantageous as the establishment of the chemical equilibrium due to selective rebinding is a time-consuming step. Absorption experiments and binding isotherms were subsequently performed in order to establish a new mathematical model for the understanding and describing of the whole separation process by the composite membrane [50]. According to the authors, this should allow prediction of the most favorable configuration for the imprinted composite membrane, and thus allow an optimal performance.

More recently, Tan and Tong have attempted the imprinting of the protein ribonuclease A using mini-emulsion polymerization [51, 52]. In these studies, the preparation of protein surface-imprinted nanoparticles was described, with diameters ranging from approximately 40 to 80 nm. The major difficulty with the imprinting of proteins was optimization of the polymerization conditions in order to avoid protein denaturation [51]. Although, the high-shear homogenization demonstrated negligible disruption to the protein conformation, the initiators and the surfactants frequently used for mini-emulsion polymerization were singled out as possible sources of template denaturation, and suitable conditions were therefore investigated. The imprinted nanoparticles prepared under optimized, nondenaturing conditions (using a redox initiator and poly(vinyl alcohol) as surfactant) displayed a good imprinting efficiency that was absent from the imprinted polymer prepared through the conventional mini-emulsion polymerization using thermal or UV initiation and sodium dodecylsulfate as surfactant.

Tan and Tong also used a variation of this approach, where mini-emulsion polymerization was used in a core–shell approach [52]. In this case, larger particles of regular shape and 700–800 nm diameter were produced, which comprised a Fe_3O_4 magnetite core and a surface-imprinted shell. The imprinted particles exhibited significant recognition and selectivity from aqueous solution, in addition to easy and efficient particle separation as a result of the magnetic core being present.

20.2.2
Precipitation Polymerization

In precipitation polymerization – unlike emulsion methods – the polymeric reaction begins in a homogeneous phase where the monomers, crosslinkers, and initiators are present in a dilute solution of porogenic solvent. As the polymerization proceeds, the expanding polymer becomes insoluble and aggregates into particles, which are stabilized against coagulation, either sterically or by their rigid crosslinked surfaces. Dispersion polymerization is another method for obtaining polymeric materials. Although the technique shows some differences

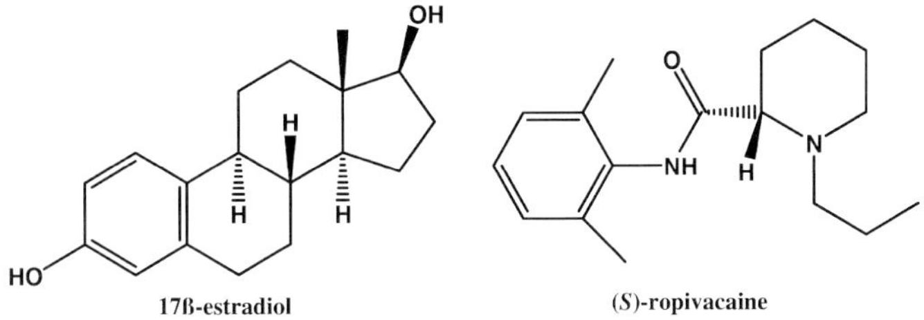

Figure 20.8 Chemical structures of 17β-estradiol and *(S)*-ropivacaine.

from precipitation polymerization, these are not significant enough to justify discrimination between the two in the context of this chapter. More detailed information on this subject is available elsewhere, however [28, 46, 53].

The first reported synthesis of imprinted nanoparticles by precipitation polymerization was by Mosbach and coworkers [54]. In these studies, MAA and trimethylolpropane trimethacrylate (TRIM) were polymerized at high dilution in the presence theophylline and 17β-estradiol (Figure 20.8). These templates were chosen to demonstrate the applicability of the method toward targets with very different hydrophobicities. Imprinted nanoparticles, with yields of >85% and an average diameter of 300 nm, were isolated by centrifugation of the polymerization solution. Following template removal, the recognition properties were characterized using a radioligand-binding assay; the binding specificity of the polymers was found to be extremely high, with <1% crossreactivity between the two target molecules. Although, the imprinted nanoparticles bound three to four times more template than the corresponding reference material, a sixfold amount of imprinted polymer was necessary for effective absorption of the more hydrophobic template (17β-estradiol) in comparison with theophylline. This effect was, however, the result of a lower affinity constant between 17β-estradiol and the polymer when rebinding in an organic solvent.

The same group also studied the effect of crosslinker type and content on the preparation of imprinted nanoparticles by precipitation polymerization [55]. The results showed that nanoparticles prepared with TRIM (with three crosslinking vinyl groups) rather than with EGDMA (two crosslinking vinyl groups) allowed the incorporation of greater amounts of template and functional monomer, without compromising the rigidity of the particles. These polymers demonstrated higher load capacities, as a result of the increased amounts of functional monomer, but without any loss of specificity due to insufficient crosslinking. More recently, Ye and coworkers studied new synthetic conditions to obtain imprinted beads with controllable size in the nanometer to micrometer range [56]. A variation of particle size, while maintaining good recognition properties, was achieved by altering the ratio of the two different crosslinking monomers, in essentially the same precipitation polymerization system.

20.2.2.1 Applications and Variations

By using the precipitation method of imprinted nanoparticle formation, Ye and coworkers developed a new sensing approach using molecular imprinting and proximity scintillation [57, 58]. The imprinting of (S)-propranolol was performed in the presence of a scintillation monomer, which fluoresces in the proximity of tritium-labeled target molecules. The authors showed that an enantioselective competitive binding assay was possible, without removal of the unbound ligand present in solution.

Nanoparticles produced by precipitation polymerization have also been used for various other applications, including the separation of enantiomers. Spégel *et al.* described the preparation of nanoparticles, which were used for separations in capillary electrochromatography [59, 60]. Two approaches for the preparation of the imprinted nanoparticles were used. The first was based on the mixing of two types of imprinted nanoparticle [(S)-propranolol and (S)-ropivacaine; Figure 20.8], while the second was based on the incorporation of two different templates during the preparation of imprinted nanoparticles. The imprinted nanoparticles were suspended in solutions of analyte, and both approaches resulted in a separation of the propranolol and ropivacaine enantiomers in one single chromatographic run.[60]

Zhu *et al.* also used the precipitation approach for the synthesis of 17β-estradiol-imprinted nanoparticles for use in chromatographic separations (high-performance liquid chromatography; HPLC) [61]. The functional monomer used in this case was 2-(trifluoromethyl)acrylic acid, and particles were obtained in the range of 300 to 1500 nm. The imprinted polymers, when packed into a column, demonstrated the separation of α- and β-estradiol. Unfortunately, the report did not include any discussion of the general applicability of nanoparticles to HPLC, and the their implications in terms of flow rates and pressures.

Ciardelli and coworkers described the preparation of theophylline-imprinted nanospheres, with a view to developing materials with combined properties of drug delivery and rebinding, for clinical applications [62]. The result was that, by varying the percentage of MAA and MMA (methyl methylacrylate) monomers, the properties of release and recognition of print molecules could be modulated. In subsequent investigations, the same group investigated an innovative approach to increase the binding and selective behavior of imprinted nanoparticles in aqueous media [63–65]. The recognition factor for theophylline and caffeine in physiological solution were found to be increased substantially when the imprinted nanoparticles were immobilized in an acrylic membrane [63]. It was suggested by the authors that the membrane had created a microenvironment that enhanced the affinity of the analyte for the imprinted nanosphere.

In a similar approach, as an alternative to incorporating the imprinted nanoparticles into a membrane, Ye *et al.* encapsulated within the polymer nanofibers that had been produced using an electrospinning technique [66]. The imprinted nanoparticles initially obtained were dissolved in dichloromethane with poly(ethylene terephthalate) (PET), which formed the matrix of the nanofiber. This solution was spun at high voltage to produce nanofibers with an average diameter of 150–300 nm. More recently, Ye and colleagues described an alternative precipitation method to prepare imprinted nanoparticles in the absence of organic solvents [67]. Using this approach, the noncovalent imprinting of propranolol was demonstrated under high-dilution conditions in supercritical carbon dioxide. The overall binding performance of the imprinted nanoparticles was comparable to that of imprinted polymers prepared in conventional organic solvents.

20.2.2.2 Microgel/Nanogel Polymerization

Precipitation polymerization can be optimized to produce microgels and/or nanogels in the size range of 10 to 600 nm [68]. One characteristic of microgels/nanogels is that they are prepared in a suitable solvent system, based on solubility parameters, and produce a low-viscosity colloidal solution that never reaches the point of precipitation. The molecular mass may be varied in a simple, controllable manner from the low thousands (nanogels) to many millions (microgels), simply by the choice of concentration at which they are prepared. A number of research groups have recently shown interest in imprinting in microgels/nanogels given their unique properties, including their solubility [69–73].

Although, a number of groups have taken steps in this direction [74], Wulff and coworkers were the first to report investigations into the preparation of suitably crosslinked microgels with molecular recognition properties [69]. Covalently imprinted microgels were successfully synthesized with 70% EGDMA and 30% MMA in cyclohexanone, cyclopentanone and N,N-dimethylformamide at 1–4 wt% monomer concentrations. The microgels were characterized using gel permeation chromatography, viscometry, and membrane osmometry, and found to be highly crosslinked macromolecules with a molecular weight comparable to that of proteins. Although rebinding selectivities were low compared to the results achieved with insoluble crosslinked polymers, the success of this approach represented a step towards the development of "artificial enzymes."

Although, at about the same time, Mosbach *et al.* produced theophylline-imprinted microgel spheres using a noncovalent approach [70], Resmini *et al.* were the first to report the preparation of imprinted soluble microgels, which acted as an enzyme mimic and displayed hydrolytic catalytic activity (Figure 20.9) [71, 72]. In these studies, a phosphate transition state analogue (TSA) was imprinted, by using two polymerizable amino acids (arginine and tyrosine) as functional monomers, in order to mimic the catalytic mechanism of hydrolytic antibodies and hydrolase enzymes with carbonate substrates. Imprinted microgels containing 70% crosslinker, and a monomer concentration of 1.5%, were found to display the highest rate enhancements of about an order of magnitude, over the uncatalyzed reaction.

More recently, Wulff and coworkers prepared phosphate-(TSA)-imprinted nanogels, with an average diameter of 20 nm, that were capable of carbonate hydrolysis, and where the k_{cat}/k_{uncat} value reached 2990 [73]. The group succeeded in imitating the natural enzymes by producing soluble nanogels that contained an average of one catalytically active site per polymeric macromolecule. Although, nanoparticles were produced with a single active site, higher-molecular-weight particles with a greater crosslink density and approximately 95 sites per particle demonstrated the greatest enhancement in catalytic activity. These results emphasized that the opposing factors which often are so critical in molecular imprinting are polymer rigidity and recognition/active site accessibility. Although, the polymer must

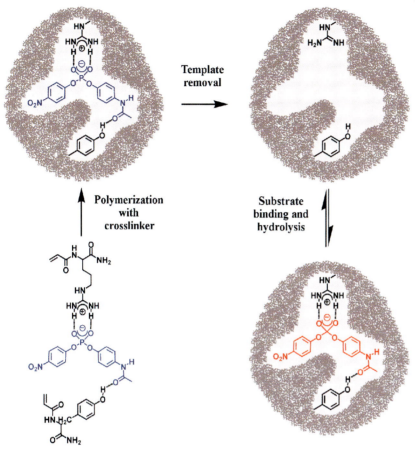

Figure 20.9 Noncovalent imprinting of a phosphate transition-state analogue, using polymerizable arginine and tyrosine as functional monomers [71].

display sufficient flexibility to allow a rapid access to the binding sites, too much flexibility would compromise its rigidity and hence its recognition properties. It was for this reason that Haupt and colleagues prepared larger nanogels of approximately 180 nm, imprinted with 2,4-dichlorophenoxyacetic acid, which were used for a (pseudo-immuno) binding assay, where the requirements were different than for catalytic applications [75].

20.2.3
Silica Nanoparticles

To date, only a relatively small number of reports have been made describing successful imprinting in silica nanoparticles, and these all used different methods to form the recognition sites at the surface of the nanoparticle [76–80]. Markowitz *et al.* used a template-directed method to imprint an α-chymotrypsin TSA at the surface of silica nanoparticles (400–600 nm diameter) prepared from tetrae-thoxysilane and a number of organically modified silanes, for the incorporation of functionality [76]. Silica particle formation was performed in a microemulsion, where a mixture of a nonionic surfactant and the acylated chymotrypsin TSA (with the TSA acting as the headgroup at the surfactant–water interface) were used as a means of creating a cavity capable of hydrolyzing benzoyl-D-arginine-*p*-nitroanalyde, a trypsin substrate. The k_{cat} and turnover number for the nanoparticles were, however, not reported, as the authors had no means of calculating the active site concentrations. As an alternative, V_{max}/K_m was used as an estimate of the relative catalytic activity, which showed an increase in line with the increasing amounts of template used. Unexpectedly, the particles were highly selective for the D-isomer of the substrate, even though the imprint molecule had the L-isomer configuration.

In a subsequent study, Markowitz and coworkers investigated the effect that the addition of functional silanes had on the catalytic activity of the surface-imprinted nanoparticles [77]. It was suggested that a variation in the basicity of the functional monomers would affect the initial rates of hydrolysis, and that imprinted particles prepared with mixtures of functional monomers would show a cooperative effect promoting catalytic activity. Subsequently, the same group used a template-directed method for the imprinting of a hydrolysis product of soman (a nerve agent) at the surface of silica nanoparticles [78]. Again, a number of different functionalized silane precursors were used, and the binding characteristics of the imprinted particles investigated. The results showed the imprinted nanoparticles to display a significantly higher degree of specificity for the imprint molecule than did the structurally related organophosphates. It was also reported that variations in functionality incorporated into the particles had a definite effect on both the porosity and absorption capacity of the polymers.

Gao *et al.* reported an alternative surface functional monomer-directing strategy for the imprinting of trinitrotoluene (TNT) at the surface of silica nanoparticles [79]. The method employed a core–shell method in which monodisperse silica cores of 100 nm diameter were prepared using the Stöber process. The surface of the core particles was initially functionalized with aminopropyl groups which were, in turn, converted to acrylamide functions. An acrylate shell, composed of acrylamide and EGDMA in the presence of the template, was subsequently synthesized, around the silica particles. The acrylamide had a strong noncovalent, charge-transfer complexing interaction with the electron-deficient aromatic ring of the template molecule, which resulted in a significant shift in the UV-visible spectrum and allowed detection of the explosive. One interesting result of this synthetic method was the ability to control shell thickness between 10 and 30 nm by varying the total quantity of polymer precursors added during the shell preparation.

Kim and colleagues also used a core–shell approach, where a covalently imprinted aromatic polyimide layer of approximately 100 nm was coated to the surface of large silica spheres (~10 μm diameter) [80]. The shell film adhered to the silica spheres through electrostatic interactions between the carboxylic groups of the polymer chains and amino functional groups at the surface of the silica. The imprinted particles were packed into a column and used as a stationary phase in the HPLC separation of estrone and structural analogues.

20.2.4
Molecularly Imprinted Nanoparticles: Miscellaneous

A number of interesting examples of imprinted nanoparticles have been reported that do not belong to the above-discussed categories, and these have been included in this section. One such approach, reported by Salam and Ulbricht, involved the imprinting of Boc-phenylalanine in "nanomonolithic" particles, that are formed by *in situ* polymerization in the nanosized pores of a polymeric membrane [81]. The authors claimed that the imprinted monoliths had a higher binding capacity and a higher enantioselectivity for the template than the reference monoliths, and suggested that the "nanomonolith" composite membranes might be used for continuous molecular-level separations with predetermined perm-selectivity.

Li *et al.* described a novel method in which uracil- and thiamine-imprinted polymeric nanospheres were prepared by diblock copolymer self-assembly [82]. Initially, a diblock copolymer was synthesized with one block containing functional groups for both hydrogen bond formation and crosslinking. In the presence of the template, the block copolymer was allowed to self-assemble to form spherical micelles in a selective solvent. This polymeric structure was then crosslinked, resulting in imprinted nanospheres of approximately 50 nm diameter, which were extracted to remove the template molecules. When compared to traditional monolithic-imprinted polymers, these imprinted nanospheres demonstrated a better solvent dispersibility, a higher capacity, and comparable selectivity. One possible drawback of this system was the presence of a hydrophobic shell around the imprinted core; however, the authors believed that hydrolysis of the shell would provide water-dispersible nanospheres with potential sensing and bioapplications.

Another novel method, which has been reported, is the preparation of covalently imprinted polymeric nanocapsules by microemulsion polymerization [83]. The polymerization of styrene and divinylbenzene in the presence of a monomer–template complex (a polymerizable derivative of estrone) was performed in oil-in-water microemulsion droplets. Following polymerization and phase separation in the micelle, the surfactant was removed, resulting in a hollow polymeric nanocapsule with diameters in the range of 20 to 25 nm. The template was thermally removed to produce highly accessible recognition sites that displayed moderate selectivity over structural analogues. However, the major interest here involved the use of these nanocapsules for controlled-release drug delivery. When the nanocapsules were incubated with a fluorescent probe (pyrene), prior to template removal, transfer of the pyrene into the interior of the capsule was not evident. Following template removal, however, a transfer of pyrene to the capsule interior was observed, confirming that the imprinted site had acted as a gateway to the interior of the capsule, and could be opened and closed by template removal and rebinding.

20.3
Molecular Imprinting with Diverse Nanomaterials

In recent years, research into molecularly imprinted materials has focused on reducing the dimensions of these materials from the micro range to the nano. As a consequence, considerable effort has aimed at the development of new polymerization methods capable of producing imprinted nanoparticles. A number of research groups have, however, studied the application of the molecular imprinting approach to other types of nanomaterials, such as nanowires [84–87], nanotubes [86], nanofibers [88], quantum dots [89, 90], fullerene [91, 92], and dendrimers [93–95]. The most significant examples of these will be reviewed in the following sections.

20.3.1
Nanowires, Nanotubes, and Nanofibers

Wang and coworkers were the first to successfully prepare molecular recognition sites at the surface of nanowires using molecular imprinting technology [84]. In this approach, a commercially available nanoporous alumina membrane with a 100 nm pore diameter was used, with a sol–gel template synthesis being used to deposit silica nanotubes inside the pores of the alumina membranes. Initially, a silane precursor with aldehyde functionality was attached to the silica nanotubes, and the template – in this case glutamic acid – was immobilized on the inner walls of the nanotubes. The nanopores were subsequently filled with the monomer mixture (pyrrole in this case), polymerized, and both the alumina membrane and the silica nanotubes removed by chemical dissolution; this left behind polypyrrole nanowires with glutamic acid binding sites situated at the surface. The selectivity of the imprinted nanowires towards glutamic acid over phenylalanine and arginine was high, and similar to that observed from bulk polymers. However, a very high rate of analyte uptake was observed resulting from this surface imprinting technique.

The same group later used a similar protocol for the surface imprinting of a variety of proteins, including albumin, hemoglobin, and cytochrome c in nanowires [85]. On this occasion, acrylamide and N,N'-methylenebisacrylamide were used for the polymerization. There was an approximate sevenfold difference between rebinding of the template to the imprinted and control nanowires, which was complemented by a large binding capacity, observed as a result of the high surface area of the nanowires. Although the imprinted nanowires could not distinguish between bovine and horse cytochrome c, there was a definite distinction between bovine and human hemoglobin. In a subsequent report, the same group described the preparation of surface-imprinted nanowires toward theophylline, which were magnetic in nature [86]. Here, the nanopores were filled with a prepolymerization

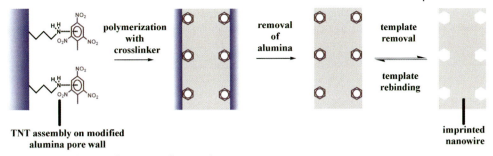

Figure 20.10 Schematic illustration of noncovalent imprinting of TNT at the surface of acrylate-based nanowires. Adapted from Ref. [87].

mixture containing the superparamagnetic $MnFe_2O_4$ nanocrystallites. Unfortunately, the behavior or potential applications of the magnetic characteristics of the nanowires were not discussed to any great extent.

Xie *et al.* described the preparation of TNT surface-imprinted nanowires (Figure 20.10), where alumina membranes with pore diameters of 70 nm were prepared by electrochemical anodization [87]. In these studies, the authors also reported the first imprinted nanotubes, which were prepared in similar fashion to the nanowires, the main differences being a reduction in the monomer concentration and an increase in the quantity of initiator used. The binding capacities of the nanotubes and nanowires were almost 2.5-fold and threefold that of normal imprinted particles with 2–3 μm diameter, respectively. This increase in binding capacity was attributed to the high ratio of surface-imprinted sites, the large surface-to-volume ratios, and the complete removal of TNT templates.

Ye and coworkers also described the incorporation of imprinted nanoparticles into electrospun nanofibers (as discussed previously in Section 20.2.2.1) [66]. The same group, however, also achieved the generation of artificial molecular recognition sites in the nanofiber itself by molecular imprinting [88]. The electrospun nanofibers were prepared from a solution mixture of PET and polyallylamine (which acts as a functional polymer) in the presence of a template molecule, 2,4-dichlorophenoxyacetic acid. Following template removal, the recognition properties of the imprinted nanofibers were identified using a radioligand-binding assay. The imprinted nanofibers displayed favorable binding characteristics in aqueous solution, where analyte binding to the imprinted nanofibers was fivefold that of the reference fibers. The authors predicted that this type of imprinted nanofiber would become highly applicable in the future for affinity-related separations.

20.3.2
Quantum Dots

Lin and coworkers are, to date, the only group to report the synthesis of molecularly imprinted polymers in conjunction with fluorescent semiconductor nanoparticles also known as quantum dots (QDs) [89, 90]. The first step of this approach was the preparation of the CdSe/ZnS semiconductor nanoparticles. In order to ensure that the nanoparticles were incorporated covalently into the polymer, the surfaces of the QDs were functionalized with 4-vinylpyridine. The functionalized nanoparticles were polymerized with MAA and EGDMA in the presence of uracil, following spin coating of the initiated solution, to form a molecularly imprinted thin film [89]. In a subsequent study, the QDs were incorporated into bulk polymers, which were imprinted with a number of template molecules including caffeine, uric acid, L-cysteine, and estriol [90]. The authors showed that binding to recognition sites in the vicinity of the fluorescent nanoparticles caused quenching of the photoluminescence emission, presumably due to the fluorescence resonance energy transfer between the QDs and the template molecules. The imprinted polymers demonstrated good selectivity towards the template when compared to structurally related compounds and, according to the authors, the control polymers exhibited no response to the analyte.

20.3.3
Fullerene

The first "homogeneous nanoscale imprinting system," wherein a saccharide was used as template and covalently bound by two boronic acid groups onto the surface of [60]fullerene, was described by Shinkai and coworkers [91]. This strategy was used in order to produce one single binding site per molecule and to avoid the creation of numerous binding sites with varying recognition properties (as illustrated in Figure 20.11) [92]. Initially, a template monomer complex, based on D,L-threitol, was synthesized which could undergo a regioselective and chiroselective double [4 + 2] cycloaddition with [60]fullerene. Subsequent cleavage of the template resulted in two optically active *cis*-3 *bis* adducts with opposite chirality in a ratio of 72.28. Competitive rebinding studies between D,L-threitol and the imprinted fullerene demonstrated that L-threitol-imprinted (^{f}A)-4 and D-threitol-

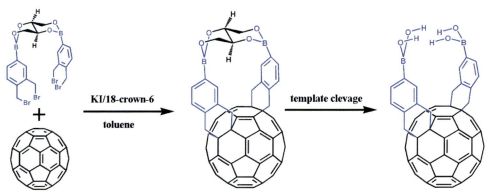

Figure 20.11 Structural representation of a monomolecular covalent imprinting strategy with fullerene for the recognition of sugars. Adapted from Ref. [91].

imprinted (^{f}C)-4 rebound the original templates in diastereoselective fashion. Although, the selectivity in this system was not extraordinary, it should be emphasized that this was the first instance in which molecular imprinting was applied to a system with a soluble nanosized matrix in homogeneous solution.

20.3.4
Dendrimers

Zimmerman *et al.* more recently used a different strategy, wherein a single porphyrin template was dynamically imprinted into a single macromolecule (dendrimer), in order to prepare molecularly homogeneous hosts each with a single recognition site [93]. For this, the authors used the covalent approach, where the porphyrin acted as the core and was covalently bound through ester linkages to eight third-generation dendrons to form the dendrimer. The porphyrin was cleaved by hydrolysis, following the crosslinking of homoallyl end groups at the exterior of the dendrimer by Grubbs ring-closing metathesis (Figure 20.12). Complexation

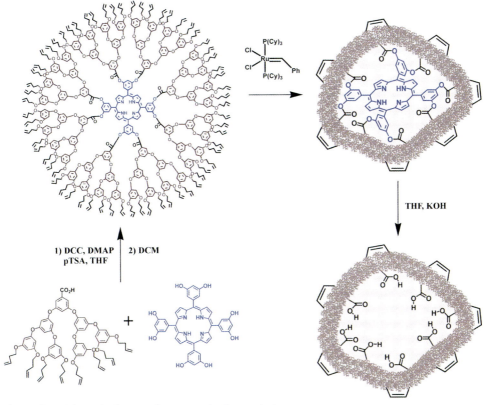

Figure 20.12 Schematic diagram of a monomolecular covalent imprinting strategy with dendrimers for the recognition of porphyrins. Adapted from Ref. [96].

studies (where rebinding occurred noncovalently) demonstrated that >95% of the imprints were effective, and their binding properties homogeneous [96]. The imprinted dendrimer was also shown to be selective, and to bind porphyrins with at least four hydrogen bond donor/acceptor sites, in toluene as solvent. The imprinted cavity was also size-selective, and although the template itself was too large to bind, porphyrins of smaller physical size and appropriate functionality were able to complex strongly.

In a subsequent study, an imprinted dendrimer was prepared around a porphyrin with four reactive alcohol groups, which was covalently bound to four fourth-generation dendrons [94]. More recently, Zimmerman's group reported the preparation of an imprinted dendrimer with a less-symmetrical template molecule, with a view to producing an amine-selective sensor [95]. Here, very high affinity binding was perceived, from a three-point interaction, which included two hydrogen bonds and one covalent linkage to a reporter group that transduced binding by color change. This monomolecular imprinting approach resulted in properties that were difficult to achieve by conventional polymerization methods, such as high-efficiency imprinting, quantitative removal of the template, solubility of the imprinted material in common organic solvents, and possibility of separation of imperfectly assembled binding sites. However, the limitations of this approach were the need for a multistep synthesis of the dendrimers, the high dilution conditions required for the ring-closing metathesis reaction, and the incompatibility of many templates with this imprinting approach.

20.4
Conclusions and Future Prospects

Nanomaterials represent one of the most important targets of the developing field of nanotechnology. The advanced scientific knowledge acquired for the synthesis and physical–chemical characteristics of these materials has led to a considerable increase in the number of their applications and commercial exploitations. The application of imprinting technologies to nanomaterials has provided a new insight, offering the possibility of creating tailored cavities with specific recognition properties.

Over the past decade, considerable effort has been devoted to developing novel synthetic strategies, and this has led to imprinted materials in a variety of formats. Attention has been mainly focused on the development of approaches to producing imprinted nanoparticles, and to providing reproducibility and efficiency among the polymerization methodologies used. Although, significant advances have been made in this area, the widespread commercialization of these materials has yet to be realized. Yet, ongoing developments in the field of nanomaterials have led to a range of promising polymeric formats, including soluble crosslinked imprinted dendrimers, microgels and nanogels, which more closely mimic the physical characteristics of enzymes and antibodies. Additional formats, such as nanofibers, nanowires and nanotubes, have arisen and have been used successfully in conjunction with imprinting technologies. These materials exploit the advantages of high surface-to-volume ratios and often also the surface-imprinting techniques that allow an easy access of target molecules to recognition sites. The results obtained to date in this area have been the outcome of an increasing interest in molecular imprinting, by both academic and industrial enterprises, and will surely lead to their significant application in the near future.

References

1 Haupt, K. and Mosbach, K. (2000) *Chem. Rev.*, **100**, 2495–2504.

2 Sellergren, B. (2003) *Molecularly Imprinted Polymers. Man-Made Mimics of Antibodies and Their Application in Analytical Chemistry*. Elsevier.

3 Polyakov, M.V. (1931) *Zh. Fiz. Khim.*, **2**, 799–805.

4 Polyakov, M.V., Stadnik, P., Paryckij, M., Malkin, I. and Duchina, F. (1933) *Zh. Fiz. Khim.*, **4**, 454–456.

5 Polyakov, M.V., Kuleshina, L. and Neimark, I. (1937) *Zh. Fiz. Khim.*, **10**, 100–112.

6 Alexander, C., Andersson, H.S., Andersson, L.I., Ansell, R.J., Kirsch, N., Nicholls, I.A., O'Mahony, J. and Whitcombe, M.J. (2006) *J. Mol. Recognit.*, **19**, 106–180.

7 Pauling, L. (1940) *J. Am. Chem. Soc.*, **60**, 2643–2657.

8 Dickey, F.H. (1949) *Proc. Natl Acad. Sci. USA*, **35**, 227–229.

9 Haldeman, R.G. and Emmett, P.H. (1955) *J. Phys. Chem.*, 1039–1043.

10 Beckett, A.H. and Anderson, P. (1957) *Nature*, **179**, 1074–1075.

11 Wulff, G. and Sarhan, A. (1972) *Angew. Chem. Int. Ed. Engl.*, **11**, 341.

12 Andersson, L., Sellergren, B. and Mosbach, K. (1984) *Tetrahedron Lett.*, **25**, 5211–5214.

13 Wulff, G., Vesper, W., Grobeeinsler, R. and Sarhan, A. (1977) *Macromol. Chem. Phys.*, **178**, 2799–2816.

14 Wulff, G. and Schauhoff, S. (1991) *J. Org. Chem.*, **56**, 395–400.

15 Wulff, G. and Vietmeier, J. (1989) *Macromol. Chem. Phys.*, **190**, 1727–1735.

16 Shea, K.J. and Sasaki, D.Y. (1989) *J. Am. Chem. Soc.*, **111**, 3442–3444.

17 Arshady, R. and Mosbach, K. (1981) *Macromol. Chem. Phys. Makromol. Chem.*, **182**, 687–692.

18 Mosbach, K. (1994) *Trends Biochem. Sci.*, **19**, 9–14.

19 Kempe, M., Fischer, L. and Mosbach, K. (1993) *J. Mol. Recognit.*, **6**, 25–29.

20 Yu, C. and Mosbach, K. (1998) *J. Mol. Recognit.*, **11**, 69–74.

21 Strikovsky, A.G., Kasper, D., Grun, M., Green, B.S., Hradil, J. and Wulff, G. (2000) *J. Am. Chem. Soc.*, **122**, 6295–6296.

22 Turkewitsch, P., Wandelt, B., Darling, G.D. and Powell, W.S. (1998) *Anal. Chem.*, **70**, 2025–2030.

23 Vlatakis, G., Andersson, L.I., Muller, R. and Mosbach, K. (1993) *Nature*, **361**, 645–647.

24 Wulff, G. and Schönfeld, R. (1998) *Adv. Mater.*, **10**, 957–959.

25 Mayes, A.G. and Whitcombe, M.J. (2005) *Adv. Drug. Deliv. Rev.*, **57**, 1742–1778.

26 Sellergren, B. and Andersson, L. (1990) *J. Org. Chem.*, **55**, 3381–3383.

27 Whitcombe, M.J., Rodriguez, M.E., Villar, P. and Vulfson, E.N. (1995) *J. Am. Chem. Soc.*, **117**, 7105–7111.

28 Pérez-Moral, N. and Mayes, A.G. (2001) *Bioseparation*, **10**, 287–299.

29 Brüggemann, O., Haupt, K., Ye, L., Yilmaz, E. and Mosbach, K. (2000) *J. Chromatogr. A*, **889**, 15–24.

30 Sellergren, B. (2001) *J. Chromatogr. A*, **906**, 227–252.

31 Wulff, G. (2002) *Chem. Rev.*, **102**, 1–27.

32 Kido, H., Miyajima, T., Tsukagoshi, K., Maeda, M. and Takagi, M. (1992) *Anal. Sci.*, **8**, 749–753.

33 Murata, M., Hijiya, S., Maeda, M. and Takagi, M. (1996) *Bull. Chem. Soc. Jpn*, **69**, 637–642.

34 Yoshida, M., Uezu, K., Goto, M., Furusaki, S. and Takagi, M. (1998) *Chem. Lett.*, **27** (9), 925–926.

35 Eshuis, A., Leendertse, H.J. and Thoenes, D. (1991) *Colloid Polym. Sci.*, **269**, 1086–1089.

36 He, W.D., Cao, C.T. and CaiYuan, P. (1996) *J. Appl. Polym. Sci.*, **61**, 383–388.

37 Ferguson, C.J., Russell, G.T. and Gilbert, R.G. (2002) *Polymer*, **43**, 6371–6382.

38 Pérez, N., Whitcombe, M.J. and Vulfson, E.N. (2000) *J. Appl. Polym. Sci.*, **77**, 1851–1859.

39 Pérez, N., Whitcombe, M.J. and Vulfson, E.N. (2001) *Macromolecules*, **34**, 830–836.

40 Pérez-Moral, N. and Mayes, A.G. (2002) *Molecularly Imprinted Materials-Sensors and Other Devices*, Vol. **723** (eds K.J. Shea and M.J. Roberts), Materials Research Society, Warrendale, PA, pp. 61–66.

41 Pérez-Moral, N. and Mayes, A.G. (2004) *Langmuir*, **20**, 3775–3779.

42 Carter, S.R. and Rimmer, S. (2002) *Adv. Mater.*, **14**, 667–670.

43 Carter, S., Lu, S.Y. and Rimmer, S. (2003) *Supramol. Chem.*, **15**, 213–220.

44 Carter, S.R. and Rimmer, S. (2004) *Adv. Funct. Mater.*, **14**, 553–561.

45 Tovar, G.E.M., Kräuter, I. and Gruber, C. (2003) *Molecularly Imprinted Polymer Nanospheres as Fully Synthetic Affinity Receptors*, Springer, pp. 125–144.

46 Schillemans, J.P. and van Nostrum, C.F. (2006) *Nanomedicine*, **1**, 437–447.

47 Vaihinger, D., Landfester, K., Kräuter, I., Brunner, H. and Tovar, G.E.M. (2002) *Macromol. Chem. Phys.*, **203**, 1965–1973.

48 Weber, A., Dettling, M., Brunner, H. and Tovar, G.E.M. (2002) *Macromol. Rapid. Commun.*, **23**, 824–828.

49 Lehmam, M., Brunner, H. and Tovar, G.E.M. (2002) *Desalination*, **149**, 315–321.

50 Lehmann, M., Dettling, M., Brunner, H. and Tovar, G.E.M. (2004) *J. Chromatogr. B Analyt. Technol. Biomed. Life Sci.*, **808**, 43–50.

51 Tan, C.J. and Tong, Y.W. (2007) *Langmuir*, **23**, 2722–2730.

52 Tan, C.J. and Tong, Y.W. (2007) *Anal. Chem.*, **79**, 299–306.

53 Arshady, R. (1992) *Colloid Polym. Sci.*, **270**, 717–732.

54 Ye, L., Cormack, P.A.G. and Mosbach, K. (1999) *Anal. Commun.*, **36**, 35–38.

55 Ye, L., Weiss, R. and Mosbach, K. (2000) *Macromolecules*, **33**, 8239–8245.

56 Yoshimatsu, K., Reimhult, K., Krozer, A., Mosbach, K., Sode, K. and Ye, L. (2007) *Anal. Chim. Acta*, **584**, 112–121.

57 Ye, L. and Mosbach, K. (2001) *J. Am. Chem. Soc.*, **123**, 2901–2902.

58 Ye, L., Surugiu, I. and Haupt, K. (2002) *Anal. Chem.*, **74**, 959–964.

59 Spégel, P. and Nilsson, S. (2002) *Am. Lab.*, **34**, 29–33.

60 Spégel, P., Schweitz, L. and Nilsson, S. (2003) *Anal. Chem.*, **75**, 6608–6613.

61 Zhu, Q.J., Tang, J., Dai, J., Gu, X.H. and Chen, S.W. (2007) *J. Appl. Polym. Sci.*, **104**, 1551–1558.

62 Ciardelli, G., Cioni, B., Cristallini, C., Barbani, N., Silvestri, D. and Giusti, P. (2004) *Biosens. Bioelectron.*, **20**, 1083–1090.

63 Silvestri, D., Borrelli, C., Giusti, P., Cristallini, C. and Ciardelli, G. (2005) *Anal. Chim. Acta*, **542**, 3–13.

64 Silvestri, D., Barbani, N., Cristallini, C., Giusti, P. and Ciardelli, G. (2006) *J. Membr. Sci.*, **282**, 284–295.

65 Ciardelli, G., Borrelli, C., Silvestri, D., Cristallini, C., Barbani, N. and Giusti, P. (2006) *Biosens. Bioelectron.*, **21**, 2329–2338.

66 Chronakis, I.S., Jakob, A., Hagström, B. and Ye, L. (2006) *Langmuir*, **22**, 8960–8965.

67 Ye, L., Yoshimatsu, K., Kolodziej, D., Francisco, J.D. and Dey, E.S. (2006) *J. Appl. Polym. Sci.*, **102**, 2863–2867.

68 Graham, N.B. and Cameron, A. (1998) *Pure. Appl. Chem.*, **70**, 1271–1275.

69 Biffis, A., Graham, N.B., Siedlaczek, G., Stalberg, S. and Wulff, G. (2001) *Macromol. Chem. Phys.*, **202**, 163–171.

70 Ye, L., Cormack, P.A.G. and Mosbach, K. (2001) *Anal. Chim. Acta*, **435**, 187–196.

71 Maddock, S.C., Pasetto, P. and Resmini, M. (2004) *Chem. Commun.*, 536–537.

72 Pasetto, P., Maddock, S.C. and Resmini, M. (2005) *Anal. Chim. Acta*, **542**, 66–75.

73 Wulff, G., Chong, B.O. and Kolb, U. (2006) *Angew. Chem. Int. Ed. Engl.*, **45**, 2955–2958.

74 Ohkubo, K., Funakoshi, Y. and Sagawa, T. (1996) *Polymer*, **37**, 3993–3995.

75 Hunt, C.E., Pasetto, P., Ansell, R.J. and Haupt, K. (2006) *Chem. Commun.*, 1754–1756.

76 Markowitz, M.A., Kust, P.R., Deng, G., Schoen, P.E., Dordick, J.S., Clark, D.S. and Gaber, B.P. (2000) *Langmuir*, **16**, 1759–1765.

77 Markowitz, M.A., Kust, P.R., Klaehn, J., Deng, G. and Gaber, B.P. (2001) *Anal. Chim. Acta*, **435**, 177–185.

78 Markowitz, M.A., Deng, G. and Gaber, B.P. (2000) *Langmuir*, **16**, 6148–6155.

79 Gao, D.M., Zhang, Z.P., Wu, M.H., Xie, C.G., Guan, G.J. and Wang, D.P. (2007) *J. Am. Chem. Soc.*, **129**, 7859–7866.

80 Kim, T.H., Do Ki, C., Cho, H., Chang, T.Y. and Chang, J.Y. (2005) *Macromolecules*, **38**, 6423–6428.

81 Salam, A. and Ulbricht, M. (2006) *Desalination*, **199**, 532–534.

82 Li, Z., Ding, J.F., Day, M. and Tao, Y. (2006) *Macromolecules*, **39**, 2629–2636.

83 Ki, C.D. and Chang, J.Y. (2006) *Macromolecules*, **39**, 3415–3419.

84 Yang, H.H., Zhang, S.Q., Tan, F., Zhuang, Z.X. and Wang, X.R. (2005) *J. Am. Chem. Soc.*, **127**, 1378–1379.

85 Li, Y., Yang, H.H., You, Q.H., Zhuang, Z.X. and Wang, X.R. (2006) *Anal. Chem.*, **78**, 317–320.

86 Li, Y., Yin, X.F., Chen, F.R., Yang, H.H., Zhuang, Z.X. and Wang, X.R. (2006) *Macromolecules*, **39**, 4497–4499.

87 Xie, C.G., Zhang, Z.P., Wang, D.P., Guan, G.J., Gao, D.M. and Liu, J.H. (2006) *Anal. Chem.*, **78**, 8339–8346.

88 Chronakis, I.S., Milosevic, B., Frenot, A. and Ye, L. (2006) *Macromolecules*, **39**, 357–361.

89 Lin, C.I., Joseph, A.K., Chang, C.K. and Lee, Y.D. (2004) *Biosens. Bioelectron.*, **20**, 127–131.

90 Lin, C.I., Joseph, A.K., Chang, C.K. and Lee, Y.D. (2004) *J. Chromatogr. A.*, **1027**, 259–262.

91 Ishi-i, T., Nakashima, K. and Shinkai, S. (1998) *Chem. Commun.*, 1047–1048.

92 Ishi-i, T., Iguchi, R. and Shinkai, S. (1999) *Tetrahedron*, **55**, 3883–3892.

93 Zimmerman, S.C., Wendland, M.S., Rakow, N.A., Zharov, I. and Suslick, K.S. (2002) *Nature*, **418**, 399–403.

94 Zimmerman, S.C., Zharov, I., Wendland, M.S., Rakow, N.A. and Suslick, K.S. (2003) *J. Am. Chem. Soc.*, **125**, 13504–113508.

95 Beil, J.B. and Zimmerman, S.C. (2004) *Chem. Commun.*, 488–489.

96 Zimmerman, S.C. and Lemcoff, N.G. (2004) *Chem. Commun.*, 5–14.

21
Near-Field Raman Imaging of Nanostructures and Devices

Ze Xiang Shen, Johnson Kasim, and Ting Yu

21.1
Introduction

Significant advances have been made in recent years in nanoscience and nanotechnology. Many applications which involve the use of nanomaterials and devices, and utilize their unique properties, have now been realized and more are expected in the near future. Despite these achievements, our understanding of nanomaterials and devices remains limited, due partly to the lack of suitable characterization techniques.

In Raman spectroscopy, molecular vibrations are measured that are determined by the structure and chemical bonding – as well as the masses – of the constituent atoms/ions. Raman spectra, as with infrared (IR) spectra, are unique in their chemical and structural identifications. The technique of Raman spectroscopy is nondestructive, sample preparation-free, there is no requirement for vacuum application, and it can be carried out in an aqueous/liquid environment. The technique can also easily be carried out at a different temperature, pressure, and electrical and magnetic fields. If it were possible to use Raman spectroscopy for nanocharacterization, it would provide critically important material-specific properties such as composition, chemical bonding, crystal and electronic structures and strain/stress, as well as supplementary information to other nanotechniques, such as scanning electron microscopy (SEM), atomic force microscopy (AFM), and transmission electron microscopy (TEM).

Conventional (far-field) micro-Raman spectroscopy has a spatial resolution of approximately 0.5 μm, which is governed by the theoretical diffraction limit, whereas IR spectrometry has at best a spatial resolution of approximately 10 μm, due to the longer wavelengths being utilized. Hence, the application of these techniques in nanoscience and nanotechnology is severely limited by their poor spatial resolution. Consequently, extensive efforts have recently been made to characterize structural information at the nanometer scale, utilizing near-field scanning optical microscopy (NSOM), the different operational modes of which are shown in Figure 21.1.

Advanced Nanomaterials. Edited by Kurt E. Geckeler and Hiroyuki Nishide
Copyright © 2010 WILEY-VCH Verlag GmbH & Co. KGaA, Weinheim
ISBN: 978-3-527-31794-3

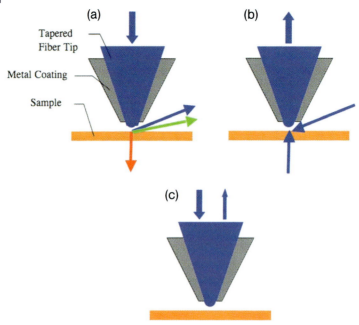

Figure 21.1 Different operational modes of near-field scanning optical microscopy. (a) Illumination mode; (b) Collection mode; (c) Illumination and collection mode.

In this chapter, presented data have been acquired using both far-field and near-field Raman imaging. The aim is to demonstrate the importance of Raman imaging in materials and device research.

21.2
Near-Field Raman Imaging Techniques

There are three main approaches to near-field scanning Raman microscopy (NSRM). The first approach is based on the principle of NSOM [1–3], which is also known as "aperture near-field Raman." It is embodied through an aperture NSOM, whereby an optical fiber tip with a small aperture (50–200 nm) is used to deliver laser light, while the fiber tip is held at a close distance (some tens of nanometers) above the sample surface. The Raman signal is collected in far-field through either a microscope objective or a lens [4–6]. The scattered light (Raman signal) is coupled to a Raman spectrometer, where the Raman spectra are recorded at each point on the sample by scanning the fiber tip across the sample surface. An NSRM image is constructed using these spectra.

The laser emerging from such a fiber tip is extremely weak (typically 100 nW) due to the low optical throughput of the fiber tip. Hence, aperture near-field Raman has been plagued by the poor signal-to-noise ratio (SNR) of the near-field

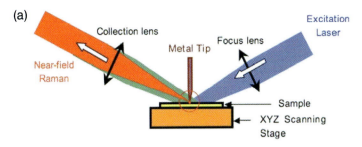

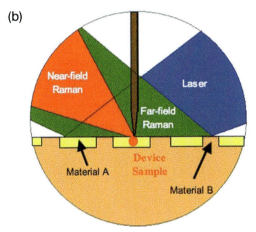

Figure 21.2 (a) Schematic diagram of tip-enhanced Raman scattering in reflection geometry. The metal tip is brought to the sample surface by tuning fork or cantilever. The Raman signals can also be collected by the focus lens instead of the collection lens; (b) Magnified view of the tip region. The far-field Raman signal comes from the whole laser-illuminated area (a few μm) and is featureless, while the near-field Raman signal comes from only the tip region (<100 nm) and carries information related to the device features.

Raman spectra. Although extensive efforts have been devoted to fabricate fiber tips with a higher throughput, the throughput is still very low, especially for tips with an aperture <200 nm. In addition, the Raman signal is intrinsically weak (typically less than 1 in 10^7 photons). A typical Raman spectrum takes several minutes to record when using the conventional near-field Raman method, which makes it prohibitive to construct a Raman image in this way. For example, when Webster *et al.* [7] used the aperture technique to study the stress distribution of a scratch on a Si crystal wafer, a period of about 9 hours was required to obtain a Raman image of 26 × 21 points.

An alternative approach to near-field Raman microscopy is to use an apertureless configuration (which is also known as tip-enhanced Raman scattering; TERS) [8–10], employing a metal or metal-coated tip (see Figure 21.2). In this configuration, the laser impinges on the apex of the metal tip, inducing a strong local electric field enhancement in the vicinity of the metal tip apex. In this case, the Raman enhancement is a result of local field enhancement by the tip, or increased polarizability of the sample due to an interaction between the tip and sample, which is similar to surface-enhanced Raman scattering. As the laser is delivered to the sample surface by an objective lens rather than by a metal-coated fiber tip, the low optical throughput of conventional fiber tips can be overcome. With a suitable metal tip (usually Ag or Au), the Raman signal from the sample close to the tip can be enhanced by several orders of magnitude.

Besides the higher overall Raman throughput, TERS has several other merits over the apertureless configuration:

1. **A higher spatial resolution:** At least in principle, it is much easier to fabricate a sharp tip than a small aperture. A spatial resolution <10 nm has been achieved using TERS.

2. **Fewer topographic artifacts:** In aperture configuration the tip is large, although its aperture may be small; therefore, the tip cannot follow the sample surface exactly. This is normally not a problem, and is less severe in TERS.

3. **A variety of potential tip materials:** In addition to noble metals, silicon and its oxides, nitrides, nanowires and nanotubes may represent good candidates for tips.

4. **A wider spectral response range:** Unlike the aperture technique, which usually functions in the visible region due to a need to use optical fibers to deliver the laser, TERS can extend its applications to a much wider range, for example from ultraviolet to IR.

Unfortunately, TERS also has some major disadvantages, the main challenge being to obtain reproducible near-field images. It is very difficult to achieve a consistent near-field enhancement in TERS, due to difficulties in controlling the geometry of the metal tip. The technique also encounters problems of wear and tear, oxidation, and contamination of the tip apex (e.g., by carbon) that may have severe adverse effects on the near-field enhancement. Another problem is that the laser spot, when focused on the tip apex, causes an intense background (far-field signal) that should be eliminated in order to achieve a better SNR [11, 12]. As a result, TERS is rarely used for routine characterization.

A third approach, developed by the present authors' group, uses a dielectric microsphere to focus the laser to a spot size that is smaller than the diffraction limit. Besides being used as the excitation source for Raman spectroscopy, the incident laser beam (linearly polarized Gaussian TEM00 mode) is also used to trap the microsphere just above the sample surface, using the well-known "optical tweezers" mechanism. In these experiments, the sample and polystyrene microspheres (3 μm diameter) were placed in a sample cell filled with deionized water. One microsphere was trapped at the center of the laser beam and was in contact with the surface of the sample during scanning. The experimental set-up is shown schematically in Figure 21.3a and b.

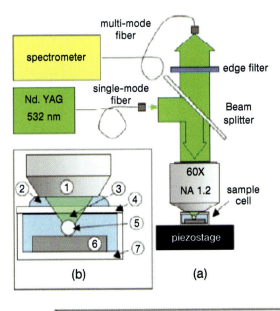

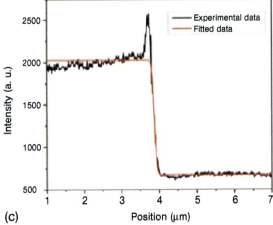

Figure 21.3 (a) Schematic diagram of the near-field Raman microscope with microsphere; (b) Detailed description of the sample cell: 1 = water immersion lens (60× NA = 1.2 water immersion), 2 = water, 3 = focused laser, 4 = cover glass, 5 = polystyrene microsphere (3 μm), 6 = sample, 7 = sample cell' (c) Typical Si–Si Raman intensity versus position with the fitted data to determine the laser spot size.

21.3
Visualization of Si–C Covalent Bonding of Single Carbon Nanotubes Grown on Silicon Substrate

In this section, it is shown that Raman microscopy can be used to characterize nanoscale samples, using Si–C bonding between the carbon nanotubes (CNTs) and the Si substrate as an example. Recently, CNTs have attracted a great deal of interest on the basis of their extraordinary electrical, chemical, thermal, and mechanical properties [13–15], all of which make them promising materials for applications in nanometer-scale electronics and devices. For example, a unique property of CNTs is that they may be either metallic or semiconducting, depending on their chirality and diameters [16]. The tuning of electrical properties of single CNTs by inducing covalent bonding between the CNTs and the Si substrate has been predicted [17–22]. Such direct integration of CNTs into well-developed silicon technologies offers a new strategy of building nanoscale, or even single-molecule, electronic devices by selectively forming covalent bonding between the CNTs and the Si substrate.

Several groups have achieved Si–CNT bonding in ceramics with good mechanical properties by annealing Si nanoparticles together with CNTs, which served as a mechanical reinforcement [23]. The direct observation of Si–C covalent bonding on single CNTs on Si substrate samples has not been reported, due partly to a lack of suitable characterization techniques. However, it has been predicted that the interaction between Si and the CNT results in a rich variety of changes in the electronic structures of the CNT [19]. For example, when Miwa *et al.* studied the single-walled CNT adsorbed onto partially and fully hydrogenated Si (001) surfaces [19, 20], it could be shown that the removal of a few H-atoms along the adsorption sites would enhance the metallic character of the CNT. However, the removal of *all* H-atoms would transform the CNT to the semiconducting state. By using first-principle calculations, Peng *et al.* were able to determine the binding energies of CNTs adsorbed onto different sites on Si (001) surfaces [22]. In fact, their results showed that the absorption sites at the surface trench for CNTs parallel to the Si dimmer rows, and between Si dimers for CNTs perpendicular to the Si dimer rows, were stable.

Such stability was also shown to depend heavily on the size (diameter) of the CNTs. When preparing these Si–CNT devices the first step is to fabricate Si–CNT covalent bonding on a large scale. Irrespective of which technique is used to characterize the sample, it must be capable of identifying those CNTs with Si–C bonds. In these studies, a micro-Raman imaging technique [24] was used that was capable of probing down to hundreds of nanometers, to investigate isolated single CNTs and their interaction with the Si substrate. The CNTs were grown on a silicon substrate at high temperature to achieve Si–CNT bonding, with Raman imaging showing that some of the single CNTs were attached covalently to the silicon wafer surface.

The CNTs were grown on the Si substrate using a thermal chemical vapor deposition (CVD) technique, the substrate having been cleaned using trichloroethylene, acetone, and ethanol in turn, by ultrasonic agitation. An ultrathin Fe film to be used as the catalyst was deposited *in vacuo* (10^{-5} Torr) by ion beam sputtering. The substrates were pre-annealed at 850 °C in a hydrogen-filled quartz tube furnace for 90 min. The CNTs were then grown at 1000 °C for 20 min using methane, hydrogen, and argon with flow rates of 75, 20, and 100 sccm (standard $cm^3 min^{-1}$), respectively. The sample was cooled below 300 °C before being exposed to ambient air, in order to avoid any damage caused by rapid cooling.

A JEOL JSM-6700F scanning electron microscope was used to study the surface morphology of samples and locations of interest for further Raman studies. The Raman spectroscopy and imaging were carried out using a WITec CRM200 confocal Raman system fitted with an Olympus microscope objective lens (100× NA = 9.5). A double-frequency Nd:YAG laser (532 nm, 100 mW laser; CNI Laser) was used as the excitation source. The Raman scattered light was dispersed using a grating (1800 grooves per millimeter), and detected using a thermoelectrically cooled charge-coupled-device (CCD) cooled to −64 °C. A piezo-stage was used to scan the sample for the Raman imaging. The spatial resolution of the Raman imaging was approximately 250 nm.

The SEM image of the sample is shown in Figure 21.4a. It can be seen that all the CNTs were grown horizontally on the Si substrate, with random orientations. The various diameters of the CNTs may be due to the different sizes of the Fe catalyst. With assistance from markings on the Si substrate, Raman imaging was performed in the same region shown in the SEM image. A careful analysis of the Raman spectra in the Raman images revealed that the Raman spectra of the CNTs shown in Figure 21.4a could be indexed as two groups: (i) those consisting of only normal CNT peaks; and (ii) those with additional peaks, which can be assigned to Si–C bonds (this point is discussed later in the chapter). For simplicity, these two types of spectra are referred to as CNT spectra and Si–C spectra, respectively. Figure 21.4b shows the typical Raman spectra of the two groups. The CNT spectrum shows only Raman bands of normal CNTs: the D band ($1344 cm^{-1}$), G band, and 2D band ($2681 cm^{-1}$) [25]. The G band can be further identified as the G− band at 1570^{-1} and the G+ band at $1590 cm^{-1}$. For certain CNTs (such as the CNT highlighted by the broken line in Figure 21.4a), the Raman spectrum shows extra bands belonging to Si–C bonds at approximately $1510 cm^{-1}$ (band A), $1692 cm^{-1}$ (band B), $1921 cm^{-1}$ (band C), $1981 cm^{-1}$ (band D), $2392 cm^{-1}$ (band E), and $2620 cm^{-1}$ (band F). While several regions were tested in this study, it should be noted that the extra bands (A–F) were observed only for certain CNTs, which suggested that the Si–C bonds formed only for certain CNTs.

Raman imaging was performed in the sample region shown in Figure 21.4a with a scanning step size of 50 nm, and with the incident laser polarized in the plane of the substrate. Raman images constructed using different peak intensities are shown in Figure 21.5a–d. Figure 21.5a was generated by the G band (ca. $1580 cm^{-1}$) intensity of the CNTs. Brighter regions represent CNTs, which show an excellent correlation with those obtained using SEM. Images were also constructed using the intensity of the extra band A; the resultant image is shown in Figure 21.5b. The shape and position of the bright region matched perfectly the

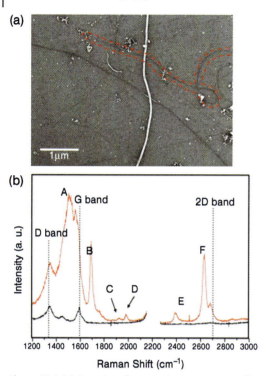

Figure 21.4 (a) Scanning electron microscope image showing the areas of interest. Raman bands belonging to Si–C bonds can be detected on a carbon nanotube (CNT), as indicated by the broken red line; (b) Comparison of Raman spectra obtained on normal CNTs (black) and the special CNT (red). The dashed lines represent the peak positions of the CNTs' D band, G band, and 2D band.

CNT, emphasized by the dashed curve in the SEM image (Figure 21.4a). Images generated using bands B and F (not shown) exhibited near-identical patterns to that of band A, clearly indicating that the three bands were all correlated to the same CNT marked in the SEM image.

In order to confirm that peaks A–F in Figure 21.4b were indeed related to Si–C covalent bonds, the Raman spectra were compared with those of bulk SiC, which have been well studied [26, 27]. Bulk SiC has several polytypes, which can give rise to the following first-order Raman peaks: TA mode between 200–300 cm^{-1}, LA mode at around 610 cm^{-1}, TO mode at around 790 cm^{-1}, and LO mode at near 970 cm^{-1} [28]. The TO mode at 790 cm^{-1} is by far the strongest Raman peak in all polytypes, which is absent in the present sample (the reasons are given later). The peak positions and relative intensities of the SiC second-order Raman bands varied for different polytypes, with generally two characteristic bands: a broad band

(a)   (b)

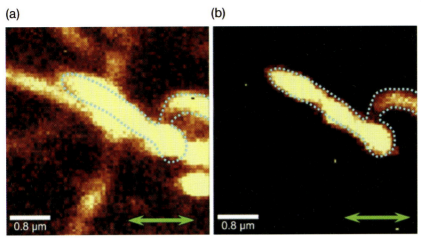

Figure 21.5 Raman images constructed using peak intensity of: (a) G band of CNTs and (b) Si–C band A at 1510 cm^{-1}. The Raman images generated using bands at 1692 cm^{-1} and 2620 cm^{-1} are identical to those in panel (b). The green arrow indicates the polarization of the incident laser. The blue dotted lines serve as a guide to the eye, and show the position and shape of the special CNT.

between 1510 cm^{-1} and 1540 cm^{-1}, and a narrower band around 1710 cm^{-1}. Both bands corresponded to multibands, resulting from the overlapping of different overtone phonons of the SiC vibration [27, 29, 30]. As shown in Figure 21.4b, both band A (1510 cm^{-1}) and band B (1692 cm^{-1}) observed in these Raman spectra showed a good agreement in shape and intensity with those two SiC characteristic bands. The slight shift to lower wavenumbers ($\Delta\omega \sim 10$ cm^{-1}) may be due to the small-size effect [31]. In addition, the weaker bands E and F also showed an agreement with the broad band of SiC at 1930 cm^{-1}. As there have been no reports on the Raman band of SiC above 2000 cm^{-1}, it was not possible to compare band F with that of bulk SiC. However, the excellent agreement between the Raman image using band F and those of band A and B showed that band F also belonged to the Si–C bonds in this sample.

The first-order Raman peak at 790 cm^{-1} is the strongest among all the Raman peaks in all polytypes of SiC, but was absent from these experiments. This can be explained by the substantial structural difference between the bulk SiC and the present CNT/Si system, where all the Si–C covalent bonds were formed at the Si–CNT interface. Such an unique an arrrangement deviates considerably from that in the bulk SiC, where the Si–C covalent bonds are arranged in three dimensions with a long-range order; in contrast, the bonds in the CNT/Si system showed a long-range order in only one dimension – that is, along the long axis of the CNTs.

21.4
Near-Field Scanning Raman Microscopy Using TERS

The NSRM system used in these experiments consisted of a standard Nanonics NSOM system and a Renishaw micro-Raman spectrometer. The Nanonics NSOM system uses a bent optic fiber tip (cantilevered tip), which makes the scanning head very compact. The compact design allows the easy integration of the spectrometer and NSOM by placing the scanning head under the microscope objective of the Renishaw system [32]. An argon ion laser (488 nm line) is focused onto the metal tip, which is made from tungsten wire through electrochemical etching [similar to the method used in scanning tunneling microscopy (STM) tip preparation], and then coated with silver through radiofrequency (RF)-sputtering [33].

The interaction between the tip and the laser enhances the local electrical field near the tip, which in turn enhances the Raman signal nearby. The same objective lens collects the Raman signal in the backscattering geometry, which is then directed into the Renishaw spectrometer by the coupling optics. In the configuration described above, both the near-field and far-field Raman signals are collected. In order to reduce the far-field component, the laser beam must be focused tightly to form a spot on the sample which is as small as possible. The collection efficiency of the optical system should also be high in order to obtain a good SNR, since the Raman signal is intrinsically weak. These two aspects require a lens or objective with a large numerical aperture (NA). In addition, the polarization direction and incident angle have strong effects on near-field enhancement, and these should be tuned carefully. The polarization direction is tuned by a λ/2 plate, which is placed just before the laser and adjusted to be perpendicular to the beam. As shown by computer simulations, the near-field enhancement is not maximum when the laser beam is parallel to the tip axis. To maximize the near-field enhancement, two approaches can be used: the first approach is to set the tip oblique to the sample and use backscattering geometry; the second is to use a perpendicular tip and an oblique incident laser.

To observe the near-field enhancement, an experiment on single crystal silicon was performed. For this, the single crystal silicon was placed on the piezo scanning stage of the NSOM system, and the silicon Raman peak at $520\,cm^{-1}$ recorded for two tip positions: (i) with the tip "touching" the silicon surface to record the Raman spectrum, which includes both the near-filed and far-field components; and (ii) with the tip lifted from the surface so as to record the far-field Raman spectrum. These are shown as spectra in Figure 21.6a and b, respectively. The near-field component is approximately 50% of the far-field signal.

It should be noted that the near-field Raman signal comes from a small region on the sample, while the far-field signal comes from a much larger volume (about 2.5 μm in diameter for our instrumental setup and the laser penetration of Si is about 0.5 μm). Assuming that the diameter of the metal tip is several tens nanometers, say 70 nm, and the near-field enhancement occurs at a depth

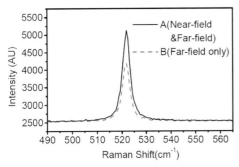

Figure 21.6 The silicon Raman spectra with the tip "touching" the sample surface (spectrum A) and the tip withdrawn (spectrum B).

of 20 nm of the Si sample. The far-field signal derives from a volume of $V_{far} = \pi\left(\dfrac{2500}{2}\right)^2 \times 500 \text{ nm}^3$, while the near-field signal is derived from a much smaller volume $V_{near} = \pi\left(\dfrac{70}{2}\right)^2 \times 20 \text{ nm}^3$. The enhancement factor can be estimated as $\dfrac{I_{near}}{I_{far}} \times \dfrac{V_{far}}{V_{near}}$, where $I_{near} = 0.35I$ and $I_{far} = (1 - 0.35)I$ are the measured near- and far-field Raman intensities, respectively. Such a simple estimation shows that the near-field enhancement factor is more than 10^4. In another experiment, the improvement in spatial resolution was demonstrated by using a featured sample: a silicon device, which consists of SiO_2 lines of 380 nm width, with 300 nm separations. The SiO_2 lines were 30 nm higher than the silicon substrate (Figure 21.7a). By comparing the width of the SiO_2 strips in the AFM profile and the real width (380 nm), the topographic spatial resolution of the set-up was derived as 34 nm (Figure 21.7b). Raman mapping (128×20 points) was performed on a strip, and the Raman image constructed using the Raman peak intensity, as shown in Figure 21.7c. The Raman spectra were collected with a 1 s integration time. In this experiment, the Raman spectrum contained both far-field and near-field components when the tip was on a Si-single crystal, but had only the far-field component when the tip was on SiO_2, where the SiO_2 prevented the tip from coming close enough to the Si to excite its near-field Raman signal. Hence, the intensity variation of the Si Raman peak represented the near-field contribution, while the far-field contributed a constant background. The Raman intensity image shown in Figure 21.7c has a clear correlation with the Si device, as might be expected. This was the first two-dimensional (2-D) near-field Raman image of a real nanometer-sized device. The 1 s integration time required to record a good quality Raman spectrum is short enough to make NSRM a useful technique for imaging purposes. It should be noted that the SiO_2 lines in the NSRM image are not straight, due to hysteresis and creep of the piezoelectric scanner, which is

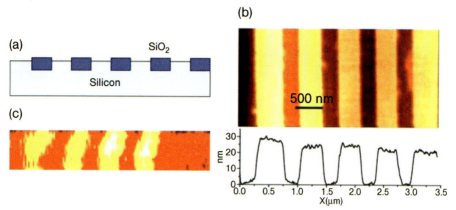

Figure 21.7 (a) Schematic cross-sectional diagram of the Si device sample; (b) Atomic force microscopy image of the Si device sample; (c) Raman intensity image of the Si device sample.

calibrated and linearized in normal AFM scanning. In Raman mapping, however, the scanning rate is much lower than that in normal AFM scanning, as the scanner must remain at each point for several seconds while the spectrometer records the Raman spectrum. Thus, the creep will become more obvious during the mapping process.

21.5
Near-Field Raman Imaging Using Optically Trapped Dielectric Microsphere

Recently, a new approach to near-field imaging has been developed, whereby the laser is focused to a spot size which is smaller than diffraction limit, by using a dielectric microsphere. In addition to being used as the excitation source for Raman spectroscopy, the incident laser beam (linearly polarized Gaussian TEM00 mode) is also used to trap the microsphere just above the sample surface, through the well-known "optical tweezers" mechanism [34, 35]. Simulation studies have shown that subdiffraction-limited focusing can be achieved when the diameter of the dielectric microsphere is comparable to the wavelength of laser [36, 37]. In the experiments, the sample and polystyrene microspheres (3 μm diameter) in solution were placed in a sample cell filled with deionized water. One microsphere was trapped at the center of the laser beam, and was in contact with the surface of the sample during scanning. The schematic diagram of the experimental set-up is shown in Figure 21.3a. The near-field Raman microscopy set-up with polystyrene microsphere is based on the WITec CRM200 confocal Raman microscopy system

(25 μm pinhole) with Olympus microscope objective (60× NA = 1.2 water immersion). A double-frequency Nd:YAG laser (532 nm, 100 mW; CNI Laser) is used as the excitation laser, and coupled into a 3.5 μm-core diameter single-mode fiber. The linearly polarized Gaussian beam (TEM00) that is used to excite the Raman signal is also used to trap the microsphere. The laser beam is incident on the sample through the microsphere. The sample is placed in a sample cell with diluted polystyrene microspheres in deionized water; the sample cell is then placed on a translation stage, which can be moved coarsely along the *x*- and *y*-axes (it can also be finely moved by using a piezostage). The Raman scattered light was directed to a 1800 grooves mm^{-1} grating, and detected using a TE-cooled CCD. Scanning electron microscopy images were recorded using field emission SEM (JEOL JSM-6700F).

This technique has many advantages over the previous near-field techniques. The Raman signal collected with microsphere using the present technique is always much stronger than that without microsphere, typically two- to fivefold depending on the objective lens used. This is a critical advantage over the aperture near-field technique, which is orders of magnitude weaker than the corresponding far-field signal. As the laser light is focused on the sample through the microsphere, there is no background far-field signal in this experimental configuration, which has been one of the major problems in TERS. There is also no requirement to use a metal or metal-coated probe (e.g., a metal-coated AFM tip) to perform the experiment. The strong near-field Raman signal, and the simplicity in performing the experiment, make this technique attractive, easy, and fast. The reproducibility is also excellent, at near-100% level. Moreover, the technique can also be used on different types of sample.

The laser spot size of the near-field Raman technique was determined using a scanning-edge method [38]. For this, the trapped microsphere was scanned across a Si/SiO$_2$ structure and the scanning spectrum (Figure 21.3c) fitted with Equation 21.1:

$$I(x) = \frac{P}{2}\left\{1 + erf\left(\frac{\sqrt{2}\,(x - x_o)}{w}\right)\right\} \tag{21.1}$$

where *P* is the total power contained in the laser beam, *x* is the position of the scanning edge, x_0 is the center of the beam, and *w* is the $1/e^2$ half-width. The spot size of the beam, the full width at half maximum (FWHM), was calculated to be ~ 80 nm, from the following relationship: FWHM = $\sqrt{2\ln 2}\,w$.

The system was used to study the SiGe/Si device sample with poly-Si gate and SiGe stressors, as shown in the electron micrograph in Figure 21.8. The patterned wafers used in this study were prepared using 65 nm device technology. After spacer formation and Si recess etching, the wafers were cleaned and the epitaxial SiGe growth was performed using a commercially available low-pressure chemical vapor deposition (LPCVD) system. The capability of the technique was also demonstrated by studying the strain on the channel below the poly-Si gates, which is compressively strained by the SiGe stressors.

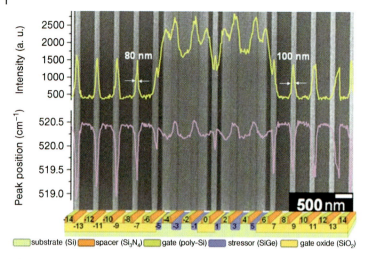

Figure 21.8 Scanning electron microscopy image with a cross-sectional diagram of the sample. The line scan of Raman Si intensity is shown in yellow, and the peak position in purple. The scanning step size was 20 nm with a 1 s integration time. The line width measured from the Si intensity profile was between 80 and 100 nm.

Straining the silicon can suppress the inter-valley scattering and reduce the effective carrier mass; the result is an improvement of the effective carrier mobility in the Si channel. In the past, the semiconductor industry has used mechanical strain as an alternative to physical scaling to improve transistor performance [39]. An appropriate strain, when applied to the channel region, can lead to a significant improvement in transistor performance. However, in the complementary metal-oxide-semiconductor (CMOS) transistor, the negative (n)-MOS and positive (p)-MOS must be strained differently. A compressive strain is known to be beneficial for p-MOS, while a tensile strain is known to improve n-MOS performance [40, 41]. It is for this reason that a technique to characterize strain with sub-100 nm resolution, on a reliable basis, is in such high demand.

Micro-Raman spectroscopy has long been a popular tool for strain measurements, because it is both nondestructive and quantitative [42–44]. Compressive strain shifts the Raman peak to higher frequency, while the tensile strain results in a red shift [45]. However, the spatial resolution of micro-Raman makes it impossible to be used for strain characterization in sub-100 nm semiconductor devices. Converging beam electron diffraction (CBED) in TEM can be used to characterize the strain with a nanometer-scale resolution [46, 47]. However, destructive and complicated sample preparations (which may alter the original strain field) have made this technique undesirable for large-scale strain characterization. Consequently, whilst the development of a reliable, nondestructive, quantitative assessment of strain on the nanometer-scale is critical, there is at present no

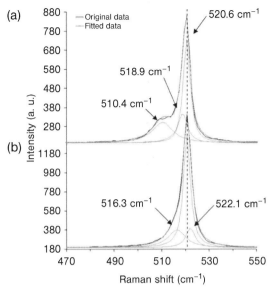

Figure 21.9 Raman spectra from SiGe and poly-Si lines with fitted peaks using Lorentzian function. (a) The Raman peaks correspond to Si–Si phonon vibrations from the SiGe ($\sim$510 cm^{-1}), tensile-strained Si just below the SiGe ($\sim$519 cm^{-1}), and the Si substrate below ($\sim$520 cm^{-1}), respectively; (b) The Raman peaks correspond to Si–Si phonon vibrations poly-Si and bulk Si below ($\sim$516 cm^{-1} and $\sim$520 cm^{-1}), and compressively strained Si in the channel region ($\sim$522 cm^{-1}).

such characterization technique available on the market. Here, we show the strain measurement on the 65 nm device lines with a much improved repeatability and SNR.

Figure 21.8 shows the SEM image as the background and the schematic cross-section diagram of the sample. This figure also shows the Raman line-scan Si intensity (yellow) and peak position (purple) profiles across the sample, with a scanning step size of 20 nm and a 1 s integration time. Both lines show an excellent correlation with the structure. Higher values of the intensity line scan correspond to positions of the poly-Si lines. It is important to note that the poly-Si lines of 45 nm width and 400 nm separations can easily be distinguished. The line-width measured from the FWHM of the Si-peak intensity is between 80 and 100 nm. The Si–Si peak position in regions below the poly-Si lines (purple line) is at a higher frequency, reflecting the fact that Si is under compressive strain exerted by the SiGe stressors. In other regions with SiGe lines (yellow line), the Si–Si peak position is at a lower frequency than that of the Si substrate, as would be expected for unstrained poly-Si lines. The difference between the center poly-Si line (marked 0) and its neighboring lines is also apparent, showing the high resolving power of the technique.

Figure 21.9a and b show the Raman spectra recorded in the SiGe line region and on top of the poly-Si line, respectively. Each spectrum was fitted with three

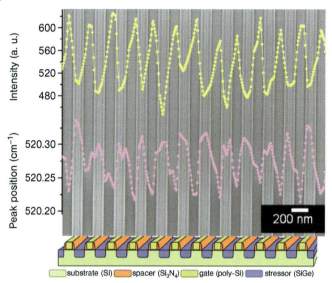

substrate (Si)   spacer (Si₃N₄)   gate (poly-Si)   stressor (SiGe)

Figure 21.10 Scanning electron microscopy image with a cross-sectional diagram of periodic poly-Si lines and SiGe stressors. The line scan of Raman Si–Si intensity from SiGe is shown in yellow, and the Si–Si peak position in purple. The line scans show excellent correspondence with the structure, and a good SNR.

Lorentzian peaks. In Figure 21.9a, the Raman peaks from SiGe line correspond to Si–Si phonon vibrations from the SiGe (~510 cm⁻¹), tensile-strained Si just below the SiGe (~519 cm⁻¹), and the Si substrate below (~520 cm⁻¹), respectively. Similarly, in Figure 21.9b, the Raman peaks correspond to the Si–Si phonon vibrations of poly-Si and bulk Si below (~516⁻¹ and ~520 cm⁻¹), and another one is from the compressively strained Si in the channel region (~522 cm⁻¹).

Figure 21.10 shows the SEM image of the device sample, together with the detailed illustration diagram. From this figure, the line profile of the intensity of Si–Si phonon vibrations from the SiGe (yellow), and the Si–Si peak position (purple), can be seen. The results show an excellent correspondence to the device structure, with a good SNR. The line profile data (intensity and peak positions) are extracted from an area of Raman mapping of $4.0 \times 1.3 \, \mu m^2$ (100 × 32 pixels), as shown in Figure 21.11, which took about 6 min to complete.

Figure 21.11a shows the Raman image from the intensity of Si–Si phonon vibrations from the SiGe of the structure shown in Figure 21.10. For comparison, Figure 21.11b shows the image from confocal Raman set-up (far-field imaging). It is clear that the present near-field technique can resolve the periodic lines with excellent repeatabiliy that far-field technique cannot resolve. The three-dimensional (3-D) Raman image in Figure 21.11c is based on the peak position of Si–Si phonon vibration from Si at ~520 cm⁻¹. From this image, it is possible to study the higher compressive strain regions, which are under the poly-Si lines and

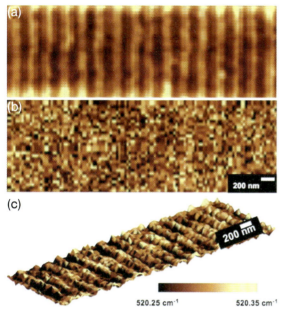

Figure 21.11 Raman images ($4.0 \times 1.3\,\mu m^2$) of the periodic poly-Si lines and SiGe stressors obtained in approximately 6 min. (a) Near-field Raman image from the Si–Si peak intensity from SiGe; (b) Confocal Raman image from the Si–Si peak intensity from SiGe; (c) 3-D near-field Raman image from the Si–Si peak position, showing the relative strain at different regions.

are compressively strained by the SiGe stressors. The fact that it took only about 6 min to carry out the mapping means that the high-resolution Raman imaging using this technique can be performed within a reasonable time.

Apart from the device sample, in order to show the capability of the technique for other types of sample, Raman mapping was also performed on CNTs and gold nanopatterns. The purified HiPCO single-walled carbon nanotubes (SWNTs) were purchased from Carbon Nanotechnologies. The HiPCO SWNTs (10 mg) were dispersed in 100 ml D_2O solution with 1 wt% sodium dodecyl sulfate (SDS). The dispersion was then treated by probe sonication (Sonics & Materials Inc.; Model VCX 130) at a power level of 250 W for 30 min, followed by ultracentrifugation at 140 000×g for 4 hours. The supernatant was drop-cast on a Si substrate and dried in air. Individual (or slightly bundled) tubes on Si were then obtained after rinsing the Si substrate with pure water. A gold nanopattern was also fabricated on a silicon substrate. For this, polystyrene microspheres (diameter 0.5 µm) in solution were dispersed on a silicon substrate, using the drop-coating method. The polystyrene microspheres self-assembled to form a monolayer, and a thin film of gold, of ~100 nm thickness, was deposited onto the Si substrate, using DC magnetron sputtering. The sample was sonicated for 1 min in chloroform to remove the

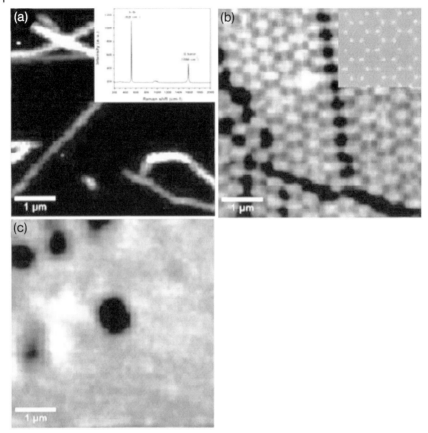

Figure 21.12 Raman images (5.0 × 5.0 μm² of the carbon nanotubes (CNTs) and gold nanopatterns. (a) Near-field Raman image from G band of CNTs; (b) Near-field Raman image from the gold nanopatterns; (c) Confocal Raman image from the gold nanopatterns. The inset in (a) is the Raman spectrum of the CNTs; the inset in (b) shows the SEM image of the gold nanopatterns, where the nanopattern size is ~100 nm.

polystyrene microspheres, and then annealed in argon ambient at 400 °C for 30 min. The size of the gold nanopattern was ~100 nm.

Figure 21.12a–c show the 5.0 × 5.0 μm² (100 × 100 pixels) Raman images of the CNTs and gold nanopattern. Figure 21.12a shows the Raman image from the G band (~1590 cm⁻¹) of CNTs; the inset shows the Raman spectrum from the CNTs on a silicon substrate. Figure 21.12b shows the Raman image from the Si–Si peak intensity, where the darker regions correspond to the gold nanopatterns. It is clear that the ~100 nm gold nanopattern can be clearly resolved using the new technique. The near-field Raman image corresponds well to the electron micrograph, as shown in the inset of Figure 21.12b. Figure 21.12c shows the confocal Raman image of the gold nanopatterns, which show no pattern details. Hence, the capability of this technique for various samples is proven.

21.6
Conclusions

Characterization tools in the nanometer regime are very important to reveal and utilize the unique properties of nanomaterials, and Raman imaging is indeed a versatile tool for studying both materials and devices. In this chapter, we have illustrated the value of Raman imaging by using three techniques:

- *Micro-Raman imaging* for the detection of Si C covalent bonding of CNTs grown on a Si substrate. The technique is simple and easy to perform, but is limited in its application to nanomaterials and devices by its limited spatial resolution.

- *Apertureless near-field Raman imaging*, where laser light is focused to the apex of a metal or metal-coated tip. This method is much superior to the aperture technique in terms of spatial resolution and SNR, but faces problems of wear and tear, oxidation, and contamination of the tip apex (e.g., by carbon). These may lead to severe adverse effects on the near-field enhancement. Far-field signals may also be problematic.

- *Microsphere near-field Raman imaging* [48], where laser is focused to a spot size smaller than the diffraction limit through a dielectric microsphere that is trapped using the optical tweezers principle. This method has many advantages over previous near-field techniques: the Raman signal is stronger; there is no background far-field signal; and the use of a metal or metal-coated tip is not required. The strong near-field Raman signal and simplicity in carrying out the experiments make this technique attractive, easy, and fast. The method's reproducibility is also excellent (close to 100%), and it can also be used on different types of sample.

References

1 Pohl, D.W., Denk, W. and Lanz, M. (1984) Optical stethoscopy: image recording with resolution λ/20. *Appl. Phys. Lett.*, **44**, 651–653.

2 Frey, H.G., Bolwien, C., Brandenburg, A., Ros, R. and Aselmetti, D. (2006) Optimized apertureless optical near-field probes with 15 nm optical resolution. *Nanotechnology*, **17**, 3105–3110.

3 Kim, J.H. and Song, K.B. (2007) Recent progress of nano-technology with NSOM. *Micron*, **38**, 409–426.

4 Tsai, D.P., Othonos, A., Moskovits, M. and Uttamchandani, D. (1994) Raman spectroscopy using a fiber optic probe with subwavelength aperture. *Appl. Phys. Lett.*, **64**, 1768–1770.

5 Grausem, J., Humbert, B., Spajer, M., Courjon, D., Burneau, A. and Oswalt, J. (1999) Near-field Raman spectroscopy. *J. Raman Spectrosc.*, **30**, 833–840.

6 Hecht, B., Sick, B., Wild, U.P., Deckert, V., Zenobi, R., Martin, O.J.F. and Pohl, D.W. (2000) Scanning near-field optical microscopy with aperture probes: fundamentals and applications. *J. Chem. Phys.*, **112**, 7761–7764.

7 Webster, S., Batchelder, D.N. and Smith, D.A. (1998) Submicron resolution measurement of stress in silicon by near-field Raman spectroscopy. *Appl. Phys. Lett.*, **72**, 1478–1480.

8 Sun, W.X. and Shen, Z.X. (2003) Near-field scanning Raman microscopy using apertureless probes. *J. Raman Spectrosc.*, **34**, 668–676.

9 Anderson, N., Hartschuh, A. and Novotny, L. (2005) Near-field Raman microscopy. *Mater. Today*, **May**, 50–54.

10 Saito, Y., Motohashi, M., Hayazawa, N., Iyoki, M. and Kawata, S. (2006) Nanoscale characterization of strained silicon by tip-enhanced Raman spectroscope in reflection mode. *Appl. Phys. Lett.*, **88**, 143109.

11 Lee, N., Hartschuh, R.D., Mehtani, D., Kisliuk, A., Maguire, J.F., Green, M., Foster, M.D. and Sokolov, A.P. (2007) High contrast scanning nano-Raman spectroscopy of silicon. *J. Raman Spectrosc.*, **38**, 789–796.

12 Ossikovski, R., Nguyen, Q. and Picardi, G. (2007) Simple model for the polarization effects in tip-enhanced Raman spectroscopy. *Phys. Rev. B*, **75**, 045412.

13 Iijima, S. (1991) Helical microtubules of graphitic carbon. *Nature*, **354**, 56–58.

14 Tans, S.J., Devoret, M.H., Dai, H.J., Thess, A., Smalley, R.E., Geerligs, L.J. and Dekker, C. (1997) Individual single-wall carbon nanotubes as quantum wires. *Nature*, **386**, 474–477.

15 Overney, G., Zhong, W. and Tomanek, D. (1993) Structural rigidity and low-frequecny vibrational modes of long carbon tubules. *Z. Phys. D*, **27**, 93–96.

16 Saito, R., Fujita, M., Dresselhaus, G. and Dresselhaus, M.S. (1992) Electronic structure of chiral graphene tubules. *Appl. Phys. Lett.*, **60**, 2204–2206.

17 Albrecht, P.M., Barraza-Lopez, S. and Lyding, J.W. (2007) Scanning tunneling spectroscopy and *ab initio* calculations of single-walled carbon nanotubes interfaced with highly doped hydrogen-passivated Si(100) substrates. *Nanotechnology*, **18**, 095204.

18 Albrecht, P.M. and Lyding, J.W. (2007) Local stabilization of single-walled carbon nanotubes on Si(100)-2 × 1:H via nanoscale hydrogen desorption with an ultrahigh vacuum scanning tunneling microscope. *Nanotechnology*, **18**, 125302.

19 Miwa, R.H., Orellana, W. and Fazzio, A. (2005) Substrate-dependent electronic properties of an armchair carbon

nanotube adsorbed on H/Si(001). *Appl. Phys. Lett.*, **86**, 213111.

20 Miwa, R.H., Orellana, W. and Fazzio, A. (2005) Carbon nanotube adsorbed on hydrogenated Si(001) surfaces. *Appl. Surf. Sci.*, **244**, 124–128.

21 de Brito Mota, F. and de Castilho, C.M.C. (2006) Carbon nanotube adsorbed on a hydrogenated Si-rich β-SiC(100) (3 × 2) surface: First-principles pseudopotential calculations. *Phys. Rev. B*, **74**, 165408.

22 Peng, G.W., Huan, A.C.H., Liu, L. and Feng, Y.P. (2006) Structural and electronic properties of 4A carbon nanotubes on Si(001) surfaces. *Phys. Rev. B*, **74**, 235416.

23 Wang, Y., Voronin, G.A., Zerda, T.W. and Winiarski, A. (2005) SiC-CNT nanocomposites: high pressure reaction synthesis and characterization. *J. Phys.: Condens. Matter*, **18**, 275–282.

24 Mews, A., Koberling, F., Basche, T., Philipp, G., Duesberg, G.S., Roth, S. and Burghard, M. (2000) Raman imaging of single carbon nanotubes. *Adv. Mater.*, **12**, 1210–1214.

25 Hiura, H., Ebbesen, T.W., Tanigaki, K. and Takahashi, H. (1993) Raman studies of carbon nanotubes. *Chem. Phys. Lett.*, **202**, 509–512.

26 Nakashima, S., Katahama, H., Nakakura, Y. and Mitsuishi, A. (1986) Relative Raman intensities of the folded modes in SiC polytypes. *Phys. Rev. B*, **33**, 5721–5729.

27 Olego, D. and Cardona, M. (1982) Pressure dependence of Raman phonons of Ge and 3C-SiC. *Phys. Rev. B*, **25**, 1151–1161.

28 Nakashima, S. and Harima, H. (1997) Raman investigation of SiC polytypes. *Phys. Status Solidi A*, **162**, 39–64.

29 Burton, J.C., Sun, L., Long, F.H., Feng, Z.C. and Ferguson, I.T. (1999) First-and second-order Raman scattering from semi-insulating 4H-SiC. *Phys. Rev. B*, **59**, 7282–7284.

30 Windl, W., Karch, K., Pavone, P., Schutt, O., Strauch, D., Weber, W.H., Hass, K.C. and Rimai, L. (1994) Second-order Raman spectra of SiC: experimental and theoretical results from *ab initio* phonon calculations. *Phys. Rev. B*, **49**, 8764–8767.

31 Sasaki, Y., Nishina, Y., Sato, M. and Okamura, K. (1989) Optical-phonon states of SiC small particles studied by Raman scattering and infrared absorption. *Phys. Rev. B*, **40**, 1762–1772.

32 Sun, W.X. and Shen, Z.X. (2003) Apertureless near-field scanning Raman microscopy using reflection scattering geometry. *Ultramicroscopy*, **94**, 237–244.

33 Sun, W.X., Shen, Z.X., Cheong, F.C., Yu, G.Y., Lim, K.Y. and Lin, J.Y. (2002) Preparations of cantilevered W tips for atomic force microscopy and apertureless near-field scanning optical microscopy. *Rev. Sci. Instrum.*, **73**, 2942–2947.

34 Ashkin, A. (1980) Applications of laser radiation pressure. *Science*, **210**, 1081–1088.

35 Ashkin, A. (1997) Optical trapping and manipulation of neutral particles using lasers. *Proc. Natl Acad. Sci. USA*, **94**, 4853–4860.

36 Li, X., Chen, Z.G., Taflove, A. and Backman, V. (2005) Optical analysis of nanoparticles via enhanced backscattering facilitated by 3-D photonic nanojets. *Opt. Express*, **13**, 526–533.

37 Lecler, S., Takakura, Y. and Meyrueis, P. (2005) Properties of a three-dimensional photonic jet. *Opt. Lett.*, **30**, 2641–2643.

38 Veshapidze, G., Trachy, M.L., Shah, M.H. and DePaola, B.D. (2006) Reducing the uncertainty in laser beam size measurement with a scanning edge method. *Appl. Opt.*, **45**, 8197–8199.

39 Chidambaram, P.R., Bowen, C., Chakravarthi, S., Machala, C. and Wise, R. (2006) Fundamentals of silicon material properties for successful exploitation of strain engineering in modern CMOS manufacturing. *IEEE Trans. Electron. Dev.*, **53** (**23**), 944–964.

40 Paul, D.J. (2004) Si/SiGe heterostructures: from material and physics to devices and circuits. *Semicond. Sci. Technol.*, **19**, R75–R108.

41 Wu, S.L., Lin, Y.M., Chang, S.J., Lu, S.C., Chen, P.S. and Liu, C.W. (2006) Enhanced CMOS performances using substrate strained-SiGe and mechanical strained-Si technology. *IEEE Electron. Device Lett.*, **27**, 46–48.

42 Wolf, I.D., Maes, H.E. and Jones, S.K. (1996) Stress measurements in Si devices through Raman spectroscopy: bridging the gap between theory and experiment. *J. Appl. Phys.*, **79**, 7148–7156.

43 Nakashima, S., Yamamoto, T., Ogura, A., Uejima, K. and Yamamoto, T. (2004) Characterization of Si/Ge_xSi_{1-x} structures by micro-Raman imaging. *Appl. Phys. Lett.*, **84**, 2533–2535.

44 Mitani, T., Nakashima, S., Okumura, H. and Ogura, A. (2006) Depth profiling of strain and defects in $Si/Si_{1-x}Ge_x/Si$ heterostructures by micro-Raman imaging. *J. Appl. Phys.*, **100**, 073511.

45 Anastassakis, E. and Liarokapis, E. (1987) Polycrystalline Si under strain: Elastic and lattice-dynamical considerations. *J. Appl. Phys.*, **62**, 3346–3352.

46 Senez, V., Armigliato, A., Wolf, I.D., Carnevale, G., Balboni, R., Frabboni, S. and Benedetti, A. (2003) Strain determination in silicon microstructures by combined convergent beam electron diffraction, process simulation, and micro-Raman spectroscopy. *J. Appl. Phys.*, **94**, 5574–5583.

47 Liu, H.H., Duan, X.F., Qi, X.Y., Xu, Q.X., Li, H.O. and Qian, H. (2006) Nanoscale strain analysis of strained-Si metal-oxide-semiconductor field effect transistors by large angle convergent-beam electron diffraction. *Appl. Phys. Lett.*, **88**, 263513.

48 Kasim, J., Yu, T., You, Y.M., Liu, J.P., See, A., Li, L.J. and Shen, Z.X. (2008) Near-field Raman imaging using optically trapped dielectric microsphere. *Opt. Express*, **16**, 7976.

22
Fullerene-Rich Nanostructures

Fernando Langa and Jean-François Nierengarten

22.1
Introduction

In recent years, the rapid advances in dendrimer synthetic chemistry have moved towards the creation of functional systems with increased attention to potential applications [1]. Among the large number of molecular subunits used for dendrimer chemistry, C_{60} has proven to be a versatile building block and fullerene-functionalized dendrimers – that is, *fullerodendrimers* [2] – have generated significant research activities in recent years [3, 4]. In particular, the peculiar physical properties of fullerene derivatives make fullerodendrimers attractive candidates for a variety of interesting features in supramolecular chemistry and materials science [5]. C_{60} itself is a convenient core for dendrimer chemistry [3], and the functionalization of C_{60} with a controlled number of dendrons dramatically improves the solubility of the fullerenes [6]. Furthermore, variable degrees of addition about the fullerene core are possible, and its almost spherical shape leads to globular systems, even with low-generation dendrons [7]. On the other hand, specific advantages are brought about by the encapsulation of a fullerene moiety in the middle of a dendritic structure [8]. The shielding effect resulting from the presence of the surrounding shell has been found useful for optimizing the optical limiting properties of C_{60} derivatives [9], to obtain amphiphilic derivatives with good spreading characteristics [10], or to prepare fullerene-containing liquid crystalline materials [11]. Use of the fullerene sphere as a photoactive core unit has also been reported [12]. In particular, the special photophysical properties of C_{60} have been used to evidence dendritic shielding effects [13] and to prepare dendrimer-based, light-harvesting systems [14]. Whereas, the majority of the fullerene-containing dendrimers reported to date have been prepared with a C_{60} core, dendritic structures with fullerene units at their surface, or with C_{60} spheres in the dendritic branches, have been essentially ignored. This is mainly associated with difficulties related to the synthesis of fullerene-rich molecules [4]. Indeed, the two major problems for the preparation of such dendrimers are the low solubility of C_{60} and its chemical reactivity, which limits the range of reactions that can be used for the synthesis of

Advanced Nanomaterials. Edited by Kurt E. Geckeler and Hiroyuki Nishide
Copyright © 2010 WILEY-VCH Verlag GmbH & Co. KGaA, Weinheim
ISBN: 978-3-527-31794-3

branched structures bearing multiple C$_{60}$ units. The most recent developments on the molecular engineering of covalent and noncovalent fullerene-rich dendrimers will be presented in the chapter, the aim of which is not to provide an exhaustive review on such systems, but rather to present significant examples that illustrate the current state of fullerene chemistry for the development of new nanostructures.

22.2
Fullerene-Rich Dendritic Branches

Over the past years, Nierengarten and coworkers have developed synthetic methodologies allowing for the preparation of dendrons substituted with several fullerene moieties [15]. As a typical example, the convergent preparation of dendritic branches with fullerene subunits at the periphery is depicted in Figures 22.1–22.3. The starting fullerene derivative, G1CO$_2$tBu, is easily obtained on a multi-gram scale, and is highly soluble in common organic solvents, based on the presence of two long alkyl chains [16]. The iterative reaction sequence used in the preparation of the subsequent dendrimer generations relies upon successive cleavage of a *t*-butyl ester moiety under acidic conditions, followed by a *N,N'*-dicyclohexylcarbodiimide (DCC)-mediated esterification reaction with the A$_2$B building block **1** possessing two benzylic alcohol functions and a protected carboxylic acid function [16].

Selective cleavage of the *t*-butyl ester group in G1CO$_2$tBu was achieved by treatment with an excess of CF$_3$COOH (TFA) in CH$_2$Cl$_2$ to afford G1CO$_2$H in a quantitative yield. Reaction of the diol **1** with carboxylic acid G1CO$_2$H under esterification conditions using DCC, 4-4-dimethylaminopyridine (DMAP), and 1-hydroxybenzotriazole (HOBt) in CH$_2$Cl$_2$ gave the protected dendron of second-generation G2CO$_2$tBu in 90% yield. Hydrolysis of the *t*-butyl ester moiety under acidic conditions then afforded the corresponding carboxylic acid G2CO$_2$H in a

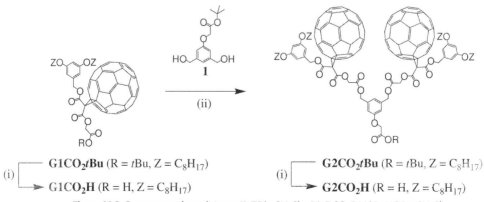

Figure 22.1 Reagents and conditions: (i) TFA, CH$_2$Cl$_2$; (ii) DCC, DMAP, HOBt, CH$_2$Cl$_2$.

G2CO₂H

(i)

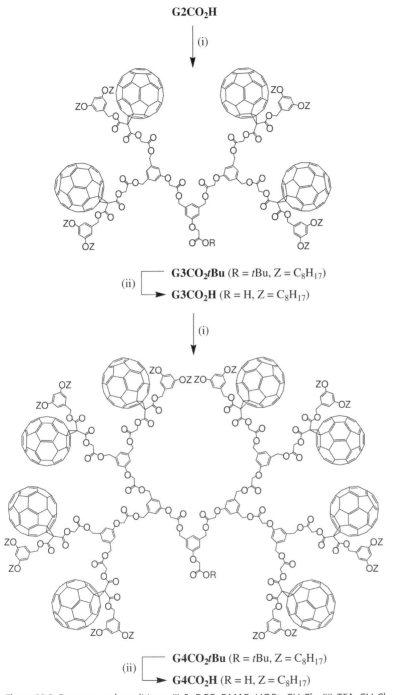

(ii) ⎡ **G3CO₂*t*Bu** ($R = t$Bu, $Z = C_8H_{17}$)
     └→ **G3CO₂H** ($R = H$, $Z = C_8H_{17}$)

(i)

(ii) ⎡ **G4CO₂*t*Bu** ($R = t$Bu, $Z = C_8H_{17}$)
     └→ **G4CO₂H** ($R = H$, $Z = C_8H_{17}$)

Figure 22.2 Reagents and conditions: (i) **1**, DCC, DMAP, HOBt, CH₂Cl₂; (ii) TFA, CH₂Cl₂.

G4CO₂H

(i)

(ii) ⎡ **G5CO₂tBu** (R = *t*Bu, Z = C₈H₁₇)

└─► **G5CO₂H** (R = H, Z = C₈H₁₇)

Figure 22.3 Reagents and conditions: (i) **1**, DCC, DMAP, HOBt, CH₂Cl₂; (ii) TFA, CH₂Cl₂.

quantitative yield. Esterification of G2CO₂H with diol **1** (DCC, HOBt, DMAP) afforded the *t*-butyl-protected fullerodendron G3CO₂*t*Bu in 87% yield (Figure 22.2). Selective hydrolysis of the *t*-butyl ester under acidic conditions afforded acid G3CO₂H in 99% yield. Subsequent reaction of G3CO₂H with the branching unit

1 in the presence of DCC, HOBt and DMAP afforded fullerodendron G4CO$_2$*t*Bu (95%), which after treatment with CF$_3$CO$_2$H gave G4CO$_2$H (97%).

By repeating the same reaction sequence from G4CO$_2$H, the fifth generation derivatives G5CO$_2$*t*Bu and G5CO$_2$H were also prepared (Figure 22.3). Compounds G1-5CO$_2$*t*Bu and G1-5CO$_2$H are highly soluble in common organic solvents such as CH$_2$Cl$_2$, CHCl$_3$, or tetrahydrofuran (THF), and complete spectroscopic characterization was easily achieved.

The unequivocal characterization of these compounds also required their mass spectrometric analysis; this is quite difficult, as no structural features allow for easy protonation of the molecules. Several ionization techniques such as matrix-assisted laser desorption/ionization (MALDI), or fast-atom bombardment (FAB) have been applied for the characterization of fullerodendritic species [17]. These tools are, however, not always well adapted for C$_{60}$ derivatives of high molecular weight, such as G1-5CO$_2$*t*Bu and G1-5CO$_2$H, as they lead to high levels of fragmentation. In contrast, electrospray mass spectrometry (ES-MS) has the ability to desolvate ions that preexist in solution, thus offering the possibility of transferring such ions into the gas phase without notable fragmentation [18]. Indeed, supramolecular cationic dendritic structures resulting from the self-assembly of fullerodendrons possessing an ammonium function at the focal point and bis-crown-ether receptors have been successfully characterized using ES-MS [19]. Unfortunately, fullerodendrons G1-5CO$_2$*t*Bu and G1-5CO$_2$H are uncharged in solution, and therefore are not suitable for ES-MS analysis, at least in principle. However, it has been shown that an *in situ* reduction of fullerenes in the injection capillary of the electrospray mass spectrometer offers an interesting possibility to analyze neutral C$_{60}$ derivatives [17]. In other words, fullerene radical anions can be generated during the electrospray process, owing to the ability of the electrospray source to behave like an electrolysis cell [17]. Indeed, fullerodendrons G1-5CO$_2$*t*Bu and G1-5CO$_2$H could be characterized by applying this technique. The ES mass spectra of the second-generation derivative G2CO$_2$*t*Bu is depicted in Figure 22.4. The mass spectrum obtained under mild conditions ($V_c = 150$ V) is dominated by the doubly charged ion peak at *m/z* 1343.2, which can be assigned to the radical di-anion of G2CO$_2$*t*Bu (calculated *m/z* 1343.43). The experimental isotopic pattern of this ion is in perfect agreement with that calculated for the doubly reduced compound. A singly charged ion is also observed as minor signal at *m/z* 2686.3, and corresponds to the singly reduced G2CO$_2$*t*Bu-structure. Finally, the minor peak observed at *m/z* 1226.1 is ascribed to the fragment G1CO$_2^-$ resulting from the cleavage of a benzylic ester unit in G2CO$_2$*t*Bu. When the ES spectrum of G2CO$_2$*t*Bu was recorded under more harsher conditions ($V_c = 300$ V), the characteristic peaks corresponding to the radical mono- and di-anions were still observed, but an additional signal could also be detected at *m/z* 1124. This characteristic fragment was the same as that observed in the spectra of G1CO$_2$*t*Bu.

The analysis of the higher-generation compounds has been more difficult. Effectively, the response factor of the anions is reduced as their molecular weight is increased. In addition, the number of ester functions is higher, thus leading to more fragmentation. The resulting fragments having a lower molecular weight – and

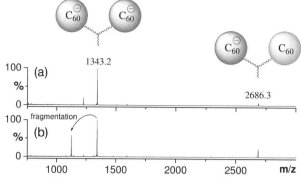

Figure 22.4 ES-MS spectra of G2CO$_2$tBu recorded with V_c values of 150 (a) and 300 V (b).

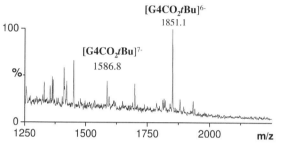

Figure 22.5 ES-MS spectrum of G4CO$_2$tBu recorded with a V_c value of 150 V.

therefore a higher response factor – gave signals with significant intensities in all the spectra, regardless of the V_c value. However, peaks corresponding to nonfragmented ions could be clearly observed. As shown in Figure 22.5, two characteristic molecular ion peaks are observed at m/z 1851.1 and 1586.8 in the ES-MS spectrum of G4CO$_2$tBu. These correspond to the hexa- and hepta-anions of the fullerodendron, in which six or seven C$_{60}$ units are reduced, respectively. Several fragment ions are also present in both cases, the low (100 V) and high (300 V) voltages. Importantly, their intensity relative to the molecular ion peaks is clearly increased when the V_c value was raised, thus showing that these additional signals result from a fragmentation of the dendrimer. Within this series of compounds, a general observation is a decrease in the absolute intensity of the signals arising from the fullerodendrons when the generation number is increased. As mentioned above, this is most likely due to the increased number of labile ester functions giving rise to more fragmentations, and to a decrease in the response factor when the molecular weight is increased.

22.3
Photoelectrochemical Properties of Fullerodendrons and Their Nanoclusters

The absorption spectra of G1-5CO$_2$tBu in a toluene/acetonitrile (1/6, v/v) mixed solvent exhibit structureless broad absorption in the range of 300 to 800 nm. These

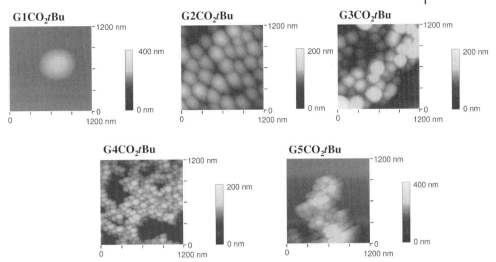

Figure 22.6 Atomic force microscopy images of (G1-5CO$_2$tBu)$_m$ on mica.

results suggest that such compounds aggregate and form large clusters in the mixed solvents [20]. The particle size of these clusters in the mixed solvent was measured using dynamic light scattering (DLS). In a mixture (1/6) of toluene/ acetonitrile at an incubation period of 15 min after injection of the toluene solution into acetonitrile, the size distribution of G1-5CO$_2$tBu was found to be relatively narrow, with different mean diameters of 790 nm for (G1CO$_2$tBu)$_m$, 210 nm for (G2CO$_2$tBu)$_m$, 170 nm for (G3CO$_2$tBu)$_m$, 100 nm for (G4CO$_2$tBu)$_m$, and 90 nm for (G5CO$_2$tBu)$_m$. The order of the mean diameters was not consistent with that of their molecular sizes. Both, dendrimers and dendrons are known to become compact, rigid structured as the generation number is increased [20]. This suggests that, in the process of cluster formation with the higher dendrimer generation, each dendritic branch is subject to interactions with branches belonging to the same dendrimer molecule (*intra*molecular), rather than to other molecules (*inter*molecular). This results in the formation of densely packed dendrimer clusters with a small, compact size (i.e., 90–100 nm). In other words, in the lower dendrimer generations, *inter*molecular interactions among branches prevail, leading to the formation of poorly packed dendrimer clusters with a large size. In order to assess the shape and morphology of (G1-5CO$_2$tBu)$_m$ clusters, atomic force microscopy (AFM) measurements have been carried out. The AFM images of the clusters prepared by spin-coating of the (G1-5CO$_2$tBu)$_m$ cluster solutions on mica surface are depicted in Figure 22.6. To remove solvent, the resulting substrates were heated under reduced pressure. The (G1-5CO$_2$tBu)$_m$ clusters were spherical, with an average horizontal diameter of 900 nm for (G1CO$_2$tBu)$_m$, 200 nm for (G2CO$_2$tBu)$_m$, 200 nm for (G3CO$_2$tBu)$_m$, 100 nm for (G4CO$_2$tBu)$_m$, and 100 nm for (G5CO$_2$tBu)$_m$. The size of the clusters agreed well with the values obtained from the DLS measurements. The vertical size of the clusters on mica correlated largely with the horizontal size of the clusters, except in the case of (G1CO$_2$tBu)$_m$. The

AFM image of the $(G1CO_2tBu)_m$ cluster revealed a rather disc-like structure with an average diameter of 900 nm and an average maximum thickness of 170 nm. Namely, the vertical size of the $(G1CO_2tBu)_m$ cluster is much smaller than the horizontal size. Although detailed structure at the molecular level is not yet clear, self-organization of $G1CO_2tBu$ in the mixed solvent would lead to the formation of multilayer vesicle. In such a case, solvent evaporation from the inner space of the multilayer vesicle may yield the disk-like structure. Similar proposed structures were reported for amphiphilic fullerene derivatives in water, mixed organic solvents, and cast films [21].

In order to determine if differences observed in the size of the nanoclusters obtained from $G1-5CO_2tBu$ could have any influence on their macroscopic properties, photoelectrochemical cells have been prepared. For this, the clusters were deposited electrophoretically onto ITO/SnO_2 electrodes by applying a DC voltage to the electrode [20]. After the electrophoretic deposition, the ITO/SnO_2 electrode turned brown in color, whereas discoloration of the cluster solution took place. As a result, cluster films with a thickness of 4–5 μm could be obtained. AFM was used to evaluate the surface morphology of the films deposited on the electrodes. The $ITO/SnO_2/(G1-5CO_2tBu)_m$ films were composed of closely packed clusters with a size of 800 nm ($G1CO_2tBu$), 200 nm ($G2CO_2tBu$), 200 nm ($G3CO_2tBu$), 100 nm ($G4CO_2tBu$), and 100 nm ($G5CO_2tBu$). The size of the clusters largely agreed with the diameters of the clusters on the mica obtained from the cluster solutions. These results also confirmed that fullerene dendrimer clusters could be successfully transferred onto the nanostructured SnO_2 electrodes. Photoelectrochemical measurements were performed in deaerated acetonitrile containing 0.5 M LiI and 0.01 M I_2 with $ITO/SnO_2/(G1-5CO_2tBu)_m$ as a working electrode, a platinum wire as a counterelectrode, and an I^-/I^{3-} reference electrode. The photocurrent responses were prompt, steady, and reproducible during repeated on/off cycles of visible light illumination. Blank experiments conducted with a bare ITO/SnO_2 electrode yielded no detectable photocurrent under similar experimental conditions. A series of photocurrent action spectra was recorded to evaluate the response of fullerodendron clusters towards photocurrent generation. The photocurrent action spectra of $ITO/SnO_2/(G1-5CO_2tBu)_m$ devices are shown in Figure 22.7. Incident photon-to-photocurrent efficiency (IPCE) was calculated by normalizing the photocurrent density for incident light energy and intensity using the expression:

$$IPCE\ (\%) = 100 \times 1240 \times i/(W_{in} \times \lambda)$$

where i is the photocurrent density (A cm^{-2}), W_{in} is the incident light intensity (W cm^{-2}), and λ is the excitation wavelength (nm). The action spectra of $ITO/SnO_2/(G1-5CO_2tBu)_m$ devices largely agreed with the absorption spectra on ITO/SnO_2, supporting the involvement of the C_{60} moieties for photocurrent generation. These results were consistent with the photoelectrochemical properties of the clusters of fullerene derivatives on SnO_2 electrodes [21]. When the IPCE values of $ITO/SnO_2/(G1-5CO_2tBu)_m$ devices were compared under the same conditions, the IPCE value

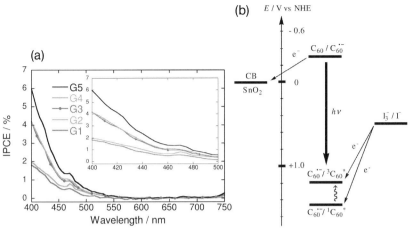

Figure 22.7 (a) Photocurrent action spectra of ITO/SnO₂/
(G1-5CO₂tBu)ₘ. Applied potential = +0.11 V versus SCE; 0.5 M
LiI and 0.01 M I₂ in acetonitrile; (b) Photocurrent generation
diagram.

at an excitation wavelength of 400 nm were seen to increase with increasing the generation number: 1.7% for ITO/SnO₂/(G1CO₂tBu)ₘ, 1.9% for ITO/SnO₂/(G2CO₂tBu)ₘ, 4.1% for ITO/SnO₂/(G3CO₂tBu)ₘ, 4.1% for ITO/SnO₂/(G4CO₂tBu)ₘ, and 6.0% for ITO/SnO₂/(G5CO₂tBu)ₘ devices.

Based on previous studies on a similar photoelectrochemical system of C_{60} and C_{60} derivatives [22], photocurrent generation diagram is illustrated in Figure 22.7. The primary step in the photocurrent generation is initiated by photoinduced electron transfer from I^- (I^{3-}/I^-, 0.5 V versus NHE) in the electrolyte solution to the excited states of fullerodendrimer clusters ($C_{60}{}^{\cdot-}/{}^1C_{60}{}^*$ = 1.45 V versus NHE, $C_{60}{}^{\cdot-}/{}^3C_{60}{}^*$ = 1.2 V versus NHE) [20]. The electron transfer rate is controlled by the diffusion of I^- (~109 s⁻¹) in the electrolyte solution. The resulting reduced C_{60} ($C_{60}{}^{\cdot-}/C_{60}$ = −0.3 V versus NHE) injects an electron directly into the SnO₂ nanocrystallites (ECB = 0 V versus NHE), or the electron is injected into the SnO₂ nanocrystallites through an electron-hopping process between the C_{60} molecules [20]. The electron transferred to the semiconductor is driven to the counterelectrode via an external circuit so as to regenerate the redox couple. It is noteworthy that the IPCE values are dependent on the dendritic generation. With increasing the dendritic generation the IPCE value increases. The structural investigation on the fullerodendrimers revealed that the higher dendrimer generation led to the formation of densely packed clusters with a smaller, compact size (*vide supra*). Such structures of fullerene dendrimer clusters on ITO/SnO₂ in the higher generation would make it possible to accelerate the electron injection process from the reduced C_{60} to the conduction band of SnO₂ via the more efficient electron hopping through the C_{60} moieties, where the average distance between the C_{60} moieties is smaller.

22.4
Fullerene-Rich Dendrimers

The results described in the previous section represent a powerful driving force to develop a new, efficient synthetic strategy for the preparation of large fullerene-rich dendritic molecules. In this respect, the self-assembly of dendrons using noncovalent interactions is particularly well suited to the preparation of fullerene-rich macromolecules [23, 24]. Indeed, the synthesis itself is restricted to the preparation of dendrons, and self-aggregation leads to the dendritic superstructure, thus avoiding tedious final synthetic steps with precursors incorporating potentially reactive functional groups such as C_{60}. For example, Nierengarten and coworkers have shown that C_{60} derivatives bearing a carboxylic acid function undergo self-assembly with *n*-butylstannonic acid (*n*BuSn(O)OH) to produce fullerene-rich nanostructures with a stannoxane core in near-quantitative yields [25]. The reaction conditions for the self-assembly of the stannoxane derivatives were first adjusted with model compound **2** (Figure 22.8). Under optimized conditions, a mixture of **2** (1 equiv.) and *n*BuSn(O)OH (1 equiv.) in benzene was refluxed for 12 h using a Dean–Stark trap. After cooling, the solution was filtered and evaporated to dryness to afford the hexameric organostannoxane derivative **3** in 99% yield. The drum-like structure of this compound, composed of a prismatic Sn_6O_6 core, was confirmed by its ^{119}Sn nuclear magnetic resonance (NMR) spectrum recorded in C_6D_6, which showed a single resonance at −479.1 ppm. This chemical shift is characteristic of a drum-shaped structure with six equivalent Sn atoms coordinated by five oxygens and one carbon [26]. Crystals suitable for X-ray crystal-structure analysis were obtained by the slow diffusion of Et_2O into a C_6H_6 solution of **3**. Despite the disorder resulting from one of the flexible butyl chains, the central Sn_6O_6 stannoxane core and the six peripheral 2-phenoxyacetate units of the structure were well resolved. As shown in Figure 22.8, the prismatic tin cage is formed by two, six-membered $(SnO)_3$ rings joined together by six Sn–O bonds [27]. The side faces of the cluster thus comprise six, four-membered $(SnO)_2$ rings, each of which is spanned by a carboxylate group that forms a bridge between two Sn atoms. A detailed observation of the stannoxane framework revealed that the six-membered rings had a chair-like conformation, with each Sn atom being bonded to three framework oxygen atoms, while the Sn–O bonds all had comparable lengths ranging from 208 to 210 pm. The six Sn atoms were seen to be hexacoordinated, with the coordination sphere being completed by a *n*-butyl group and two oxygen atoms from two different carboxylate groups. The Sn–O bonds to the bridging carboxylate atoms were longer than the core bonds, and ranged from 215 to 218 pm.

The reaction conditions used to prepare **3** from 2-phenoxyacetic acid were applied to the fullerene building blocks G1-3CO₂H. The organostannoxane derivatives **4–6** were thus obtained in almost quantitative yields (Figures 22.9 and 22.10). These compounds are highly soluble in common organic solvents such as CH_2Cl_2, $CHCl_3$, C_6H_6 or toluene, and complete spectroscopic characterization was easily achieved. The ^1H and ^{13}C NMR spectra of **4–6** clearly revealed the characteristic

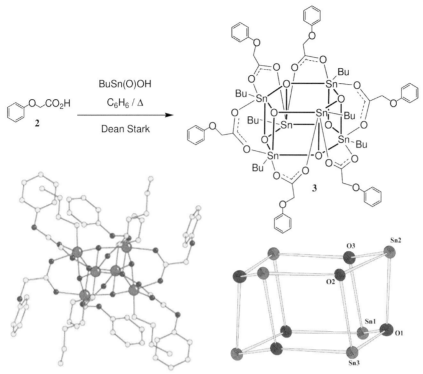

Figure 22.8 Top: Preparation of compound **3**. Bottom: X-ray crystal structure of **3** (C = pale gray, O = black, Sn = gray) and detailed view of the Sn_6O_6 core. Selected bond lengths: Sn(1)-O(1): 2.084(3) Å, Sn(2)-O(1): 2.075(4) Å, Sn(2)-O(2): 2.097(3), Sn(2)-O(3): 2.093(3) Å, Sn(3)-O(1): 2.101(3) Å.

signals of the starting carboxylic acid precursors **4–6**, as well as the expected additional resonances arising from the *n*-butyl chains. Importantly, the spectra clearly showed that all the peripheral fullerene subunits were equivalent in **4-6**, as might be expected for a sixfold symmetric assembly with a drum-shaped organostannoxane core. In addition, a single resonance was observed at approximately −480 ppm in the ^{119}Sn NMR spectra of **4–6** recorded in C_6D_6. This characteristic signature of tin-drum clusters provided definitive evidence for the formation of **4–6**.

The absorption spectra obtained from CH_2Cl_2 solutions of compounds **4–6** were identical to those recorded for the corresponding starting carboxylic acid precursors; this demonstrated an absence of any significant influence of the stannoxane core on the electronic properties of the fullerene moieties. To further confirm that the characteristics of the fullerene subunits would be maintained for **4–6**, their electrochemical properties were investigated using cyclic voltammetry (CV). For the sake of comparison, electrochemical measurements were also carried out with **3** and G1CO$_2$*t*Bu. All of these experiments were conducted at room temperature

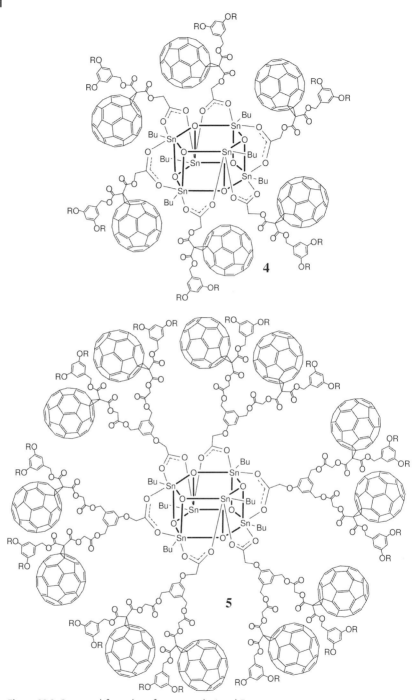

Figure 22.9 Structural formulae of compounds **4** and **5**.

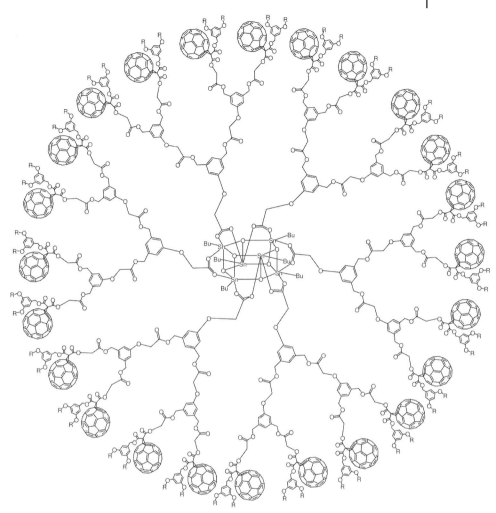

Figure 22.10 Structural formula of compound **6**.

in CH$_2$Cl$_2$ solutions containing tetra-*n*-butylammonium tetrafluoroborate (0.1 *M*) as the supporting electrolyte, with a Pt wire as the working electrode and a saturated calomel electrode (SCE) as a reference. The potential data for all of these compounds are collected in Table 22.1.

In the anodic region, all of the studied compounds presented at least one irreversible peak at approximately +1.7–1.9 V versus SCE that could likely be attributed to oxidation of the dialkyloxyphenyl and/or alkyloxyphenyl units. Whereas, model compound **3** was found to be electrochemically silent in the cathodic region, the fullerene-substituted stannoxane derivatives **4–6** revealed the typical electrochemical response of fullerene derivatives [28]. Indeed, the electrochemical behavior of **4–6** appeared to be similar to that of the model compound G1CO$_2$*t*Bu. This

Table 22.1 Electrochemical data on the reduction of **4–6** and G1CO$_2$tBu determined by cyclic voltammetry on a Pt working electrode in CH$_2$Cl$_2$ + 0.1 M nBu$_4$NBF$_4$ at room temperature. Values shown are for $(E_{pa} + E_{pc})/2$ (in Volts) versus SCE and ΔE_{pc} (in milliVolts) (in parentheses) at a scan rate of 0.1 V s^{-1}.

	E_1	E_2	E_3
G1CO$_2$tBu	−0.51 (75)	−0.89 (75)	−1.34 (65)
4	−0.53 (85)	−0.91 (80)	−1.35 (110)
5	−0.54 (75)	−0.92 (90)	−1.33 (110)
6	−0.54 (80)	−0.92 (90)	−1.34 (110)

indicated that all methanofullerene moieties in **4–6** behaved as independent redox centers, and that their electrochemical properties were not affected by the stannoxane core.

22.5
Conclusions

The preparation of covalent, fullerene-rich dendrimers is difficult and involves a significant number of synthetic steps, thus limiting their accessibility and therefore their applications. Recent obtained results on the self-assembly of fullerene-containing components, by using supramolecular interactions rather than covalent bonds, have shown clearly that this strategy represents an attractive alternative for their preparation. Indeed, fullerene-rich derivatives are thus easier to produce, and the range of systems that can be prepared is not severely limited by the synthetic route. In this way, in-depth investigations of their properties are possible, and the use of fullerene-rich materials for specific applications can be envisaged. Despite some remarkable recent achievements, it is clear that the examples discussed in this chapter represent only the first steps towards the design of fullerene-rich molecular assemblies that can display functionality at the macroscopic level. Clearly, further research is required in this area to fully explore the possibilities offered by these materials, for example, in nanotechnology or in photovoltaics.

Acknowledgments

These research studies were supported by the CNRS. The authors warmly thank all of their coworkers and collaborators for their outstanding contributions; their names are cited in the references.

References

1 (a) Newkome, G.R., Moorefield, C.N. and
Vögtle, F. (2001) *Dendrimers and
Dendrons: Concepts, Syntheses, Applications,*
Wiley-VCH Verlag GmbH, Weinheim.
(b) Newkome, G.R., Moorefield, C.N. and
Vögtle, F. (2001) *Dendrimers and Other
Dendritic Polymers* (eds J.M.J. Fréchet and
D.A. Tomalia), John Wiley & Sons, Ltd,
Chichester.

2 Nierengarten, J.-F. (2000) *Chem. Eur. J.*, **6**,
3667.

3 Hirsch, A. and Vostrowsky, O. (2001) *Top.
Curr. Chem.*, **217**, 51.

4 Nierengarten, J.-F. (2003) *Top. Curr.
Chem.*, **228**, 87.

5 (a) Nierengarten, J.-F. (2004) *New J.
Chem.*, **28**, 1177.
(b) Imahori, H. and Fukuzumi, S. (2004)
Adv. Funct. Mater., **14**, 525.
(c) Martin, N. (2006) *Chem. Commun.*,
2093.

6 (a) Wooley, K.L., Hawker, C.J., Fréchet,
J.M.J., Wudl, F., Srdanov, G., Shi, S., Li,
C. and Kao, M. (1993) *J. Am. Chem. Soc.*,
115, 9836.
(b) Hawker, C.J., Wooley, K.L. and
Fréchet, J.M.J. (1994) *J. Chem. Soc. Chem.
Commun.*, 925.
(c) Nierengarten, J.-F., Habicher, T.,
Kessinger, R., Cardullo, F., Diederich, F.,
Gramlich, V., Gisselbrecht, J.-P., Boudon,
C. and Gross, M. (1997) *Helv. Chim. Acta*,
80, 2238.
(d) Brettreich, M. and Hirsch, A. (1998)
Tetrahedron Lett., **39**, 2731.
(e) Rio, Y., Nicoud, J.-F., Rehspringer,
J.-L. and Nierengarten, J.-F. (2000)
Tetrahedron Lett., **41**, 10207.

7 (a) Camps, X. and Hirsch, A. (1997)
J. Chem. Soc. Perkin Trans. 1, 1595.
(b) Camps, X., Schönberger, H. and
Hirsch, A. (1997) *Chem. Eur. J.*, **3**, 561.
(c) Herzog, A., Hirsch, A. and
Vostrowsky, O. (2000) *Eur. J. Org. Chem.*,
171.

8 Nierengarten, J.-F. (2003) *C. R. Chimie*, **6**,
725.

9 Rio, Y., Accorsi, G., Nierengarten, H.,
Rehspringer, J.-L., Hönerlage, B.,
Kopitkovas, G., Chugreev, A., Van

Dorsselaer, A., Armaroli, N. and
Nierengarten, J.-F. (2002) *New J. Chem.*,
26, 1146.

10 (a) Cardullo, F., Diederich, F., Echegoyen,
L., Habicher, T., Jayaraman, N., Leblanc,
R.M., Stoddart, J.F. and Wang, S. (1998)
Langmuir, **14**, 1955.
(b) Zhang, S., Rio, Y., Cardinali, F.,
Bourgogne, C., Gallani, J.-L. and
Nierengarten, J.-F. (2003) *J. Org. Chem.*,
68, 9787.

11 Chuard, T. and Deschenaux, R. (2002)
J. Mater. Chem., **12**, 1944 and references
therein.

12 Nierengarten, J.-F., Armaroli, N., Accorsi,
G., Rio, Y. and Eckert, J.-F. (2003) *Chem.
Eur. J.*, **9**, 36.

13 (a) Kunieda, R., Fujitsuka, M., Ito, O., Ito,
M., Murata, Y. and Komatsu, K. (2002)
J. Phys. Chem. B, **106**, 7193.
(b) Murata, Y., Ito, M. and Komatsu, K.
(2002) *J. Mater. Chem.*, **12**, 2009.
(c) Rio, Y., Accorsi, G., Nierengarten, H.,
Bourgogne, C., Strub, J.-M., Van
Dorsselaer, A., Armaroli, N. and
Nierengarten, J.-F. (2003) *Tetrahedron*, **59**,
3833.

14 (a) Armaroli, N., Barigelletti, F., Ceroni,
P., Eckert, J.-F., Nicoud, J.-F. and
Nierengarten, J.-F. (2000) *Chem.
Commun.*, 599.
(b) Segura, J.L., Gomez, R., Martin, N.,
Luo, C.P., Swartz, A. and Guldi, D.M.
(2001) *Chem. Commun.*, 707.
(c) Accorsi, G., Armaroli, N., Eckert, J.-F.
and Nierengarten, J.-F. (2002) *Tetrahedron
Lett.*, **43**, 65.
(d) Langa, F., Gómez-Escalonilla, M.J.,
Díez-Barra, E., García-Martínez, J.C., de la
Hoz, A., Rodríguez-López, J., González-
Cortés, A. and López-Arza, V. (2001)
Tetrahedron Lett., **42**, 3435.
(e) Guldi, D.M., Swartz, A., Luo, C.,
Gomez, R., Segura, J.L. and Martin, N.
(2002) *J. Am. Chem. Soc.*, **124**,
10875.
(f) Pérez, L., Garcia-Martinez, J.C.,
Diez-Barra, E., Atienzar, P., Garcia, H.,
Rodriguez-Lopez, J. and Langa, F. (2006)
Chem. Eur. J., **12**, 5149.

(g) Armaroli, N., Accorsi, G., Clifford, J.N., Eckert, J.-F. and Nierengarten, J.-F. (2006) *Chem. Asian J.*, **1**, 564.

15 (a) Nierengarten, J.-F., Felder, D. and Nicoud, J.-F. (1999) *Tetrahedron Lett.*, **40**, 269.
(b) Nierengarten, J.-F., Felder, D. and Nicoud, J.-F. (2000) *Tetrahedron Lett.*, **41**, 41.
(c) Nierengarten, J.-F., Eckert, J.-F., Rio, Y., Carreon, M.P., Gallani, J.-L. and Guillon, D. (2001) *J. Am. Chem. Soc.*, **123**, 9743.
(d) Gutiérrez-Nava, M., Accorsi, G., Masson, P., Armaroli, N. and Nierengarten, J.-F. (2004) *Chem. Eur. J.*, **10**, 5076.

16 Hahn, U., Hosomizu, K., Imahori, H. and Nierengarten, J.-F. (2006) *Eur. J. Org. Chem.*, 85.

17 Herschbach, H., Hosomizu, K., Hahn, U., Leize, E., Van Dorsselaer, A., Imahori, H. and Nierengarten, J.-F. (2006) *Anal. Bioanal. Chem.*, **386**, 46.

18 Felder, D., Nierengarten, H., Gisselbrecht, J.-P., Boudon, C., Leize, E., Nicoud, J.-F., Gross, M., Van Dorsselaer, A. and Nierengarten, J.-F. (2000) *New J. Chem.*, **24**, 687.

19 (a) Elhabiri, M., Trabolsi, A., Cardinali, F., Hahn, U., Albrecht-Gary, A.-M. and Nierengarten, J.-F. (2005) *Chem. Eur. J.*, **11**, 4793.
(b) Nierengarten, J.-F., Hahn, U., Trabolsi, A., Herschbach, H., Cardinali, F., Elhabiri, M., Leize, E., Van Dorsselaer, A. and Albrecht-Gary, A.-M. (2006) *Chem. Eur. J.*, **12**, 3365.

20 Hosomizu, K., Imahori, H., Hahn, U., Nierengarten, J.-F., Listorti, A., Armaroli, N., Nemoto, T. and Isoda, S. (2007) *J. Phys. Chem. C*, **111**, 2777.

21 Zhou, S., Burger, C., Chu, B., Sawamura, M., Nagahama, N., Toganoh, M., Hackler, U.E., Isobe, H. and Nakamura, E. (2001) *Science*, **291**, 1944.

22 (a) Kamat, P.V., Barazzouk, S., Thomas, K.G. and Hötchandani, S. (2000) *J. Phys. Chem. B*, **104**, 4014.
(b) Hotta, H., Kang, S., Umeyama, T., Matano, Y., Yoshida, K., Isoda, S. and Imahori, H. (2005) *J. Phys. Chem. B*, **109**, 5700.

23 Hahn, U., Cardinali, F. and Nierengarten, J.-F. (2007) *New J. Chem.*, **31**, 1128.

24 (a) Hahn, U., González, J.J., Huerta, E., Segura, M., Eckert, J.-F., Cardinali, F., de Mendoza, J. and Nierengarten, J.-F. (2005) *Chem. Eur. J.*, **11**, 6666.
(b) van de Coevering, R., Kreiter, R., Cardinali, F., van Koten, G., Nierengarten, J.-F. and Klein Gebbink, R.J.M. (2005) *Tetrahedron Lett.*, **46**, 3353.

25 Hahn, U., Gégout, A., Duhayon, C., Coppel, Y., Saquet, A. and Nierengarten, J.-F. (2007) *Chem. Commun.*, 516.

26 Chandrasekhar, V., Schmid, C.G., Burton, S.D., Holmes, J.M., Day, R.O. and Holmes, R.R. (1987) *Inorg. Chem.*, **26**, 1050.

27 Chandrasekhar, V., Gopal, K., Nagendram, S., Steiner, A. and Zacchini, S. (2006) *Cryst. Growth Des.*, **6**, 267.

28 (a) Nierengarten, J.-F., Habicher, T., Kessinger, R., Cardullo, F., Diederich, F., Gramlich, V., Gisselbrecht, J.-P., Boudon, C. and Gross, M. (1997) *Helv. Chim. Acta*, **80**, 2238.
(b) Boudon, C., Gisselbrecht, J.-P., Gross, M., Isaacs, L., Anderson, H.L., Faust, R. and Diederich, F. (1995) *Helv. Chim. Acta*, **78**, 1334.

23
Interactions of Carbon Nanotubes with Biomolecules: Advances and Challenges

Dhriti Nepal and Kurt E. Geckeler

23.1
Introduction

Carbon nanotubes (CNTs) are allotropes of carbon, composed up of rolled-up graphene sheets with a unique combination of properties that include excellent electrical, optical, and mechanical properties [1]. Such properties have opened up a great range of new possibilities for these fascinating materials in a wide range of research, including electronics, optics, high-performance fibers, composites, and biotechnology, as well as other fields of materials science.

23.2
Structure and Properties

The chemical bonding of nanotubes is composed entirely of sp^2 bonds, similar to those of graphite. The strength of the sp^2 carbon–carbon bonds gives the CNTs amazing mechanical properties, as these bonds are much stronger than the sp^3 bonds found in diamonds. Such properties, when coupled with the low specific weight of CNTs, provides these materials with great potential over a wide range of applications. The extraordinary electronic properties of CNTs are due to the quantum confinement effect, which is based on a unique structure that allows electrons to propagate only along the nanotube axis. Depending on how the two-dimensional (2-D) graphene sheet is "rolled up," the CNTs may be either metals or semiconductors. Three types of nanotubes are possible, referred to as *armchair, zigzag,* and *chiral* nanotubes.

Despite the great potential that CNTs offer, many fundamental challenges remain to be met before such potential can be fully utilized. This includes the purification, dispersion, debundalization, and chiral-selective separation of nanotubes. The major hurdle results from the natural tendency of CNTs to remain in highly aggregated states, due to their intrinsic van der Waals interactions. In

Advanced Nanomaterials. Edited by Kurt E. Geckeler and Hiroyuki Nishide
Copyright © 2010 WILEY-VCH Verlag GmbH & Co. KGaA, Weinheim
ISBN: 978-3-527-31794-3

fact, under normal conditions this property not only makes it practically impossible to solubilize the materials in any type of solvent but also hampers their electrical, optical, and mechanical properties. Consequently, the functionalization of nanotubes holds much promise as it provides processability and also helps when studying the materials' properties using solution-based techniques. In order to explore the potential of nanotubes, they have been functionalized in many ways, including direct covalent reaction with a variety of reagents, indirect reaction with different functional moieties starting from oxidized precursors, via organometallic routes, and noncovalent interactions. The reactions of nanotubes with polymers, whether by grafting or by using the nanotubes as an anchor to initiate polymerization, has led to the production of a range of nanotube-based polymer composites [2, 3]. Functionalization with well-defined chemistry has also helped in the design of nanotube-based "smart" devices, by modifying their properties.

Today, single-walled carbon nanotubes (SWNTs) are attracting increasing attention in biomedical research [4–6]. Because of their unique physical and chemical properties, they offer the promise for the development of novel diagnostic and therapeutic methods, high-sensitivity biosensors, and biofuel cells. When SWNTs absorb energy from near-infrared (NIR) light they emit heat—a property which can play important role in the design of novel biomedical devices or novel therapeutic agents. In developing such applications, the challenges of *in vivo* detection and biocompatibility are the most essential. The functionalization of biomolecules not only offers biocompatibility, but also provides new possibilities in the material world for the advancement of future technology. DNA and proteins represent major classes of biopolymers with extensive potential in both the natural as well as the synthetic world. In this chapter, an attempt has been made to summarize recent developments on the functionalization of CNTs using these polymers, with special attention on noncovalent functionalization. The aim also is to highlight some of the key challenges, and future prospects.

23.3
Debundalization

Due to their inherent van der Waals interactions and their special geometric structures, CNTs normally aggregate to form large bundles. Unfortunately, this process of "bundalization" causes the most outstanding characteristics of CNTs—namely their electrical [7], mechanical [8], and optical [9] properties—to be diminished. As an example, it has been shown experimentally that an individual SWNT has a tensile strain of ~280% and can undergo superplastic deformation by reducing its diameter 15-fold before breaking [10]. Yet, bundled SWNTs [11] and multi-walled carbon nanotubes (MWNTs) [12] will break at strains of less than ~6% and 12%, respectively. This provides a clear picture of the importance of including debundalized nanotubes in nanocomposites. Moreover, the development of efficient techniques to produce stable dispersions of debundalized nanotubes is a clear prerequisite to many applications, ranging from molecular electronics to nanocomposites.

In an effort to overcome these problems, many research groups worldwide have investigated the debundalization or individualization of SWNTs. The debundalization of SWNTs has been achieved via two routes, namely covalent [13] or noncovalent [14] modification/functionalization [3], with the former approach causing inevitable damage to the sidewalls of the nanotubes that altered their chemical and physical properties [15]. As a consequence, a range of studies has focused on noncovalent approaches to obtaining individual nanotubes by using different agents and polymers.

Previous studies have been focused on using synthetic polymers or surfactants. For example, Smalley and coworkers [9] obtained individual nanotubes by encasing each one in cylindrical micelles. This was made possible by ultrasonically agitating an aqueous dispersion of raw SWNTs in sodium dodecyl sulfate (SDS), followed by ultracentrifugation to remove tube bundles and residual catalyst. The absorbance spectra hold the key role to determining the degree of debundalization. The quasi-one-dimensionality of the SWNTs produces sharp van Hove peaks in the density of electronic states [16]. In this respect, for example in the case of HiPCO-SWNTs, the optical absorption spectrum consists of a series of transitions: first van Hove transitions of *met*-SWNTs (metallic-SWNT) (M_{11}) (~400 and ~600 nm); second van Hove transitions of *sem*-SWNTs (semimetallic-SWNT) (S_{22}) (~550 to 900 nm); and first van Hove transitions of *sem*-SWNTs (S_{11}) (~800 to ~1600 nm). The well-resolved absorbance feature is a key indication of the debundalization, as otherwise the spectra are broadened (Figure 23.1).

In the bulk state, the individualization can be further confirmed by recording fluorescence spectra, which have been observed directly across the band gap of *sem*-SWNTs. A fluorescence spectrum can only be obtained in the debundalized state, because otherwise the presence of *met*-SWNTs will quench the electronic excitation on adjacent *sem*-SWNTs in the bundle, which prevents luminescence. This approach of direct fluorescence observation from individual SWNTs in solution has great importance for the development of SWNT-based optical d evices. Different surfactants [17], synthetic polymers, and also biopolymers [3] were each found to be excellent dispersing agents for SWNTs, especially for making them disperse individually in aqueous media. A study was conducted to identify the best dispersing agent for SWNTs by comparing various anionic, cationic, and nonionic surfactants and polymers [17]. Each of these agents was compared with respect to their ability to suspend individual SWNTs, and the quality of the absorption and fluorescence spectra. For ionic surfactants, sodium dodecylbenzene sulfonate (SDBS) was found to provide well-resolved spectral features, and this correlated directly with the most favorable ionic surfactant for debundalization. For the nonionic systems, surfactants with a higher molecular mass could suspend more SWNT material and had more pronounced spectral features.

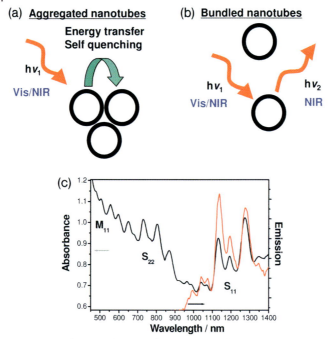

Figure 23.1 Schematic diagram of the dispersion-dependent energy transfer in nanotubes, and evidence of isolated nanotubes by optical spectroscopy. (a) Model of energy-transfer quenching in a matrix with bundled nanotubes; (b) Model of fluorescent emission, if nanotubes are individually isolated; (c) Absorbance and emission spectra of individually isolated nanotubes in a protein solution.

23.4
Noncovalent Functionalization

Covalent functionalization weakens the strength of all interband transitions in both the semiconducting and metallic SWNTs, because the saturated bonds function as defects. This perturbs the periodicity of the pseudo-one-dimensional (1-D) lattice and eventually destroys the electronic band structure altogether. However, the largest effects of covalent functionalization occur in metallic SWNTs, because the metallic state should be extremely sensitive to the occurrence of defects introduced by saturation of the delocalized electronic structure. In particular, these effects should be most strongly manifested in the transitions at E_F because these transitions are characteristic of a metal, which is a species without a gap in the energy band spectrum.

On the other hand, the noncovalent functionalization of CNTs represents one of the best alternatives for preserving the sp^2-structure of nanotubes and, thus, their electronic characteristics. The advantage is the possibility of attaching chemical functionalities to the surface, without disrupting the bonding network of the nanotubes. Thus, the noncovalent functionalizaton of CNT with different polymers or organic molecules has attracted great attention.

23.5
Dispersion of Carbon Nanotubes in Biopolymers

Synergetic effects with biomolecules on CNTs has attracted much attention for a wide range of biotechnological applications, including sensors, cancer therapy [18], drug delivery [19], tissue engineering, and antimicrobial surfaces [20]. The results of recent studies have shown CNTs to be compatible with mammalian cells and neurons [21], with CNT-based composites having been identified as excellent materials for scaffolds for neuron growth [22] and bone proliferation [23], as well as having potential for tissue engineering [24]. These types of bio-focused composites have been prepared using a variety of natural biopolymers, including different polysaccharides, polypeptides, and polynucleotides (Figure 23.2) [2]. Interestingly, the biopolymers have shown much promise, since "mother Nature" has provided an abundance of well-designed polymers for defined applications. The exploration of these molecules in conjunction with nanotubes not only facilitates the creation of suitable products from abundantly available resources, but also assists in mimicking their smart functions for the advancement of nanotube-based devices. For example, a degree of success has been achieved in preparing a unique supramolecular conjugate of nanotubes and lysozyme (LSZ), one of the most common proteins. In this way, novel properties could be induced (i.e., pH-sensitive dispersion and debundalization of the nanotubes) that directly gave their signal simply by switching photoluminescence. A similar success was achieved when nanotubes were individually dispersed, using different proteins.

Recently, biopolymer integration has attracted much attention due to their increased compatibility with biological systems compared to synthetic materials. Consequently, a variety of biopolymers was chosen in order to monitor their efficiency for the dispersion of CNTs. For example, Rozen *et al.* [25] used the water-soluble polysaccharide *gum arabica* to obtain a stable dispersion of individual, full-length tubes in an aqueous solution from as-produced SWNTs. Wenseleers *et al.* [26] reported an efficient isolation of pristine SWNTs in bile salts, such as the sodium salts of deoxycholic acid (DOC) and their taurine-substituted analogue

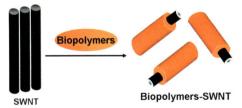

Figure 23.2 Schematic diagram of the wrapping of single-walled carbon nanotubes by biopolymers.

taurodeoxycholic acid (TDOC), simply by stirring at room temperature. When using this process it is possible to avoid damage induced on the walls of the SWNTs during ultrasonication. The efficiency of debundalization of the SWNTs was monitored using NIR, fluorescence, and Raman spectroscopies. Most interestingly, a dramatically improved resolution of the radial breathing modes (RBMs) in the Raman spectra with multi-peaks was obtained with the bile salts, in contrast to the single peak obtained in most previous studies with arc discharge tubes. When the ability of the salts to disperse the SWNTs was compared with other common surfactants, by using NIR spectroscopy, among the most common surfactants, both DOC and TDOC (i.e., anionic, nonionic, cationic) showed the highest dispersion powers (five- to 20-fold higher).

A water-soluble product also has great potential in the design of an electrode for bioelectrochemical sensors, by taking advantage of the noncovalent interactions of chitosan with CNTs [27]. Hasegawa *et al.* showed that biopolymers such as Schizophyllan (s-SPG) and curdlan were capable of wrapping SWNTs, creating a "periodical" helical structure that reflected the helical nature of the SPG main-chain on the SWNT surface [28]. Kim *et al.* [28] reported a simple, but efficient, process for the solubilization of SWNTs with amylose in aqueous dimethylsulfoxide (DMSO), by using sonication. The former step separated the SWNT bundles, while the latter step provided a maximum cooperative interaction of SWNTs with amylose, leading to an immediate and complete solubilization. Both, scanning electron microscopy (SEM) and atomic force microscopy (AFM) images of the encapsulated SWNTs appeared as loosely twisted ribbons wrapped around the SWNTs, which were locally intertwined as a multiple twist; however, no clumps of the host amylose were seen on the SWNT capsules.

The potential of CNTs for the development of novel bioelectronic devices has been realized and, indeed, biomolecules have been immobilized on CNTs [29]. The major benefit in choosing biomolecules is to take advantage of their amphoteric nature to render the nanotubes processable, while simultaneously utilizing their unique properties to design novel artificial systems. In this respect, a variety of biomolecules hold great promise for this type of development.

Here, attention is focused on DNA and proteins for the interaction with CNTs. The major motivation to use these polymers includes: (i) their unique features and typical characteristics, as well as their diversity, which controls almost every living system in Nature; and (ii) the potential to bring about novel possibilities by uniting them, as this can have huge impact on both basic and applied research.

23.6
Interaction of DNA with Carbon Nanotubes

The interaction of DNA with CNTs has been a major focus of recent research [30–32], due mainly to the unique structure of DNA (Figure 23.3) and its properties with regard to both biological and nonbiological applications. For instance, recent reports have already focused on several applications as a new material, including

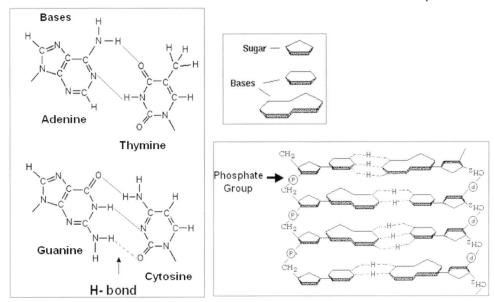

Figure 23.3 Schematic showing the chemical structure of DNA, which is composed of the four different bases, sugar (ribose), and phosphate groups.

drug delivery [33, 34], molecular electronics, nanoscale robotics [35], computation [36], self-assembly, 1-D electron conduction [34], and photonics [37]. When considering the huge potential of CNTs, a combination of DNA with CNTs in the quest to identify new and advanced performances for a wide range of synthetic systems has been of vast interest. Yet, the major attraction for DNA has been based on a recent breakthrough which described its ability to individually disperse CNTs in aqueous solutions, under controlled conditions [32, 38, 39], and this has in turn led to many other possibilities.

The first report on the direct interaction of DNA with CNT was made by Tsang *et al.*, in a study which was based on the visualization of CNT by platinated oligo-nucleotides [40]. The major breakthrough in this field was based on the ability of DNA to disperse CNTs individually in water [30, 38], while others reported on a DNA-assisted dispersion of CNTs [39, 41]. All of these studies were based on the noncovalent functionalization of DNA with nanotubes. The covalent bonding of DNA with SWNTs [42] has been also reported, while field effect transistors [43] using such materials have also been studied.

Previously, the use of mechanochemical reactions has been suggested [44] as a generally efficient technique for various processes and, by using the same approach, supramolecular adducts of DNA–CNT conjugates have been reported [39]. Highly reactive centers generated on the CNTs by the mechanical energy in the solid state helped to promote good interactions with DNA (Figure 23.4), which in turn led to a short-length, highly water-soluble, CNT-based product in a single step. This

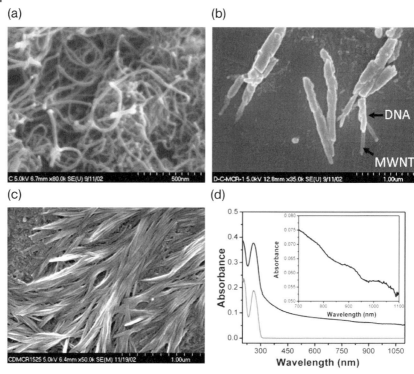

Figure 23.4 Scanning electron microscopy images. (a) Pristine CNT; (b) CNT–DNA conjugate; (c) Self-assembly of the CNT–DNA conjugate; (d) UV-visible-NIR absorbance spectra of DNA (gray) and DNA–MWNT dispersion (black). The inset shows the van Hove transitions of the semiconducting nanotubes.

technique has also been used successfully to prepare both fullerene derivatives [45] and alcohol-functionalized CNTs [46] Interestingly, this soluble CNT product appeared fully wrapped with DNA and seemed to differ from that described in previous reports on mechanochemical reactions for covalently functionalized products [46].

Zheng *et al.* obtained individual SWNTs in a solution of single-stranded DNA (ssDNA) [30, 38, 47]. Notably, removal of the free DNA by either anion-exchange column chromatography or nuclease digestion did not cause any flocculation of the nanotubes, indicating that binding of the DNA to the SWNTs was very strong. In order to obtain putative binding structures and to approximately quantify the thermodynamics of binding, Zheng's group simulated interactions between ssDNA molecules and nanotubes with a (10,0)-chiral vector. As a consequence, it appeared that the short ssDNA strands could bind in many ways to the nanotube surface, including helical wrapping with right- and left-handed turns, or simply by surface adsorption with a linearly extended structure. The bases could extend from the backbone and stack onto the nanotube, whereas the sugar-phosphate backbone could lead to solubilization of the SWNTs. The phosphate groups of the product provided a negative charge density on the surface and affected the surface charge of the nanotubes, which was directly related to the electronic properties of the SWNTs. In this respect, DNA-*met*-SWNTs were found to have a lower surface charge than DNA-*sem*-SWNTs, due to the greater positive charge created in the *met*-SWNTs. This led to a successful separation of *met*-SWNTs from *sem*-SWNTs by using ion-exchange liquid chromatography. Moreover, by measuring the electronic absorption spectra for all fractions, a remarkable difference was noted between the early fractions, which were enriched with *met*-SWNTs, and the later fractions, which were enriched with *sem*-SWNTs.

23.7
Interaction of Proteins with Carbon Nanotubes

The main focus of preparing debundalized SWNTs with the aid of proteins was motivated by the outstanding properties that each of these proteins display in the natural environment. This offers the possibility to bring functionality into synthetic systems, if the correct technology were to be developed. A broad range of different types are available in Nature, which allows complicated synthetic procedures to be avoided and their compatibility with biological systems to be enhanced, when compared to synthetic materials. In addition, the intrinsic smart function in a synthetic system may be mimicked [2, 48]. Functionalization is possible by both the covalent and noncovalent approaches. For example, CNTs may be functionalized by the protein bovine serum albumin (BSA) via a diimide-activated amidation, under ambient conditions [49]. The nanotube–BSA conjugates thus obtained are highly water-soluble, and form dark-colored aqueous solutions. Results from characterizations using AFM, thermal gravimetric analysis (TGA), Raman spectroscopy and gel electrophoresis showed that the conjugate samples indeed contained both CNTs and BSA proteins, and that the protein species were intimately associated with the nanotubes. When the bioactivity of the nanotube-bound proteins was evaluated using a total protein microdetermination assay, the results showed that 90% of the protein species present in the nanotube–BSA conjugates remained bioactive [49].

Zorbas *et al.* [50] isolated long, individual SWNTs wrapped with a specially designed peptide in aqueous solution by sonication and centrifugation. Based on AFM studies, it was observed that the product contained SWNTs with an average length of $1.2 \pm 1.1\,\mu m$ and an average diameter of $2.4 \pm 1.3\,nm$. Furthermore, the peptide–peptide interactions apparently assisted the assembly of the SWNT structures, specifically in the formation of Y-, X-, and intraloop junctions. In this way, apart from the debundalization and biocompatibility of the peptide–SWNT product, their properties showed great potential for the development of SWNT-based, higher-ordered structures. Similarly, the importance of aromatic groups of the amino acid residues on the interaction with CNTs has been observed [51]. Various systematically designed peptides with different types of amino acid showed a stronger selectivity for CNTs than did peptides with a higher aromatic residue content.

A study was conducted to evaluate the affinity of aromatic amino acids toward SWNTs in the absence of complications from peptide folding or self-association, by synthesizing a series of surfactant peptides [52]. Each surfactant peptide was designed with a lipid-like architecture: two Lys residues at the C-terminus as a hydrophilic head; five Val residues to form a hydrophobic tail; and the testing amino acid at the N-terminus. Both, Raman and circular dichroism (CD) spectroscopic studies revealed that the surfactant peptides had a large, unordered structural component, which was independent of the peptide concentration. This suggested that the peptides had undergone minimal association under experimental conditions, thus removing this interference from any interpretation of the peptide–CNT interactions. A lack of peptide self-association was also indicated by sedimentation equilibrium data. The optical spectroscopy of the peptide–CNT dispersions indicated that, among the three aromatic amino acids, tryptophan had the highest affinity for CNTs (both bundled and individual states), when incorporated into a surfactant peptide, while the Tyr-containing peptide was more selective for individual CNTs. A protein-assisted solubilization of nanotubes was also reported [53]. Although these recent findings have opened novel possibilities for nanobiohybrid systems, it remains an overt issue as to whether there is any selectivity and any effect of the proteins on the nanotubes. In addition, it is very important to note that these natural polyampholytes are highly pH-sensitive, and that their solubility and ionizability can easily be tuned with respect to pH. This, simultaneously, should influence the nanotube dispersion, which might offer novel options for the precise control of selectivity towards nanotubes.

The present authors studied highly dispersed and debundalized CNTs (Figure 23.5), which had been prepared in an aqueous solution of LSZ, using a combina-

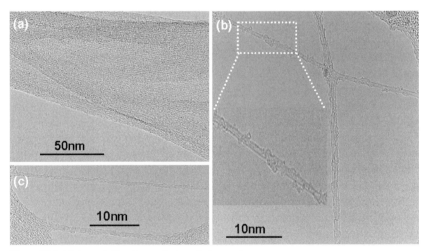

Figure 23.5 High-resolution transmission electron microscopy images of (a) SWNT and (b, c) L-SWNT. The inset in (b) shows the magnified image of a nanotube.

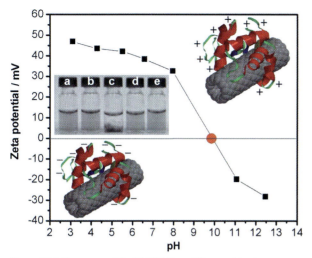

Figure 23.6 Zeta potential of *L*-SWNT at different pH values. The inset shows vials containing the products at different pH values: (a) 2.95; (b) 6.5; (c) 9.0; (d) 11.15; (e) 12.47.

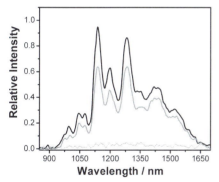

Figure 23.7 Photoluminescence spectrum of *L*-SWNT, showing the pH-dependent reversibility in emission. The black (solid), black (dotted), and gray (solid) curves are from the solution at pH 12.5, same solution adjusted to pH 9.0, and again readjusted to pH 12.5, respectively.

tion of ultrasonication and ultracentrifugation [14]. The product showed a pH-sensitive dispersion, which remained in a highly dispersed state at pH < 8 and pH > 11, and in an aggregated state at pH = 8–11 (Figure 23.6). Photoluminescence measurements showed that, by changing the pH, a reversible conversion of the highly dispersed state to the aggregated state could be observed, or *vice-versa* (Figure 23.7).

It is important to understand the nature of the interactions involved between the LSZ and the SWNTs. LSZ has long been considered a structurally stable

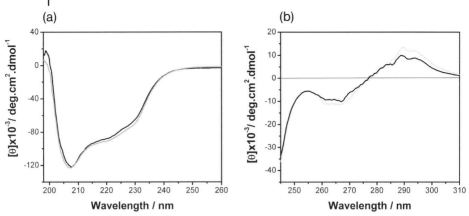

Figure 23.8 Far-UV-CD (a) and near-UV-CD (b) spectra of
L-SWNT (black) and lysozyme (gray).

protein; hence, in order to appreciate the state of LSZ in the product, a CD analysis
of the samples was carried out in the far- and near-UV regions (Figure 23.8). It
has been well established that shorter wavelengths reflect the secondary structure,
whereas longer wavelengths arise from the tertiary structure of proteins. A strong
negative band was appeared in the range of 200–260 nm. The signal intensity at
208 nm was greater than at 222 nm, which was characteristic for an $\alpha + \beta$ protein.
These data confirmed that the secondary structure had remained intact in the
product.

In an attempt to understand the efficiency of common proteins for SWNT dispersion, a selection of eight common proteins (histone [HST], hemoglobin [HBA],
myoglobin [MGB], ovalbumin [OVB], bovine serum albumin [BSA], trypsin [TPS],
glucose oxidase [GOX] and lysozyme [LSZ]), with a variety of molecular masses
and isoelectric points, was selected. The basic physical data of the proteins are
listed in Table 23.1. In a typical experiment, the SWNTs were ultrasonicated in an
aqueous solution of the proteins, followed by a double-step ultracentrifugation
process. This resulted in highly dispersed and debundalized SWNTs in the
aqueous system. Schematic diagrams of the protein-stabilized SWNTs are shown
in Figure 23.9.

The vials containing products from the different proteins are shown in Figure
23.10, each at three different pH values (intrinsic, acidic, and basic). It is clear
from the figure that the color of the solution varies with respect to pH, ranging
from a colorless-transparent liquid to a dark-gray solution. This color change corresponds directly to variations in the yields of nanotubes in the solutions.

Microscopic studies of the dispersed nanotubes using high-resolution transmission electron microscopy (HR-TEM) showed that the starting material consisted
of bundled ropes assembled from a few tens to a hundred tubes, whereas HR-TEM
images of the product HST–SWNT showed debundalized tubes, well separated

Table 23.1 Comparison of the dispersion limits and yields of debundalized single-wall nanotubes (SWNTs) with different proteins, and some physical parameters of the proteins. The proteins are arranged in alphabetic order.

Protein	Code	Molar mass (kg mol^{-1})^a	Isoelectric point	Initial pH^b	Final pH^c	Dispersion limit (mg l^{-1})^d	Debundalized SWNTs (mg l^{-1})^e	Debundalization degree (DD) (%)^f
Bovine serum albumin	BSA	69.293	4.7	1.8	1.5	2.44	0.98	1
				7.4	6.7	25.44	6.38	10
				11.4	11.2	132.71	43.41	65
Glucose oxidase	GOX	131.2764	4.2	1.6	1.8	0.33	0	0
				6.9	4.5	0.14	0	0
				11.2	11.3	1.39	0.01	0
Hemoglobin	HBA	61.933	9.4	1.7	1.5	179.96	36.46	55
				6.4	5.9	7.09	0.74	1
				11.5	11.1	134.52	27.09	41
Histone	HST	15.316	11.7	1.9	1.8	67.00	11.66	17
				5.7	4.2	196.78	66.70	100
				11.6	11.5	17.59	2.85	4
Lysozyme	LSZ	14.313	10.7	1.4	1.5	197.8	54.03	81
				6.5	4.7	255.3	58.97	88
				11.6	11.2	2.19	0.49	1
Myoglobin	MGB	16.953	7.2	1.9	1.6	91.93	28.96	43
				6.6	5.8	42.58	15.54	23
				11.6	11.5	119.42	37.12	56
Ovalbumin	OVB	42.861	4.6	1.6	1.6	67.78	20.53	31
				6.9	6.6	143.99	29.12	44
				11.2	11.2	153.40	45.49	68
Trypsin	TPS	23.783	9.2	1.8	1.7	0.13	0.01	0
				4.2	4.1	9.98	2.65	4
				11.8	11.6	42.78	12.56	19

a   Molar masses were calculated based on the sequence downloaded from the protein data bank.
b   pH of the solution before sonication.
c   pH of the solution after sonication.
d   Concentration of SWNTs in the supernatant solution from the first step (centrifugation for 3 h at 18 000×g).
e   Concentration of SWNTs in the supernatant solution from the final step (centrifugation for 4 h at 120 000×g).
f   HST-SWNT (which has the highest DD at the intrinsic pH) was considered as 100% DD; values for other products were calculated with respect to this value.

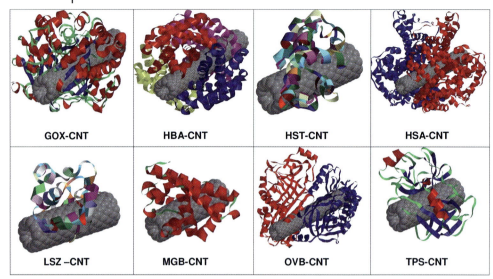

Figure 23.9 Schematic representations of the single-walled carbon nanotube (SWNT) adducts with different proteins: glucose oxidase (GOX-SWNT), hemoglobin (HBA-SWNT), histone (HST-SWNT), human serum albumin (HSA-SWNT), lysozyme (LSZ– SWNT), myoglobin (MGB-SWNT), ovalbumin (OVB-SWNT), trypsin (TPS-SWNT). Color-coding: secondary structure for GOX, MGB, and TPS; amino acid residue sequence for HST and LSZ; and strand for HAS, HBA, and OVB.

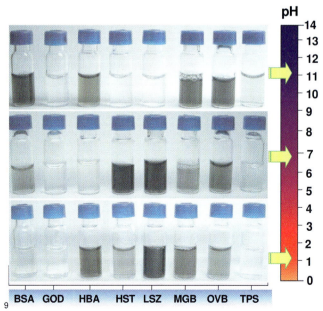

Figure 23.10 Vials containing the SWNT products obtained from different proteins at different pH values. From left: bovine serum albumin (BSA), glucose oxidase (GOX), hemoglobin (HBA), histone (HST), lysozyme (LSZ), myoglobin (MGB), ovalbumin (OVA), and trypsin (TPS). The arrow in each row indicates the pH range of the respective solutions.

from the bundles. The majority of the tubes were found to be individual, although small bundles of two to four tubes also existed. The irregular coating seen around the nanotube surface was the protein, as confirmed by the detection of nitrogen in the energy-dispersive X-ray (EDX) spectrum. Likewise, this confirmed that the nanotubes had become well dispersed in the protein solution. The AFM images of the product showed a nanotube diameter distribution of between 1 and 5 nm, with a majority of tubes being >3 nm in size. Similar observations were made for the other products. These data illustrate clearly that, during sonication, the nanotubes became simultaneously debundalized due to strong interactions of the proteins with the nanotube surface, and this resulted in highly dispersed protein–nanotube adducts in the aqueous systems. During high-speed ultracentrifugation, the bundled adducts were precipitated due to their higher density, and this resulted in highly individual or less-bundalized protein–nanotubes in the supernatant.

In order to quantify the yield and to check the efficiency of the proteins, the concentrations of nanotubes in the products were measured using absorbance spectroscopy at 500 nm (Figure 23.11). The concentrations were calculated using optical absorbance data fitted to a Beer–Lambert plot. The values (which are summarized in Table 23.1) showed clearly that the highest yield of debundalized nanotubes was obtained from HST at the intrinsic pH, followed by LSZ. However, the yield was exclusively dependent on the pH. In the cases of HBA, OVB, BSA, and MGB, the yields were much higher at basic pH. In contrast, under similar conditions TPS showed a slight improvement, but for GOX the value was almost zero.

In a bid to understand these diverse phenomena, the primary structure of these proteins was investigated. Numerous forces, including van der Waals, electrostatic, hydration, steric, hydrophobic, and chemically specific interactions, are known to act between the colloidal particles in solution. Proteins serve as a rich source of different functionalities due to the presence of different types of amino acid residues (Table 23.2) and the different interactions that are responsible for their interaction with substrates. Hence, the coordinates of these proteins were downloaded from a protein data bank and the different types of residues present in each sequence calculated (Table 23.3). For a better understanding of the major interactions responsible for the dispersion, an attempt was made to correlate the efficiency of proteins for the nanotube dispersion with the residues. The correlation of the debundalization degree (DD) of SWNTs at the intrinsic pH with respect to percentage of hydrophobic residues (HR), aromatic residues (ArR), polar residues (PR), acidic residues (AcR), basic residues (BR), and charged residues (CR) contained in each protein is shown as a three-dimensional plot in Figure 23.12.

23.8
Technology Development Based on Biopolymer-Carbon Nanotube Products

A variety of novel materials has been developed by combining biopolymers with nanotubes. Some important contributions of these materials are outlined in the following sections.

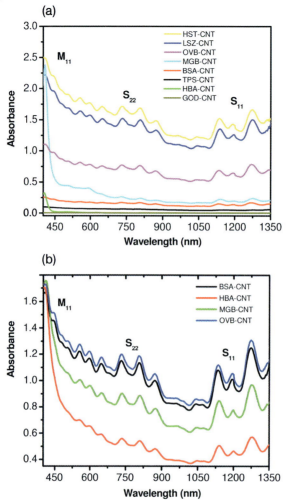

Figure 23.11 NIR absorbance spectra of the SWNT adducts with different proteins at (a) intrinsic pH, and (b) basic pH. At basic pH, the enhanced intensity of the nanotubes confirmed the high yield of nanotubes in the products. S_{11} and S_{22} correspond to the first and second interband transitions of semiconducting nanotubes; M_{11} corresponds to the first transition of metallic nanotubes.

Table 23.2 Amino acid residues and their basic chemical and physical parameters.

Amino Acid	Symbol	Structure	Type(s)	Isoelectric point[a]
Alanine	A, Ala	$^+H_3N-\overset{\overset{\textstyle COO^-}{\textstyle \|}}{\underset{\underset{\textstyle CH_3}{\textstyle \|}}{C}}-H$	Hydrophobic	6.01
Arginine	R, Arg	$^+H_3N-\overset{\overset{\textstyle COO^-}{\textstyle \|}}{C}-H$ CH_2 CH_2 CH_2 NH $C=NH_2^+$ NH_2	Basic Hydrophilic Polar	10.76
Asparagine	N, Asn	$^+H_3N-\overset{\overset{\textstyle COO^-}{\textstyle \|}}{C}-H$ CH_2 $H_2N-C\!\!=\!\!O$	Hydrophilic Polar	5.41
Aspartic acid	D, Asp	$^+H_3N-\overset{\overset{\textstyle COO^-}{\textstyle \|}}{C}-H$ CH_2 COO^-	Acidic Polar Hydrophilic	2.85
Cysteine	C, Cys	$^+H_3N-\overset{\overset{\textstyle COO^-}{\textstyle \|}}{C}-H$ CH_2 SH	Hydrophobic Nonpolar	5.05
Glutamine	Q, Gln	$^+H_3N-\overset{\overset{\textstyle COO^-}{\textstyle \|}}{C}-H$ CH_2 CH_2 $H_2N-C\!\!=\!\!O$	Hydrophilic Polar	5.65

Table 23.2 Continued.

Amino Acid	Symbol	Structure	Type(s)	Isoelectric point[a]
Glutamic acid	E, Glu	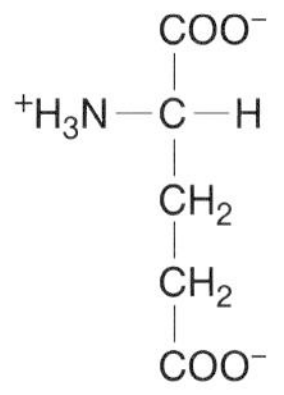	Acidic Hydrophilic Polar	3.15
Glycine	G, Gly	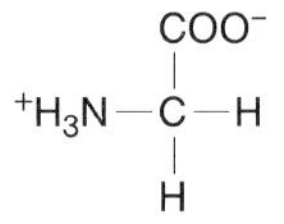	Hydrophobic	6.06
Histidine	H, His	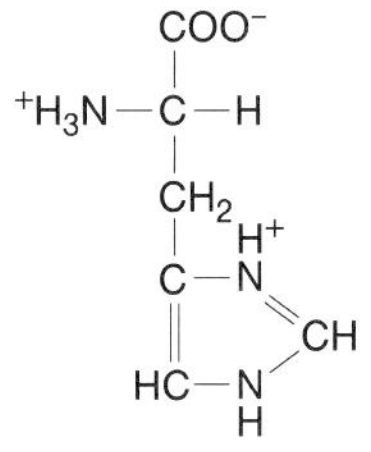	Basic Hydrophilic Polar	7.60
Isoleucine	I, Ile	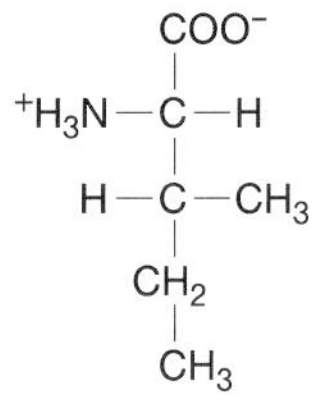	Hydrophobic	6.05
Leucine	L, Leu	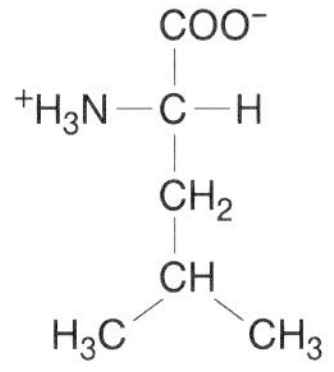	Hydrophobic	6.01
Lysine	K, Lys	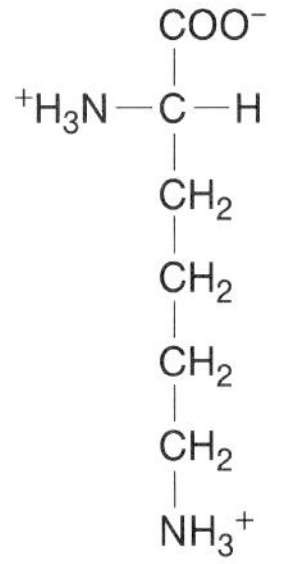	Basic Hydrophilic Polar	9.60

Glutamic acid:
$$COO^-$$
$$^+H_3N-C-H$$
$$CH_2$$
$$CH_2$$
$$COO^-$$

Glycine:
$$COO^-$$
$$^+H_3N-C-H$$
$$H$$

Histidine:
$$COO^-$$
$$^+H_3N-C-H$$
$$CH_2$$
$$C-N \overset{H^+}{}$$
$$\|\diagdown CH$$
$$HC-N$$
$$H$$

Isoleucine:
$$COO^-$$
$$^+H_3N-C-H$$
$$H-C-CH_3$$
$$CH_2$$
$$CH_3$$

Leucine:
$$COO^-$$
$$^+H_3N-C-H$$
$$CH_2$$
$$CH$$
$$H_3C \diagup \diagdown CH_3$$

Lysine:
$$COO^-$$
$$^+H_3N-C-H$$
$$CH_2$$
$$CH_2$$
$$CH_2$$
$$CH_2$$
$$NH_3^+$$

Table 23.2 Continued.

Amino Acid	Symbol	Structure	Type(s)	Isoelectric point[a]
Methionine	M, Met		Hydrophobic Nonpolar	5.74
Phenylalanine	F, Phe		Hydrophobic Aromatic Nonpolar	5.49
Proline	P, Pro		Hydrophobic Nonpolar	6.30
Serine	S, Ser		Hydrophilic Polar	5.68
Threonine	T, Thr		Hydrophilic Polar	5.60
Tryptophan	W, Trp		Hydrophobic Aromatic Nonpolar	5.89

Table 23.2 Continued.

Amino Acid	Symbol	Structure	Type(s)	Isoelectric point[a]
Tyrosine	Y, Tyr	COO⁻ structure	Aromatic Polar Hydrophilic	5.64
Valine	V, Val	COO⁻ structure	Hydrophobic	6.00

a http://en.wikipedia.org/wiki/Amino_acid; Doolittle, R.F. (1989) Redundancies in protein sequences, in *Predictions of Protein Structure and the Principles of Protein Conformation* (G.D. Fasman, ed.) Plenum Press, New York, pp. 599–623; Nelson, D.L. and Cox, M.M. (2000) *Lehninger, Principles of Biochemistry*, 3rd ed., Worth Publishers.

Table 23.3 Calculation of different types of residue content in different proteins.

Protein	Hydrophobic residue (V, I, L, M, F, W, C)	Aromatic residue (F, W, Y, H)	Basic residue (R, K, H)	Acidic residue (E, D)	Charged residue (R, K, E, D)	Polar residue (R, K, D, E, N, Q)
HST	28.9	2.8	23.1	17.3	39	39
LSZ	30.4	10.1	14	7	20.2	33.4
OVB	34.5	10.4	10.9	12.1	21.2	29.5
MGB	28.8	13.1	19.6	14.4	28.1	34.7
HBA	31.5	14.9	16.4	9.4	19.2	24.1
BSA	31.5	11.7	17	16.3	30.5	36.1
TPS	30	10.3	9	5.3	12.5	23.2
GOD	28.2	12.6	9.6	11	17.3	26.6

23.8.1
Diameter- or Chirality-Based Separation of Carbon Nanotubes

Another very important challenge remaining in nanotube research is the separation of nanotubes based on their electronic properties. This is feasible because, during the synthesis of nanotubes (notably of SWNTs) they may be semi-conducting, semi-metallic, or metallic, depending on the rolling vector. However, none of

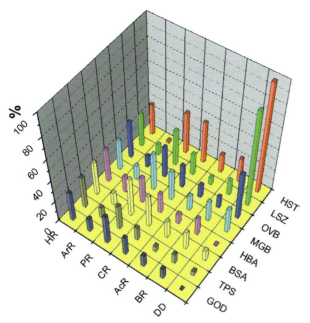

Figure 23.12 Three-dimensional plot showing the debundalization degree (DD) with respect to the percentage of hydrophobic residues (HR), aromatic residues (ArR), polar residues (PR), charged residues (CR), acidic residues (AcR), and basic residues (BR) contained in each protein.

the synthetic techniques reported to date can be used produce a single type of SWNT; rather, a heterogeneous mixture of SWNTs is usually obtained. More importantly, as the electrical and optical properties depend heavily on the chirality and diameter of the nanotubes, a separation of these tubes according to type is a prerequisite for further development of the technology. For example, when fabricating optical devices based on photoluminescence, *sem*-SWNTs are essential, but the presence of *met*-SWNTs leads to a quenching of these properties. Moreover, the problem is worsened due to the nature of nanotubes remaining in aggregation or bundalization, based on the strong van der Waals interactions of the parallel tubes. Although efforts have been made to overcome this problem, an effective and simple technique, which preferably could easily be scaled up, has yet to be developed. One of the major achievements when using DNA for the dispersion of nanotubes is its property to separate the nanotubes selectively, based on their electronic properties [47]. Different experiments have confirmed that the DNA-based approach using chromatography (both, ion-exchange and size exclusion) is highly effective for metal/semiconducting CNT separations, and single-chirality CNT enrichment for certain small-diameter tubes. Such separation is possible due to the selective interaction ability of the designed polynucleotide being synthesized with a specific sequence. For example, an improved metal/semiconductor separation was possible with shorter (guanine/thymine)$_n$ sequences.

Table 23.4 Enrichment factors of metallic (*met*-SWNTs) and semiconducting (*sem*-SWNTs) single-wall carbon nanotubes (SWNTs) of small and large diameters in the products using the 532 nm and 632.8 nm Raman excitation laser lines.

Product	Enrichment factor[a]				*met:sem*[a]
	(met)$_s$[b]	*(met)*$_L$[c]	*(sem)*$_s$[c]	*(sem)*$_L$[b]	
BSA-SWNT[d]	1.71	1.31	0.44	1.09	1.08:1
HBA-SWNT[e]	1.42	1.31	0.53	1.23	1.68:2
HBA-SWNT[d]	1.24	1.54	0.5	1.55	1.57:2
HST-SWNT[f]	1.74	1.08	0.47	1	1:1
LSZ-SWNT[e]	1.45	1.62	0.34	1.36	1.12:1
LSZ-SWNT[f]	1.63	2.38	0.34	1.09	1.61:1
MGB-SWNT[e]	1.47	1.38	0.41	1.18	1.06:1
MGB-SWNT[f]	1.53	1.62	0.5	1.23	1.01:1
MGB-SWNT[d]	1.55	1	0.47	1.14	1.71:2
OVB-SWNT[f]	1.68	1.46	0.38	1.09	1.23:1
OVB-SWNT[d]	1.71	1.39	0.5	1.14	1.01:1
TRP-SWNT[d]	1.34	2	0.63	1	1.04:1

a   Calculated according to the method of Samsonidze, G.G., Chou, S.G., Santos, A.P. *et al.* (2004) *Appl. Physics Lett.* **85** (6), 1006–1008. *(met)*$_s$ = small diameter metallic SWNTs; *(met)*$_L$ = large diameter metallic SWNTs; *(sem)*$_s$ = small diameter semiconducting SWNTs; *(sem)*$_L$ = large diameter semiconducting SWNTs.
b   Calculated from 633 nm laser.
c   Calculated from 532 nm laser.
d   Basic pH.
e   Acidic pH.
f   Intrinsic pH.

By taking advantage of natural proteins under controlled conditions, it was possible to obtain metallic-enriched nanotubes in the final products, which ranged from ~33% (pristine nanotubes) to ~63% (final product) (Table 23.4). This unique and simple technique to enrich metallic nanotubes has had a major impact on metallic-semiconducting nanotube separation in bulk, on a large scale.

23.8.2
Fibers

Recently, DNA-dispersed nanotube products have been used to sort nanotubes based on chirality [30], to synthesize polymers [54], as well as serving as templates for nanoparticles [55]. In addition, these products have been shown to prepare liquid crystals [56], fibers [57], and coatings [20]. Whilst the future prospects of CNTs in biomedical research continue to rise, the preparation of CNT-based biofibers has been introduced only recently [57–60]. DNA-based SWNT fibers were first produced via wet spinning using poly(vinyl alcohol) in the coagulation bath [57], this being possible only due to the excellent dispersion ability of DNA for the SWNTs. Typically, a concentration of 1 wt% SWNTs was able to disperse very well in 1 wt% salmon DNA, which made it feasible to produce DNA–SWNT fibers via wet spinning. Of note, this is almost impossible using surfactants, where ratios of surfactant to SWNTs of at least 2:1 [61] or 3:1 [62] are necessary. The fibers exhibited excellent mechanical properties, with a Young's modulus of up to 14–19 GPa – much higher than for the fibers obtained using sodium dodecyl sulfate (5 GPa), one of the most commonly used surfactants for CNT dispersion. Recently, other fibers based on heparin–SWNT, chitosan–SWNT, and hyaluronic acid–SWNT fibers have also been processed [58]. In this way, the unique nature of each of these biopolymers, and its effect on the CNTs, may open wide possibilities for future exploration. Indeed, combining the excellent conductivity and mechanical strength of CNTs with the inherent properties of different biomolecules might one day bring about new and exciting products.

23.8.3
Sensors

The size-dependent properties of nanomaterials have made them attractive for the development of highly sensitive sensors and detection systems. This is especially the case in the biological sciences, where the efficiency of a detection system is reflected by the size of the detector and the amount of sample required for detection. In this respect, CNTs hold much promise due to their typical size, which offers a great potential for enhancing the sensitivity of detection and diagnostics, while reducing the sample size (which may consist of a few individual proteins and antibodies). As carbon atoms dominate the surface of the SWNTs, the binding of proteins or antibodies to their surfaces can alter the surface state, and this in turn may result in a varied electrical and optical functionality. DNA-coated CNTs have a great potential in the design of a variety of sensors, and several research groups are currently working on this direction with electrochemical sensors [63]. In addition, protein- or DNA-incorporated CNTs can be used as a base for detecting biological surface reactions in a single protein or antibody attached to the CNT surface.

   The other main advantages of nanotubes include their optical activity, as long as they are neither aggregated nor chemically altered. Under appropriate conditions, semiconducting SWNTs show a bandgap fluorescent emission in the NIR spectral region at wavelengths characteristic of their specific (n,m)-structure. As natural biomolecules are relatively transparent and nonemissive in this range, the sharp spectra of SWNTs can be detected even in a complex biological environment. By taking advantage of these properties different sensors have been designed, including those used to detect glucose [15].

23.8.4
Therapeutic Agents

Today, functionalized CNTs are emerging as a new family of nanovectors for the delivery of different types of therapeutic molecules. A recent study proved that functionalized CNTs can penetrate into cells, with the CNTs being loaded with bioactive molecules via stable covalent bonds or supramolecular assemblies based on noncovalent interactions. When the cargos have been carried into the different cells, tissues, and organs, they are able to express their biological function. Recently, Cheruki and coworkers [64] injected debundled, chemically pristine SWNTs into rabbits and monitored their characteristic NIR fluorescence. The nanotube concentration in the blood serum decreased exponentially, with a half-life of 1 h.

The ability of CNTs to convert NIR light into heat provides an opportunity to create a new generation of immunoconjugates for cancer phototherapy, with high performance and efficacy. The fact that biological tissue is relatively transparent to NIR suggests that targeting CNTs to tumor cells, followed by noninvasive exposure to NIR light, would cause tumor ablation within the range of accessibility of the NIR light. Recently, it was shown that antibody-functionalized SWNTs operated very efficiently to kill targeted tumor cells [65], this being achieved by a specific binding of antibody-coupled CNTs to tumor cells *in vitro*, followed by their highly specific ablation with NIR light. For this, biotinylated polar lipids were used to prepare the stable, biocompatible, and noncytotoxic CNT dispersions that were then attached to antibodies. Following exposure to NIR light, the CNT–antibody construct was shown to have killed the targeted cells with great specificity [65].

23.9
Conclusions

The functionalization of nanotubes with biopolymers has achieved good progress in a relatively short time. When compared to the programmed properties and functions they possess in a natural environment, the technological development achieved to date remains in its infancy. It may be premature at this stage to claim advancements of the different fields using these materials, and especially for the biotechnological sector, where issues of toxicity have not yet been fully resolved. Nonetheless, the vast potential for these materials will surely offer rapid research and development in this direction. There is also no doubt in anticipating that this field will bring future breakthroughs for the advancement of humankind.

Acknowledgments

This work was supported by the Ministry of Education ("Brain Korea 21" project) and a grant from the Institute of Medical System Engineering at GIST.

References

1 Haddon, R.C. (2002) Carbon nanotubes. *Acc. Chem. Res.*, **35**, 997.

2 Nepal, D. (2006) PhD thesis. Dispersion, Individualization and Novel Supramolecular Adducts of Carbon Nanotubes. Department of Materials Science and Engineering 140, Gwangju Institute of Science and Technology (GIST), Gwangju.

3 Nepal, D. and Geckeler, K.E. (2006) Functional Nanomaterials. In: *Functionalization of Carbon Nanotubes*, (eds K.E. Geckeler and E. Rosenberg), American Scientific Publishers, Valencia, pp. 57–79.

4 Li, S.S., He, H., Jiao, Q.C. and Chuong, P.H. (2008) Applications of carbon nanotubes in drug and gene delivery. *Prog. Chem.*, **20**, 1798–1803.

5 Kim, S.N., Rusling, J.F. and Papadimitrakopoulos, F. (2007) Carbon nanotubes for electronic and electrochemical detection of biomolecules. *Adv. Mater.*, **19**, 3214–3228.

6 Prato, M., Kostarelos, K. and Bianco, A. (2008) Functionalized carbon nanotubes in drug design and discovery. *Acc. Chem. Res.*, **41**, 60–68.

7 Kovtyukhova, N.I., Mallouk, T.E., Pan, L. and Dickey, E.C. (2003) Individual single-walled nanotubes and hydrogels made by oxidative exfoliation of carbon nanotube ropes. *J. Am. Chem. Soc.*, **125**, 9761–9769.

8 Ajayan, P.M., Schadler, L.S., Giannaris, C. and Rubio, A. (2000) Single-walled carbon nanotube-polymer composites: strength and weakness. *Adv. Mater.*, **12**, 750–753.

9 O'Connell, M.J., Bachilo, S.M., Huffman, C.B., Moore,V.C., Strano, M.S., Haroz, E.H., Rialon, K.L., Boul, P.J., Noon, W.H., Kittrell, C., Ma, J., Haugl, R.H., Weisman, R.B., and Smalley, R.E. (2002) Band gap fluorescence from individual single-walled carbon nanotubes. *Science*, **297**, 593–596.

10 Huang, J.Y., Chen, S., Wang, Z.Q. *et al.* (2006) Superplastic carbon nanotubes. *Nature*, **439**, 281.

11 Yu, M.-F., Files, B.S., Arepalli, S. and Ruoff, R.S. (2000) Tensile loading of ropes of single wall carbon nanotubes and their mechanical properties. *Phys. Rev. Lett.*, **84**, 5552–5555.

12 Yu, M.-F., Lourie, O., Dyer, M.J., Moloni, K., Kelly, T.F. and Ruoff, R.S. (2000) Strength and breaking mechanism of multiwalled carbon nanotubes under tensile load. *Science*, **287**, 637–640.

13 Hudson, J.L., Casavant, M.J. and Tour, J.M. (2004) Water-soluble, exfoliated, nonroping single-wall carbon nanotubes. *J. Am. Chem. Soc.*, **126**, 11158–11159.

14 Nepal, D. and Geckeler, K.E. (2006) pH-sensitive dispersion and debundling of single-walled carbon nanotubes: lysozyme as a tool. *Small*, **2**, 406–412.

15 Barone, P.W., Baik, S., Heller, D.A. and Strano, M.S. (2005) Near-infrared optical sensors based on single-walled carbon nanotubes. *Nat. Mater.*, **4**, 86–92.

16 Saito, R., Dresselhaus, G. and Dresselhaus, M.S. (2000) Trigonal warping effect of carbon nanotubes. *Phys. Rev. B*, **61**, 2981.

17 Moore, V.C., Strano, M.S., Haroz, E.H., Hauge, R.H. and Smalley, R.E. (2003) Individually suspended single-walled carbon nanotubes in various surfactants. *Nano Lett.*, **3**, 1379–1382.

18 Zhang, Z.H., Yang, X.Y., Zhang, Y., Zeng, B., Wang, Z.J., Zhu, T.H., Roden, R.B.S., Chen, Y.S. and Yang, R.C. (2006) Delivery of telomerase reverse transcriptase small interfering RNA in complex with positively charged single-walled carbon nanotubes suppresses tumor growth. *Clin. Cancer Res.*, **12**, 4933–4939.

19 Hillebrenner, H., Buyukserin, F., Stewart, J.D. and Martin, C.R. (2006) Template synthesized nanotubes for biomedical delivery applications. *Nanomedicine*, **1**, 39–50.

20 Nepal, D., Balasubramanian, S., Simonian, A.L. and Davis, V.A. (2008) Strong antimicrobial coatings: single-walled carbon nanotubes armored with biopolymers. *Nano Lett.*, **8**, 1896–1901.

21 Dubin, R.A., Callegari, G.C., Kohn, J. and Neimark, A.V. (2008) Carbon nanotube fibers are compatible with mammalian

cells and neurons. *IEEE Trans. Nanobioscience*, **7**, 11–14.

22 Hu, H., Ni, Y., Mandal, S.K., Montana, V., Zhao, B., Haddon, R.C. and Parpura, V. (2005) Polyethyleneimine functionalized single-walled carbon nanotubes as a substrate for neuronal growth. *J. Phys. Chem. B*, **109**, 4285–4289.

23 Usui, Y., Aoki, K., Narita, N., Murakami, N., Nakamura, I., Nakamura, K., Ishigaki, N., Yamazaki, H., Horiuchi, H., Kato, H., Taruta, S., Kim, Y.A., Endo, M. and Saito, N. (2008) Carbon nanotubes with high bone-tissue compatibility and bone-formation acceleration effects. *Small*, **4**, 240–246.

24 Harrison, B.S. and Atala, A. (2007) Carbon nanotube applications for tissue engineering. *Biomaterials*, **28**, 344–353.

25 Bandyopadhyaya, R., Nativ-Roth, E., Regev, O. and Yerushalmi-Rozen, R. (2002) Stabilization of individual carbon nanotubes in aqueous solutions. *Nano Lett.*, **2**, 25–28.

26 Wenseleers, W., Vlasov, I.I., Goovaerts, E., Obraztsova, E.D., Lobach, A.S. and Bouwen, A. (2004) Efficient isolation and solubilization of pristine single-walled nanotubes in bile salt micelles. *Adv. Funct. Mater.*, **14**, 1105–1112.

27 Zhang, M., Smith, A. and Gorski, W. (2004) Carbon nanotube-chitosan system for electrochemical sensing based on dehydrogenase enzymes. *Anal. Chem.*, **76**, 5045–5050.

28 Hasegawa, T., Fujisawa, T., Numata, M., Umeda, M., Matsumoto, T., Kimura, T., Okumura, S., Sakurai, K., and Shinkai, S. (2004) Single-walled carbon nanotubes acquire a specific lectin-affinity through supramolecular wrapping with lactose-appended schizophyllan. *Chem. Commun.*, 2150–2151.

29 Star, A., Steuerman, D.W., Heath, J.R. and Stoddart, J.F. (2002) Starched carbon nanotubes. *Angew. Chem. Int. Ed. Engl.*, **41**, 2508–2512.

30 Zheng, M., Jagota, A., Strano, M.S., Santos, A.P., Barone, P., Chou, S.G., Diner, B.A., Dresselhaus, M.S., McLean, R.S., Onoa, G.B., Samsonidze, G.G., Semke, E.D., Usrey, M., and Walls, D.J. (2003) Structure-based carbon nanotube sorting by sequence-dependent DNA assembly. *Science*, **302**, 1545–1548.

31 Cathcart, H., Quinn, S., Nicolosi, V., Kelly, J.M., Blau, W.J. and Coleman, J.N. (2007) Spontaneous debundling of single-walled carbon nanotubes in DNA-based dispersions. *J. Phys. Chem. C*, **111**, 66–74.

32 Liu, Y., Chen, J., Anh, N.T., Too, C.O., Misoska, V. and Wallace, G.G. (2008) Nanofiber mats from DNA, SWNTs, and poly(ethylene oxide) and their application in glucose biosensors. *J. Electrochem. Soc.*, **155**, K100–K103.

33 Jonganurakkun, B., Liu, X.D., Nodasaka, Y., Nomizu, M. and Nishi, N. (2003) Survival of lactic acid bacteria in simulated gastrointestinal juice protected by a DNA-based complex gel. *J. Biomater. Sci., Polym. Ed.*, **14**, 1269–1281.

34 Fink, H.-W. and Schonenberger, C. (1999) Electrical conduction through DNA molecules. *Nature*, **398**, 407–410.

35 Yan, H., Zhang, X., Shen, Z. and Seeman, N.C. (2002) A robust DNA mechanical device controlled by hybridization topology. *Nature*, **415**, 62–65.

36 Turberfield, A. (2003) *Phys. World*, **16** (3), 43–46.

37 Steckl, A.J. (2007) DNA – a new material for photonics? *Nat. Photonics*, **1**, 3–5.

38 Zheng, M., Jagota, A., Semke, E.D., Diner, B.A., Mclean, R.S., Lustig, S.R., Richardson, R.E. and Tassi, N.G. (2003) DNA-assisted dispersion and separation of carbon nanotubes. *Nat. Mater.*, **2**, 338–342.

39 Nepal, D., Sohn, J.-I., Aicher, W.K., Lee, S. and Geckeler, K.E. (2005) Supramolecular conjugates of carbon nanotubes and DNA by a solid-state reaction. *Biomacromolecules*, **6**, 2919–2922.

40 Tsang, S.C., Guo, Z., Chen, Y.K., Green, M.L.H., Hill, H.A.O., Hambley, T.W. and Sadler, P.J. (1997) Immobilization of platinated and iodinated oligonucleotides on carbon nanotubes. *Angew. Chem. Int. Ed. Engl.*, **36**, 2198–2200.

41 Nakashima, N., Okuzono, S., Murakami, H., Nakai, T. and Yoshikawa, K. (2003) DNA dissolves single-walled carbon nanotubes in water. *Chem. Lett.*, **32**, 456–457.

42 Baker, S.E., Cai, W., Lasseter, T.L., Weidkamp, K.P. and Hamers, R.J. (2002) Covalently bonded adducts of

deoxyribonucleic acid (DNA) oligonucleotides with single-wall carbon nanotubes: synthesis and hybridization. *Nano. Lett.*, **2**, 1413–1417.

43 Keren, K., Berman, R.S., Buchstab, E., Sivan, U. and Braun, E. (2003) DNA-templated carbon nanotube field-effect transistor. *Science*, **302**, 1380–1382.

44 Drexler, K. (1992) *Nanosystems: Molecular Machinery, Manufacturing, and Computation*, John Wiley & Sons, Inc., New York.

45 Wang, G.-W., Komatsu, K., Murata, Y. and Shiro, M. (1997) Synthesis and X-ray structure of dumb-bell-shaped C120. *Nature*, **387**, 583–586.

46 Pan, H., Guo, Z.-X., Dai, L., Zhang, F., Zhu, D., Czerw, R. and Carroll, D.L. (2003) Carbon nanotubols from mechanochemical reaction. *Nano Lett.*, **3**, 29–32.

47 Tu, X. and Zheng, M. (2008) A DNA-based approach to the carbon nanotube sorting problem. *Nano Res.*, **1**, 185–194.

48 Kane, R.S. and Stroock, A.D. (2007) Nanobiotechnology: protein-nanomaterial interactions. *Biotechnol. Prog.*, **23**, 316–319.

49 Huang, W.J., Taylor, S., Fu, K.F., Lin, Y., Zhang, D.H., Hanks, T.W., Rao, A.M. and Sun, Y.P. (2002) Attaching proteins to carbon nanotubes via diimide-activated amidation. *Nano Lett.*, **2**, 311–314.

50 Zorbas, V., Ortiz-Acevedo, A., Dalton, A.B., Yoshida, M.M., Dieckmann, G.R., Draper, R.K., Baughman, R.H., Jose-Yacaman, M., and Musselman, I.H. (2004) Preparation and characterization of individual peptide-wrapped single-walled carbon nanotubes. *J. Am. Chem. Soc.*, **126**, 7222–7227.

51 Zorbas, V., Smith, A.L., Xie, H., Ortiz-Acevedo, A., Dalton, A.B., Dieckmann, G.R., Draper, R.K., Baughman, R.H., and Musselman, I.H. (2005) Importance of aromatic content for peptide/single-walled carbon nanotube interactions. *J. Am. Chem. Soc.*, **127**, 12323–12328.

52 Xie, H., Becraft, E.J., Baughman, R.H., Dalton, A.B. and Dieckmann, G.R. (2008) Ranking the affinity of aromatic residues for carbon nanotubes by using designed surfactant peptides. *J. Peptide Sci.*, **14**, 139–151.

53 Karajanagi, S.S., Yang, H., Asuri, P., Sellitto, E., Dordick, J.S. and Kane, R.S. (2006) Protein-assisted solubilization of single-walled carbon nanotubes. *Langmuir*, **22**, 1392–1395.

54 Ma, Y., Chiu, P.L., Serrano, A., Ali, S.R., Chen, A.M. and He, H. (2008) The electronic role of DNA-functionalized carbon nanotubes: efficacy for in situ polymerization of conducting polymer nanocomposites. *J. Am. Chem. Soc.*, **130**, 7921–7928.

55 Han, X., Li, Y. and Deng, Z. (2007) DNA-wrapped single walled carbon nanotubes as rigid templates for assembling linear gold nanoparticle arrays. *Adv. Mater.*, **19**, 1518–1522.

56 Badaire, S., Zakri, C., Maugey, M., Derre, A., Barisci, J.N., Wallace, G. and Poulin, P. (2005) Liquid crystals of DNA-stabilized carbon nanotubes. *Adv. Mater.*, **17**, 1673–1676.

57 Barisci, J.N., Tahhan, M., Wallace, G.G., Badaire, S., Vaugien, T., Maugey, M. and Poulin, P. (2004) Properties of carbon nanotube fibers spun from DNA-stabilized dispersions. *Adv. Funct. Mater.*, **14**, 133–138.

58 Lynam, C., Moulton, S.E. and Wallace, G.G. (2007) Carbon-nanotube biofibers. *Adv. Mater.*, **19**, 1244–1248.

59 Razal, J.M., Gilmore, K.J. and Wallace, G.G. (2008) Carbon nanotube biofiber formation in a polymer-free coagulation bath. *Adv. Funct. Mater.*, **18**, 61–66.

60 Shin, S.R., Lee, C.K., So, I.S. *et al.* (2008) DNA-wrapped single-walled carbon nanotube hybrid fibers for supercapacitors and artificial muscles. *Adv. Mater.*, **20**, 466–470.

61 Muñoz, E., Suh, D.-S., Collins, S., Selvidge, M., Dalton, A.B., Kim, B.G., Razal, J.M., Ussery, G., Rinzler, A.G., Martínez, M.T. and Baughman, R.H. (2005) Highly conducting carbon nanotube/polyethyleneimine composite fibers. *Adv. Mater.*, **17**, 1064–1067.

62 Kozlov, M.E., Capps, R.C., Sampson, W.M., Ebron, V.H., Ferraris, J.P. and Baughman, R.H. (2005) Spinning solid and hollow polymer-free carbon nanotube fibers. *Adv. Mater.*, **17**, 614–617.

63 Yogeswaran, U., Thiagarajan, S. and Chen, S.M. (2008) Recent updates of DNA incorporated in carbon nanotubes

and nanoparticles for electrochemical sensors and biosensors. *Sensors*, **8**, 7191–7212.

64 Cherukuri, P., Gannon, C.J., Leeuw, T.K., Schmidt, H.K., Smalley, R.E., Curley, S.A. and Weisman, R.B. (2006) Mammalian pharmacokinetics of carbon nanotubes using intrinsic near-infrared fluorescence. *Proc. Natl Acad. Sci. USA*, **103**, 18882–18886.

65 Chakravarty, P., Marches, R., Zimmerman, N.S., Swafford, A.D.E., Bajaj, P., Musselman, I.H., Pantano, P., Draper, R.K., and Vitetta, E.S. (2008) Thermal ablation of tumor cells with anti body-functionalized single-walled carbon nanotubes. *Proc. Natl Acad. Sci. USA*, **105**, 8697–8702.

24
Nanoparticle-Cored Dendrimers and Hyperbranched Polymers: Synthesis, Properties, and Applications

Young-Seok Shon

24.1
Introduction

Functionalized nanoparticles have attracted remarkable academic and industrial interest due to their unique chemical, biological, catalytic, optical, and electronic properties, as well as their potential applications in electronic devices, sensors, molecular reagents, and catalysis [1–3]. Nanoparticles protected by a polymeric ligand have been employed to prepare nanoscale building blocks for functional materials. Polymeric stabilizers such as thiolated polymers [4–6] and electrostatic polymer multilayers [7–9] have been used for the stabilization of metal nanoparticles. Nanoparticles have also been modified with polymer chains covalently bound to the surface using "grafting from" reactions, such as atom transfer radical (ATR) polymerization, living cationic ring-opening polymerization, and surface-immobilized ring-opening metathesis polymerization (ROMP) [10–12].

Dendrimers have attracted an intense interest because of their well-defined structures with interior cavities and chemical versatility [13–15]. Dendrimers also often have provided some of the most versatile tools to building blocks used in the construction of three-dimensional (3-D) organic/inorganic nanostructures [16]. The current research on dendrimers is mostly focused on the development of functionalized dendrimers with an active core or external surface groups [17–19]. These functional dendrimers have shown a variety of applications in the fields of medicinal chemistry (e.g., magnetic resonance imaging and drug delivery), host–guest chemistry, and catalysis [17–19]. Significant research progress has also been made in the use of dendritic frameworks to surround functional core molecules such as rotaxane, porphyrin, and fullerene [20–23].

Due to the interesting properties of metal nanoparticles [1], the incorporation of nanoparticles into the dendrimer templates has been attempted [24–33]. For example, monometallic nanoparticles of gold, palladium, platinum, copper, and silver were incorporated into the dendrimers by reduction of their metal salts with $NaBH_4$ in the presence of "Tomalia-type" poly(amidoamine) dendrimers [24–26]. These materials were synthesized using a template approach in which metal ions

Advanced Nanomaterials. Edited by Kurt E. Geckeler and Hiroyuki Nishide
Copyright © 2010 WILEY-VCH Verlag GmbH & Co. KGaA, Weinheim
ISBN: 978-3-527-31794-3

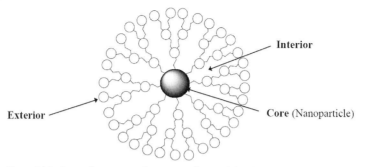

Figure 24.1 General structure of nanoparticle-cored dendrimers (NCDs).

were first extracted into the interior of dendrimers and then subsequently reduced by chemical reactants to yield quite monodisperse particles having dimensions of less than 5 nm. Both, bimetallic and semiconductor nanoparticles have also been incorporated into the dendrimers by similar routes [28, 29]. These materials have a number of applications, but are especially useful in catalysis (e.g., hydrogenations, Heck coupling, Stille, and Suzuki reactions) [30–33]. Several reviews have been prepared detailing the synthesis, characterization, and application of dendrimer-encapsulated nanoparticles [23–25].

Structurally well-defined, inorganic, cluster-cored dendrimers have also attracted interest. Dendrimers supported by the $[Re_6Se_8]^{2+}$ metal cluster were synthesized by Zheng and coworkers [34], this being the first example of utilizing metal clusters to construct cluster-cored metallodendrimers. These dendrimers exhibited dramatic color changes with only slight variations in the structure of dendrons, which suggested their interesting–and potentially important–optical tunability. Gorman *et al.* reported the synthesis and characterization of iron–sulfur $[Fe_4S_4]$ cluster-cored dendrimers [35, 36]. Here, it was recognized that the iron–sulfur cluster employed as the core unit in these dendrimers displayed redox potentials that were very sensitive to environments such as the generation of dendrons and the solvent media. Dendrimers based on polyhedral oligomeric silsesquioxane (POSS) cores were also prepared, and provided an increased selectivity in hydroformylation reactions due to a positive dendrimer effect [37].

Nanoparticle-cored dendrimers represent an organic–inorganic hybrid nanostructure with a nanoparticle core and well-defined dendritic wedges (Figure 24.1) [38–58]. The preparation of nanoparticle-cored dendrimers represents an important advance in the control and preparation of new organized nanostructures. Dendritic scaffolds offer an enhanced stability and well-defined property to the materials, and can provide some segregation of external groups and internal nanoparticle core. While the external functionality mainly controls the overall solubility and reactivity of nanoparticle-cored dendrimers (NCDs), internal functionality provides a completely different environment for the nanoparticle core from that of the bulk solution and the exterior. Another important structural feature of NCDs is the large void spaces present inside; these may act as guest cavities and also function as "reaction vessels" for catalytic reactions.

In this chapter we present a concise review of synthetic strategies for nanoparticle-cored dendrimers, along with their properties and applications. The most popular synthetic approach – the *direct method* – uses a modified Schiffrin reaction with dendrons containing thiols or disulfides [38–49]. In contrast, the *indirect method* involves two-step reactions, namely the synthesis of monolayer-stabilized nanoparticles followed by the ligand-place exchange of thiolated dendrons [47, 50–54]. A third synthetic strategy is also described, in which single or multistep organic reactions are employed to build dendritic architectures around a monolayer-protected metal nanoparticle [55–58]. Finally, the synthesis of nanoparticle-cored hyperbranched polymers is introduced as the convenient alternate for nanoparticle-cored dendrimers. In this approach, well-defined hyperbranched polymers are grafted onto a nanoparticle by self-organization [59–61]. The chemical and physical properties, and also the technological applications, of nanoparticle-cored dendrimers and hyperbranched polymers are described with regards to the synthetic methods for these nanostructures.

24.2
Synthesis of Nanoparticle-Cored Dendrimers via the Direct Method, and their Properties and Application

When, in 1994, Schiffrin *et al.* reported the convenient two-phase synthesis of isolable and soluble alkanethiolate-protected gold nanoparticles, this had a major impact on nanoparticle research and development [1]. In this reaction, $AuCl_4^-$ was transferred to toluene, using tetraoctylammonium bromide as the phase-transfer reagent. The addition of alkanethiol to organic-phase $AuCl_4^-$, followed by reduction with $NaBH_4$, generated alkanethiolate-protected gold nanoparticles. The first examples of nanoparticle-cored dendrimers were synthesized using a modified, two-phase Schiffrin protocol (Scheme 24.1).

Shultz, Linderman, Feldheim, and their coworkers reported that the alkene-terminated tripodal ligand (**1**; Figure 24.2), which contained a branched structure with a single thiolate group on one end and three alkene-terminated hydrocarbon chains on the other end, could be used for the stabilization of gold nanoparticles that were approximately 5 nm in diameter [38]. The alkene-terminated tripodal ligand was used in order to maximize polymeric crosslinking, which led to the formation of structurally rigid hollow capsules following removal of the gold nanoparticle. The dimensions of the capsules (~10 ± 2.5 nm) agreed well with the diameter expected based on the size of the template gold nanoparticle and the length of the alkylthiol spacer. Crosslinking was carried out by olefin metathesis of the thiol-derivatized nanoparticles using a catalytic quantity of Grubbs catalyst [62].

Similarly, poly(oxymethyl) dendrons with a single thiol group (**2**; Figure 24.2) at the focal point were used as the stabilizer, and produced dendron-stabilized gold

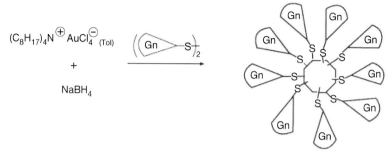

Scheme 24.1 Synthesis and structure of nanoparticle-cored dendrimers (NCDs), where Gn represents a dendritic wedge of variable generation number. Reproduced with permission from Ref. [41]; © 2003, The American Chemical Society.

nanoparticles [39]. This approach afforded highly stable gold nanoparticles with average diameters of 2.4 to 3.1 nm. Generation-one [G-1], two [G-2], and three [G-3] dendron-stabilized gold nanoparticles were highly stable, both in solution and in the solid state. These NCDs remained unchanged after standing overnight at elevated temperatures (at 50 °C in solution, and at 160 °C in the solid state). The high stability of NCDs protected by dendrons with a thiol group was a clear improvement over that of dendrimer-encapsulated nanoparticles. The core size and size distribution of the nanoparticle-cored dendrimers, when examined with high-resolution transmission electron microscopy (HR-TEM), were not significantly affected by the use of different molar ratios of dendrons/HAuCl₄ in this study [39].

Torigoe *et al.* also used poly(oxymethylphenylene) dendrons (PPDs) (**3**; Figure 24.2) of generations 1–4 functionalized with a thiol group at the focal point as capping ligands for gold nanoparticle-cored dendrimers [40]. The average particle size was 1.5 and 1.4 nm for [G-1] and [G-2], respectively with a relatively small standard deviation. The particle size was increased for higher-generation dendrons (2.2 nm for [G-3] and 3.8 nm for [G-4]), with a much broader size distribution. The spontaneous formation of one-dimensional (1-D) arrays of [G-1] and [G-2] NCDs was observed over a larger area. This suggested that the dendron ligands had a strong tendency to crystallize due to an intermolecular π–π stacking interaction of thiol-terminated PPDs, as in the case of the pyridine derivative system in which two-dimensional (2-D) superlattices have been observed [63]. The formation of 1-D arrays over a larger area would provide more elaborate nanoelectronic systems.

The systematic characterization of stable gold nanoparticle-cored dendrimers with different generations ([G-1] to [G-5]) prepared by the Schiffrin reaction using polyaryl ether dendritic disulfide as a capping reagent, was reported by Whitesell and Fox in 2003 (**4**; Figure 24.2) [41]. Results obtained with transmission electron microscopy (TEM) suggested that the nanoparticle core of NCDs exhibited a relatively high polydispersity. The average core sizes were increased from [G-1] NCDs to [G-4] NCDs, and then showed a decrease as the dendron wedge sizes increased to [G-5]. The structure and composition of NCDs were more carefully studied

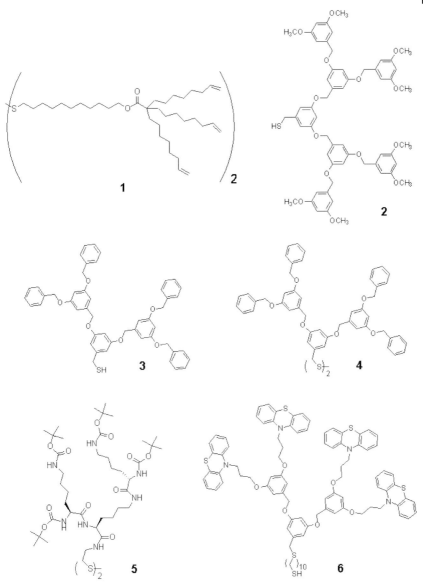

Figure 24.2 Dendron thiols and disulfides used for the synthesis of nanoparticle-cored dendrimers.

using ultraviolet (UV), Fourier transform infrared (FT-IR) and nuclear magnetic resonance (NMR) spectrosocopies in this study. The surface plasmon bands (~520 nm) of these NCDs were similar to those for arenethiolate-protected clusters reported elsewhere [1, 3]. The IR spectra of NCDs and their dendritic disulfide precursors were almost identical. In the ^{1}H NMR spectra, the peaks appeared very similar for dendrons and NCDs, except for broadened peaks in the bases. The

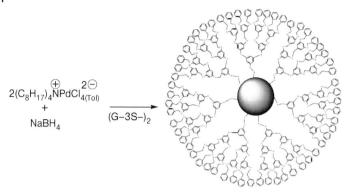

$$2(C_8H_{17})_4\overset{\oplus}{N}PdCl_4\overset{2\ominus}{(Tol)}$$
$$+$$
$$NaBH_4 \quad \xrightarrow{\text{(G–3S–)}_2}$$

Scheme 24.2 Synthesis of Pd-G-3 NCDs. Reproduced with permission from Ref. [42]; © 2003, The American Chemical Society.

number of dendron units on a nanoparticle core, which was obtained via TEM and thermogravimetric analysis (TGA), decreased with the increased generation of the dendrons, with values falling from 2.18 nm^{-2} for [G-2] NCDs to 0.27 nm^{-2} for [G-5] NCDs. Therefore, for the higher-generation NCDs a large fraction of the surface area of the metal nanoparticle was not passivated and thus available for the catalytic reaction. Based on these characteristics, transition metal-based NCDs were recognized as being good catalysts for many organic reactions.

Whitesell and Fox prepared [G-3] Pd NCDs (as shown in Scheme 24.2) and investigated their catalytic properties [42]. A combination of data acquired using TEM and TGA showed that, for the Pd NCDs, more than 90% of the nanoparticle surface was unpassivated. If most of the unpassivated surface atoms were to be available for catalysis, and the required reagents could penetrate the dendritic exterior to reach the metal surface, then these NCDs might serve as excellent catalysts for many organic reactions. The Heck and Suzuki reactions (Scheme 24.3), both of which are normally carried out on Pd-phosphine catalysts, were tested on these Pd NCDs. The turnover number (TON; calculated as moles of product per mole catalyst) and turnover frequency (TOF; calculated as moles of product per mole catalyst per hour) for both the Heck and Suzuki reactions were very high. In addition, the absence of any side product other than the starting compound, products, and solvent for both reactions indicated the high efficiency of Pd NCDs as a new catalytic system.

Whitesell and Fox also reported the synthesis of NCDs stabilized by polyaryl ether dendrons with ester groups on the periphery, using the Schiffrin reaction [43]. The ester-terminated NCDs were converted to the corresponding carboxylate-functionalized NCDs by treatment with sodium hydroxide (Scheme 24.4). The sodium salts of carboxylate-terminated NCDs were soluble in water and exhibited micelle-like properties in solution; this was an important point because dendritic micelles can be used as catalysts in water and as drug-delivery agents. The micellar properties of NCDs were probed by the spectral shifts of dye molecules (Figure

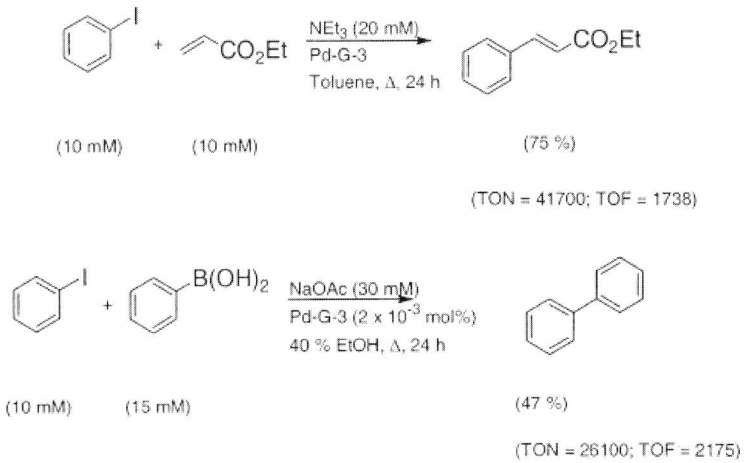

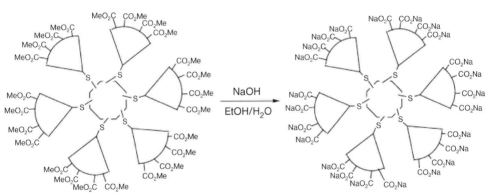

Scheme 24.3 Heck reaction (top) and Suzuki reaction (bottom) with Pd-G-3 NCDs. TON = turnover number; TOF = turnover frequency. Reproduced with permission from Ref. [42]; © 2003, The American Chemical Society.

Scheme 24.4 Basic hydrolysis of ester-terminated NCDs to produce unimolecular micelles. Reproduced with permission from Ref. [43]; © 2003, The American Chemical Society.

24.3). In aqueous solution, pinacyanol chloride (PC) showed absorption bands at 549 and 601 nm, but these were red-shifted to 564 and 610 nm, respectively, in NCD micelles. Such a result suggested that NCDs could act as micellar hosts and, potentially, also as water-soluble catalysts.

Chechik, Smith, and coworkers reported the synthesis of L-lysine based dendron-stabilized gold nanoparticles using the modified one-phase Schiffrin reaction (5; Figure 24.2) [44]. The size of the nanoparticle core was found to decrease with increasing dendron generation, a finding which contrasted with data reported for other NCD systems [39, 41, 45]. These differences suggested that the structure and

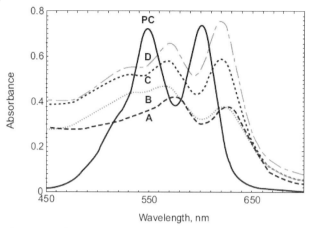

Figure 24.3 Absorption spectra of pinacyanol chloride
(1×10^{-5} M) in water (PC) and in aqueous solutions of
(a) Au-G1(CO$_2$Na), (b) Au-G2(CO$_2$Na), (c) Au-G3(CO2Na),
and (d) Au-G4(CO2Na). [Au-Gn(CO$_2$Na)] were ~0.1 mg ml^{-1}
water. Reproduced with permission from Ref. [43]; © 2003,
The American Chemical Society.

functionality of the dendron ligands must play an important role in determining
the characteristics of the gold core. Such a size relationship could be explained in
terms of steric effects, as the much more bulky higher-generation dendritic system
would be expected to pack more efficiently around a small core, thus favoring a
smaller particle size. The thermal stability of the NCDs in solution was governed
by the extent of branching in the surrounding dendron ligands. A different order
of thermal ripening was observed at 120 °C, which increased in size in the order
[G-1] > [G-2] > [G-3].

Astruc and colleagues synthesized redox-active NCDs using the Schiffrin reaction from a 1:1 mixture of dodecanethiol and nonaferrocenyl thiol dendrons
(Scheme 24.5) [46, 47]. However, this approach resulted in NCDs without any open
catalytic sites on the nanoparticle core. Smaller dodecanethiols could be coassem-
bled on the nanoparticle surface, so that the integrity and solubility of NCDs was
suitable for their applications in the electrochemical sensing of anions. The gold
nanoparticle core was seen to be surrounded by ~360 silylferrocenyl units, such
that the general structure of these NCDs closely resembled that of large metallodendrimers. These redox-active NCDs were capable of selectively recognizing
H$_2$PO$_4^-$ anions and adenosine-5′-triphosphate (ATP^{2-}) with a positive dendritic
effect, even in the presence of other anions such as Cl$^-$ and HSO$_4^-$. The titration
of NCDs using anions resulted in a shift of the cyclic voltammetry (CV) wave to
a less positive potential. The nonaferrocenyl thiol dendron-functionalized nanoparticles could be deposited on the Pt electrode by dipping it into the NCD solution. This modified electrode was quite robust and was capable of recognizing
different anions. Moreover, the salt of these anions could be easily removed simply

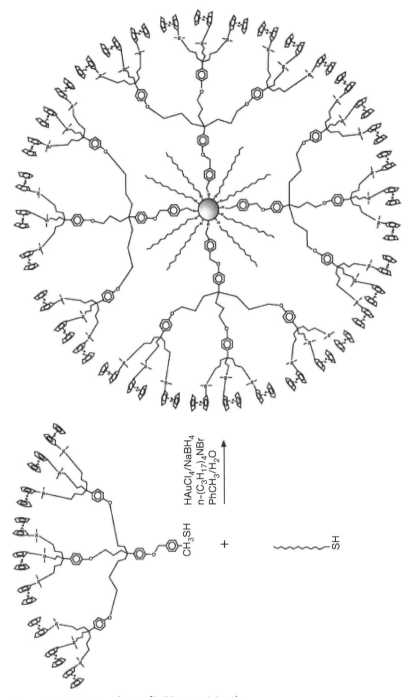

Scheme 24.5 Direct synthesis of NCDs containing the nonaferrocenyl thiol dendron. Reproduced with permission from Ref. [47]; © 2003, The American Chemical Society.

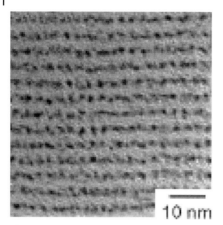

Figure 24.4 Transmission electron microscopy image of one-dimensional arrays of phenothiazine-terminated NCDs. Reproduced with permission from Ref. [48]; © 2006, The Royal Society of Chemistry.

by dipping the electrode in CH_2Cl_2, the consequence being that the electrode could be re-used many times.

Phenothiazine-terminated NCDs were prepared by Fujihara and coworkers using the two-phase Schiffrin reaction (**6**; Figure 24.2) [48]. The gold nanoparticles with a higher-generation dendron bearing a long chain alkanethiol at the focal point had a smaller core size with a narrow size distribution. These NCDs underwent the spontaneous formation of 1-D arrays (Figure 24.4), and exhibited an interesting one-electron transfer behavior. The intermolecular π–π stacking interaction of the thiol-terminated phenothiazine is believed to be the driving force for this self-organization. The 1-D assembly of redox-NCDs not only provides a good model to study the size-dependent electronic and optical properties of metal nanoparticles, but may also play an important future role in nanoelectronics.

Dendrons with another focal group that is capable of metal complexation were reported by Zheng *et al.* [45]. The Oct_4N^+-$AuCl_4$ in toluene was reduced by $NaBH_4$ in the presence of 4-pyridone-functionalized dendrons (**7**; Figure 24.5) as a capping reagent, and this resulted in stable, gold nanoparticle-cored dendrimers. There is no known example of monolayer formations of pyridone derivatives on bulk gold surfaces due to the chemical instability of such monolayers. Therefore, these results suggests that the high surface curvature of the gold nanoparticles can support the high-density packing of the capping ligands, which have only weak metal surface-binding properties. An examination using TEM highlighted the correlation between the particle size and the generation of dendrons, with higher-generation dendrons producing larger particles ([G-1], [G-2], [G-3] = 2.0, 3.3, 5.1 nm, respectively). However, the [G-3] NCDs were less stable than [G-1] and [G-2] NCDs, because the larger dendrons led to an increased open space between the ligands. The resultant weaker force between the dendrons and the particle caused a more rapid particle agglomeration of [G-3] NCDs.

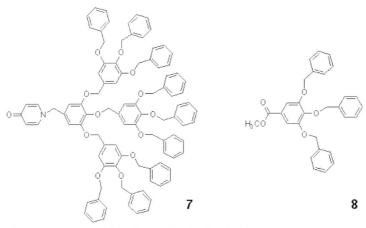

7   **8**

Figure 24.5 Structure of dendrons with other functional groups used for the synthesis of nanoparticle-cored dendrimers.

Dendrons with an ester focal point were also used for the synthesis of polyaryl dendron-protected Pd nanoparticles (**8**; Figure 24.5) [49]. Briefly, H_2PdCl_4 was phase-transferred into the organic phase using tetraoctylammonium bromide. The dendrons were then added to the reaction mixture before addition of N_2H_4 as reducing agent. The resultant mixture was stirred under dry argon for an additional 24 h at room temperature. The Pd nanoparticle-cored dendrimers formed had mostly a spherical shape, and an average core size which ranged from ~10 to ~70 nm. The Pd core size was seen to increases with the decreasing molar ratio of dendrons to metal ions.

24.3
Synthesis of Nanoparticle-Cored Dendrimers by Ligand Exchange Reaction, and their Properties and Applications

Nanoparticles with protecting monolayers composed of thiolate ligands can be functionalized by a partial ligand-replacement [1]. In this exchange reaction, the incoming ligands replace the thiolate ligands on nanoparticles by an associative reaction, while the displaced thiolate becomes a thiol. Ligand-exchange reactions of dodecanethiolate-protected gold nanoparticles with triferrocenyl thiol dendrons in dichloromethane were first reported by Astruc *et al.* (Scheme 24.6) [46, 47]. Although an excess of functional thiol dendrons was used, the percentages of dendron thiols introduced as ligands in dodecanethiolate-protected gold nanoparticles were less than 5% of the overall ligands. The limit of the ligand-exchange reaction was even greater with larger dendrons. Attempts to synthesize NCDs with nonaferrocenyl thiol dendrons resulted in the incorporation of very small amounts of dendrons (less than one per nanoparticle). As the rate of ligand-exchange

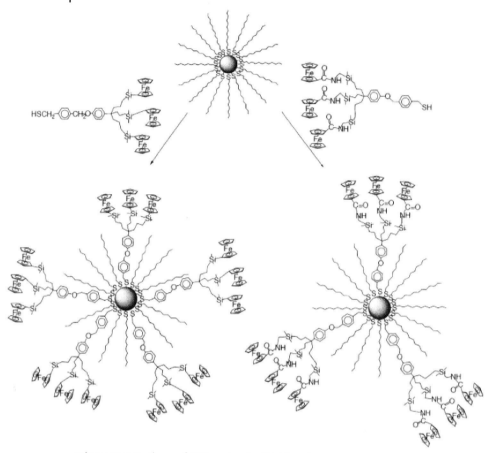

Scheme 24.6 Synthesis of NCDs using the thiol-ligand exchange reaction. Reproduced with permission from Ref. [47]; © 2003, The American Chemical Society.

between thiols is dependent on the chain length and/or steric bulk of the initial monolayers on nanoparticles, the bulky or very small incoming ligands are difficult to replace the original ligands on the nanoparticle surface due to either kinetic (e.g., steric hindrance) or thermodynamic effects, respectively. Thus, the main short-coming of the ligand-exchange reaction, when using larger dendrons, was in fact quite consistent with previous results on ligand-place exchange reactions of alkanethiolate-protected nanoparticles [1, 3]. In contrast, the exchange between ligands having different functional groups was much more efficient.

Peng *et al.* synthesized hydrophilic NCDs by the ligand-exchange of citrate-capped gold nanoparticles using hydroxy-functionalized dendron thiols (**9**; Figure 24.6) [50]. The citrate reduction of HAuCl$_4$ in water led to citrate-capped gold nanoparticles with core sizes that ranged from 10 to 150 nm [3]. Such nanoparticles

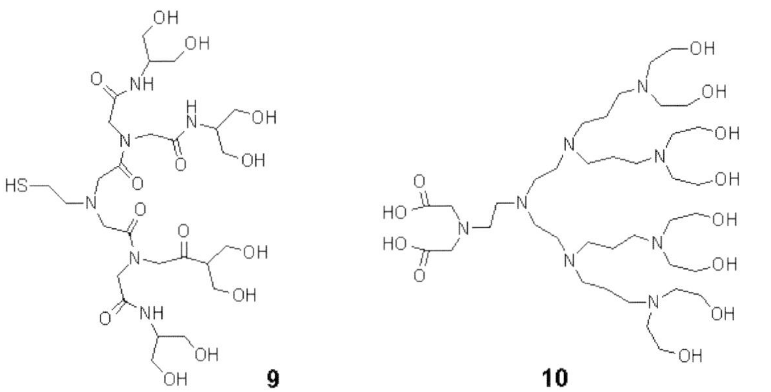

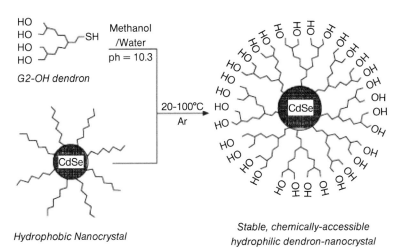

Figure 24.6 Structure of hydroxy-functionalized dendrons used for the synthesis of nanoparticle-cored dendrimers.

Scheme 24.7 Schematic process for converting hydrophobic semiconductor nanoparticles into hydrophilic and chemically processable NCDs. Reproduced with permission from Ref. [50]; © 2002, The American Chemical Society.

are typically useful when a rather loose shell of ligands is required around the gold core for ligand-exchange [64]. Peng also performed ligand-exchange reactions of trioctylphosphine oxide (TOPO)-capped CdSe nanoparticles with hydroxy-func-tionalized dendron thiols (Scheme 24.7) [50, 51]. Dendron thiols can rather easily replace the loosely bound TOPO groups, which have a weak binding property with CdSe nanoparticles. The chemistry related to CdSe NCDs can be applied for devel-oping photoluminescence-based labeling reagents for biomedical applications. The photochemical, thermal, and chemical stability of both CdSe and Au NCDs was exceptionally good compared to that of the corresponding alkanethiolate-

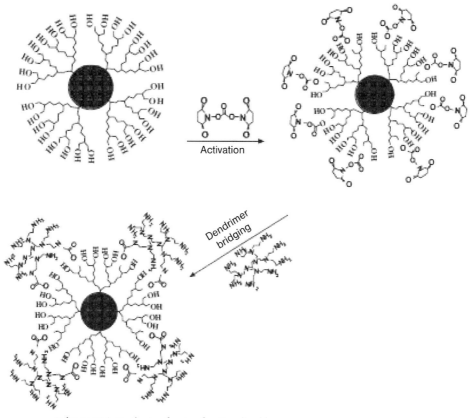

Scheme 24.8 Synthesis of amine-functionalized box nanoparticles. Reproduced with permission from Ref. [51]; © 2003, The American Chemical Society.

protected nanoparticles. The stability of CdSe NCDs could be further enhanced by a global crosslinking of the dendron ligands around a nanoparticle core [51]. Such crosslinking of alkene-terminated CdSe NCDs was achieved via ring-closing metathesis, and led to each nanoparticle being sealed in a dendron box. Another strategy–*dendrimer bridging*–was also reported by Peng for the simultaneous formation and functionalization of a biocompatible and bioaccessible dendron box with a semiconductor nanoparticle core [52] (Scheme 24.8). The dendrimer bridging involved an activation of the hydroxyl terminal groups of CdSe NCDs with a homobifunctional crosslinker, N,N-disuccinimidyl carbonate, followed by a crosslinking reaction with [G-2] polyamidoamine (PAMAM) dendrimers. The resulting amine box nanoparticles were very stable chemically, thermally, and photochemically. In addition, the amine groups on the surface of the box nanoparticles provided reactive sites for the conjugation of biological entities.

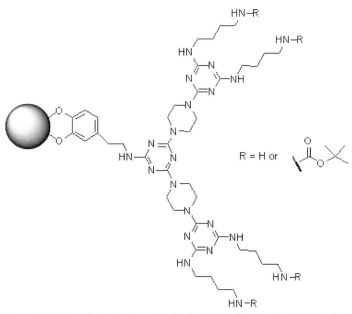

Figure 24.7 Simanek (melamine)-type dendron-coated iron oxide nanoparticles.

Bioconjugation and biodetection using box nanoparticles were each successfully demonstrated with the avidin–biotin system [52].

Liu and Peng reported the synthesis of highly luminescent water-soluble CdSe/CdS core–shell NCDs by ligand-exchange reactions [53]. A dendron ligand with eight hydroxyl terminal groups and two carboxylate anchoring groups was found to replace alkylamine ligands on CdSe/CdS nanoparticles in toluene, and to convert the particles to water-soluble NCDs (**10**; Figure 24.6). The resultant water-soluble NCDs retained 60% of the photoluminescence value of the alkylamine-protected CdSe/CdS core–shell nanoparticles in toluene. Compared to the dendron thiol-protected NCDs, the photoluminescence value of these NCDs was much higher (about sixfold). Moreover, the UV-brightened photoluminescence could be retained for several months.

Dendron (with a shell of Simanek)-fuctionalized superparamagnetic (iron oxide) nanoparticles were prepared by Gao *et al.* using ligand-exchange methods (Figure 24.7) [54]. The resulting Fe_2O_3 NCDs exhibited a switchable solubility in a variety of solvents, depending on the terminal groups of the dendritic shells, and were examined as soluble matrices for supporting magnetically recoverable homogeneous Pd catalysts for Suzuki crosscoupling reactions in organic solvents. For this, Gao immobilized a Pd-triphenyl phosphine moiety to the termini of dendron-protected iron oxide nanoparticles for catalysis. The Fe_2O_3 NCDs were also investigated as potential contrast agents for magnetic resonance imaging (MRI) in aqueous media.

24.4
Synthesis of Nanoparticle-Cored Dendrimers by Dendritic Functionalization, and their Properties and Applications

The direct synthesis of NCDs was somewhat problematic because it required a large excess of dendronized thiols or disulfides, especially for the synthesis of NCDs with a small and monodispersed nanoparticle core [39–41]. The direct method also provided little control over the nanoparticle core dimension. NCDs with different core size were produced, when dendrons with different sizes (or generations) were used. In addition, the NCDs synthesized by the direct method often contained small amounts of trapped tetraoctylammonium bromide that could not be removed completely, even by repeated extraction with solvent. The indirect method using the ligand-exchange reaction was also somewhat limited considering a low exchange rate, especially for the synthesis of metal nanoparticle-cored dendrimers, as described in Section 24.3. A more convenient and cost-efficient synthetic methodology for the synthesis of NCDs with controlled particle core sizes, generations, and dendritic wedge density has been considered highly desirable for the basic understanding of structure–property relationships of these nanostructures.

The present authors' approach is based on a strategy in which the synthesis of monolayer-protected nanoparticles is followed by building the dendrimer architecture on the nanoparticle surface, using either the convergent or divergent approach (Scheme 24.9). A convergent synthesis of NCDs (**11**) can be accomplished by the coupling of reactive nanoparticles with functionalized dendrons. As the reaction only takes place at the functional groups on the surface of nanoparticles, this approach eliminates the need for a large excess of functionalized dendrons, and also maintains an intact core size for the synthesis of NCDs with different interior layers (generations) and dendritic wedge densities. The divergent approach

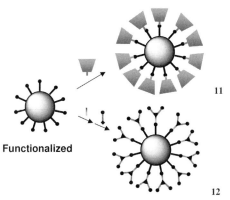

Functionalized

11

12

Scheme 24.9 General reaction schemes for the synthesis of nanoparticle-cored dendrimers by (**11**) convergent and (**12**) divergent approaches.

involves an iterative sequence of reaction steps, which lead to the addition of a branch (or branches) to the monolayer-protected cluster (MPC) framework (**12**).

24.4.1
Nanoparticle-Cored Dendrimers by the Convergent Approach

The convergent synthesis of NCDs (Scheme 24.10) was accomplished by the ester coupling reaction of 11-mercaptoundecanoic acid-functionalized gold nanoparticles with hydroxy-functionalized "Fréchet-type" dendrons ([G-1]-OH, [G2]-OH, and [G-3]-OH) in the presence of ester coupling reagents (1,3-dicycohexylcarbodiimide [DCC] and 4-dimethylaminopyridine [DMAP]) [56–58]. This synthetic approach provided an access to functionalized NCDs, which contain multiple reactive (COOH) functional groups inside dendritic wedges, even after the coupling of dendrons. These unreacted COOH groups resulted from the steric repulsion of dendrons during the synthesis of NCDs. The appearance of a new band at ~1735 cm^{-1} in FT-IR spectroscopy after the reaction indicated a formation of ester bonds due to a successful coupling reaction between 11-mercaptoundecanoic acids and [G-n]-OH. The ^{1}H NMR results showed the same peak-broadening effect, as do monolayers of alkanethiolate-protected nanoparticles generated from alkanethiols. This peak-broadening effect also confirmed that [G-n]-OH was reacted with 11-mercaptoundecanoic acid-functionalized gold nanoparticles and bonded onto the surface of nanoparticles. The increased organic fractions were also observed by TGA.

The dendritic encapsulation of a functional molecule provides a site isolation, which mimics principles from biomaterials because dendritic scaffolds can provide the segregation of external and internal functionality. The incorporation of electroactive compounds onto the surfaces of nanoparticles in dendritic architecture should provide an opportunity to tune their electrochemical properties. The incorporation of redox-active groups in NCDs was achieved by a simple coupling reaction of ferrocene methanol with the remaining COOH groups inside dendritic wedges of NCDs ([G-2]$_{conv}$ NCDs), as shown in Scheme 24.11 [56, 58]. After an ester coupling reaction between [G-2]$_{conv}$ NCDs and ferrocene methanol, the IR spectrum of the NCD-encapsulated ferrocene (Fc@NCDs) showed an increase in the intensity of a band at 1735 cm^{-1} (ester), in addition to the strong band around 1450 cm^{-1} which indicated the presence of cyclopentadienyl groups. The NMR spectrum of Fc@NCDs showed additional broad bands at ~4.10 ppm, which was corresponding to the signals from cyclopentadienyl groups. Cyclic voltammetry was employed to compare electrochemical behaviors of ferrocene methanol and Fc@NCDs. Both, ferrocene methanol and Fc@NCDs exhibited well-defined voltammetric peaks corresponding to ferrocenyl groups. From the voltammogram of ferrocene methanol, it could be seen that there was a pair of voltammetric waves with a peak splitting of 233 mV at 100 mV s^{-1}, indicating a quasi-reversible electron-transfer processes. Interestingly, the peak splitting was found to be 112 mV at 100 mV s^{-1} for Fc@NCDs, which was only half of that for the ferrocene methanol; this was most likely due to either multielectron transfer or to the adsorption of particles onto the electrode.

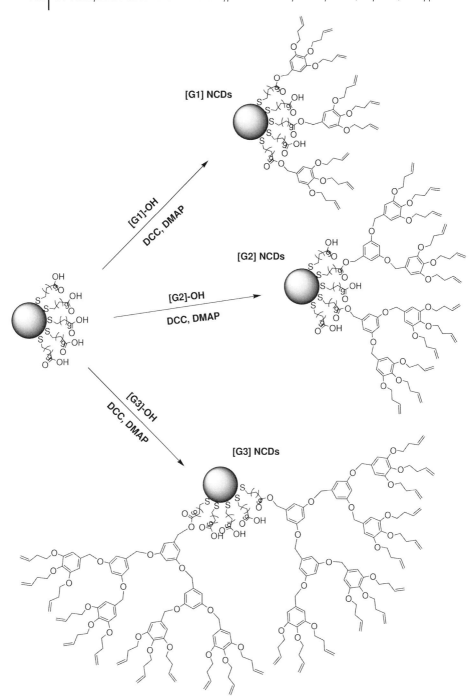

Scheme 24.10 Synthesis of [G-n]$_{conv}$ NCDs by an ester coupling reaction.

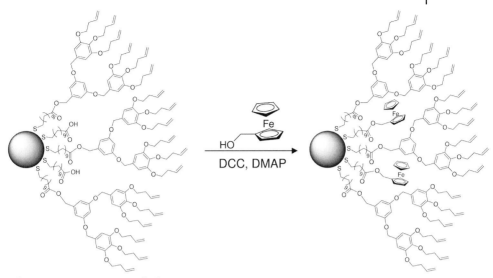

Scheme 24.11 Incorporation of a ferrocene moiety in NCDs.

24.4.2
Nanoparticle-Cored Dendrimers by the Divergent Approach

The NCDs (**6**) were grown in a stepwise manner from a central nanoparticle involving the amide coupling reaction with 1,2-diaminoethane ($NH_2CH_2CH_2NH_2$) in the presence of DCC and the Michael addition reaction with methyl acrylate ($CH_2CH\text{-}CO_2CH_3$) [11]. The repetition of the two reactions in the divergent approach was attempted for the synthesis of higher-generation NCDs (Scheme 24.12) [55, 57]. The IR spectra of 11-mercaptoundecanoic acid (MUA) MPCs showed a strong carbonyl (C=O) stretching band at ~$1710\,cm^{-1}$, the disappearance of which and the appearance of a new band at ~$1650\,cm^{-1}$, after completion of the first reaction, indicated the formation of amide bonds. This was evidence of the successful coupling reaction between the COOH groups and 1,2-diaminoethane. The excess (10/1 = [1,2-diaminoethane]/[COOH of MUA MPCs]) of 1,2-diaminoethane was used to prevent the intermolecular crosslinking of COOH groups on the clusters. These amine-functionalized MPCs ([G-0.5]$_{div}$ NCDs) were reacted with excess methyl acrylate, such that the conjugate addition of the amine moiety occurred in "cascade" form (2:1 molar ratio of methyl acrylate to amine functional group) to generate [G-1]$_{div}$ NCDs with terminal ester groups. The appearance of an ester band at ~$1730\,cm^{-1}$ was a clear indication of the successful addition reaction. The amide bands at ~$1650\,cm^{-1}$ indicated the presence of internal amide groups. Following the amide coupling reaction of [G-1]$_{div}$ NCDs with 1,2-diaminoethane, the disappearance of the ester band at $1730\,cm^{-1}$ was clearly observed, indicating the

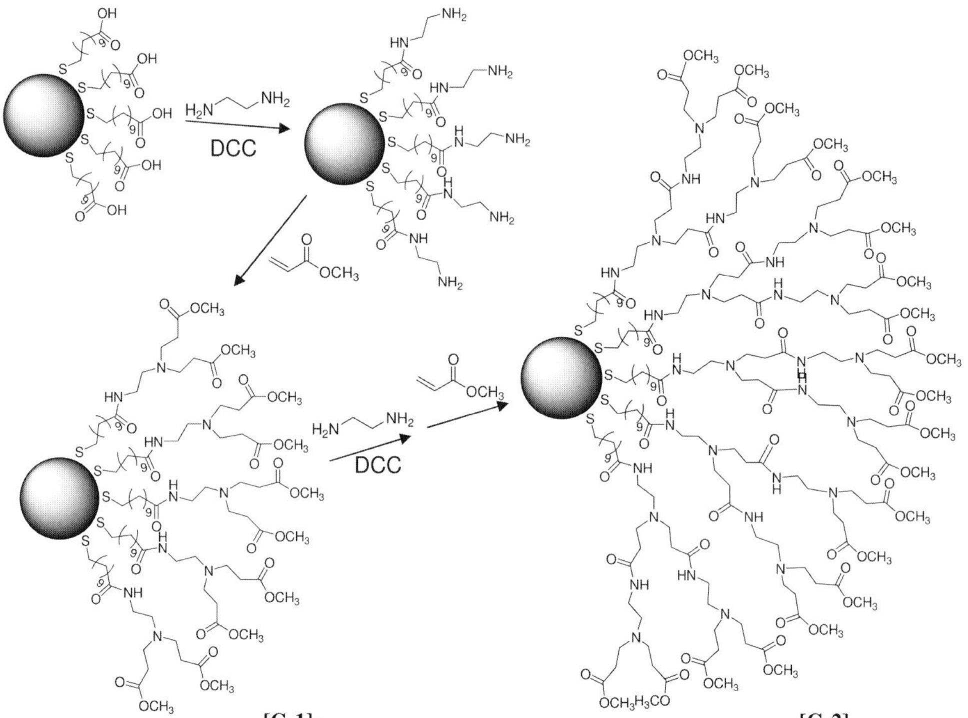

Scheme 24.12 Synthesis of [G-1]$_{div}$ NCDs and [G-2]$_{div}$ NCDs.

formation of amine-functionalized [G-1.5]$_{div}$ NCDs. The reaction was again followed by the Michael addition reaction of [G-1.5]$_{div}$ NCDs; this resulted in the reappearance of ester bands in the IR spectra of [G-2]$_{div}$ NCDs. However, the relatively low intensity of this band suggested that the reaction had not been completed. Further iterative reactions of [G-2]$_{div}$ NCDs were no longer effective and did not involve incorporation of methyl acrylate, as evidenced by an absence of ester bands in the IR spectra of [G-2]$_{div}$ NCDs. A TEM image of the [G-1] NCDs showed the presence of slightly fused domains upon evaporation of the solvent. Additional TEM images of higher-generation NCDs prepared by the divergent approach showed that the multistep reactions of NCDs had resulted in the formation of large aggregates. Such aggregate formation prevented or minimized any further dendritic branching reactions for the higher-generation NCDs. This aggregation behavior of "Tomalia-type" NCDs during the iterative reactions probably resulted from the presence of strong inter-cluster hydrogen-bonding interactions.

24.5
Synthesis of Nanoparticle-Cored Hyperbranched Polymers by Grafting on Nanoparticles

Further studies regarding the synthesis and characterization of NCDs are important for a fundamental understanding of nanostructures, and could also generate interesting technological applications in areas such as catalysis and sensing. However, one weakness of investigations involving NCDs is that multiple reaction steps and purification processes must be included in the synthesis of functionalized dendrons. As hyperbranched polymers may have properties and structures which are quite similar to those of dendrimers (even though the synthetic procedure requires only a direct, one-pot grafting reaction), the substitution of dendron-capping ligands with hyperbranched polymer ligands would definitely benefit in realizing some of the proposed technological applications in the real world, and the commercialization of these nanostructures. Some examples of grafting hyperbranched polymers on either silica or zinc oxide nanoparticle surfaces have recently been reported [59–61].

The first example of nanoparticle-cored hyperbranched polymers (NCHPs) was prepared by self-condensing vinyl polymerization via atom transfer radical polymerization (ATRP) from OH-functionalized silica nanoparticle surfaces (Scheme 24.13) [59]. For polymerization, the ATRP initiators were covalently attached to the

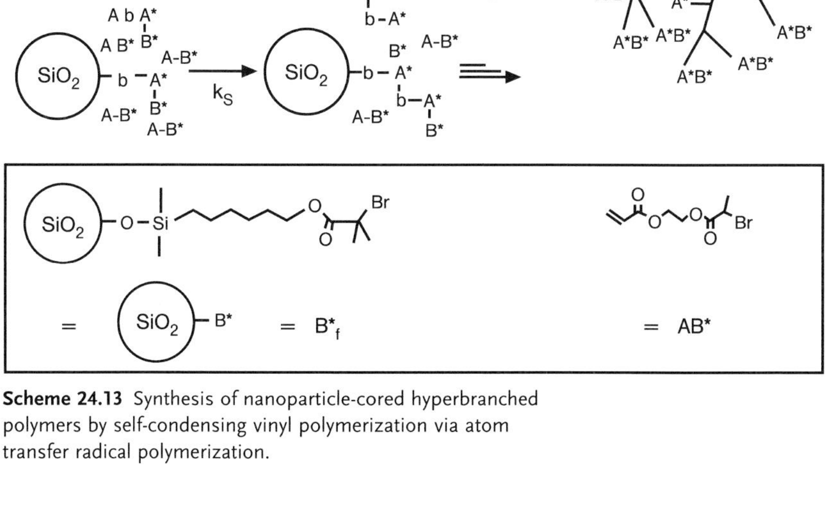

Scheme 24.13 Synthesis of nanoparticle-cored hyperbranched polymers by self-condensing vinyl polymerization via atom transfer radical polymerization.

surface of silica nanoparticles. Well-defined hyperbranched polymer chains with multifunctional bromoester end groups were confirmed by gel permeation chromatography (GPC), GPC/viscosity, and NMR. These NCHPs can be designed to have a fairly open structure, allowing the functional materials (such as metal ions) to penetrate the ligands more easily than in conventional linear polymer layers. Therefore, a well-controlled synthesis for these NCHPs might lead to the creation of novel core–shell hybrids that are controllable on the nanoscopic scale, and have chemically sensitive interfaces.

A similar ATRP reaction involving a surface-initiated self-condensing vinyl polymerization of *p*-chloromethyl styrene was used for the preparation of NCHP from NH_2-functionalized silica nanoparticles [60]. The surface Br atom of the NCHP could initiate another surface-initiated ATRP of methyl methacrylate and produce a dendritic-graft copolymer grafted onto silica nanoparticles. The same strategy was applied for the synthesis of ZnO nanoparticle-cored hyperbranched polymers with multifunctional chlorobenzyl functional end groups [61].

24.6
Conclusions and Outlook

Nanoparticle-cored dendrimers or hyperbranched polymers can be prepared using four different synthetic strategies: (i) a direct method, using a modified Schiffrin reaction; (ii) an indirect method, involving a ligand-place exchange reaction; (iii) the dendritic functionalization of nanoparticles; and (iv) the surface grafting of a hyperbranched polymer. The direct method is especially useful for the synthesis of nanoparticle-cored dendrimers with a large fraction of unpassivated metal nanoparticle surface, which is ideal for catalytic reaction. The indirect method, using a ligand-place exchange reaction, is ideal for the loading of functionalized dendrons on the surface of nanoparticle core. Dendritic functionalization is most versatile in terms of controlling the structure and composition of nanoparticle-cored dendrimers, while surface grafting of hyperbranched polymers might be the most convenient method as it can eliminate the multistep organic reactions involved in the synthesis of dendrons. Today, the synthetic procedures developed for nanoparticle-cored dendrimer preparation permits near-complete control over the critical molecular design parameters, such as core size, dendron wedge density, surface/interior chemistry, flexibility/rigidity, and solubility/micelle-like properties. The availability of effective methods to prepare distinct nanoparticle-cored dendrimers and hyperbranched polymers will allow investigators to exploit these materials in a variety of ways, including dendritic catalysis, energy storage, sensors, and drug delivery.

Acknowledgment

These investigations were supported by a start-up grant and an SCAC grant from California State University, Long Beach, California, USA.

References

1 Astruc, D. and Daniel, M.-C. (2004) *Chem. Rev.*, **104**, 293–346.

2 Pasquato, L., Pengo, P. and Scrimin, P. (2004) *J. Mater. Chem.*, **14**, 3481–3487.

3 Shon, Y.-S. (2004) Metal nanoparticles protected with monolayers: synthetic methods, in *Dekker Encyclopedia of Nanoscience and Nanotechnology*, Marcel Dekker, New York, pp. 1–11.

4 Zhu, M.-Q., Wang, L.-Q., Exarhos, G.J. and Li, A.D.Q. (2004) *J. Am. Chem. Soc.*, **126**, 2656–2657.

5 Lowe, A.B., Sumerlin, B.S., Donovan, M.S. and McCormick, C.L. (2002) *J. Am. Chem. Soc.*, **124**, 11562–11563.

6 Shimmin, R.G., Schoch, A.B. and Braun, P.V. (2004) *Langmuir*, **20**, 5613–5620.

7 Schneider, G. and Decher, G. (2004) *Nano Lett.*, **4**, 1833–1839.

8 Gittins, D.I. and Caruso, F. (2001) *J. Phys. Chem. B*, **105**, 6846–6852.

9 Caruso, F. and Möhwald, H. (1999) *J. Am. Chem. Soc.*, **121**, 6039–6046.

10 Werne, T. and Patten, T.E. (2001) *J. Am. Chem. Soc.*, **123**, 7497–7505.

11 Jordan, R., West, N., Ulman, A., Chou, Y.-M. and Nuyken, O. (2001) *Macromolecules*, **34**, 1606–1611.

12 Watson, K.J., Zhu, J., Nguyen, S.T. and Mirkin, C.A. (1999) *J. Am. Chem. Soc.*, **121**, 462–463.

13 Zeng, F. and Zimmerman, S.C. (1997) *Chem. Rev.*, **97**, 1681–1712.

14 Bosman, A.W., Janssen, H.M. and Meijer, E.W. (1999) *Chem. Rev.*, **99**, 1665–1688.

15 Fischer, M. and Vögtle, F. (1999) *Angew. Chem. Int. Ed. Engl.*, **38**, 884–905.

16 Tully, D.C. and Fréchet, J.M.J. (2001) *Chem. Commun.*, 1229–1239.

17 Fréchet, J.M.J. and Hecht, S. (2001) *Angew. Chem. Int. Ed. Engl.*, **40**, 74–91.

18 Adronov, A. and Fréchet, J.M.J. (2000) *Chem. Commun.*, 1701–1710.

19 Balzani, V., Ceroni, P., Gestermann, S., Kauffmann, C., Gorka, M. and Vögtle, F. (2000) *Chem. Commun.*, 853–854.

20 Amabilino, D.B., Ashton, P.R., Balzani, V., Brown, C.L., Credi, A., Fréchet, J.M.N., Leon, J.W., Raymo, F.M., Spencer, N., Stoddart, J.F. and Venturi, M. (1996) *J. Am. Chem. Soc.*, **118**, 12012–12020.

21 Stapert, H.R., Nishiyama, N., Jiang, D.L., Aida, T. and Kataoka, K. (2000) *Langmuir*, **16**, 8182–8188.

22 Herzog, A., Hirsch, A. and Vostrowsky, O. (2000) *Eur. J. Org. Chem.*, **2000**, 171–180.

23 Smith, D.K. (2006) *Chem. Commun.*, 34–44.

24 Crooks, R.M., Zhao, M., Sun, L., Chechik, V. and Yeung, L.K. (2001) *Acc. Chem. Res.*, **34**, 181–190.

25 Scott, R.W.J., Wilson, O.M. and Crooks, R.M. (2005) *J. Phys. Chem. B*, **109**, 692–704.

26 Ye, H., Scott, R.W.J. and Crooks, R.M. (2004) *Langmuir*, **20**, 2915–2920.

27 Floriano, P.N., Noble, C.O. IV, Scoonmaker, J.M., Poliakoff, E.D. and McCarley, R.L. (2001) *J. Am. Chem. Soc.*, **123**, 10545–10553.

28 Wilson, O.M., Scott, R.W.J., Garcia-Martinez, J.C. and Crooks, R.M. (2005) *J. Am. Chem. Soc.*, **127**, 1015–1024.

29 Lemon, B.I. and Crooks, R.M. (2000) *J. Am. Chem. Soc.*, **122**, 12886–12887.

30 Niu, Y., Yeung, L.K. and Crooks, R.M. (2001) *J. Am. Chem. Soc.*, **123**, 6840–6846.

31 Narayanan, R. and El-Sayed, M.A. (2004) *J. Phys. Chem. B*, **108**, 8572–8580.

32 Garcia-Martinez, J.C., Lezutekong, R. and Crooks, R.M. (2005) *J. Am. Chem. Soc.*, **127**, 5097–5103.

33 Rahim, E.H., Kamounah, F.S., Frederiksen, J. and Christensen, J.B. (2001) *Nano Lett.*, **1**, 499–501.

34 Wang, R. and Zheng, Z. (1999) *J. Am. Chem. Soc.*, **121**, 3549–3550.

35 Gorman, C.B. and Smith, J.C. (2000) *J. Am. Chem. Soc.*, **122**, 9342–9343.

36 Chasse, T.L., Sachdeva, R., Li, Q., Li, Z., Petrie, R.J. and Gorman, C.B. (2003) *J. Am. Chem. Soc.*, **125**, 8250–8254.

37 Ropartz, L., Morris, R.E., Foster, D.F. and Cole-Hamilton, D.J. (2001) *Chem. Commun.*, 361–362.

38 Wu, M., O'Neill, S.A., Brousseau, L.C., McConnell, W.P., Shultz, D.A., Linderman, R.J. and Feldheim, D.L. (2000) *Chem. Commun.*, 775–776.

39 Kim, M.-K., Jeon, Y.-M., Jeon, W.-S., Kim, H.-J., Hong, S.G., Park, C.G.

and Kim, K. (2001) *Chem. Commun.*, 667–668.

40 Nakao, S., Torigoe, K. and Kon-No, K. (2002) *J. Phys. Chem. B*, **106**, 12097–12100.

41 Gopidas, K.R., Whitesell, J.K. and Fox, M.A. (2003) *J. Am. Chem. Soc.*, **125**, 6491–6502.

42 Gopidas, K.R., Whitesell, J.K. and Fox, M.A. (2003) *Nano Lett.*, **3**, 1757–1760.

43 Gopidas, K.R., Whitesell, J.K. and Fox, M.A. (2003) *J. Am. Chem. Soc.*, **125**, 14168–14180.

44 Love, C.S., Chechik, V., Smith, D.K. and Brennan, C. (2004) *J. Mater. Chem.*, **14**, 919–923.

45 Wang, R., Yang, J., Zheng, Z., Carducci, M.D., Jiao, J. and Seraphin, S. (2001) *Angew. Chem. Int. Ed. Engl.*, **40**, 549–552.

46 Astruc, D., Daniel, M.-C. and Ruiz, J. (2004) *Chem. Commun.*, 2637–2649.

47 Daniel, M.-C., Ruiz, J., Nlate, S., Blais, J.-C. and Astruc, D. (2003) *J. Am. Chem. Soc.*, **125**, 2617–2628.

48 Komine, Y., Ueda, I., Goto, T. and Fujihara, H. (2006) *Chem. Commun.*, 302–304.

49 Jiang, G., Wang, L., Chen, T., Yu, H. and Wang, J. (2004) *Nanotechnology*, **15**, 1716–1719.

50 Wang, Y.A., Li, J.J., Chen, H. and Peng, X. (2002) *J. Am. Chem. Soc.*, **124**, 2293–2298.

51 Guo, W., Li, J.J., Wang, Y.A. and Peng, X. (2003) *J. Am. Chem. Soc.*, **125**, 3901–3909.

52 Guo, W., Li, J.J., Wang, Y.A. and Peng, X. (2003) *Chem. Mater.*, **15**, 3125–3133.

53 Liu, Y., Kim, M., Wang, Y., Wang, Y.A. and Peng, X. (2006) *Langmuir*, **22**, 6341–6345.

54 Duanmu, C., Saha, I., Zheng, Y., Goodson, B.M. and Gao, Y. (2006) *Chem. Mater.*, **18**, 5973–5981.

55 Shon, Y.-S., Cutler, E.C., Chinn, L.E. and Garabato, B.D. (2004) *Polym. Mater. Sci. Eng.*, **91**, 1068.

56 Choi, D., Chinn, L.E. and Shon, Y.-S. (2005) *Polym. Mater. Sci. Eng.*, **93**, 739–740.

57 Cutler, E.C., Lundin, E., Garabato, B.D., Choi, D. and Shon, Y.-S. (2007) *Mater. Res. Bull.*, **42**, 1178–1185.

58 Shon, Y.-S. and Choi, D. (2006) *Chem. Lett*, **35**, 644–645.

59 Mori, H., Seng, D.C., Zhang, M. and Müller, A.H.E. (2002) *Langmuir*, **18**, 3682–3693.

60 Mu, B., Wang, T. and Liu, P. (2007) *Ind. Eng. Chem. Res.*, **46**, 3069–3072.

61 Liu, P. and Wang, T. (2007) *Poly. Eng. Sci.*, **47**, 1296–1301.

62 Beil, J.B., Lemcoff, N.G. and Zimmerman, S.C. (2004) *J. Am. Chem. Soc.*, **126**, 13576–13577.

63 Teranishi, T., Haga, M., Shiozawa, Y. and Miyake, M. (2000) *J. Am. Chem. Soc.*, **122**, 4237.

64 Watson, K.J., Zhu, J., Nguyen, S.B.T. and Mirkin, C.A. (1999) *J. Am. Chem. Soc.*, **121**, 462–463.

25
Concepts in Self-Assembly

Jeremy J. Ramsden

25.1
Introduction

This chapter begins with some basic definitions, followed by a review of the early studies conducted by Ashby and von Foerster. Some actual systems recognized as exemplars of self-assembly are then examined, after which self-assembly as a valid approach to nanofacture is considered; in this context it is usually known as "bottom-up" manufacture. There follow some brief notes on useful ideas for understanding the self-assembly processes. The aim is not to discuss ordering related to phase transitions [1] (this will be detailed in a forthcoming article), or reaction-diffusion systems leading to striation and other regular patterns, nor phenomena such as spinodal dewetting [2] and other types of segregation (e.g., in block copolymers).

Assembly means gathering things together and arranging them (fitting them together) to produce an organized structure. A boy assembling a mechanism from "Meccano" captures the meaning, as does an assembly line where the workers and machines progressively build a complicated product such as a motor-car from simpler components; or when an editor assembles a newspaper from copy provided by journalists. In all of these examples the component parts are subject to constraints – otherwise they would not fit together – and their entropy S must necessarily decrease. Since the free energy $\Delta G = \Delta H - T\Delta S$ must then necessarily increase, in general the process will not, by itself, happen spontaneously. On the contrary, a segregated arrangement will tend to become homogeneous. Hence, in order for self-assembly to become a reality, something else must be included. Typically, enthalpy H is lost through the formation of connections (bonds) between the parts, and provided that $|\Delta H|$ exceeds $|T\Delta S|$ there is at least the possibility that the process can occur spontaneously – which is presumably what is meant by "self-assembly." The final result is generally considered to be in some sort of equilibrium.

As *entropy* is always multiplied by the temperature T in order to compute the free energy, it should be noted that the relevant temperature is not necessarily what is measured with a thermometer. Rather, only the (typically small number

Advanced Nanomaterials. Edited by Kurt E. Geckeler and Hiroyuki Nishide
Copyright © 2010 WILEY-VCH Verlag GmbH & Co. KGaA, Weinheim
ISBN: 978-3-527-31794-3

of) relevant degrees of freedom should be taken into account; the temperature is the mean energy per degree of freedom.

If slightly sticky granules are thrown into a rotating drum, or stirred in a mixing bowl, they will all eventually clump together to form a "random" structure (e.g., Figure 25.1a), but one which is evidently less random than the initial collection of freely flowing granules. An alternative example might be the dry ingredients of concrete (sand and cement) when thrown into a "cement mixer," to which a small amount of water is added. Indeed the entropy is lowered, but each time the experiment is repeated the result will be different in detail. Thus, if the system were really ergodic, and if there were sufficient time to make all the repetitions, the ergodicity would only be broken because of insufficient time.

The goal of an ideal self-assembly process is for the same structure to be formed each time the constituent particles are mixed together (Figure 25.1b), as was imagined by von Foerster [3]. Of course, energy must be put into the system; in the "purest" form of self-assembly, the energy is thermal (random), but it could also be provided by an external field (e.g., electric or magnetic).

If we are satisfied by the constituent particles being merely joined together in a statistically uniform fashion and, moreover, the process happens spontaneously, then it is more appropriate to speak of "self-joining" or "self-connecting." The mixing of concrete referred to above is, at least at first sight, an example of such a self-connecting process; more generally, one can refer to "gelation." On approaching the phenomenon from a practical viewpoint, it is clear that gelation is almost always triggered by a change of external conditions imposed upon the system, such as a change of temperature, or dehydration, and the spontaneity implied by the prefix "self-" is absent. The only action that can be imposed upon the system without violating the meaning of "self-" is that of bringing the constituent particles

(a)   (b)

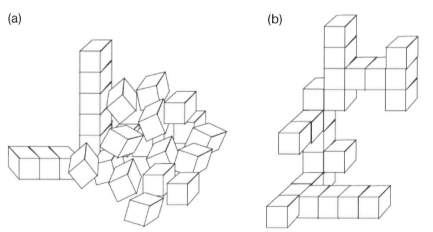

Figure 25.1 (a) The result of mixing isotropically sticky cubelets; (b) Putative result of mixing selectively sticky cubelets.

together. Note that even here we diverge from the everyday meaning of the term "assembly," which includes the gathering together of people, either spontaneously as when a group of people wish to protest against some new measure introduced by an authoritarian government, or by decree as in the case of the event with which the day is begun in many British schools.

If the process in which we are interested is deemed to begin at the instant the constituent particles are brought together, then we can place the mixing of concrete into the category of self-joining, because we could (although it is not usually done in this way) bring the wetted particles of sand and cement together, whereupon they would indeed spontaneously join together to form a mass. The meaning of self-joining (of which self-connecting is a synonym) is then the property possessed by certain particles of spontaneously linking together with their neighbors when they are brought to within a certain separation. One can also imagine there being kinetic barriers to joining, which can be overcome given enough time. Note that the particles each need more than one valency (unit of combining capacity), otherwise dimers would be formed and the process would then stop. A good example is when steam condenses to form water. We can suppose that the steam is first supercooled, which brings the constituent particles (H_2O molecules) together; the transition to liquid water is actually a first-order phase transition that requires an initial nucleus of water to be formed spontaneously. "Gelation" then occurs by the formation of weak hydrogen bonds (a maximum of four per molecule) throughout the system.

Strictly speaking, it is not necessary for all the constituent particles to be brought together instantaneously, as implied above. Once the particles are primed to be able to connect themselves to their neighbors, they can be brought together one by one. This is the model of diffusion-limited aggregation (DLA) [4]. In Nature, this is how biopolymers are formed – the monomers (e.g., nucleic acids or amino acids) are joined sequentially by strong covalent bonds to form a gradually elongating linear chain. The actual self-assembly into a compact three-dimensional (3-D) structure involves additional weak hydrogen bonds between neighbors that may be distant according to their positions along the linear chain (see Section 25.3.7); some of the weak bonds formed early are broken before the final structure is reached.

In the chemical literature, "self-assembly" is often used as a synonym of self-organization. A recapitulation of the examples already discussed shows, however, that the two terms cannot really be considered synonymous. The diffusion-limited aggregate is undoubtedly assembled, but can scarcely be considered to be organized, not least because every repetition of the experiment will lead to a result that is different in detail, and only the same when considered statistically. "Organized" is an antonym of "random"; therefore, the entropy of a random arrangement is high while that of an organized arrangement is low. It follows that inverse entropy may be taken as a measure of the degree of organization, and this notion will be further refined in the following section.

The diffusion-limited aggregate differs from the heap of sand only insofar as the constituent particles are connected to each other. An example of organization is shown in Figure 25.1b. The impossibility of self-organization is proved by Ashby, as will be described in Section 25.2.2.

Before discussing self-organization, we must first discuss organization, of which self-organization is a part. If the elements in a collection (here we shall not say "system," because that already implies a degree of organization) are organized to some degree, that implies they are in some way connected to each other, which can be considered as a kind of communication, and are hence subject to certain constraints. In other words, if the state of element B is dependent on the state of A to some extent, then we can say that B's state is conditional on that of A. Likewise, the relationship between A and B may be conditional on the state of C. Whenever there is conditionality, there is *constraint*: B is not as free to adopt states as it would be in a totally unorganized system [5].

25.2
Theoretical Approaches to Self-Organization

25.2.1
Thermodynamics of Self-Organization

Consider a universe U comprising a system S and its environment E; that is, $U = S \cup E$ (Figure 25.2). Self-organization (of S) implies that its entropy spontaneously diminishes; that is:

$$\delta S_S / \delta t < 0. \tag{25.1}$$

Accepting the Second Law of Thermodynamics, such a spontaneous change can only occur if, concomitantly,

$$\delta S_E / \delta t > 0, \tag{25.2}$$

with some kind of coupling to ensure that the overall change of entropy is greater than or equal to zero. If all processes were reversible, the changes could exactly

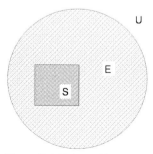

Figure 25.2 Universe U comprising system S and its environment E.

balance each other, but since (inevitably, we may suppose) some of the processes involved will be irreversible, overall

$$\delta S_U/\delta t > 0. \tag{25.3}$$

Therefore, although the system itself has become more organized, overall it has generated more disorganization than the organization created, and it is more accurate to call it a *self-disorganizing system* [3]. Hence, the "system" must properly be expanded to include its environment: it is evidently intimately connected with it, since without it there could be no organization. Despite its true nature as a self-disorganizing system having been revealed, nevertheless we can still speak of a self-organizing *part* S of the overall system that consumes order (and presumably energy) from its environment. It follows that this environment must necessarily have structure itself, otherwise there would be nothing to be usefully assimilated by the self-organizing part.

The link between entropy (i.e., its inverse) and organization can be made explicit with the help of relative entropy R (called *redundancy* by Shannon); this is defined by

$$R = 1 - S/S_{max}, \tag{25.4}$$

where S_{max} is the maximum possible entropy. With this new quantity R, self-organization implies that $\delta R/\delta t > 0$. Differentiating Equation 25.4, we obtain

$$\frac{dR}{dt} = \frac{S(dS_{max}/dt) - S_{max}(dS/dt)}{S_{max}^2}; \tag{25.5}$$

our criterion for self-organization (that R must spontaneously increase) is plainly

$$S\frac{dS_{max}}{dt} > S_{max}\frac{dS}{dt}. \tag{25.6}$$

The implications of this inequality can be seen by considering two special cases [3]:

1. The maximum possible entropy S_{max} is constant; therefore $dS_{max}/dt = 0$ and $dS/dt < 0$. Now, the entropy S depends on the probability distribution of the constituent parts (at least, those that are to be found in certain distinguishable states); this distribution can be changed by rearranging the parts, which von Foerster supposed could be accomplished by an "internal demon" (see Ref. [6] for an updated description of James Clerk Maxwell's original invention)

2. The entropy S is constant; therefore $dS/dt = 0$ and the condition that $dS_{max}/dt > 0$ must hold; that is, the maximum possible disorder must increase. This could be accomplished, for example, by increasing the number of elements, but care must be taken to ensure that S then indeed remains constant, which

probably needs an "external" demon. Looking again at inequality (Equation 25.6), we see how the labor is divided among the demons: dS/dt represents the internal demon's efforts, and S is the result; dS_{max}/dt represents the external demon's efforts, and S_{max} is the result. There is therefore an advantage (in the sense that labor may be spared) in cooperating. For example, if the internal demon has worked hard in the past, the external demon can get away with putting in a bit less effort in the present. Yet, one should not underestimate the burden placed on the demons – particularly the external one – which must evidently possess an almost divine omnipotence.

25.2.2
The "Goodness" of the Organization

Examining again Figure 25.1, it can be asserted that both putative results of mixing slightly sticky cubelets together are organized, although most people would not hesitate to call the structure in (b) better organized than that in (a). Evidently, there is some meaning in the notion of "good organization," even though it seems difficult to formulate an unambiguous definition. Can a system spontaneously (automatically) change from a bad to a good organization? This would be a reasonable interpretation of "self-organization," but has been proved to be formally impossible [5]. Consider a device that can be in any one of three states, A, B or C, and the device's operation is represented by some transformation, for example:

$$\downarrow \begin{array}{ccc} A & B & C \\ B & C & A \end{array}$$

Now suppose that we can provide an input f to the device, and that the output is determined by the value of f, for example:

$\downarrow$	A	B	C
f_A	B	C	A
f_B	A	A	A
f_C	A	B	C

Spontaneous (automatic) operation means that the device is able to autonomously select its input. The different possible input values are here represented by a subscript indicating the state of the device on which the input now depends. However, this places severe constraints on the actual operation, because $f_A(B)$ (for example) is impossible; only $f_A(A)$, $f_B(B)$ and $f_C(C)$ are possible, hence the operation necessarily reduces to the simple transform, lacking any autonomy:

$$\downarrow \begin{array}{ccc} A & B & C \\ B & A & C \end{array}$$

Any change in f must therefore come from an external agent.

25.2.3
Programmable Self-Assembly

The results summarized in the two preceding subsections might well engender a certain pessimism regarding the ultimate possibility of realizing true self-assembly. Yet in biology, numerous examples are known (see also Section 25.3.5): the final stages of assembly of bacteriophage viruses [7], of ribosomes [8], and of microtubules [9], which occur not only *in vivo*, but which can also be demonstrated *in vitro* by simply mixing the components together in a test-tube. As apparent examples of what might be called "passive" self-assembly, in which objects possessing certain asymmetric arrangements of surface affinities are randomly mixed and expected to produce ordered structures [3], they seem to contradict the predictions of Sections 25.2.1 and 25.2.2.

It has long been known that biomolecules are constructions [10]; that is, they have a small number of macroscopic (relative to atomic vibrations) degrees of freedom, and can exist in a small number (≥ 2) of stable conformations [11]. Without these properties, the actions of enzymes, active carriers such as hemoglobin, and the motors that power muscle, for example, are not understandable. Switching from one conformation to another is typically triggered by the binding or dissociation of a small molecule; for example, the "substrate" of an enzyme, or adenosine triphosphate (ATP) [12]. The initial collision of two particles is followed by a conformational change in one or both of them; for example:

$$A + B + C \rightarrow A - B^* + C \rightarrow A - B^* - C, \tag{25.7}$$

where the asterisk denotes a changed conformation induced by binding to A; C has no affinity for B, but binds to B* [7]. This process is illustrated in Figure 25.3, and is called "programmable self-assembly." It has recently been modeled by *graph grammar*, which can be thought of as a set of rules encapsulating the outcomes of

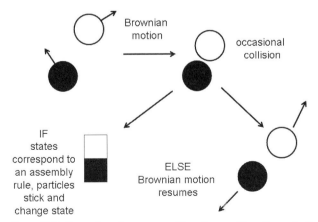

Figure 25.3 Illustration of programmable self-assembly, with a primitive local rule.

interactions between the particles [13, 14] (cf. stigmergic assembly; Section 25.3.5). The concept of graph grammar has brought about a significant advance in the formalization of programmable self-assembly, including the specification of minimal properties that must be possessed by a self-assembling system (e.g., the result implying that no binary grammar can generate a unique stable assembly [15]).

25.3
Examples of Self-Assembly

25.3.1
The Addition of Particles to the Solid/Liquid Interface

A chemically and morphologically unstructured surface of the solid substratum is prepared and brought into contact with a fluid medium. Self-assembly is initiated by replacing the pure fluid by a suspension of particles, the buoyancy of which in the fluid is such that they move purely by diffusion (Brownian motion) and are not influenced by gravity. Occasionally, those particles in the vicinity of the interface will strike it; the materials of substratum (1), fluid medium (2) and particle (3) are chosen such that the interfacial free energy $\Delta G_{123}^{\parallel}$ is negative.[1] In some cases $\Delta G_{123}^{\parallel}$ is positive, but $\Delta G_{13}^{\parallel}$ (i.e., when all the intervening fluid is eliminated) is negative. This signifies that there is an energy barrier hindering adsorption of the particle to the solid substratum. The rate of arrival of the particles at the substratum is proportional to the product of particle concentration c_b in the suspending medium and the diffusion coefficient D of a particle, the constant of proportionality depending on the hydrodynamic régime; this rate will be reduced by a factor $1/\int_{\ell_o}^{\infty}[\exp(\Delta G_{123}(z)/k_B T)-1]\mathrm{d}z$ in the presence of an energy barrier; the reduction factor could easily be a hundred- or a thousand-fold.

Once a particle of radius r adheres to the substratum, evidently the closest the center of a second particle can be placed to the first one without overlapping it is at a distance $2r$ from the center of the first; in effect, the first particle creates an *exclusion zone* around itself (Figure 25.4). If the particles interact with each other apart from via hard body (Born) repulsion, then the effective radius must be increased to the distance at which the particle–particle interaction energy $\Delta G_{323}(z)$ equals the thermal energy $k_B T$ [17]. A corollary of the existence of exclusion zones

---

1) Please refer to Ref. [16] for details of the calculation of $\Delta G_{123}^{\parallel}$. The subscript $\parallel$ refers to the interfacial free energy between infinite parallel surfaces of materials 1 and 3 separated by the closest distance of approach ℓ_0, when Born repulsion sets in (it is considered to be about 0.16 nm) and medium 2. The magnitude of the interfacial free energy between an infinite planar surface of 1 and a sphere of 3 separated by an arbitrary distance z can be calculated from $\Delta G_{123}^{\parallel}$ using the Derjaguin approximation [16]; for many purposes this might not even be necessary, especially if only the sign of the interfacial interaction is really needed.

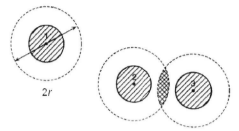

Figure 25.4 Diagram to illustrate the concept of exclusion zone. The projected area of a spherical particle is hatched; the area enclosed by the dashed circles represents the exclusion zone, with twice the radius of the actual particle. The cross-hatched area marks the overlap of the exclusion zones of particles 2 and 3.

is that the interface will be jammed (i.e., unable to accept a further particle) at a surface coverage of substantially less than 100%. The actual value of the jamming limit depends on the shape of the particle; for spheres, it is about 54% of the complete surface coverage [18].

This process is known as random sequential addition (RSA). Although a random dispersion of particles in three dimensions is thereby reduced to a two-dimensional (2-D) layer, the positions of the particles remain random: the radial distribution function is totally unstructured [19]. Even if the particles can move laterally, allowing the interface to equilibrate in a certain sense, it is still jammed at a surface coverage of well below 100%. If, however, the particles can adhere to each other on the interface, the possibility for organizing arises. This has been very clearly demonstrated when lateral mobility was expressly conferred on the particles by covering the surface with a liquid-crystalline lipid bilayer and anchoring the particles (large spherical proteins) in the bilayer through a hydrophobic "tail" [20]. The particles structure themselves to form a 2-D ordered array. When such an affinity exists between the particles trapped at the interface, the exclusion zones are annihilated. From this fact alone (which can be very easily deduced from the kinetics of addition [20]), one cannot distinguish between random aggregation forming a diffusion-limited aggregate [4] (cf. reaction-limited aggregation [21]) and 2-D crystallization; these can generally be distinguished, however, by the fact that the crystal unit cell size is bigger than the projected area of the particle.[2] The process of 2-D crystallization has two characteristic time scales, namely the interval τ_a between the addition of successive particles to the interface

$$\tau_a = 1/[aF\phi(\theta)] \tag{25.8}$$

where a is the area per particle, F is the flux of particles to an empty surface (proportional to the bulk particle concentration and some power <1 of the coefficient

---

2) This is especially to be expected if the particle is a protein; typically about 70% of the volume of three-dimensional protein crystals is solvent.

of diffusion in three dimensions), and ϕ is the fraction of the surface available for addition, which is some function of θ, the fractional surface coverage of the particles at the interface; and the characteristic time τ_D for rearranging the surface by lateral diffusion (with a diffusion coefficient D_2)

$$\tau_D = a/(D_2\theta). \tag{25.9}$$

If $\tau_D \gg \tau_a$, then lateral diffusion is encumbered by the rapid addition of fresh particles before self-organization can occur and the resulting structure is indistinguishable from that of random sequential addition. Conversely, if $\tau_a \gg \tau_D$ there is time for 2-D crystallization to occur. Note that some affinity-changing conformational change must be induced by the interface; otherwise the particles would already aggregate in the bulk suspension. In the example of the protein with the hydrophobic tail, when the protein is dissolved in water the tail is buried in the interior of the protein, but partitions into the lipid bilayer when the protein arrives at its surface.

An intriguing example of interfacial organization is the heaping into cones of the antifreeze glycoprotein (AFGP) consisting of repeated alanine–alanine–glycosylated threonine triplets added to the surface of a solid solution of nanocrystalline $Si_{0.6}Ti_{0.4}O_2$ [22]. Under otherwise identical conditions, on mica the glycoprotein adsorbs randomly sequentially. It is indeed possible that in this, as in other cases, we are seeing examples of programmable self-assembly (Sections 25.2.3 and 25.3.7), but current limitations in resolving the shapes of nanometer-sized biomolecules adsorbed at interfaces mean that it can only be inferred (e.g., from the overall kinetics of the assembly process), rather than observed directly.

25.3.1.1 Numerically Simulating RSA

The process is exceptionally easy to simulate. For each addition attempt, one selects a point at random: if it is further than $2r$ from the center of any existing particle then a new particle is added (the available area for less symmetrical shapes may have to be computed explicitly), but if it is closer then the attempt is abandoned. The success of this simple algorithm is due to the fortuitous cancellation of two opposing processes: correlation and randomization. In reality, if a particle cannot be added at a selected position because of the presence of a previously added one, it will make another attempt in the vicinity of the first, because of the Rabinowitch ("cage") effect [23]; successive attempts are strongly positionally correlated. On the other hand, as a particle approaches the interface through the bulk fluid, it experiences *hydrodynamic friction*, which exerts a randomizing effect; the two effects happen to cancel out each other [24].

25.3.2
Self-Assembled Monolayers (SAMs)

If particles randomly and sequentially added to the solid–liquid interface are asymmetrical (i.e., elongated) and have an affinity for the solid substratum at only one

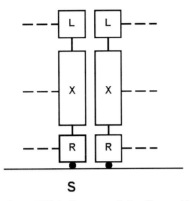

Figure 25.5 A (fragment of a) self-assembled monolayer (SAM). The component molecules have the general formula LXR, where X is an apolar chain (e.g., alkyl), and R is a reactive group capable of binding to the substratum, S. X can be functionalized at the end opposite from R with a group L to form molecules L–XR; the nature of L can profoundly change the wetting properties of the SAM.

end (but an ability to move laterally at the interface), and with the "tail" (the rest) poorly solvated by the liquid, they will tend to adsorb in a compact fashion, by strong lateral interaction between the tails (Figure 25.5).

Self-assembled monolayers were discovered by Bigelow *et al.* [25]. The original example was eicosyl alcohol ($C_{20}H_{41}OH$) dissolved in hexadecane ($C_{16}H_{34}$) adsorbing onto silica glass. Later, molecules with R = –SH (thiol or mercaptan), which bind strongly to metals such as Au, Ag, Cu, Hg, and so on, were investigated [26, 27], as well as organosilanes, which bind strongly (covalently) to silica. If the tail moiety X is poorly solvated by the liquid, its flexibility may enable it to be compactified while the molecule is in the bulk solution, tending to prevent self-aggregation, and only unfurling itself after R has attached to the solid surface; this can be seen as a rudimentary form of programming.

SAMs provide a very convenient way to change the wettability of a solid surface. Bigelow *et al.*'s monolayers were both hydrophobic and oleophobic. An octadecanethiol (L = –H) film adsorbed onto gold would be both oil- and water-repellent; if L = –OH it would be hydrophilic. Equilibrium wetting is quantified by Young's equation; if wetting is complete, then the contact angle of a droplet of liquid L on solid S in the presence of vapor V is zero, and Young's equation becomes

$$0 = \gamma_{SV} - \gamma_{SL} - \gamma_{LV} \tag{25.10}$$

where γ_{LV} is the interfacial tension of liquid L (or an epitaxially grown solid) in contact with its vapor, and *mutatis mutandis* for the other symbols. Out of equilibrium, the spreading coefficient S introduced by Cooper and Nuttall [28] is useful:

$$S = \gamma_{SV} - \gamma_{SL} - \gamma_{LV} \tag{25.11}$$

where $\gamma_{\bar{S}V}$ is the interfacial tension of a *dry* (unwetted) solid S in contact with vapor V. Three régimes can be defined:

1. S > 0. This corresponds to $\gamma_{\bar{S}V} > \gamma_{SV}$; that is, the wetted surface has a lower energy than the unwetted one. Hence, wetting takes place spontaneously. The thickness h of the film is greater than monomolecular if $S \ll \gamma_{LV}$. The difference $\gamma_{\bar{S}V} - \gamma_{SV}$ can be as much as $300\,\text{mJ}\,\text{m}^{-2}$ for water on metal oxides. Such systems therefore show enormous hysteresis between advancing and receding contact angles.

2. S = 0. This occurs if $\gamma_{\bar{S}V}$ practically equal to γ_{SV}, as is typically the case for organic liquids on molecular solids.

3. S < 0. This is partial wetting. Films thinner than a certain critical value, usually ~1 mm, break up spontaneously into droplets.

More elaborate groups L can be incorporated into SAMs; however, if they are bulky the functionalized molecules should be mixed with unfunctionalized ones so as to avoid packing defects. Mixtures with different chain lengths (e.g., X = C_{12} and C_{22}) produce liquid-like SAMs.

The biologically ubiquitous lipid bilayer membrane could be considered to belong to this category. The component molecules are of type XR, where X is an alkyl chain as before, and R is a rather polar "head group," the volume of which is typically roughly equal to that of X. Placing XR molecules in water and gently agitating the mixture will lead spontaneously to the formation of spherical bilayer shells called *vesicles*. The vesicles will coat a planar hydrophilic substratum with a lipid bilayer when brought into contact with it [29].

If the substratum is electrified (via the Gouy–Chapman mechanism) and the dissolved molecule is a polyion with an electrostatic charge of opposite sign, then it will adsorb onto the surface and invert the charge; the strong correlations within the polymeric ion render the Gouy–Chapman mechanism invalid [30]. The polyion-coated substratum can then be exposed to a different polyion of opposite sign, which will in turn be adsorbed and again invert the charge; this process can be repeated *ad libitum* to assemble thick films [31].

25.3.3
Quantum Dots (QDs)

Whereas SAMs can be prepared with very simple equipment, a somewhat analogous process – molecular beam epitaxy (MBE) – which was developed at the AT & T Bell Laboratories during the late 1960s, takes place in ultrahigh vacuum and requires elaborate and expensive equipment for its realization. In this process, the material to be assembled is evaporated from a store and beamed onto the substratum. A very slow deposition (a fraction of a nanometer per second) is the key to achieving epitaxy; ultrathin layers with atomically sharp interfaces are capable of being deposited. Equation 3.11 also applies. If $S > 0$, we have the Frank–van der Merwe régime, where the substratum is wet and layer-by-layer growth takes place. If $S < 0$, we have the Volmer–Weber régime, where there is no wetting and three-dimensional (3-D) islands grow. But, if $S > 0$ – for the first layer at least – and if there is a geometric mismatch between the lattices of the substratum and the deposited layer, then strain will build up in the latter, this being subsequently relieved by the spontaneous formation of monodisperse islands (the Stranski–Krastanov régime); it can be thought of as "frustrated wetting" [42]. The main application of this process is to fabricate QDs for lasers [43].[3] In this application, the main difficulty is to ensure that the dots comprising a device are uniformly sized. If not, the density of states is smeared out and the behavior reverts to bulk-like. Initially, the QDs were prepared by conventional semiconductor processing [44], but it proved to be very difficult to eliminate defects and impurities. The Stranski–Krastanov self-assembly process does not introduce these problems.

25.3.4
Crystallization and Supramolecular Chemistry

It is a remarkable fact that many (or even most) organic molecules, despite their usually very complicated shapes, are able spontaneously to form close-packed crystals. Perhaps only familiarity with the process prevents it from occupying a more prominent place in the world of self-assembly. In seeking to better understand this remarkable phenomenon, Kitaigorodskii formulated an Aufbau principle [32], according to which the self-assembly of complicated structures takes place in a hierarchical fashion in the following sequence:

- at stage zero, we have a single molecule (or a finite number of independent molecules)
- in stage one, single molecules join up to form linear chains
- in stage two, the linear chains are bundled to form 2-D monolayers
- in stage three, the 2-D monolayers are stacked to form the final 3-D crystal.

Many natural structures are evidently hierarchically assembled. Wood, for example, derives its structural strength from glucose polymerized to form cellulose chains, which are bundled to form fibrils, that are in turn glued together using lignin to form robust fibers. It should be noted that the interactions between the molecular components of the crystal may be significantly weaker than the covalent chemical bonds that hold the atomic constituents of the molecules together. Such weakness enables defects to be annealed by local melting of the partially formed structure.

There is an obvious analogy between "crystallization" in two dimensions and tiling a plane. Since tiling is connected to computation, self-assembly – which can

3) Quantum dot lasers are a development of quantum well lasers. The carriers are confined in a small volume and population inversion occurs more easily than in large volumes, leading to lower threshold currents for lasing. The emission wavelength can be readily tuned by simply varying the dimensions of the dot (or well).

perhaps be regarded as a type of generalized crystallization – has in turn been linked to computation [33].

Just as QDs containing several hundred atoms can in some sense (e.g., with regard to their discrete energy levels) be regarded as "superatoms," so can supramolecular assemblies (typically made up from very large and elaborate molecules [34]) be considered as "supermolecules." The obtainable hierarchically assembled structures provide powerful demonstrations of the Kitaigorodskii Aufbau principle [35–39], and an enormous literature has subsequently accumulated. The use of metal ions as organizing centers in these assemblies has been a particularly significant practical development [35, 40, 41].

25.3.5
Biological Examples

The observation that preassembled bacteriophage components (head, neck, and legs) could be mixed in solution [7] provided profound inspiration for the world of "shake and bake" advocates. These components are essentially made up of proteins – heteropolymers made from irregular sequences chosen from the 20 natural amino acids of the general formula $H_2N-C^{(\alpha)}HR\text{-}COOH$, where R is naturally one of 20 different side chains (residues), ranging from R=H in glycine, the simplest amino acid, to elaborate heterocycles such as $R=CH_2-[C_8NH_6]$ in tryptophan. The conformation – and hence affinity – of a protein depends on environmental parameters such as the pH and ionic strength of the solution in which it is dissolved [7]. Indeed, this is one mechanism for achieving programmability in assembly, as the local pH and ion concentration around a protein molecule will depend on the amino acids present at its surface. These factors also determine the conformation of the highly elongated, so-called "fibrous" proteins such as fibronectin, which are now known to consist of a large number of rather similar modules strung together [45]. Some other examples of biological self-assembly have already been mentioned in Section 25.2.3. A further example is provided by the remarkable S-layers with which certain bacteria are coated [46]. DNA is discussed in Section 25.3.6. Mention should also be made here of the oligopeptides found in fungi and the stings of bees and wasps, which self-assemble into pores when introduced into a bilayer lipid membrane (see, e.g., Ref. [47]). Yet, biological self-assembly and self-organization is by no means limited to the molecular scale: the development of an embryo consisting of a single cell into a multicellular organism is perhaps the most striking example of self-organization in the living world. The process of cell differentiation into different types can be very satisfactorily simulated on the basis of purely local rules enacted by the initially uniform cells (see Ref. [48] for the modeling of neurogenesis). On a yet larger scale, it is likely that the construction of nests by social insects such as ants and wasps relies on simple rules held and enacted by individual agents (the insects), according to local conditions [49]; this process has been called *stigmergy* and is evidently conceptually the same as the programmable self-assembly modeled by graph grammars (Section 25.2.3).

25.3.6
DNA

The importance of deoxyribonucleic acid (DNA), both as the quasi-universal carrier of genetic information in living creatures and as a sufficiently robust material to serve as the basis for synthetic self-assembling systems, warrants it having a subsection of its own. Similar to proteins, DNA is a heteropolymer, the monomers of which comprise an invariant backbone [in this case, of alternating sugar (deoxyribose) and phosphate units] and a variety of four "bases", adenine (A), cytosine (C), guanine (G) and thymine (T). The rather specific base pairings between A and T, and between C and G, which are due to respectively two and three interbase hydrogen bonds [50], are responsible for the famous "double helix" [51] in which genetic information is stably stored in living cells and many viruses. The combinatorial uniqueness even of base strings of fairly modest length is exploited in the fabrication of "gene chips" [52] used to identify genes and genomes. In these devices, the sample to be identified (e.g., the nucleic acids extracted from bacteria found in the bloodstream of a patient) is dispersed over the surface of the chip, which comprises an array of contiguous microzones containing known oligomers of nucleic acids complementary to the sought-for sequences (e.g., the fragment GATTACA is complementary to CTAATGA).

Seeman was the first to point out that the specific complementary base pairing could form the basis of employing DNA as a structural nanomaterial [53, 54]: oligonucleotides could be artificially synthesized according to what are now straightforward, routine procedures and randomly stirred together, assembling in a unique fashion (Figure 25.6). This field has now grown enormously to encompass very elaborate constructions (e.g., Refs [55, 56]) and is connected with tile assembly and computation [57, 58]. Fragments of DNA have also been fastened to nanoparticles to confer selective affinity upon them (e.g., Ref. [59]).

25.3.7
RNA and Proteins

Although ribonucleic acid (RNA) differs only very slightly from DNA, heteropolymers formed from its four bases (the same as in DNA, except that uracil (U)

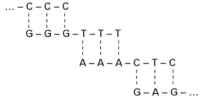

Figure 25.6 Four oligonucleotides, which can only assemble in the manner shown. Long dashes represent strong covalent bonds; short dashed lines represent weak hydrogen bonds.

replaces thymine (T)) do not assemble into a double helix, but rather "fold." That is, the linear polymer chain – each monomer of which is bonded covalently only to its two nearest neighbors – typically assembles into a compact 3-D structure (though not all sequences will do so) with additional base-pairing (A=U and C≡G) bonds between distant monomers. Predicting the final 3-D structure is *prima facie* a difficult problem. Energetics are clearly involved, because bonds between distant monomers form spontaneously (if geometric constraints are satisfied), releasing enthalpy and hence lowering the free energy. On the other hand, this raises the entropy because the chain becomes more constrained. At one time it was believed that the observed stable folded configurations were the result of minimizing the free energy of the molecule, and were found by systematically searching configuration space. It was, however, quickly realized that for a large molecule with thousands of atoms this was a practically impossible task that might take longer than the age of the universe. Both, *in vivo* and *in vitro*, it can be observed that biopolymers can typically fold within an interval of the order of seconds, and the only way that this can be achieved is for the system to find a pathway expeditiously leading to the goal (which is evidently some form of energy minimum, even if not the global one). This in turn suggests that a least-action principle might be operating. The action is the integral of the Lagrangian $\mathcal{L}$ ($= L - F$ for conservative systems, where L and F are respectively the kinetic and potential energies). The most expedient path is found by minimizing the action. This is, in fact, an inerrant principle for finding the correct solution of any dynamical problem; the difficulty lies in the fact that there is no general recipe for constructing $\mathcal{L}$, and it can often be found only by "fiddling around," as has been remarked by Feynman.

Folded RNA contains certain characteristic structures, notably loops and "hairpins". Loop closure is considered to be the most important folding event. F (the potential) is identified with the enthalpy; that is, the number n of base pairings (contacts), and L corresponds to the entropy [60]. At each stage in the folding process, as many as possible new favorable intramolecular interactions are formed, while minimizing the loss of conformational freedom (the principle of sequential minimization of entropy loss, SMEL [60]). The entropy loss associated with loop closure is ΔS_{loop} (and the rate of loop closure $\sim \exp(\Delta S_{\text{loop}})$); the function to be minimized is therefore $\exp(-\Delta S_{\text{loop}}/R)/n$, where R is the universal gas constant. A quantitative expression for ΔS_{loop} can be found by noting that the N monomers in an unstrained loop ($N \geq 4$) have essentially two possible conformations, pointing either inwards or outwards. For loops smaller than a critical size, N_0, the inward loops are in an apolar environment, as the nano-enclosed water no longer has bulk properties [61], while the outward loops are in polar bulk water. For $N < N_0$, $\Delta S_{\text{loop}} = -RN \ln 2$ (for $N > N_0$, the Jacobson–Stockmayer approximation based on excluded volume yields $\Delta S_{\text{loop}} \sim R \ln N$ [60]).

A similar approach can be applied to proteins [62], although in proteins the main intramolecular structural connector (apart from the covalent bonds between successive amino acid monomers) is the backbone hydrogen bond. This form of hydrogen bond is responsible for the appearance of characteristic structures such as the alpha helix but, being single, is necessarily weaker than the double and triple hydrogen bonds in DNA and RNA. They must therefore be protected from competition for hydrogen bonding by water, and this can be achieved by bringing amino acids with apolar residues to surround the hydrogen bonds [63]. This additional feature, coupled with the existence of multiple conformational states already referred to in Section 25.2.3, means that proteins are particularly good at engaging in programmable self-assembly – a possibility that is, of course, made abundant use of in Nature.

25.4
Self-Assembly as a Manufacturing Process

Practical, industrial interest in self-assembly is strongly driven by the increasing difficulty of reducing the feature sizes that can be fabricated by semiconductor processing technology [44]. Self-assembly is positioned as a rival to the so-called "top-down" processes, which also include precision engineering (Figure 25.7) [64]. For certain engineering problems, it may already be useful to manufacture regular structures such as dots or stripes over an indefinitely large area; passive self-assembly might be able to produce such structures with feature sizes at the molecular scale of a few nanometers. These could be used either as masks for photolithography (as a cheaper and faster method than electron beam writing for creating such small features), or they might be useful in their own right (e.g., as nanoporous membranes for separating vapors [65]). Self-assembly is a particularly attractive option for the fabrication of molecular electronic components [66, 67], for in this case – almost by definition – conventional semiconductor processing is useless.

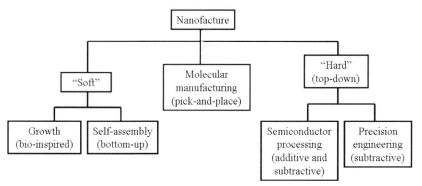

Figure 25.7 Different possible modes of nanomanufacture (nanofacture) [64].

25.5
Useful Ideas

25.5.1
Weak Competing Interactions

It is known from biological self-assembly (e.g., of compact RNA structures) that bonds must be broken and reformed before the final structure is achieved. This is particularly apparent because these molecules are synthesized as a linear polymer, which begins to fold spontaneously as soon as a few tens of monomers have been connected. As the chain becomes longer, some of these earlier bonds must be broken so as to allow connections between points more distant along the chain to be made. Since hydrogen bonds have only about one-tenth of the strength of ordinary covalent bonds, they have an appreciable probability of being melted (broken) even at room temperature. Furthermore, the polymer is surrounded by water, each molecule of which is potentially able to participate in four hydrogen bonds (although at room temperature only about 10% of the maximum possible number of hydrogen bonds in water are broken; see also Section 25.5.4). Hence, there is ceaseless competition between the intramolecular and intermolecular hydrogen bonds.

If the competing interactions have a different sign and range, then ordered structures can assemble spontaneously [68]. Consider nanoparticles suspended in water and weakly ionized, such that they all carry the same electrostatic charge. When the suspension is stirred, suppose that the repulsive electrostatic force is too weak to overcome the attractive van der Waals force when two particles happen to collide. Therefore, every collision will lead to sticking, and aggregates will slowly form. The van der Waals force is, however, very short ranged and can only act between nearest neighbors. The electrostatic force, on the other hand, has a much longer range, and can therefore be summed over the entire aggregate. Ultimately, the aggregate will become large enough for the summed electrostatic repulsion to exceed the van der Waals nearest-neighbor attraction; the result is monodisperse "superspheres" (i.e., aggregates of small, perhaps spherical, particles).

25.5.2
Percolation

Percolation can be considered to be a formalization of gelation. Initially, in a flask we have isolated sol particles, which are gradually connected to each other in nearest-neighbor fashion until the gel is formed. Although the particles can be placed anywhere in the medium, subject only to the constraint of hard-body exclusion, it is convenient to consider them placed on a 2-D square lattice (imagine a Gō board). Two particles are considered to be connected if they share a common side (this is called "site percolation"). Alternatively, neighboring lattice points are connected if they are bonded together ("bond percolation").

In principle the lattice is infinite, but in reality it is merely very large. Percolation occurs if one can trace the continuous path of connections from one side to the other. Initially, all the particles are unconnected. In site percolation, the lattice is initially considered to be empty, and particles are added. In bond percolation, initially all particles are unconnected and bonds are added. The problem is to determine what fraction of sites must be occupied, or how many bonds must be added, in order for percolation to occur.

In the remainder of this subsection we shall consider site percolation. Let the probability of a site being occupied be p (and of being empty, $q = 1 - p$). The average number of singlets per site is $n_1(p) = pq^4$ for the square lattice, as each site is surrounded by four shared sides. The average number of doublets per site is $n_2(p) = 2p^2q^6$, as there are two possible orientations. A triplet can occur in two shapes, straight or bent, and so on. Generalizing,

$$n_s(p) = \sum_t g(s, t) p^s q^t \tag{25.12}$$

where $g(s, t)$ is the number of independent ways that an s-tuplet can be put on the lattice, and t counts the different shapes. If there is no "infinite" cluster (i.e., one spanning the lattice from side to side) then

$$\sum_s sn_s(p) = p. \tag{25.13}$$

The first moment of this distribution gives the mean cluster size

$$S(p) = \frac{\sum s^2 n_s(p)}{p}. \tag{25.14}$$

Writing this as a polynomial in p using Equation 25.12, it will be noticed that for $p < p_c$ the value of the series converges, but for $p > p_c$ the value of the series diverges. The "critical" value $p = p_c$ corresponds to the formation of the "infinite" cluster. For site percolation on a square lattice, $p = p_c = 0.5$; the universal Galam–Mauger formula [69]

$$p_c = a[(d-1)(q-1)]^{-b}, \tag{25.15}$$

with $a = 1.2868$ and $b = 0.6160$ predicts, with less than 1% error, all known thresholds for lattices of connectivity q embedded in space of dimension d. The larger the lattice, the sharper the transition from not percolating to percolating. For a 3×3 lattice there can be no percolation for two particles or less, but the probability of randomly placing three particles in a straight line from edge to edge is evidently one-third.

25.5.3
Cooperativity

When considering the interactions between precursors of a self-assembly process, it has tacitly been assumed that every binding event is independent. Similarly, when considering conformational switches, it has been assumed that each molecule switches independently. This, however, is often not the case: switching or binding of one facilitates the switching or binding of neighbors, whereupon we have cooperativity (if it hinders rather than facilitates, then it is called *anticooperativity*).

A cooperative processes can be conceptualized as two subprocesses, nucleation and growth. Let our system exist in one of two states (e.g., bound or unbound, conformation A or B), which we shall label 0 and 1. We have [70]

$$\text{nucleation:} \quad \cdots 000 \cdots \xrightleftharpoons{\sigma S} \cdots 010 \cdots \tag{25.16}$$

and

$$\text{growth:} \quad \cdots 001 \cdots \xrightleftharpoons{S} \cdots 011 \cdots \tag{25.17}$$

Let $\{1\} = \theta$ denote the probability of finding a "1"; we have $\{0\} = 1 - \theta$. The parameter λ^{-1} is defined as the conditional probability of "00", given that we have a "0", written as (00) and equal to $\{00\}/\{0\}$. It follows that $(01) = 1 - (00) = (\lambda - 1)/\lambda$.

According to the mass action law (MAL), for nucleation we have

$$S = \frac{\{011\}}{\{001\}} = \frac{(11)}{(00)} \tag{25.18}$$

from which we derive $(11) = S/\lambda$, and hence $(10) = 1 - (11) = (\lambda - S)/\lambda$. Similarly, for growth

$$S = \frac{(01)(10)}{(00)^2} = (\lambda - 1)(\lambda - S). \tag{25.19}$$

Solving for λ gives

$$\lambda = \left[1 + S + \sqrt{(1-S)^2 + 4\sigma S} \right] / 2. \tag{25.20}$$

To obtain the sought-for relationship between θ and S, we note that $\theta = \{01\} + \{11\} = \{0\}(01) + \{1\}(11)$, which can be solved to yield

$$\lambda = \frac{1 + (S-1) / \sqrt{(1-S)^2 + 4\sigma S}}{2}. \tag{25.21}$$

25.5.4
Water Structure

So many self-assembly processes take place in water (not least all the biological ones) that it is advisable to bear some salient features of this remarkable liquid into account. Water (H–O–H) can participate in four hydrogen bonds (HB). The two-electron lone pairs (LPs) on the oxygen atom are electron donors, and hence HB acceptors. The two hydrogens at the ends of the hydroxyl groups (OH) are HB donors, and hence electron acceptors. The equilibrium

$$H_2O_{\text{fully bonded}} \rightleftharpoons OH_{\text{free}} + LP_{\text{free}} \tag{25.22}$$

is such that at room temperature about 10% of the OHs and LPs are nonbonded (i.e., free). It is especially noteworthy that the concentrations of the free species are seven to eight orders of magnitude greater than the concentrations of the perhaps more familiar entities H^+ and OH^-, and their chemical significance is correspondingly greater.

The OH moiety has a unique infrared absorption spectrum, which differs according to whether it is hydrogen-bonded or free, and can therefore be used to investigate the reaction in Equation 25.22 (e.g., the effect of adding ionic cosolutes; see Ref. [16] for further discussion).

25.6
Conclusions and Challenges

The most promising direction seems to be programmable self-assembly. Although some wholly synthetic examples have already been demonstrated [71], we can expect significant progress in the future to derive from the convergence of understanding of cooperative switching transitions in biomolecules with graph grammar or other formalisms for describing the assembly process. The final challenge will then be to reverse-engineer a device in order to specify the components (and environmental conditions) that should be stirred together for its assembly.

References

1 Bray, A.J. (2002) Phase-ordering kinetics. *Adv. Phys.*, **51**, 481–587.

2 Higgins, A.M. and Jones, R.A.L. (2000) Anisotropic spinodal dewetting as a route to self-assembly of patterned surfaces. *Nature*, **404**, 476–478.

3 von Foerster, H. (1960) On self-organizing systems and their environments, in *Self-Organizing Systems* (eds M.C. Yorvitz and S. Cameron), Pergamon Press, Oxford, pp. 31–50.

4 Meakin, P. (1994) What do we know about DLA? *Heterogen. Chem. Rev.*, **1**, 99–102.

5 Ashby, R.W. (1962) Principles of the self-organizing system, in *Principles of Self-Organization* (eds H. von Foerster and G.W. Zopf), Pergamon Press, Oxford, pp. 255–278.

6 Maruyama, K., Nori, F. and Vedral, V. (2009) The physics of Maxwell's demon and information. *Rev. Mod. Phys.*, **81**, 1–23.

7 Kellenberger, E. (1972) Assembly in biological systems, in *Polymerization in Biological Systems*, CIBA Foundation Symposium 7 (new series), Elsevier, Amsterdam.

8 Culver, G.M. (2003) Assembly of the 30S ribosomal subunit. *Biopolymers*, **68**, 234–249.

9 Holy, T.E. and Leibler, S. (1994) Dynamic instability of microtubules as an efficient way to search in space. *Proc. Natl Acad. Sci. USA*, **91**, 5682–5685.

10 Blumenfeld, L.A. (1981) *Problems of Biological Physics*, Springer, Berlin.

11 Frauenfelder, H. (1984) From atoms to biomolecules. *Helv. Phys. Acta*, **57**, 165–187.

12 Ramsden, J.J. (2001) Biophysical chemistry, in *Encyclopaedia of Chemical Physics and Physical Chemistry* (eds J.H. Moore and N.D. Spencer), IOP, Philadelphia, PA, pp. 2509–2544.

13 Klavins, E. (2004) Directed self-assembly using graph grammars, in *Foundations of Nanoscience: Self-Assembled Architectures and Devices*, Snowbird, UT.

14 Klavins, E. (2004) Universal self-replication using graph grammars. International Conference on MEMs, NANOs and Smart Systems, Banff, Canada.

15 Klavins, E., Ghrist, R. and Lipsky, D. (2004) Graph grammars for self-assembling robotic systems. Proceedings, International Conference on Robotics and Automation, New Orleans, pp. 5293–5300.

16 Cacace, M.G., Landau, E.M. and Ramsden, J.J. (1997) The Hofmeister series: salt and solvent effects on interfacial phenomena. *Q. Rev. Biophys.*, **30**, 241–278.

17 Ramsden, J.J. and Máté, M. (1998) Kinetics of monolayer particle deposition. *J. Chem. Soc., Faraday Trans.*, **94**, 783–788.

18 Schaaf, P. and Talbot, J. (1989) Surface exclusion effects in adsorption processes. *J. Chem. Phys.*, **91**, 4401–4409.

19 Schaaf, P., Voegel, J.-C. and Senger, B. (1998) Irreversible deposition/adsorption processes on solid surfaces. *Ann. Phys.*, **23**, 3–89.

20 Ramsden, J.J., Bachmanova, G.I. and Archakov, A.I. (1994) Kinetic evidence for protein clustering at a surface. *Phys. Rev. E*, **50**, 5072–5076.

21 Meakin, P. and Family, F. (1988) Structure and kinetics of reaction-limited aggregation. *Phys. Rev. A*, **38**, 2110–2123.

22 Lavalle, Ph., DeVries, A.L., Cheng, C.-H.C., Scheuring, S. and Ramsden, J.J. (2000) Direct observation of postadsorption aggregation of antifreeze glycoproteins on silicates. *Langmuir*, **16**, 5785–5789.

23 Luthi, P.O., Ramsden, J.J. and Chopard, B. (1997) The role of diffusion in irreversible deposition. *Phys. Rev. E*, **55**, 3111–3115.

24 Bafaluy, J., Senger, B., Voegel, J.C. and Schaaf, P. (1993) Effect of hydrodynamic interactions on the distribution of adhering Brownian particles. *Phys. Rev. Lett.*, **70**, 623–626.

25 Bigelow, W.C., Pickett, D.L. and Zisman, W.A. (1946) Oleophobic monolayers. *J. Colloid Sci.*, **1**, 513–538.

26 Nuzzo, R.G., Fusco, C.O. and Allara, D.L. (1987) Spontaneously organized molecular assemblies. 3. Proportion and properties of solution-adsorbed monolayers of organic disulfides at surfaces. *J. Am. Chem. Soc.*, **109**, 2358–2368.

27 Sellers, H., Ulman, A., Shnidman, Y. and Eilers, J.E. (1993) Structure and binding of alkane thiolates on gold and silver surfaces. *J. Am. Chem. Soc.*, **115**, 9389–9401.

28 Cooper, W. and Nuttall, W. (1915) The theory of wetting and the determination of the wetting power of dipping and spraying fluids containing a soap basis. *J. Agric. Sci.*, **7**, 219–239.

29 Csúcs, G. and Ramsden, J.J. (1998) Interaction of phospholipid vesicles with smooth metal oxide surfaces. *Biochim. Biophys. Acta*, **1369**, 61–70.

30 Grosberg, A.Y., Nguyen, T.T. and Shklovskii, B.I. (2002) The physics of charge inversion in chemical and biological systems. *Rev. Mod. Phys.*, **74**, 329–345.

31 (a) Ramsden, J.J., Lvov, Y.A. and Decher, G. (1995) Optical and X-ray structural monitoring of molecular films assembled via alternate polyion adsorption. *Thin Solid Films*, **254**, 246–251.

(b) Ramsden, J.J., Lvov, Y.A. and Decher, G. (1995) Optical and X-ray structural monitoring of molecular films assembled via alternate polyion adsorption. *Thin Solid Films*, **261**, 343–344.

32 Kitaigorodskii, A.I. (1961) *Organic Chemical Crystallography*, especially Chapters 3 and 4. Consultants Bureau, New York.

33 Winfree, E. (1996) On the computational power of DNA annealing and ligation, in *DNA-Based Computers* (eds R.J. Lipton and E.B. Baum), American Mathematical Society, Princeton, NJ, pp. 199–221.

34 Lehn, J.-M. (2000) Supramolecular polymer chemistry—scope and perspectives, in *Supramolecular Polymers* (ed. A. Ciferri), Dekker, New York, pp. 615–641.

35 Constable, E.C., Hannon, M.J., Edwards, A.J. and Raithby, P.R. (1994) Metallosupramolecular assembly of dinuclear double helicates incorporating a biphenyl-3,3′ diyl spacer. *J. Chem. Soc., Dalton Trans.*, 2669–2677.

36 Fujita, M., Ibukuro, F., Hagihara, H. and Ogura, K. (1994) Quantitative selfassembly of a [2]catenane from two preformed molecular rings. *Nature*, **367**, 720–723.

37 Desiraju, G.R. (1995) Supramolecular synthons in crystal engineering. *Angew. Chem. Int. Ed. Engl.*, **34**, 2311–2327.

38 Kálmán, A., Argay, G., Fábián, L. *et al.* (2001) Basic forms of supramolecular self-assembly organized by parallel and antiparallel hydrogen bonds in the racemic crystal structures of six disubstituted and trisubstituted cyclopentane derivatives. *Acta Crystallogr.*, **B57**, 539–550.

39 Lackinger, M., Griessl, S., Heckl, W.M. *et al.* (2005) Self-assembly of trimesic acid at the liquid-solid interface—a study of solvent-induced polymorphism. *Langmuir*, **21**, 4984–4988.

40 Barigelletti, F., Flamigni, L., Balzani, V. *et al.* (1994) Rigid rod-like dinuclear Ru(II)/Os(II) terpyridine-type complexes. *J. Am. Chem. Soc.*, **116**, 7692–7699.

41 Whittell, G.R. and Manners, I. (2007) Metallopolymers: new multifunctional materials. *Adv. Mater.*, **19**, 3439–3468.

42 Shchukin, V.A. and Bimberg, D. (1999) Spontaneous ordering of nanostructures on crystal surfaces. *Rev. Mod. Phys.*, **71**, 1125–1171.

43 Bimberg, D., Ledentsov, N.N., Grundman, M. *et al.* (1996) InAs-GaAs quantum pyramid lasers. *Jpn. J. Appl. Phys.*, **35**, 1311–1319.

44 Mamalis, A.G., Markopoulos, A. and Manolakos, D.E. (2005) Micro and nanoprocessing techniques and applications. *Nanotechnol. Percept.*, **1**, 63–73.

45 Rocco, M., Infusini, E., Daga, M.G., Gogioso, L. and Cuniberti, C. (1987) Models of fibronectin. *EMBO J.*, **6**, 2343–2349.

46 Pum, D., Sára, M., Schuster, B. and Sleytr, U.B. (2006) Bacterial surface layer proteins, in *Nanotechnology: Science and Computation* (eds J. Chen, N. Jonoska and G. Rozenberg), Springer, Berlin, pp. 277–290.

47 Schwarz, G., Stankowski, S. and Rizzo, V. (1986) Thermodynamic analysis of incorporation and aggregation in a membrane. *Biochim. Biophys. Acta*, **861**, 141–151.

48 Luthi, P.O., Preiss, A., Chopard, B. and Ramsden, J.J. (1998) A cellular automaton model for neurogenesis in *Drosophila*. *Physica D*, **118**, 151–160.

49 Theraulaz, G. and Bonabeau, E. (1995) Coordination in distributed building. *Science*, **269**, 686–688.

50 Ageno, M. (1967) Linee di ricerca in fisica biologica. *Accad. Naz. Lincei*, **102**, 3–58.

51 (a) Watson, J.D. and Crick, F.H.C. (1953) A structure for deoxyribose nucleic acid. *Nature*, **171**, 737–738. (b) Watson, J.D. and Crick, F.H.C. (1953) Genetical implications of the structure of deoxyribonucleic acid. *Nature*, **171**, 964–967.

52 Chumakov, S., Belapurkar, C., Putonti, C. *et al.* (2005) The theoretical basis of universal identification systems for bacteria and viruses. *J. Biol. Phys. Chem.*, **5**, 121–128.

53 Seeman, N.C. (1982) Nucleic-acid junctions and lattices. *J. Theor. Biol.*, **99**, 237–247.

54 Seeman, N.C. and Lukeman, P.S. (2005) Nucleic acid nanostructures: bottom-up control of geometry on the nanoscale. *Rep. Prog. Phys.*, **68**, 237–270.

55 Goodman, R.P., Schaap, I.A.T., Tardin, C.F. *et al.* (2005) Rapid chiral assembly of rigid DNA building blocks for molecular nanofabrication. *Science*, **310**, 1661–1665.

56 Rothemund, P.W.K. (2006) Folding DNA to create nanoscale shapes and patterns. *Nature*, **440**, 297–302.

57 Högberg, B. and Olin, H. (2006) Programmable self-assembly–unique structures and bond uniqueness. *J. Comput. Theor. Nanosci.*, **3**, 391–397.

58 Chen, J., Jonoska, N. and Rozenberg, G. (eds) (2006) *Nanotechnology: Science and Computation*, Springer, Berlin.

59 Mirkin, C.A., Letsinger, R.L., Mucic, R.C. and Storhoff, J.J. (1996) A DNA-based method for rationally assembling nanoparticles into macroscopic materials. *Nature*, **382**, 607–609.

60 Fernández, A. and Cendra, H. (1996) In vitro RNA folding: the principle of sequential minimization of entropy loss at work. *Biophys. Chem.*, **58**, 335–339.

61 Sinanoğlu, O. (1981) What size cluster is like a surface? *Chem. Phys. Lett.*, **81**, 188–190.

62 Fernández, A. and Colubri, A. (1998) Microscopic dynamics from a coarsely defined solution to the protein folding problem. *J. Math. Phys.*, **39**, 3167–3187.

63 Fernández, A. and Scott, R. (2003) Dehydron: a structurally encoded signal for protein interaction. *Biophys. J.*, **85**, 1914–1928.

64 Ramsden, J.J. (2009) *Nanotechnology*, Ventus, Copenhagen.

65 Ying, J.Y., Mehnert, C.P. and Wong, N.S. (1999) Synthesis and applications of supramolecular-templated mesoporous materials. *Angew. Chem. Int. Ed. Engl.*, **38**, 57–77.

66 Percec, V., Glodde, M., Bera, T.K. *et al.* (2002) Self-organization of supramolecular helical dendrimers into complex electronic materials. *Nature*, **417**, 384–387.

67 Grill, L., Dyer, M., Lafferentz, L. *et al.* (2007) Nano-architectures by covalent assembly of molecular building blocks. *Nat. Nanotechnol.*, **2**, 687–691.

68 Ramsden, J.J. (1987) The stability of superspheres. *Proc. R. Soc. Lond. A*, **413**, 407–414.

69 Galam, S. and Mauger, A. (1996) Universal formulas for percolation thresholds. *Phys. Rev. E*, **53**, 2177–2181.

70 Schwarz, G. (1970) Cooperative binding to linear biopolymers. *Eur. J. Biochem*, **12**, 442–453.

71 Bishop, J., Burden, S., Klavins, E., Kreisberg, R., Malone, W., Napp, N. and Nguyen, T. (2005) Programmable parts, in *International Conference on Intelligent Robots and Systems*, IEEE/RSJ Robotics and Automation Society.

26
Nanostructured Organogels via Molecular Self-Assembly

Arjun S. Krishnan, Kristen E. Roskov, and Richard J. Spontak

26.1
Introduction

Gels comprise an intermediate state of matter that is neither solid nor liquid. Generally speaking, gels are at least binary systems that consist of a crosslinked network swollen by a liquid, which constitutes the major component of the system. Due to their inherently high concentration of liquid, gels tend to be soft and possess the cohesive properties of a solid but the diffusive properties of a liquid [1]. Although they are largely liquid, gels behave mechanically as solids because the dominant liquid species is entrapped in a solid three-dimensional (3-D) network by capillary forces and adhesion. The underlying network responsible for gel behavior is stabilized by crosslink sites, and gels may be classified on the basis of the nature of their crosslinks. In *chemical gels*, the crosslinks are permanent (i.e., thermally irreversible) due to the formation of covalent bonds. Chemical gels are routinely produced by either connecting individual polymer chains with a bi- (or higher) functional crosslinking agent, or synthesizing macromolecular networks from small molecules in the presence of a crosslinking agent with at least trifunctionality [2]. In most cases, chemical gels are examples of molecular gels, wherein crosslink sites cannot be experimentally delineated from the remainder of the network.

Physical gels, the subject of this chapter, can be generated when microphase separation or weak physical interactions (e.g., hydrogen bonding, ion complexation, van der Waals forces, conformational changes and π–π stacking) promote the formation of crosslinks and, hence, contiguous networks. Because microphase separation is generally enthalpically-driven and the interactions involved in creating such crosslinks are on the order of kT (where k denotes the Boltzmann constant and T the absolute temperature), the formation of crosslinks is usually thermoreversible, which means that physical gels frequently exist as liquids at elevated temperatures and undergo reversible gelation upon cooling. One of the best-known examples of a physical gel involves the protein *gelatin* in water. Gelatin molecules possess a random coil conformation at elevated temperatures, but spontaneously adopt a triple helix conformation at 37 °C, thereby crosslinking

Advanced Nanomaterials. Edited by Kurt E. Geckeler and Hiroyuki Nishide
Copyright © 2010 WILEY-VCH Verlag GmbH & Co. KGaA, Weinheim
ISBN: 978-3-527-31794-3

individual molecules and forming a thermoreversible network that can be dissolved upon reheating above 37 °C [3, 4]. Although gelatin provides an excellent example of a system known to form a physical gel in water, only physical gels that develop in organic solvents (to yield so-called "organogels") are considered here. Moreover, attention is focused specifically on systems capable of generating discrete nanoscale crosslink sites that can be experimentally interrogated.

While physical gels are often (and sometimes mistakenly) identified on the basis of a liquid-rich system undergoing solidification (at temperatures well above the fusion temperature of the liquid) upon cooling, a more scientifically rigorous, quantitative definition of a gel remains a matter of discipline-specific debate. One of the most widely accepted definitions of a gel derives from its response to dynamic mechanical deformation, and is thus of rheological origin. In dynamic rheology, a relatively soft specimen, such as a physical gel, is subjected to oscillatory shear deformation between two metal surfaces. The magnitude and frequency of deformation can be independently adjusted and used in concert to probe structure existing within the specimen. Measurements are normally expressed in terms of the dynamic storage modulus (G') and the dynamic loss modulus (G''), which correspond to the in-phase and out-of-phase components, respectively, of the response of the specimen to oscillatory deformation [5]. In the specific case of classifying systems as gels, Kramer and coworkers [6] recommend the following criteria:

1. No experimentally discernible equilibrium modulus.
2. A storage modulus that exhibits a prominent frequency-independent plateau.
3. Frequency spectra wherein the G' plateau extends to time scales (i.e., reciprocal frequencies) on the order of seconds or more.
4. Values of G'' that are lower than those of G' by an experimentally significant difference (at least one order of magnitude) over all frequencies.

Most physical gels relax at very low frequencies, eventually causing reductions in both G' and G'', and so the applicability of the definition offered above depends on the experimental time scale. In addition, some physical gels "ripen"–that is, the extent and stability of their network improves–over time under quiescent conditions, in which case sufficient time must be permitted for such gel networks to mature fully.

This chapter is divided into two sections on the basis of the type of nanostructured network that develops spontaneously in an organic solvent due to molecular self-assembly. While homopolymer molecules can undergo a variety of conformational transitions, such as helix formation–for example, a mixture of syndiotactic and isotactic poly(methyl methacrylate) (PMMA) in a variety of solvents [7], or atactic polystyrene (PS) in carbon disulfide [8] – or crystallization in organic solvents, these systems are not considered further due to the lack of a clearly identifiable network-forming nanostructure. For this reason, we likewise exclude liquid crystalline polymer solutions that can form gels due to intermolecular association between mesogenic groups along the backbone or in side chains [9]. Instead, attention is first focused on macromolecular systems composed of chemically heterogeneous chains (e.g., block copolymers) that microphase-separate to form self-assembled micellar networks (SAMINs). A second section addresses physical organogels generated when low-molar-mass organic gelators (LMOGs) undergo specific interactions in organic solvents and organize into self-assembled fibrillar networks (SAFINs).

26.2
Block Copolymer Gels

Block copolymers are macromolecules composed of long contiguous sequences ("blocks") of chemically distinct repeating units that are covalently linked together. If the blocks are sufficiently incompatible, they can microphase-separate into a rich variety of periodic nanostructures, including spheres on a body- (or face-) centered cubic lattice, cylinders on a hexagonal lattice, bicontinuous channels exhibiting $Ia\bar{3}d$ symmetry (the "gyroid"), or alternating lamellae [10, 11]. Micro-phase separation results from the competition between enthalpically-driven block stretching to reduce repulsive contacts and entropically-driven block elasticity to minimize chain deformation. The resultant morphology is generally dictated by chain packing along the interface that divides the chemically dissimilar microdomains. In many (but not all) cases, interfacial chain packing relates to molecular composition, expressed in terms of the number fraction (f) of one of the blocks. As stated above, block copolymers microphase-separate if they are sufficiently incompatible, where thermodynamic incompatibility is given by the coupled parameter χN (χ is the temperature-dependent Flory–Huggins interaction parameter and N is the number of repeat units along the copolymer backbone [12]). As illustrated in Figure 26.1 for an AB diblock copolymer composed of A and B repeat units [13], an increase in χN for a given f_A tends to favor microphase-ordering for $\chi N > (\chi N)_{ODT}$, where ODT denotes the order–disorder transition. Similar behavior is observed for block copolymers possessing different molecular architectures. The architecture of primary interest here is the linear ABA triblock design. Unlike their simple AB analogues, ABA copolymer molecules are capable of forming supramolecular networks by depositing their A endblocks in different microdomains. More specifically, the ABA copolymers discussed hereafter possess glassy (PS or PMMA) endblocks at ambient temperature, so that the A microdomains effectively serve as physical crosslinks upon gel formation.

26.2.1
Concentration Effects

When an ABA triblock copolymer is dissolved in a B-selective organic solvent (i.e., the solvent is more compatible with, and thus preferentially swells, the B block), the A endblocks can, under propitious conditions, self-assemble to form microdomains that measure on the size scale of the endblocks [14–16]. Depending on

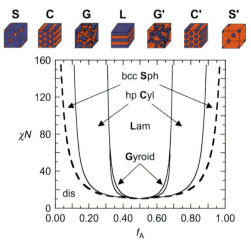

Figure 26.1 Theoretical phase diagram predicted for a solvent-free AB diblock copolymer wherein the thermodynamic incompatibility (χN) is presented as a function of copolymer composition (f_A). Ordered morphologies are labeled in the diagram and depicted schematically in order of increasing f_A above the diagram. The order–disorder transition (ODT) is denoted by the dashed bold line. Adapted from Ref. [13] and used with permission from the American Chemical Society.

factors such as the copolymer composition (f_A), copolymer chain length (N) and copolymer concentration in the solvent (w_{BC}), the microdomains will adopt a thermodynamically favored morphology at a given temperature (where T varies inversely with χ). In these studies, only solvated ABA systems composed of discrete micelles (SAMINs) with glassy, A-rich cores are considered further. At the critical micelle concentration (cmc), flower-like micelles first develop by a closed association mechanism characterized by dynamic equilibrium between unimers (u) and micelles (m) such that

$$nM^{(u)} \longleftrightarrow M^{(m)}$$

where M denotes mass and n is the molecular association number of the micelle. In the case of a flower-like micelle, both A endblocks of each copolymer molecule co-reside in the same micelle so that the molecule forms a loop, or one endblock remains in the incompatible matrix to yield a dangling end. Although thermodynamically unfavorable, the effect of dangling ends on the phase behavior and ordering of molecularly asymmetric A_1BA_2 triblock copolymers possessing A endblocks that differ in length ($N_{A_1} \neq N_{A_2}$) has been systematically investigated [17] in copolymer melts. In solvated systems consisting of molecularly symmetric ABA copolymers, dangling ends may become energetically preferable due to the entropic penalty (F_{loop}) associated with midblock looping in the micellar corona, which can be written as [18]

$$F_{\text{loop}} = \frac{1}{2}kT\ln(\pi\chi N_{\text{B}})$$ (26.1)

At copolymer concentrations beyond the cmc, copolymer chains added to a solution predominantly form micelles [10]. The solvent molecules serve to increase the effective volume fraction of the copolymer midblock in the phase diagram, and concurrently decrease the effective χ by screening repulsive A–B contacts [19]. Micellization is generally favored when the endblock size is large, and corresponding values of n may be high. Conversely, when the copolymer midblock is large, dangling ends are favored due to the high entropy of loop formation, which leads to smaller association numbers [20]. Conflicting trends of micelle association number and size with regards to block copolymer size and chemistry reported in independent studies [18, 21] can be attributed to variations in solvent–block interactions causing one of the blocks to dominate over the other. If the concentration of triblock copolymer is increased further, the number and size of micelles both increase, which results in a lower intermicellar spacing [22]. Due to their respective free energy penalties, the loops and dangling ends eventually form bridges wherein the endblocks locate in different micelles (cf. Figure 26.2), thereby generating a 3-D network stabilized by glassy micelles (SAMINs). The concentration at which a percolated SAMIN first develops is termed the critical gelation concentration (cgc) and is often significantly higher than the cmc under isothermal conditions. In the specific case of an ABA copolymer having glassy A endblocks and a rubbery B midblock, the neat copolymer is widely referred to as a thermoplastic elastomer (TPE), while the physically crosslinked, thermoreversible gel formed upon midblock-selective solvation has been previously termed [23] a thermoplastic elastomer gel (TPEG). Although TPEGs exhibit substantially lower moduli than their TPE

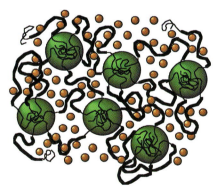

Figure 26.2 Illustration of the micellar morphology that develops due to triblock copolymer self-assembly in SAMINs. The three distinct conformations depicted include the midblock loop, midblock bridge and dangling endblock (labeled).

precursors, they tend to display remarkable elasticity, as evidenced by elongations to break in excess of 2000% strain [24].

The effect of copolymer concentration on molecular conformation and, hence, midblock bridging and network formation, is elucidated by the frequency (ω) spectrum obtained from dynamic rheology. Below the cgc, the copolymer solutions exhibit terminal responses that are characteristic of a liquid: $G' \sim \omega^2$ and $G'' \sim \omega$. At the sol–gel transition, however, $G' \sim G'' \sim \omega^m$, according to Winter and Chambon [25]. Experimental values of m commonly range from 0.4 to 0.7 [26], whereas percolation models predict values of m between 0.65 and 0.75 [27]. If the gel is self-similar, the Kramers–Kronig relationship can be invoked to yield $\tan\delta = G''(\omega)/G'(\omega) = \tan(m\pi/2)$ [26]. At copolymer concentrations above the cgc, G' ultimately becomes independent of ω, as required by the definition of a gel and evidenced in Figure 26.3. The plateau modulus G_0 increases with increasing copolymer concentration (c) as $G_0 \sim c^\alpha$. At low c ($c > $ cgc), the intermicellar distance of SAMINs is sufficiently large so that the micellar coronas remain unentangled and the modulus only incorporates contributions from bridged midblocks that behave as elastic springs in the unentangled-chain (Rouse) limit [28] with $\alpha = 1$. The corresponding plateau modulus depends on the molecular weight M_B of the bridge and is given by [29]

$$G_0^{\text{Rouse}} = \frac{RTc}{M_B} \tag{26.2}$$

At higher c, the midblock loops become entangled and α theoretically increases to 2.25 in the presence of a good B-selective solvent [30]. While several studies [29, 31, 32] have demonstrated experimentally that α can be close to this theoretical value, others [33, 34] have reported values of α between 3 and 4 for copolymer concentrations just above the cgc. This variation is attributed to the presence of a large number

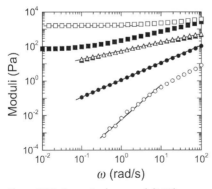

Figure 26.3 Dynamic shear moduli (G', open symbols; G'', filled symbols) provided as a function of oscillatory frequency (ω) for triblock copolymer organogels varying in copolymer concentration (in wt%): 1 (circles), 4 (triangles) and 10 (squares). The solid lines correspond to the scaling relationships discussed in the text. Adapted from Ref. [26] and used with permission from the American Chemical Society.

of clusters (or flocs) not physically attached to the network at the gel point. Roos and Creton [35] have derived an expression for the plateau modulus in this entangled regime by first considering the neat ABA triblock copolymer, which can be envisaged as an elastomer filled with hard spheres. If c_A represents the concentration of endblock microdomains ("hard spheres"), then, according to Guth [36],

$$G_0^{(neat)} = \frac{\rho RT}{M_e}(1 + 2.5c_A + 14.1c_A^2)$$

(26.3)

where ρ is the midblock density and M_e is the molecular weight between entanglements. The plateau modulus of the gel is consequently obtained by considering the copolymer in a B-selective solvent so that

$$G_0^{(gel)} = \phi^{2.25}G_0^{(neat)}$$

(26.4)

where φ is the volume fraction of copolymer. More elaborate frameworks such as the Slip Tube Network (STN) model proposed by Rubinstein and Panyukov [37] to depict polymer networks subject to uniaxial tensile deformation provide a more physically realistic representation of block copolymer gels, and its applicability to such nanostructured gels varying in concentration and molecular weight has been recently established [31], as discussed in Section 26.2.4.

Koňák *et al.* [38, 39] have confirmed the presence of block copolymer clusters in the vicinity of the cgc by probing the dynamics of a styrenic triblock copolymer in a midblock-selective solvent (*n*-heptane) with dynamic light scattering (DLS). Different dynamic modes are observed in the relaxation time distributions, $A(\tau)$ (cf. Figure 26.4), depending on whether the solution resides in the dilute or semidilute regimes, which are separated by the overlap concentration (c^*). The overlap concentration is defined as $3\,M/(4\pi N_A R_g^3)$, where the gyration radius scales as M^ν with $0.5 < \nu < 0.6$ for good solvents. At $c < c^*$, the correlation function is a simple exponential decay that indicates a single diffusive dynamic mode. This single mode at low copolymer concentrations is attributed to the translational diffusion of flower-like micelles. However, as $c \rightarrow c^*$, the correlation function becomes the sum of several simple exponentials (corresponding to flower-like micelles) and a slower stretched exponential. These dynamics reveal that, close to the overlap concentration, individual micelles, as well as polydisperse clusters of connected micelles, undergo translational diffusion, with the diffusion mode of the clusters being slower than that of the micelles. At $c > c^*$, three different dynamic modes are identified. The slow and fast modes appear diffusive in nature, while the middle dynamic mode is a relaxation mode at low temperatures but a diffusive mode at temperatures above 40 °C. The authors have hypothesized that, for sufficiently high copolymer concentrations, a gel network forms, and the fast mode reflects the collective diffusion of networked micelles. The middle mode arises due to the relaxation of nodes comprising the network. When the temperature is increased, the network begins to break down as a consequence of endblock pullout, in which case the middle dynamic mode becomes diffusive. Although the slowest

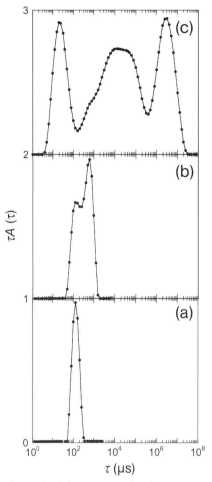

Figure 26.4 Relaxation time distributions, represented by $A(\tau)$ where τ denotes time, for triblock copolymer organogels varying in copolymer concentration (in wt%): (a) 0.49, (b) 1.50, and (c) 5.50. The solid lines correspond to model fits, as discussed in the text. Adapted from Ref. [39] and used with permission from John Wiley & Sons, Inc.

diffusive mode has been previously observed [28], and is presumed to correspond to the presence of large-scale heterogeneities (impurities), its origin nonetheless requires further investigation.

Watanabe [40, 41] has developed a dielectric technique to experimentally ascertain the fraction of midblock loops (φ_{L}) in a triblock copolymer melt or solution. Dielectric properties are dependent on the polarization (**P**) of the system under investigation, which, in turn, depends on the arrangement of the dipoles present. *Cis*-polyisoprene molecules possess "type-A" dipoles, in which case their dipoles orient parallel to the polymer backbone and **P** is proportional to the chain end-to-end vector. Thus, in a poly(styrene-*b*-isoprene-*b*-styrene) (SIS) triblock copolymer system, dielectric measurements provide information regarding the relative movement of chain ends. Since the domain boundaries are immobilized in a quiescent state at ambient temperature, the SIS copolymer is dielectrically inert [40]. If dipole inversion is introduced at the center of the midblock, **P** is proportional to the difference between the end-to-center vectors. This type of triblock (abbreviated as SIIS) is dielectrically active even under quiescent conditions. Quantitative comparison of the dielectric behavior of SIIS triblock and matched SI diblock copolymers yields [42] a simple expression for the fraction of looped midblocks (φ_L). For a triblock copolymer dissolved in n-tetradecane, φ_L is found to decrease from 0.8 to 0.6 as the copolymer concentration is increased from 20 to 50 wt%. The contributions to the overall modulus provided by midblock loops, due to osmotic effects of microlattice formation, and midblock bridges, due to physical crosslinks, measure on the same order of magnitude, with the latter being larger by a factor of 2–3×.

A block-selective solvent can induce morphological transitions in block copolymers because it serves to increase the effective volume fraction of the compatible block and reduce the effective χ of the system by screening repulsive A–B monomer contacts. Thus, gels initially possessing a lamellar nanostructure will be transformed into gels with micellar nanostructures (cylindrical or spherical) by increasing the solvent fraction [43]. Representative transmission electron microscopy (TEM) images illustrating this morphological progression are provided in Figure 26.5, in which a microphase-separated SIS triblock copolymer swollen to different extents in mineral oil exhibits lamellar (Figure 26.5a), cylindrical (Figure 26.5b), and spherical (Figure 26.5c) morphologies. In these figures, the unsaturated

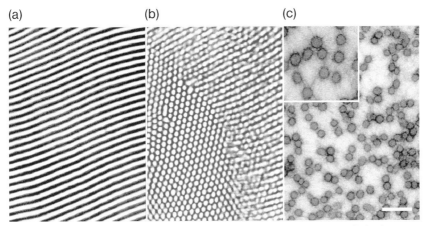

(a)   (b)   (c)

Figure 26.5 Transmission electron microscopy images of SIS triblock copolymer organogels portraying several morphologies at different copolymer concentrations. (a) Lamellar; (b) Cylindrical; (c) Spherical. In all three images, the isoprenic midblocks are selectively stained and appear dark. The inset in (c) is an enlargement that evinces the existence of fine structure in the solvent-rich matrix. Scale bar shown in (c) corresponds to 200 nm for (a) and (b), and 100 nm for (c).

isoprenic midblocks are selectively stained and appear dark. In Figure 26.5c, the coronas around the periphery of the micelles are delineated. It is important to recognize that nanoscale strands between and connecting neighboring micelles are evident in this image (and not in images wherein the styrenic endblocks are stained), which suggests that some bundles of bridged midblocks are sufficiently correlated that they become detectable upon staining. Morphological transitions in solvated triblock copolymer systems can be accompanied by abrupt changes in mechanical properties, such as the plateau or tensile moduli [44], and are readily identified by small-angle scattering patterns, such as those shown in Figure 26.6, due to signature peak ratios corresponding to various ordered morphologies [43]. While relatively few studies have systematically addressed the phase behavior of triblock copolymers in the presence of an organic solvent, Lodge and coworkers [19] have broadly scrutinized the effects of solvent concentration and selectivity on the phase behavior of diblock copolymers.

It is appropriate at this juncture to mention that several published reports [45, 46] describe diblock copolymer gels. Diblock copolymer molecules swollen in a selective solvent can self-organize into the same morphologies observed for solvated triblock copolymers, but there remains an important molecular-level difference. In a micellar morphology, for the sake of illustration, diblock copolymer molecules can only adopt a tail topology wherein the solvent-incompatible block forms the micellar core and the solvent-compatible block accounts for the solvent-swollen corona. Intermicellar bridging is not possible. At low copolymer concentrations, such systems behave as suspensions of soft particles. As the copolymer concentration is increased, the micelles order on a cubic lattice, a process that has been referred to as "microlattice structuring" [47]. Watanabe and Kotaka [48] have studied the microlattice formation of SI diblock copolymers in *n*-tetradecane, an I-selective solvent. In moderately concentrated solutions, overlapping coronas

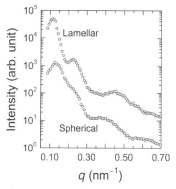

Figure 26.6 Small-angle X-ray scattering (SAXS) patterns acquired from SEPS triblock copolymer organogels varying in morphology (as labeled) due to differences in copolymer concentration at ambient temperature.

assume random conformations to maximize the conformational entropy, but are forced to maintain a uniform concentration distribution to minimize the free energy. A compromise yields mutually correlated microdomains with overlapping (and thus entangled) coronal blocks. When a small strain is applied, the local variation in concentration promotes an osmotic pressure gradient, which, in turn, generates a restoring force. Micellar solutions of diblock copolymers at moderate or high concentrations are elastic below a "yield" point. These systems, although gel-like at low strains, are not strictly considered organogels, as they undergo relaxation at ambient temperature. The presence of bridges in tri-block copolymer organogels serves to increase relaxation times, so that at suffi-ciently high copolymer concentrations ($c >$ cgc) and molecular weights, the copolymer network never relaxes at ambient temperature, which explains why G' is independent of frequency in rheological tests. Because of the copolymer network, SAMINs derived from triblock copolymers can undergo large strains and subse-quently snap back to their original shape, indicating that they possess shape memory.

26.2.2
Temperature Effects

At ambient temperature, a TPEG behaves as an elastic solid with relatively little hysteresis (i.e., nonrecoverable, or permanent, strain) induced upon cycling. When the temperature is raised above the T_g of the endblocks, the network-stabilizing crosslinks soften, and the gel transforms into a viscoelastic liquid with a distinct yield stress due to pull-out of endblocks from their microdomains upon deforma-tion. According to dynamic rheological analysis, the glass transition is manifested by a small decrease in the storage modulus (G') and a broad maximum in the loss modulus (G'') [49]. Care must be taken, however, not to confuse the endblock glass transition with other copolymer transitions, such as order–order transitions (OOTs), corresponding to morphological transformations, and the order–disorder transition (ODT), sometimes termed the lattice-disordering transition. In the latter case, long-range (lattice) order is replaced by short-range (liquid-like) order, and G' is observed to drop precipitously. This progression [50] is illustrated schemati-cally in Figure 26.7, in which cmT denotes the critical micelle temperature (i.e., the temperature at which micelles spontaneously form at constant concentration). Structural transitions such as OOTs and the ODT lie between the cmT and the endblock T_g. It must be recognized that all these transition temperatures depend on factors such as endblock size and endblock–solvent compatibility at elevated temperatures [22].

The dynamics of micelles in block copolymer gels have attracted considerable attention. In small-molecule (i.e., surfactant-based) micellar systems, micelle exchange is a well-documented phenomenon [51]. As alluded to above, under favorable conditions, block copolymer organogels can exhibit similar behavior. Watanabe and coworkers [52], for instance, have investigated network disruption and recovery of poly(butadiene-*b*-styrene-*b*-butadiene) (BSB) triblock copolymers

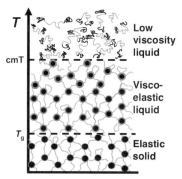

Figure 26.7 Schematic diagram showing the effect of temperature on the micellar network formed in triblock copolymer organogels. Not included here is the order–disorder transition (ODT), if it exists at the copolymer concentration of interest. This progression implicitly presumes that $c >$ cgc. Adapted from Ref. [50] and used with permission from the American Chemical Society.

in dibutyl phthalate, an S-selective solvent, using dielectric spectroscopy. Unlike conventional TPEGs with endblocks that form glassy crosslinks, solvated systems composed of BSB copolymer molecules consist of soft, deformable crosslinks that can experience chain pullout. When a large stress is applied at ambient temperature ($T \gg T_g$ of the endblocks), flow is restricted to pre-existing lattice defects. During shear, bridges across the defect plane convert to loops, which has likewise been observed [53] in triblock copolymer melts. This change in chain conformation generates a force that causes the micelles to migrate away from the defect plane. Some loops produced in this fashion mix transiently with the matrix, thereby forming dangling ends. A full recovery of the network upon cessation of shear is achieved when the thermodynamically unequilibrated dangling ends reform into bridges. Thus, the time for recovery is a function of the thermodynamic stability of the dangling end, which is a shear-independent relationship.

Another aspect of organogel dynamics that warrants close examination pertains to the kinetics of ordering and disordering at temperatures in close proximity to the ODT. Bansil *et al.* [54] have employed time-resolved small-angle X-ray scattering (SAXS) to probe the kinetics associated with the nanostructural (dis)ordering of organogels containing 20 wt% poly[styrene-*b*-(ethylene-*co*-butylene)-*b*-styrene] (SEBS) triblock copolymer. In this case, the kinetics of ordering upon cooling are found to be much slower (on the order of hours) than those of disordering upon heating (on the order of seconds). Upon slow cooling ramps, the effective volume fraction of hard spheres deduced from the Percus–Yevick scattering model [55] initially increases to a maximum of ~0.52 (indicating an increase in order) near the ODT, and then decreases slowly. At lower temperatures, Bragg peaks become discernible and increase in intensity, indicating that the system has ordered into a body-centered-cubic (bcc) lattice. This change is accompanied by a small, but abrupt, reduction in micelle size. The ordering transition has also been studied by performing a thermal quench from 140 °C (at which the gel is completely disordered) to a temperature below T_{ODT}, but above T_g. Two distinguishable stages are evident from the results, as shown in Figure 26.8. During an induction period (~1000 s or more after quenching), no change in the maximum SAXS intensity is evident, initially due to thermal equilibration and then to micelle supercooling. An analysis of the scattering curves by the Percus–Yevick model [56] indicates that the hard sphere volume fraction increases to ~0.53, which precedes the appearance of scattering reflections. Nucleation and growth of the ordered nanostructure occur during the second stage of structural development and account for an increase in scattering intensity at Bragg peak positions. The duration of the induction time depends on the depth of the thermal quench, exhibiting a minimum at an intermediate temperature between T_{ODT} and T_g. This observation suggests that a competition exists between the driving force for ordering, which increases at deeper quenches, and chain mobility, which is reduced at lower temperatures.

As alluded to earlier, block copolymer gels subjected to a thermal ramp can, under favorable conditions, also transform from one morphology to another at an OOT temperature. In this case, the initial morphology is either cylindrical or lamellar. Starting with a gel composed of a SEBS triblock copolymer swollen with a midblock-selective solvent and exhibiting the cylindrical morphology, Li *et al.* [57] have examined the transformation kinetics from hexagonally packed cylinders to bcc spheres by performing temperature-jump SAXS. Knowing the OOT and ODT temperatures (127 and 180 °C, respectively) from complementary rheological and SAXS measurements, temperature jumps between the OOT and ODT could be used to follow the mechanism of cylinder → sphere transformation. For temperatures up to ~145 °C, this structural evolution proceeds by a three-stage nucleation and growth process. In the first (incubation) stage, the cylinders retain their

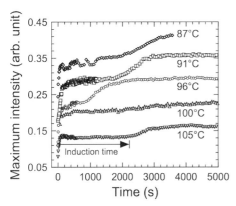

Figure 26.8 Time evolution of the maximum SAXS intensity for a triblock copolymer organogel with 20 wt% copolymer quenched from a temperature above the ODT to different temperatures (as labeled) below the ODT. The induction time (labeled for data acquired at 105 °C) corresponds to the time before the maximum intensity undergoes a nearly stepwise increase. Adapted from Ref. [54] and used with permission from the American Chemical Society.

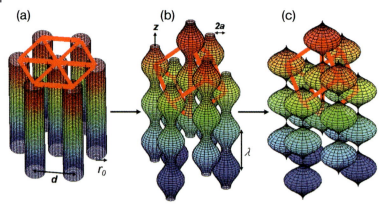

Figure 26.9 Scheme depicting the temperature-induced time evolution of hexagonally packed cylinders to bcc spheres in triblock copolymer organogels. Initial and subsequently modulated cylinders are displayed in (a) and (b), respectively, whereas cylinder break-up into spheres is portrayed in (c). The solid lines follow the accompanying crystallographic development discussed in the text. Reproduced from Ref. [57] and used with permission from the American Chemical Society.

shape, but a shift in the principal scattering peak indicates a reduction in the distance between cylinders. A critical number of cylinders rupture into modulated spheres during the second stage. These regions serve as nucleation sites to promote further rupture of cylinders and modulation of weakly correlated spheres in an outgoing ripple. Lastly, the spheres order on a bcc lattice so that the (100) plane of the hexagonal cylinders becomes the (110) plane of the bcc spheres and the cylinder axis evolves into the <111> direction of the bcc spheres, as illustrated in Figure 26.9. At temperatures beyond 150 °C, the first two stages coincide, and the mechanism commences via spinodal decomposition in which the cylinders transform directly into disordered spheres by correlated ripples. Above the ODT, the temperature jump permits formation of transient bcc spheres prior to disordering. Qualitatively similar results have been reported [58] for the transition from lamellae to cylinders in a triblock copolymer gel consisting of an endblock-selective solvent. Although no studies to date have examined the kinetics of OOTs in block copolymer gels upon cooling, corresponding efforts performed experimentally [59] and theoretically [60] on solvent-free copolymers have likewise revealed that the sphere → cylinder transformation occurs by a nucleation and growth process, the details of which are sensitive to the depth of the thermal quench used to promote the transition.

26.2.3
Microdomain Alignment

Microdomains that form by molecular self-organization in a block copolymer organogel, such as a TPEG, frequently appear deformed due to interfacial packing frustration that occurs during rapid solvent evaporation or thermal quenching. In either case, specimen processing can promote metastable conformations that become frozen-in once the endblocks undergo vitrification. It immediately follows, then, that increasing the temperature of such gels above the endblock T_g should give the endblocks ample opportunity to relax and redistribute, and the nanostructure time to equilibrate. Quenching of such ordered gels to temperatures below the endblock T_g should again lock-in the morphology, even though it is nonequilibrium at the lower temperature. Independent efforts [32, 61, 62] have explored the effect of annealing on the morphologies and properties of organogels composed of SEBS copolymers. Laurer *et al.* [32] have directly observed moderately improved long-range order in annealed gels by TEM, whereas Soenen *et al.* [61, 62] have reported an increase in the number of peaks in SAXS patterns of gels upon annealing. Although annealing in the melt can enhance the nanostructural order of block copolymer organogels, it remains unclear that long-time annealing serves to either improve their mechanical properties or alter their phase behavior.

Similar structural results (cf. Figure 26.10) have been purported by Kleppinger *et al.* [63, 64], who have also shown that SAXS patterns acquired from gels, annealed or otherwise, retain a broad structure factor maximum at high temperatures, indicating that network clusters remain. For gels possessing intermicellar distances such that the midblocks behave as statistical coils, an abrupt transformation to the bcc morphology proceeds upon annealing in the melt [65, 66]. This morphology is preserved even after cooling below T_g. For gels with higher copolymer concentrations and compressed midblocks, the bcc morphology eventually forms upon longer annealing times. In this case, a temperature increase promotes an increase in midblock gyration radius, thus improving orientational order. At lower temperatures, this morphology is nonequilibrium, which may result in lattice distortion due to frustration of close-packed spheres. An alternative route to improved microdomain orientation in block copolymer gels involves shear-

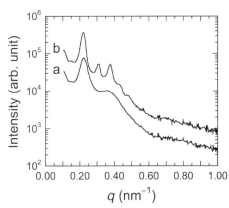

Figure 26.10 SAXS patterns obtained from triblock copolymer organogels with 20 wt% copolymer under two different thermal conditions. (a) Quenched and unannealed; (b) Quenched and annealed at 90 °C for 24 h. The higher-order Bragg diffraction peaks in (b) identify the morphology as bcc spheres. Adapted from Ref. [63] and used with permission from John Wiley & Sons, Inc.

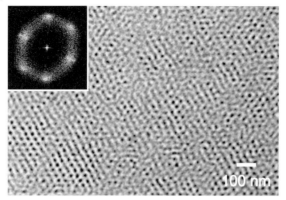

Figure 26.11 Transmission electron microscopy image of a triblock copolymer organogel composed of 20 wt% copolymer and subjected to large-amplitude oscillatory shear to induce nanostructural alignment. In contrast to Figure 26.5, the styrenic endblocks are selectively stained and appear dark. The inset is a 2-D Fourier transform of the image to confirm the high degree of order in this specimen.

induced disordering and reordering [64]. After the gel has been preannealed to form a bcc morphology, a large-amplitude oscillatory shear is applied to disrupt the nanostructure. Upon reordering, the gel exhibits a highly ordered, single-crystal, twinned bcc morphology, as depicted in Figure 26.11. Thus, a combination of thermal and shear treatment can be used to achieve single-crystal order in block copolymer gels. By surveying a range of strain amplitudes and frequencies, Mortensen *et al.* [67] have developed a morphology diagram (cf. Figure 26.12) that establishes the conditions under which block copolymer gels can be highly oriented. Orientation refinement is found to occur only in the frequency range where $G'' \approx G'$, and long shear times are required for dislocation planes to form.

26.2.4
Tensile Deformation

Reynaers and coworkers [68] have subjected disordered SEBS and poly[styrene-*b*-(ethylene-*co*-propylene)-*b*-styrene] (SEPS) gels to moderate uniaxial strains (up to 100%) and elucidated the resultant nanostructure by small-angle neutron scattering (SANS), as illustrated in Figure 26.13. Affine deformation of the supramolecular network is observed only at high copolymer concentrations, which favor entanglement of the copolymer midblocks. According to Prasman and Thomas [69], Poisson's ratio must remain constant to achieve affine deformation. At lower copolymer fractions, stretching in one direction up to 100% strain promotes the formation of distinct diffraction spots along the strain direction (Figure 26.13a) and well-defined layers of ordered micelles normal to the strain direction (Figure 26.13b). At higher strains (up to 1000%), however, the corresponding SANS patterns appear markedly different [70]. Between 200 and 400% strain, the patterns

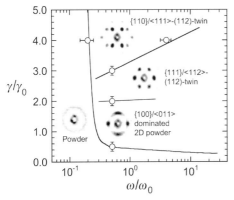

Figure 26.12 Normalized strain-frequency (γ-ω) diagram revealing the dynamic shear conditions under which a triblock copolymer organogel can be highly oriented into a twinned bcc spherical morphology. Adapted from Ref. [67] and used with permission from the American Chemical Society.

(a)   (b)

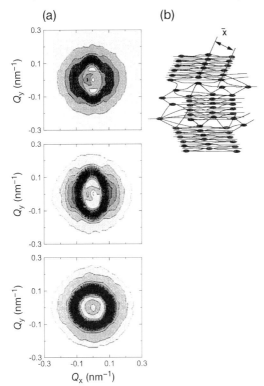

Figure 26.13 (a) 2-D SANS patterns of triblock copolymer organogels consisting of 18 wt% SEPS copolymer and subjected to different uniaxial tensile strains: (bottom) 0%, (middle) 50%, (top) 100%; (b) A clustered micelle arrangement consistent with experimental observations that SANS patterns do not change beyond a particular strain level. Adapted from Ref. [14] and used with permission from the American Chemical Society.

become more diffuse, clearly indicating distortion of the regular layers formed at lower strains. Above 400% strain, no structural changes are detectable at the length scales probed by SANS. This observation has been interpreted to mean that (i) the strained network consists of large, discrete clusters similar to those that develop at the sol–gel transition upon initial network formation; and (ii) these clusters vary in the extent of their connectivity. Cluster boundaries, for instance, possess fewer crosslink sites than the clusters themselves, and are thus more deformable than the clusters, which saturate and produce no further structural changes at high strains. Krishnan [71] has employed SAXS to examine the response of SEBS gels to biaxial tensile deformation up to 300% × 300% strain, and has found, by using the Percus–Yevick scattering model [72] incorporating a square-shoulder potential, that the micelles remain (for the most part) spherical and that the micellar coronas increasingly overlap with escalating strain. Kleppinger *et al.* [73] have likewise investigated the conditions responsible for affine behavior when SEBS gels are subjected to uniaxial strain. Affine behavior is observed in unannealed SEBS gels at high extension rates, while nonaffine behavior is dominant at low extension rates. If, however, a gel is preannealed so that the micellar nanostructure exhibits long-range order prior to deformation, then affine behavior is realized even at low extension rates.

Uniaxial tensile deformation of triblock copolymer gels deviates from the predictions of classical rubber elasticity [74], and is therefore often described in terms of the semi-empirical Mooney–Rivlin model [75, 76] that initially was developed for chemically crosslinked elastomers. A more recent attempt to relate the bulk physical behavior of triblock copolymer gels to their underlying nanoscopic network is the STN model [37], as introduced in Section 26.2.1. In this model, the measured nominal (or engineering) stress (σ) is given by

$$\sigma = \left(G_c + \frac{G_e}{g(\lambda)} \right) \left(\lambda - \frac{1}{\lambda^2} \right) f(\phi, \Phi) \tag{26.5}$$

where G_c and G_e denote the contributions of permanent crosslinks and transient entanglements, respectively, to G. Here, $g(\lambda) = 0.74\lambda + 0.61\lambda^{-1/2} - 0.35$, where λ represents the extension ratio, and $f(\phi,\Phi)$ depends on the concentration of hard filler (ϕ) – for example, glassy micelles, governed by the composition of the copolymer – and the copolymer volume fraction (Φ). Shankar *et al.* [31] have demonstrated that this model can provide valuable insight into molecular factors, such as the copolymer chain length (N), that influence gel network formation and properties.

26.2.5
Network Modifiers

26.2.5.1 Inorganic Nanofillers
Due to a growing interest in hybrid organic–inorganic nanocomposites [77], several studies [49, 78–83] have recently addressed the effects of inorganic nano-

scale fillers on morphology and property development in block copolymer gels. Electron microscopy images of gels modified with colloidal silica nanoparticles and a surface-modified organoclay are provided in Figures 26.14 and 26.15, and indicate the extent to which the additives disperse, which is of paramount importance with regard to controllable property development. Mechanical properties are generally found to improve when surface-modified silicas [49, 78] and organoclays [78] are incorporated into the gel matrix. Efforts to use carbon nanotubes (CNTs) [78] have yielded less impressive results due to challenges associated with sufficient dispersion. An enhanced modulus is achieved when the additive is more

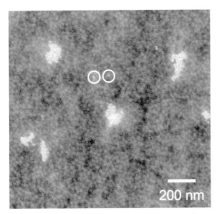

Figure 26.14 Energy-filtered TEM image of a triblock copolymer organogel composed of 10 wt% SEBS copolymer and modified with 3 wt% colloidal silica nanoparticles. The styrenic micelles, stained by the vapor of RuO$_4$(aq), appear light, whereas the siliceous nanoparticles appear bright, due to imaging at an energy loss of 200 eV. Two neighboring individual nanoparticles are circled.

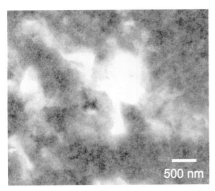

Figure 26.15 Energy-filtered TEM image of the same triblock copolymer organogel pictured in Figure 26.14, but modified with 3 wt% organoclay. The siliceous clay platelets likewise appear bright due to imaging under the same conditions as those employed in the previous figure.

compatible with the matrix (solvent and midblock), as this renders improved dispersion and, consequently, more efficient stress absorption from the soft phase [49]. Properties are observed [79] to generally improve with increasing nanofiller content, even at surprisingly high (60 wt%) loading levels. The addition of nanoparticles may also expand the mechanical performance and stability of block copolymer organogels at high temperatures, especially if (i) the attractive interactions between the nanoparticles and matrix are particularly strong [49]; or (ii) the nanoparticles themselves form a secondary, load-bearing network that remains thermally stable [78].

26.2.5.2 Polymeric Modifiers

Endblock-Selective Homopolymer In the previous section, the addition of inorganic nanofillers to a block copolymer gel resulted in a hybrid material wherein the nanofillers were highly dispersed to yield nearly discrete nanoscale particulates with an ultrahigh surface-to-volume ratio. For this reason, and to avoid macroscopic phase separation between the nanofillers and the gel, only very low nanofiller concentrations can be considered. The addition of an endblock-selective homopolymer to a block copolymer gel can likewise result in several different scenarios, depending on factors such as endblock compatibility, molecular weight disparity, and homopolymer concentration [84]. If the homopolymer is chemically identical (hA) to the endblocks of an ABA triblock copolymer, then only the molecular weight disparity ($\alpha = N_{hA}/N_A$) and hA concentration constitute key design parameters. As in solvent-free block copolymers, if α is large (>1), the hA molecules will not be physically accommodated within the brush comprising the A-rich microdomains. In this case, the brush is said to remain *dry* due to the lack of penetration of homopolymer molecules [85]. This entropic penalty favors macrophase separation between the copolymer and homopolymer molecules even at relatively low hA concentrations. In this limit, hA-rich domains measuring on the order of micrometers or larger coexist with the gel network, and the accompanying mechanical properties are largely dictated by the separating interface. The incorporation of semicrystalline syndiotactic polystyrene (sPS) into a SEBS gel, for example, results in the formation of discrete sPS crystals, which appear as filaments and sheets (cf. Figure 26.16) that greatly improve the modulus due to adhesion between the crystals and the styrenic micelles [86].

As α becomes smaller, however, due to a reduction in N_{hA} or an increase in N_A, the smaller hA molecules can locate within the A-rich microdomains and *wet* the compatible block brush. In this limit, added hA can serve to facilitate, or even induce, copolymer micellization because of the corresponding increase in the population of unfavorable A–B contacts [87], and it can likewise promote a change in interfacial curvature and, hence, gel morphology [88, 89]. Mechanical properties are found [90] to generally improve with increasing hA fraction up to a molecular-weight-dependent level, beyond which macrophase separation occurs. One way to lessen the propensity for macrophase separation and to ensure the encapsulation of a homopolymer within the endblock-rich microdomains responsible for stabiliz-

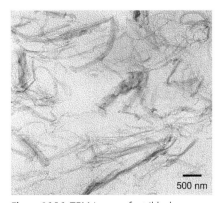

Figure 26.16 TEM image of a triblock copolymer organogel composed of 8.5 wt% SEBS copolymer and 1.5 wt% syndiotactic polystyrene (sPS). The styrenic micelles and crystalline sPS sheets and filaments are both selectively stained and appear dark. Adapted from Ref. [86] and used with permission from the American Chemical Society.

ing the gel network is to increase the homopolymer/endblock compatibility. In the case of gels composed of styrenic triblock copolymers (i.e., copolymers with PS endblocks), poly(2,6-dimethyl-1,4-phenylene oxide) (PPO) constitutes an ideal candidate in this regard, since χ between these two polymers is negative over all compositions and a large temperature range [91]. Moreover, since PPO possesses a relatively high T_g (~210 °C), it may be added to improve the service temperature of styrenic SAMINs [92]. This increase has been reported [93] to be as high as ~30 °C upon the addition of 3 wt% PPO to a SEBS gel. At higher loading levels, the mechanical properties improve substantially and morphological transitions can be expected [91].

Cosurfactant Although a variety of midblock-selective homopolymers can be blended into block copolymer gels (e.g., polyolefins such as polypropylene [94] added to gels with a primarily aliphatic solvent) to modify process or application properties, such modification normally results in the formation of macrophase-separated systems consisting of homopolymer-rich and gel-rich domains that are discrete or cocontinuous, depending on the relative concentrations. For this reason, such multicomponent systems are not considered further here. Another means by which to alter gel properties at the molecular level involves the addition of an AB diblock copolymer as a cosurfactant to the ABA triblock copolymer network. In this scenario, the AB molecules, if sufficiently incompatible, are forced to coreside with their ABA analogues, resulting in submicrodomain stratification [95]. Due to the presence of AB molecules, the ABA molecules are entropically forced to form bridges (rather than re-enter to form loops) due to coronal volume exclusion, in which case addition of an AB copolymer in small quantities can improve mechanical properties even when $c < \mathrm{cgc}$ for the parent ABA solution (cf.

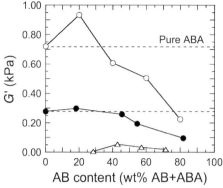

Figure 26.17 Dependence of G′ on the fraction of AB diblock copolymer added to triblock copolymer organogels and solutions varying in total copolymer concentration (in wt%): 15 (○), 11 (●), and 7 (△). Note that a frequency-independent modulus indicative of a gel network is not achieved in the system with 7 wt% copolymer until the diblock copolymer is added. The solid lines serve to connect the data. Adapted from Ref. [96] and used with permission from the American Chemical Society.

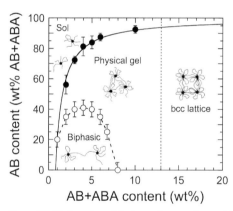

Figure 26.18 Experimental phase diagram for an ABA triblock copolymer organogel modified with an AB diblock copolymer. The lines serve as guides for the eye. Adapted from Ref. [29] and used with permission from John Wiley & Sons, Inc.

Figure 26.17) [96, 97]. Complementary SAXS studies performed by Vega *et al.* [29] reveal that the presence of an AB copolymer could help to prevent macrophase separation due to a reduction in intermicellar distance, and could, in general, be used to tune to the phase behavior of the system (cf. Figure 26.18).

26.2.6
Nonequilibrium Mesogels

Thermoplastic elastomer gels are normally prepared by mixing a triblock copolymer and a low-volatility solvent, with or without a carrier solvent (to reduce viscosity during mixing and evaporate thereafter), at elevated temperatures and then cooling the solution below the endblock T_g to induce glassy crosslinks that serve to stabilize the gel network. An alternative approach to preparing gels from the same copolymer and solvent pair is to introduce a solvent directly into the ordered copolymer by diffusion at temperatures below the endblock T_g. As the solvent-incompatible endblocks – and hence their microdomains – are glassy, they do not dissolve as the midblocks swell. Midblock swellability depends on both the solubility of the solvent in the midblock and the extent to which the midblocks stretch (which is entropically unfavorable), as shown schematically in Figure 26.19. Gels produced in this fashion from an ordered copolymer have been referred [99] to as "mesogels" because they retain the characteristics of the neat copolymer mesophase, since fabrication occurs under nonequilibrium conditions. King *et al.* [100] have observed that, on progressive swelling, such gels derived from a copolymer possessing the lamellar morphology retain highly swollen lamellae, as evidenced by the TEM image shown in Figure 26.20, even at solvent concentrations that

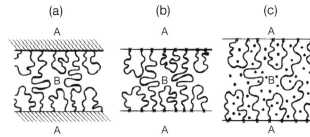

Figure 26.19 Schematic illustration of the procedure to generate lamellar mesogels from midblock-selective solvation of ABA triblock copolymers. (a) Microphase ordering of the copolymer from solvent casting or melt processing; (b) Reduction in temperature so that the A lamellae are rigid (i.e., glassy or semicrystalline); (c) Diffusion of solvent into the B lamellae to induce swelling while retaining a layered morphology. Reproduced with permission from Ref. [98]; © American Chemical Society.

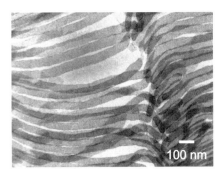

Figure 26.20 Transmission electron microscopy image of a triblock copolymer mesogel demonstrating that the procedure depicted in Figure 26.19 yields intact styrenic lamellae (stained) in a solvent-swollen midblock matrix. Reproduced with permission from Ref. [100]; © American Chemical Society.

would have otherwise induced morphological transformations. Li *et al.* [101] have expanded earlier theoretical efforts [98, 99, 102] designed to model mesogels in terms of swollen brushes by simulating the equilibrium swelling volume fraction as a function of morphology, bridge fraction and midblock–solvent interaction parameter. It is interesting to note that the extension and compression behavior of lamellar mesogels are predicted [102] to be dissimilar. During extension, chain elasticity dominates, whereas osmotic pressure governs compression. In general, however, mesogels tend to exhibit improved mechanical properties (expressed in terms of modulus [100]) relative to their equilibrium counterparts at the same gel composition.

26.2.7
Special Cases

26.2.7.1 Liquid Crystals

Generally speaking, liquid crystals (LCs) can be envisaged as anisotropic, rod-like molecules that are capable of developing orientational and/or positional order in the liquid state [103]. Of the three commonly encountered types of liquid crystal-line mesophases reported (nematic, smectic and cholesteric, or twisted nematic), the nematic, wherein the molecules align along a single direction with no positional order, constitutes the simplest [104]. *Thermotropic* LCs are temperature-sensitive, and an increase in temperature causes the nematic mesophase to disorder into an unstructured, isotropic liquid. Kornfield and coworkers [105] have successfully synthesized thermoresponsive triblock copolymer gels containing a midblock-selective LC solvent. To ensure sufficient compatibility between the copolymer and nematic solvent (4-cyano-4-*n*-pentylbiphenyl; known commercially as 5CB) and to avoid macrophase separation, the copolymer molecule is designed to have glassy (styrenic) endblocks and a midblock functionalized with a nematic side group. In this case, the copolymer midblock is soluble in both the LC and isotropic phases of 5CB. At low copolymer concentrations, the copolymer endblocks are soluble in the isotropic phase, but aggregate in the nematic phase due to endothermic mixing and a low entropy of mixing. Unlike isotropic solvents, in which the solvent quality changes gradually with temperature, this LC solvent undergoes an abrupt change in solvent quality at the relatively sharp isotropic → nematic phase transition [106]. In essence, the gel exhibits an "on–off" LC response at this transition temperature, thereby imparting the gel with added functionality. At higher copolymer concentrations (20 wt%), the endblocks also become insoluble in the isotropic phase of 5CB, in which case the gel network remains intact even at temperatures above the nematic → isotropic transition. Mesogels of LC triblock copolymers swollen by a nematic solvent have likewise been investigated [107]. An interesting finding is that the modulus of the gel can be reversibly changed by applying an electric field. The gel can also be sheared by applying a field above a certain threshold value, thus evincing quasi-piezoelectricity.

26.2.7.2 Ionic Liquids

Ionic liquids (ILs) constitute an emerging class of functional compounds that exhibit electrical conductivity and possess negligibly low vapor pressure, as well as broadly tunable physical properties [108]. Lodge and coworkers [26] have fabricated thermoreversible block copolymer gels by dissolving a poly(styrene-b-ethylene oxide-b-styrene) (SEOS) triblock copolymer in an IL at elevated temperatures, and allowing the glassy endblocks to self-organize and vitrify upon cooling. The IL used in this study remains liquid at ambient temperature. It is interesting to note that the same type of copolymer, with a polar midblock, has been used [109] to prepare mesogels in conjunction with poly(ethylene glycol) (PEG) for enhanced carbon dioxide separation. Characterization of the IL-based copolymer gel, for which the cgc is 4 wt% at 10 °C, reveals that the temperature dependence of the ionic conductivity is comparable to that of the bulk ionic liquid in the absence of the copolymer network. Similar gels have also been produced [110] with a poly(N-isopropyl acrylamide-b-ethylene oxide-b-N-isopropyl acrylamide) triblock copolymer, which possesses temperature-sensitive endblocks. Gelation in this system can be induced due to the lower critical solution temperature (LCST) behavior of the endblock in the IL solvent. Pioneering efforts such as these are charting the course for future research in the bottom-up design of conductive gels, especially as the copolymers and solvents can be further modified to improve both electrical and mechanical properties.

26.2.7.3 Multiblock Copolymers

Thus far, nanostructured organogels composed exclusively of ABA triblock copolymers have been considered. Multiblock copolymers generically designated as $(A_nB_n)_m$ copolymers are likewise expected to form stabilizing network structures in an A- or B-selective solvent. Similar to triblock copolymer gels containing cosurfactant (diblock copolymer) molecules (cf. Section 26.2.5.2), each microdomain in a multiblock copolymer consists of dangling endblocks, as well as looped and bridged midblocks, in proportions that depend only on n [111]. While the morphological and property attributes of solvent-free multiblock copolymers have received considerable attention [112, 113], few studies have explored the utility of well-defined multiblock copolymer gels. Bansil and coworkers, for instance, have examined an ABABA pentablock copolymer in two different solvents: 1,4-dioxane [114], a slightly good solvent for A and a θ solvent for B; and n-hexane [115], a strongly selective solvent for the B blocks. In 1,4-dioxane solutions, the marginally less-soluble B blocks appear to be physically connected by swollen A blocks. An interesting result is that a gel network does not develop in these solutions, even at the highest copolymer concentrations studied, due possibly to (i) an insufficiently low fraction of microphase-separated B blocks; or (ii) insufficient solvent selectivity. In the presence of n-hexane, macrophase separation occurs at low copolymer concentrations, whereas gelation is accompanied by solvent expulsion at higher concentrations. Gindy et al. [116] have attempted to explain this result by performing Monte Carlo simulations and have proposed that, in dilute solutions, macrophase separation occurs when the ratio m/n exceeds a critical value (as observed experimentally in *n*-heptane). Gelation at higher copolymer concentrations is attributed to the association of collapsed insoluble microdomains, similar to multiplets in ionomers. More complicated gel systems derived from randomly coupled multiblock copolymers possessing broad block and chain polydispersities (e.g., polyurethanes [117]) have likewise been generated and studied, but their nanostructures are typically not well-defined, which is why they are not considered further here.

26.2.7.4 Cosolvent Systems

Although ABA triblock copolymer gels normally consist of a single, low-volatility solvent that is selected to be sufficiently B-selective and A-incompatible, a mixture of miscible solvents can certainly be employed to fine-tune solvent quality and controllably alter the phase behavior and physical properties of the resultant gel [118]. More recent studies [71] have demonstrated that this strategy can likewise be used to adjust the time-responsive dynamic nature of such gels. The addition of a triblock copolymer to a saturated tackifying resin, which possesses saturated ring groups and a T_g near or slightly above ambient temperature, yields a time-dependent viscoelastic system that, upon uniaxial or biaxial deformation, slowly returns to its original shape. In this case, the elastic restoring force of the copolymer network is thwarted by the high viscosity of the solvent matrix. Results acquired from dynamic rheology confirm that both G' and G'' are strong functions of frequency, indicating that these systems are not gels according to the rheological criteria listed earlier. The addition of a low-T_g aliphatic oil to the system, however, results in a composition-dependent progressive shift of the frequency spectrum to higher frequencies. This apparent time-composition equivalence is similar in effect to time–temperature equivalence [5] and permits the construction of a superpositioned frequency spectrum over a broader range than could be measured experimentally (cf. Figure 26.21). As frequency relates to reciprocal time, the behavior of the cosolvent gel at very long or very short times can be accurately assessed at ambient temperature by simply changing the cosolvent composition.

26.3
Organic Gelator Networks

Organic gelling agents, or *gelators*, with molecular masses of less than ~2 kg mol^{-1} are referred to as low molar-mass organic gelators (LMOGs), and constitute a growing class of compounds that provides fundamental insight into molecular self-organization and practical use for applications requiring responsive materials [1, 119, 120]. In stark contrast to solvated block copolymers that form gels by microphase separation, LMOGs are generally classified according to their molecular structure and the intermolecular interactions that promote physical gelation. In this case, the organogel networks are stabilized via noncovalent physicochemical interactions such as hydrogen bonds, π–π stacking, or London dispersion forces. As with block copolymer gels, gels produced by LMOGs are thermorevers-

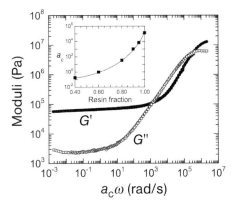

Figure 26.21 Frequency spectra shown as master curves for G' and G'' measured from triblock copolymer organogels composed of 25 wt% copolymer and a cosolvent (mineral oil/tackifying resin) that varies in composition, thereby demonstrating that a ternary SAMIN system exhibits time–composition equivalence. The composition-dependent shift factor (a_c), determined using a reference composition of 60 wt% tackifying resin, is included as a function of resin fraction in the inset.

ible, in which case the load-bearing networks dissolve into the surrounding liquid matrix upon heating above a composition-dependent dissolution temperature (T_{dis}), but reform upon cooling. Physical gels are typically generated by first heating a relatively low concentration (typically a few percent by mass) of the LMOG in an organic solvent or low-T_g polymeric liquid until all the components become a solution, or sol, and then cooling the sol to below the gelation temperature (T_{gel}). The value of T_{gel} is identified as the temperature at which flow is no longer discernible over long periods [119]. It is important to recognize that T_{gel} is generally lower than T_{dis}, since more thermal energy is required to break apart and dissolve the gel network than to form it. Conversely, gel network formation may require a finite degree of supercooling to initiate either (i) crystallization; (ii) precipitation; or (iii) aggregation of the LMOG, thereby producing a gel [121].

The resultant gels consist of 3-D self-assembled fibrillar networks (SAFINs) that are characterized by entangled nanoscale fibrils exhibiting a high surface-to-volume ratio. These networks have been visualized by a variety of imaging methods, including scanning electron microscopy (SEM), TEM and atomic force microscopy (AFM). Representative examples of SEM [122] and TEM [123] images of SAFINs are provided in Figures 26.22 and 26.23, respectively, and demonstrate that the fibrils can range in size from nanometers to micrometers in width, and from micrometers to millimeters in length. Due to the large solid–liquid interfacial area, the matrix solvent is effectively entrapped by capillary forces within the network. At the macroscopic level, the total volume of solvent is immobilized, resulting in a solid-like material [124]. Because the networks do not consist of long, elastic chains (as in block copolymer gels), LMOG-based gels tend to be exquisitely shear-sensitive, and their networks readily break apart during steady or large-amplitude

(a)   (b)

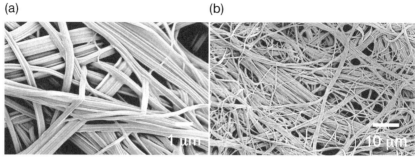

Figure 26.22 Scanning electron microscopy images collected at (a) high and (b) reduced magnification from a polycatenar organogel consisting of 3 wt% gelator, illustrating the morphology representative of SAFIN organogels. Reproduced with permission from Ref. [122]; © American Chemical Society.

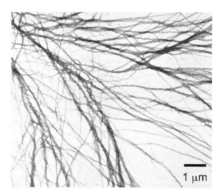

Figure 26.23 Transmission electron microscopy image acquired from a 1-acetonitrile organogel consisting of 0.07 wt% gelator, confirming the existence of nanofibrils (measuring 40–70 nm in diameter), some of which exhibit helical twist with a pitch of ~150 nm. To improve contrast, the nanofibrils have been selectively stained. Adapted from Ref. [123] and used with permission from John Wiley & Sons, Inc.

oscillatory shear, but reform upon cessation of shear. The kinetics of network healing depend on the chemistry (and interaction mechanism) of the LMOG, the concentration of the LMOG, and the quality of the solvent matrix. As gels produced with LMOGs depend on specific intermolecular interactions, this section is divided into three types of interactions that LMOGs require to promote physical gelation, namely hydrogen bonding, π–π stacking, and London dispersion forces.

26.3.1
Hydrogen Bonding

Hydrogen bonding is the attractive force that exists between an electronegative atom and a hydrogen attached to another electronegative atom, thereby imparting the hydrogen with a partial positive charge. The electronegative atom must possess one or more unshared electron pairs, and thus has a partial negative charge. Hydrogen bonding can occur *inter*molecularly, between molecules, or *intramolecularly*, between parts of the same molecule. Hydrogen bonding is weaker than covalent or ionic bonds, but stronger than van der Waals forces. Originally thought to be a random event, hydrogen bonding constitutes an example of a highly ordered occurrence and is ubiquitous throughout nature [125, 126]. Characteristics that are common to almost all gelators include [123]:

- Molecular factors, such as the presence of long alkyl substituents, that promote one-directional growth which prohibits the formation of 3-D crystal structure.

- The ability of SAFINs to branch and entangle, often due to active functional groups that promote intermolecular interactions, and thus develop a 3-D network capable of immobilizing the solvent.

- At least one chiral center within the molecule.

It should be noted that, while this list provides guidelines for known gelators, the existence of any or all of these traits in a molecule does not guarantee that it can gel organic solvents. Meléndez *et al.* [127] have suggested three common features of efficient gelators that self-assemble via hydrogen bonding: (i) the presence of one or more hydrogen-bonding groups; (ii) long alkyl substituents; and (iii) stereogenic centers within the molecule. Although these types of interaction unquestionably contribute to the ability of a molecule to induce gelation, this section focuses on the functional groups, chirality, and the self-aggregated structure of several important families of gelators. The aim here is to identify the common features that are responsible for gelation, and to classify LMOGs on the basis of their chemical similarity. As it is not possible to include all gelators at this point, several excellent reviews [119, 128, 129], each dedicated to LMOGs and their organogels, are recommended to the reader.

26.3.1.1 Amides

The presence of two electronegative atoms in amides allows them to produce highly ordered hydrogen-bonded networks that can extend in linear arrays, as well as form eight-membered ring dimers [130]. The addition of long alkyl chains to an amide can help to prevent crystallization [123], thereby resulting in an effective gelling agent. Cyclic amide derivatives possessing butyl chains, for instance, are less effective at promoting gelation than the same derivatives with chains possessing ten or more carbons [131]. The formation of extended molecular sheets due to gelation of a diaminocyclohexane constitutes another example of how hydrogen bonding promotes the entanglement of SAFINs through complementary functional groups [123]. Chirality in amides can also play a critical role when determining which enantiomeric form can induce gelation in an organic or organic-containing solvent. *Trans*-cyclohexane-1,2-diamide (see Figure 26.24) is an efficient gelator of polydimethylsiloxane (silicone) oil and liquid paraffin, whereas the *cis* enantiomer is unable to produce a gel [132]. This difference in gelling efficacy is attributed to the fact that the *trans* molecule can exist in an anti-parallel arrangement that

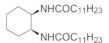

Figure 26.24 Chemical structure of *trans*-cyclohexane-1,2-diamide, an example of an efficient LMOG for SAFIN organogels.

maximizes the self-complementary interaction of hydrogen-bonding groups. Other types of amide gelators include aromatic polyamides [133], synthetic peptides [134], and perfluorinated amides [135].

26.3.1.2 **Ureas**

Ureas (also referred to as *carbamides*) have attracted much attention as another family of LMOGs. More specifically, bis-urea molecules have been used as organic gelators that assemble into thin rectangular sheets by hydrogen bonding and form thermoreversible gels [136]. Similar to many other organic gelators, only one enantiomeric form, *trans* bis-urea, is capable of promoting gelation because, unlike its *cis* analogue, it can self-aggregate into an anti-parallel arrangement and thus form linear aggregates [137]. A practical challenge with urea gelators is that the minimum temperature, at which the gels are stable, is much higher than ambient temperature (~100 °C) [127]. Hamilton [138] has, however, successfully established that a group of bis-urea–amino acid conjugates can gel solvents as low as 5 °C, which is unique for this molecule and highly advantageous for industrial applications. "Spacer" moieties such as azobenzenes, thiophenes and bithiophene groups have been incorporated between two urea molecules to synthesize new urea-based gelators [139–141]. These designer molecules self-assemble into highly ordered monolayers by hydrogen bonding between the urea groups [137].

26.3.1.3 **Sorbitols**

Sorbitol is a sugar alcohol used in diverse applications in the food and medical industries. One commercial derivative of sorbitol is 1,3:2,4-dibenzylidene sorbitol (DBS), which is an effective LMOG due to its solubility in a broad range of organic solvents and polymers at elevated temperatures and the low concentrations of DBS necessary to induce gelation. The DBS molecule (Figure 26.25) is both chiral and amphiphilic, and is often described as "butterfly-like" due to its sorbitol body and two phenyl "wings." The two hydrophobic phenyl groups are largely responsible for the solubility of DBS in various polymers and organic solvents, while the hydroxyl and acetal oxygen functionalities induce intermolecular hydrogen bonding between DBS molecules, resulting in nanofibril formation [142]. Network formation is achieved by the aggregation of DBS nanofibrillar strands or bundles, which is strongly influenced by both DBS concentration and solvent polarity [143]. The primary unit size of DBS nanofibrils has been repeatedly measured to be about 9–10 nm in diameter. Using dynamic rheology, Wilder *et al.* have investigated the time–temperature equivalence principle of DBS in polypropylene glycol and PEGs differing in endblock chemistry and, hence, polarity [144]. The extraction of activation energies from shift factors discerned during the analysis reveals

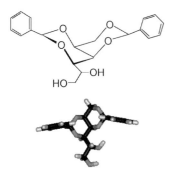

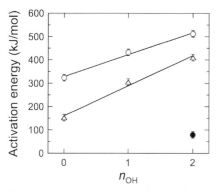

Figure 26.25 Chemical structure (top) and energy-minimized molecular configuration (bottom) of the commercial gelator 1,3:2,4 dibenzylidene sorbitol (DBS).

Figure 26.26 Activation energy discerned as a function of solvent polarity (expressed in terms of the number of hydroxyl endgroups, n_{OH}) from the time–temperature equivalence of frequency spectra collected from DBS organogels containing functionalized polyethylene glycol at two different DBS concentrations ($\bigcirc$, 2 wt%; $\triangle$, 5 wt%) and polypropylene glycol ($\bullet$) at a DBS concentration of 0.4 wt%. The solid lines are linear regressions of the data.

that DBS gelation is most energetically favored (with relatively low activation energies) in nonpolar solvents (cf. Figure 26.26).

The reason for this behavior is that, as the polarity of the solvent increases, the DBS molecules can hydrogen bond – even if only transiently – with solvent molecules rather than with other DBS molecules, thus reducing the gelation effectiveness and increasing the time required for gelation. Although through independent experimental and theoretical efforts, much insight has been gained into the macroscopic properties of DBS organogels, little is known regarding the precise molecular mechanism by which DBS induces gelation. Since D,L-DBS, a racemate of DBS with equal mixtures of D and L forms, is unable to form gels [145], the chirality of the DBS molecule appears to constitute a critical consideration in elucidating the self-assembly behavior of DBS molecules. Fourier-transform infrared (FTIR) spectroscopy has likewise confirmed that DBS molecules self-assemble via hydrogen bonding by the appearance of a broad spectral peak at 3250 cm^{-1} [145, 146], although molecular modeling [143] suggests that phenyl stacking may also play an important role. One of the reasons why DBS is singled out here, other than its commercial availability, is that it is capable of inducing gelation in both nonpolar solvents and molten polymers [147] at relatively low concentrations (ca. 0.25 wt%), which makes it attractive for scientific enquiry and technological applications.

26.3.1.4 Miscellaneous LMOG Classes

Hydrazides are hydrazine derivatives possessing a N–N covalent bond with four substituent groups, at least one of which is an acyl group. Tan *et al.* [148] have reported that long-chain substituted benzoic acid hydrazides (cf. Figure 26.27a) can form stable gels in organic solvents. Spectroscopic analysis confirms that the molecules self-assemble via hydrogen bonding due to the synergistic interactions between the carbonyl and amine groups, and the growth of the resultant SAFINs in one dimension is attributed to self-alignment of the long alkyl group. Although it is typically essential that LMOGs possess a chiral center, Bai *et al.* [149] have successfully synthesized achiral hydrazides. These compounds, composed of a core hydrazide unit with three exterior alkoxy chains of varying length, self-assemble due to intermolecular hydrogen bonding into columnar aggregates that are capable of efficiently gelling a variety of nonpolar organic solvents. *Calixarenes*, which are cup-shaped LMOGs characterized by several repeat units (each of which possesses a long acyl group at the *para* position of an aromatic ring; see Figure 26.27b), have likewise been shown [150] to generate stable gels in organic liquids such as alkanes, alcohols, carbon tetrachloride, and aromatic solvents. Physical gelation due to hydrogen bonding is attributed to the carbonyl in the acyl group, since alkyl groups in the *para* position do not gel common organic liquids. Because of the large number of possible binding sites available on calixarenes, the addition of metal ions provides a viable route by which to produce stable metallogels [151].

26.3.2
π–π Stacking

Aromatic interactions, also known as π–π stacking, refer to noncovalent intermolecular interactions between organic compounds that contain phenyl units. This type of interaction is caused by overlapping *p*-orbitals in π-conjugated systems,

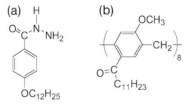

Figure 26.27 Chemical structures of two LMOGs with long alkyl-chain substitution. (a) A benzoic acid hydrazide; (b) A calixarene.

and π–π interactions are strongest for flat, polycyclic aromatic hydrocarbons such as anthracene, triphenylene or coronene, due to the numerous delocalized π-electrons residing in these molecules. Sufficiently strong interactions may develop between coplanar aromatic rings so that these groups on neighboring molecules stack, in which case the molecules consequently self-assemble to form SAFINs. One of the most diverse classes of LMOGs that rely on π–π stacking for network formation derives from the cholesterol, a steroid of which the backbone consists of hydrocarbons arranged in three six-membered rings and one five-membered ring (cf. Figure 26.28a). Although cholesterol is amphiphilic due to a polar hydroxyl head group and hydrophobic body and tail groups, it alone is incapable of gelling organic liquids. Relatively simple synthetic derivatives of cholesterol, however, constitute excellent examples of LMOGs. Lin *et al.* [152, 153] have established that molecules composed of an aromatic (A) unit connected to a steroidal (S) group through a functionalized linkage (L) serve as efficient and predictable ALS-type gelators, and numerous variations of the ALS design motif have been investigated since its introduction. The aromatic group is essential for gelation to ensure π–π stacking. On a side note, the delocalized π-electrons in the aromatic ring may likewise impart resultant organogels with electronic and optical properties that provide added functionality. While the stacking of aromatic units in ALS-type molecules is generally responsible for the formation of linear aggregates, cholesterol derivatives, in particular, tend to self-assemble so that the aromatic groups are arranged helically, facing outward, at the periphery of a columnar core, as experimentally verified by the AFM observations of Song *et al.* [154, 155].

Even if the linker between the A and S segments is relatively long, ALS-type LMOGs still produce stable organogels due to π–π interactions. The development of intercolumnar aromatic–aromatic interactions further stabilizes the gel network. Other factors affecting the stability of the gel network include the size and shape of the aromatic group. For example, large polyphyrin rings or simple phenyl rings yield relatively weak gels unless they are simultaneously stabilized by other intermolecular interactions induced by (i) coexisting functionalities to promote, for

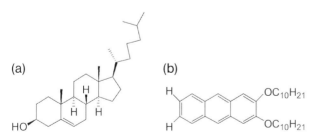

Figure 26.28 Chemical structures of two highly aromatic compounds used as LMOGs alone or modified.
(a) Cholesterol, which requires derivatization (cf. the ALS approach discussed in the text); (b) 2,3-di-*n*-alkoxyanthracene, which can gel polar solvents.

instance, hydrogen bonding [143]; or (ii) chemical additives such as metal ions, amines or nucleobases [156]. It is interesting to note that the sol–gel transition in ALS systems can be controlled by UV irradiation, as well as by temperature. The *trans*-isomer of an azobenzene ALS, for example, is able to promote gelation of organic solvents. Upon irradiation, it switches to the *cis*-isomer, which is incapable of producing a gel [157]. Over the past decade, cholesterol-based gels have become more chemically complex through the strategic incorporation of inorganic complexes [158, 159], novel linker groups [160, 161], and multiple LS moieties [124]. In the event that gelation relies exclusively on aromatic interactions, however, LMOGs such as dialkoxybenzenes (cf. Figure 26.28b) perform more effectively if they are symmetrically substituted [162, 163]. Using SANS, Terech *et al.* [164] have ascertained that the molecules which fit into this classification tend to self-assemble into bundles possessing a hexagonal or square symmetry, which suggests that the molecules are shifted radially to maximize π–π interactions. Unlike other LMOGs that rely to different extents on hydrogen bonding, these molecules form gel networks solely on the basis of nonpolar interactions and can thus gel polar organic liquids.

26.3.3
London Dispersion Forces

London dispersion forces are weak compared to other noncovalent interactions and arise from induced dipoles between two molecules, regardless of their intrinsic polarity. They are typically stronger between molecules that (i) can be easily polarized; or (ii) contain large or electron-dense atoms. While it has been suggested previously in this chapter that SAFINs require chemically complex LMOGs, simpler LMOGs based on functionalized long-chain *n*-alkanes are also capable of gelling organic liquids due to London dispersion forces. Most of these alkanes are at least partially fluorinated or possess N or S heteroatoms incorporated along the backbone [165, 166]. Perfluoroalkylalkanes with a chemical structure of the form $F(CF_2)_n(CH_2)_mH$ constitute examples of LMOGs that undergo self-assembly due to thermodynamic incompatibility between the chemically dissimilar fluorinated and hydrocarbon segments. Due to this incompatibility, the two segments microphase-separate into lamellar nanostructures and form gel networks stabilized only by London dispersion forces as a consequence of the strong dipole of the fluorinated segment [167]. The SAFINs generated by perfluoroalkylalkanes, which inherently possess a low surface energy [168], have been utilized to create superhydrophobic surfaces in conjunction with organic solvents [169].

26.3.4
Special Considerations

26.3.4.1 Biologically Inspired Gelators
Amino acid derivatives constitute another class of organogelators that are becoming increasingly important due to their ability to gel both organic and aqueous

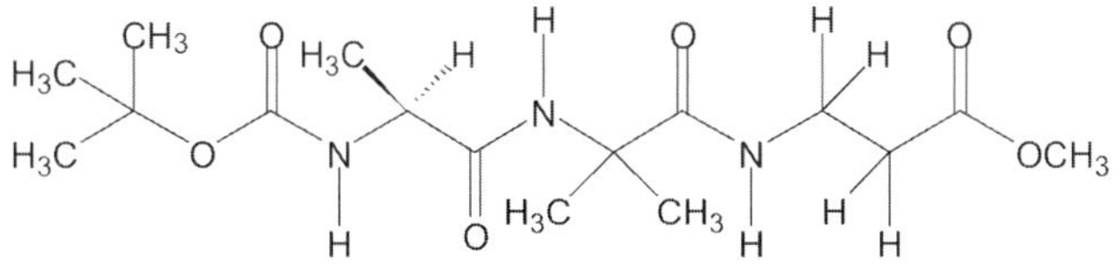

Figure 26.29 Chemical structure of the tripeptide LMOG
Boc-Ala-(α-aminoisobutyric acid)-(β-Ala)-Ala-Ome.

liquids [1]. Even in polar solvents, these compounds self-assemble via hydrogen bonding due to the large number of donor sites available per molecule, and the gelation efficacy improves as the number of peptide units increases. The thermodynamic balance between the hydrophobic alkyl substituent and the hydrophilic amide segment governs how these molecules self-assemble. According to UV-visible spectroscopy, these molecules can organize into one of three morphologies: lamellar spheres (multilamellar vesicles); helical ribbons; or tubules. Discrete tubules and ribbons become entangled as the solution approaches the sol–gel transition, which consequently promotes network formation via hydrogen bonding [170]. Lamellar spheres, on the other hand, are unstable and incapable of forming a gel network. One amino acid commonly used as an organogelator is L-lysine, which can bind with long aliphatic chains to form fibrillar micelles [171], a unique organogel nanostructure. Further studies [172] with L-Lysine have yielded binary gels. An interesting feature of this system is that, while neither α,ω-diaminoalkane alone can form a gel, together the two species can mutually self-assemble and promote network formation. This discovery of organogels derived from amino acid derivatives greatly facilitates control over the conditions responsible for gelation, including the sol–gel transition temperature.

Both, linear and cyclopeptides are capable of gelling a variety of organic solvents and aqueous systems. In the same vein as amino acid derivatives, their gelation effectiveness is due, for the most part, to the large number of hydrogen-bonding donor cites located on these molecules. Oligopeptide-based gelators and tripeptides (cf. Figure 26.29) have attracted much attention due to their ability to self-assemble through intermolecular hydrogen bonding into anti-parallel β-sheets, which can be envisaged as an entangled network of rodlike fibrils [173, 174]. Hirst *et al.* [175] have investigated mixing different molecular building blocks of dendritic peptides to identify the key factors that influence the self-assembly of such LMOGs. These authors have reported that mixtures of molecules differing in size and chirality can self-organize, whereas mixing molecules differing in shape, defined as the length of the spacer connecting the peptide head groups, reduces the propensity for supramolecular organization. Findings such as these regarding LMOG mixtures are beneficial to the rational design of organogels with tailorable properties.

26.3.4.2 Isothermal Gelation

Although all organogels discussed thus far are generated by a change in temperature, some can be formed isothermally by bubbling CO_2 and N_2 through solutions of primary *n*-alkanamines [176]. This distinctive gelation process has been shown to be chemically reversible, which allows the sol to be recovered. Another type of latent organogel is achieved by bubbling HCl through amino acid derivatives of cholesterol to protonate the amino groups [177]. Organogelation can also be performed *in situ* in the presence of highly reactive solvents. This route not only eliminates the need for thermal treatment, but also shortens the gelation time as the crystalline state is bypassed during formation of the 3-D fibrillar network [178].

26.3.4.3 Solvent Effects

Solvent polarity plays a critical role in the hydrogen-bonding (and hence gelation) efficacy of many LMOGs. The degree to which the binding sites in a gelator are solvated affects the strength of the gel network, which explains why dramatically different properties can be realized with a single LMOG in different solvents. A recent study [179] has addressed the ability of the disaccharide trehalose to gel organic solvents on the basis of the Hildebrand solubility parameter theory. Gelation is found to improve markedly as solvent–gelator interactions decrease and solvent polarity is low; this is consistent with the findings of Wilder *et al.* [143] regarding the gelation effectiveness of DBS in polyethers differing in polarity due to endgroup substitution. Zhu and Dordick [179] have likewise demonstrated that, when the solvent–gelator interactions are low, the gel network forms from thin, entangled fibrils. Yet, in the presence of strong solvent–gelator interactions, clusters of gelator molecules form thick, rigid fibrils. In addition to solvent polarity, two other criteria must be considered when matching a solvent with an appropriate LMOG to form a stable organogel:

- The boiling and melting temperatures of the solvent and pure LMOG, respectively, must be high so as to avoid undesirable loss of material due to vaporization.

- The solubility of the LMOG in the solvent of choice must be low so as to avoid bulk crystallization of the LMOG [120].

The organic solvents considered thus far in the design of SAFINs have all been isotropic liquids, but this requirement is relaxed below as we again consider organogel networks containing LCs (cf. Section 26.2.7.1).

Because the ability of LC molecules to reorient depends on external conditions such as temperature and electromechanical fields [180], they are used as active displays in watches, televisions, computer monitors, and projectors. Attempts to better control these functional liquids have shown [181] that organogels produced from LCs in the presence of LMOGs can yield fast electro-optical responses due to the low concentrations of gelator required for network formation. As in conventional SAFINs, the LMOG is first dispersed in an isotropic LC solution wherein the LMOG can, upon cooling, self-assemble into fibrils to form a thermoreversible gel network containing immobilized LCs [182]. Unlike other physical gels with only two thermoreversible states (sol and gel), however, LC gels possess three distinct states: isotropic liquid, isotropic gel, and liquid-crystalline gel [182]. Results obtained with FTIR spectroscopy reveal [183, 184] that, while hydrogen bonds are absent in the isotropic liquid state, such interactions develop at the sol–gel transi-

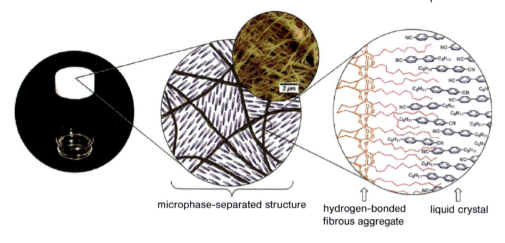

microphase-separated structure   hydrogen-bonded   liquid crystal
                                    fibrous aggregate

electric field OFF   electric field ON

light scattering   light transmission

Figure 26.30 Top: Schematic illustrations of organogels containing a nanostructured liquid crystal (LC) solvent, wherein the SAFIN entraps the anisotropic LC molecules (labeled). Bottom: Images of LC organogel films with an applied electric field off (light scattering, left) and on (light transmission, right), demonstrating the field-responsive optical nature of these organogels. Reproduced with permission from Ref. [180]; © American Association for the Advancement of Science.

tion. Deindörfer *et al.* [185] have recently reported on the formation of photosensitive gels composed of smectic LCs and capable of responding elastically to small stresses. The scattering of light from LC physical gels is also of tremendous scientific and technological interest, as the degree of transparency can be controlled electrically without using a polarizer (see Figure 26.30) [186, 187]. Such gels may likewise display anisotropic mechanical responses and recovery, which can be strongly affected by temperature [188].

26.4
Conclusions

Nanostructured organogels developed from SAMINs (through, for instance, the use of selectively solvated triblock copolymers) or SAFINs (through the use of sparingly soluble LMOGs) afford a new class of soft materials [189–191] that possess a broad range of interesting and useful properties. In the first case, supramolecular networks stabilized by nanoscale micelles are readily generated by allowing an incompatible block copolymer possessing at least one midblock to self-organize into discrete micelles in the presence of a midblock-selective solvent. If the micelle-forming blocks are glassy, the micelles effectively behave as physical crosslinks that can endow this genre of organogels with remarkable elasticity and shape memory, depending on factors such as copolymer composition and molecular weight. Because of their unique property attributes, SAMINs are currently under investigation [24, 31, 192, 193] as next-generation dielectric elastomers designed for use as synthetic muscle in microrobotics and as pumps in microfluidics. Through judicious selection of the midblock-selective solvent, SAMINs can likewise exhibit temperature-responsive optical [105] or conductivity [26] properties. In the case of SAFINs, high-melting LMOGs form (nano)fibrillar networks in various organic solvents through site-specific intermolecular interactions such as hydrogen bonds, π–π stacking or London dispersion forces. Unlike SAMINs, however, the stability of SAFINs tends to be exquisitely sensitive to mechanical deformation, as well as to temperature. Once the supramolecular network of a SAFIN is broken under shear, for instance, the solvent can flow. Upon cessation of shear, the network begins to reform so that, over a system-specific period of time, the initial SAFIN may be fully recovered. In addition to such mechanically responsive properties, SAFINs have been designed with unique optical properties and can be produced through nonconventional routes, such as exposure to light [194]. Taken together, these two families of nanostructured organogels afford tremendous versatility in the fabrication of soft materials exhibiting designer properties for use in a wide range of mature and emerging (nano)technologies [121].

Acknowledgments

These studies were supported by Eaton Corporation and the U.S. National Science Foundation. The authors thank Mr Arif O. Gozen for his editorial services, and the various authors who contributed their results to this work.

References

1 Weiss, R.G. and Terech, P. (2006) *Molecular Gels: Materials with Self-Assembled Fibrillar Networks*, Springer, Dordrecht.

2 Painter, P.C. and Coleman, M.M. (1997) *Fundamentals of Polymer Science: An Introductory Text*, 2nd edn, Technomic Pub. Co., Lancaster, PA.

3 Joly-Duhamel, C., Hellio, D., Ajdari, A. and Djabourov, M. (2002) *Langmuir*, **18**, 7158–7166.

4 Joly-Duhamel, C., Hellio, D. and Djabourov, M. (2002) *Langmuir*, **18**, 7208–7217.

5 Larson, R.G. (1999) *The Structure and Rheology of Complex Fluids*,

Oxford University Press,
New York.

6 Admal, K., Dyre, J., Hvidt, S. and
Kramer, O. (1993) *Polym. Gels
Networks*, **1**, 5–17.

7 Spěváček, J. and Schneider, B. (1987)
Adv. Colloid Interface Sci., **27**, 81–150.

8 Bui, H.S. and Berry, G.C. (2007) *J.
Rheol.*, **51**, 915–945.

9 Franse, M.W.C.P., Nijenhuis, K.T. and
Picken, S.J. (2003) *Rheol. Acta*, **42**,
443–453.

10 Hamley, I.W. (1998) *The Physics of Block
Copolymers*, Oxford University Press,
Oxford, New York.

11 Bates, F.S. and Fredrickson, G.H. (1999)
Phys. Today, **52**, 32–38.

12 Leibler, L. (1980) *Macromolecules*, **13**,
1602–1617.

13 Matsen, M.W. and Bates, F.S. (1996)
Macromolecules, **29**, 1091–1098.

14 Lindman, B. and Alexandridis, P.
(2000) *Amphiphilic Block Copolymers:
Self-Assembly and Applications*,
1st edn, Elsevier, Amsterdam, New
York.

15 Alexandridis, P. and Spontak, R.J.
(1999) *Curr. Opin. Colloid Interface Sci.*,
4, 130–139.

16 Hamley, I.W. (2005) *Block Copolymers in
Solution: Fundamentals and Applications*,
John Wiley & Sons, Ltd, Chichester,
England and Hoboken, NJ.

17 Hamersky, M.W., Smith, S.D., Gozen,
A.O. and Spontak, R.J. (2005) *Phys. Rev.
Lett.*, **95**, 168306.

18 Balsara, N.P., Tirrell, M. and Lodge,
T.P. (1991) *Macromolecules*, **24**,
1975–1986.

19 Lodge, T.P., Pudil, B. and Hanley, K.J.
(2002) *Macromolecules*, **35**, 4707–4717.

20 Giacomelli, F.C., Riegel, I.C., Pctzhold,
C.L., da Silveira, N.P. and Spěváček, P.
(2009) *Langmuir*, **25**, 731–738.

21 Quintana, J.R., Jáñez, M.D. and Katime,
I. (1998) *Polymer*, **39**, 2111–2117.

22 Seitz, M.E., Burghardt, W.R., Faber,
K.T. and Shull, K.R. (2007)
Macromolecules, **40**, 1218–1226.

23 Laurer, J.H., Bukovnik, R. and Spontak,
R.J. (1996) *Macromolecules*, **29**,
5760–5762.

24 Shankar, R., Ghosh, T.K. and Spontak,
R.J. (2007) *Adv. Mater.*, **19**, 2218–2223.

25 Winter, H.H. and Chambon, F. (1986) *J.
Rheol.*, **30**, 367–382.

26 He, Y.Y., Boswell, P.G., Bühlmann, P.
and Lodge, T.P. (2007) *J. Phys. Chem. B*,
111, 4645–4652.

27 Yu, J.M., Jérôme, R., Overbergh, N. and
Hammond, P. (1997) *Macromol. Chem.
Phys.*, **198**, 3719–3735.

28 Raspaud, E., Lairez, D., Adam, M. and
Carton, J.P. (1996) *Macromolecules*, **29**,
1269–1277.

29 Vega, D.A., Sebastian, J.M., Loo, Y.L.
and Register, R.A. (2001) *J. Polym. Sci.,
B: Polym. Phys.*, **39**, 2183–2197.

30 de Gennes, P.-G. (1979) *Scaling Concepts
in Polymer Physics*, Cornell University
Press, Ithaca, NY.

31 Shankar, R., Krishnan, A.K., Ghosh,
T.K. and Spontak, R.J. (2008)
Macromolecules, **41**, 6100–6109.

32 Laurer, J.H., Mulling, J.F., Khan, S.A.,
Spontak, R.J. and Bukovnik, R. (1998) *J.
Polym. Sci., B: Polym. Phys.*, **36**,
2379–2391.

33 Yu, J.M., Dubois, P., Teyssié, P.,
Jérôme, R., Blacher, S., Brouers, F. and
L'Homme, G. (1996) *Macromolecules*, **29**,
5384–5391.

34 Inomata, K., Nakanishi, D., Banno, A.,
Nakanishi, E., Abe, Y., Kurihara, R.,
Fujimoto, K. and Nose, T. (2003)
Polymer, **44**, 5303–5310.

35 Roos, A. and Creton, C. (2005)
Macromolecules, **38**, 7807–7818.

36 Guth, E. (1945) *J. Appl. Phys.*, **16**, 20–25.

37 Rubinstein, M. and Panyukov, S. (2002)
Macromolecules, **35**, 6670–6686.

38 Koňák, Č., Helmstedt, M. and Bansil, R.
(2000) *Polymer*, **41**, 9311–9315.

39 Koňák, Č., Fleischer, G., Tuzar, Z. and
Bansil, R. (2000) *J. Polym. Sci., B:
Polym. Phys.*, **38**, 1312–1322.

40 Watanabe, H. (1995) *Macromolecules*, **28**,
5006–5011.

41 Watanabe, H. (2001) *Macromol. Rapid
Commun.*, **22**, 127–175.

42 Watanabe, H., Sato, T. and Osaki, K.
(2000) *Macromolecules*, **33**, 2545–2550.

43 Laurer, J.H., Khan, S.A., Spontak, R.J.,
Satkowski, M.M., Grothaus, J.T., Smith,
S.D. and Lin, J.S. (1999) *Langmuir*, **15**,
7947–7955.

44 Ceauşescu, E., Bordeianu, R., Ghioca,
P., Buzdugan, E., Stancu, R. and

Cerchez, I. (1984) *Pure Appl. Chem.*, **56**, 319–328.

45 Watanabe, H. (1997) *Acta Polym.*, **48**, 215–233.

46 Hamley, I.W. (2001) *Phil. Trans. R. Soc. A*, **359**, 1017–1044.

47 Hashimoto, T., Shibayama, M., Kawai, H., Watanabe, H. and Kotaka, T. (1983) *Macromolecules*, **16**, 361–371.

48 Watanabe, H. and Kotaka, T. (1983) *J. Rheol.*, **27**, 223–240.

49 Theunissen, E., Overbergh, N., Reynaers, H., Antoun, S., Jérôme, R. and Mortensen, K. (2004) *Polymer*, **45**, 1857–1865.

50 Drzal, P.L. and Shull, K.R. (2003) *Macromolecules*, **36**, 2000–2008.

51 Zana, R. (2005) *Dynamics of Surfactant Self-Assemblies: Micelles, Microemulsions, Vesicles, and Lyotropic Phases*, Taylor & Francis, Boca Raton.

52 Tan, H., Watanabe, H., Matsumiya, Y., Kanaya, T. and Takahashi, Y. (2003) *Macromolecules*, **36**, 2886–2893.

53 Morrison, F.A. and Winter, H.H. (1989) *Macromolecules*, **22**, 3533–3540.

54 Nie, H.F., Bansil, R., Ludwig, K., Steinhart, M., Koňák, Č. and Bang, J. (2003) *Macromolecules*, **36**, 8097–8106.

55 Percus, J.K. and Yevick, G.J. (1958) *Phys. Rev.*, **110**, 1–13.

56 Kinning, D.J. and Thomas, E.L. (1984) *Macromolecules*, **17**, 1712–1718.

57 Li, M., Liu, Y., Nie, H., Bansil, R. and Steinhart, M. (2007) *Macromolecules*, **40**, 9491–9502.

58 Liu, Y.S., Li, M.H., Bansil, R. and Steinhart, M. (2007) *Macromolecules*, **40**, 9482–9490.

59 Sota, N., Sakamoto, N., Saijo, K. and Hashimoto, T. (2006) *Polymer*, **47**, 3636–3649.

60 Matsen, M.W. (2001) *J. Chem. Phys.*, **114**, 8165–8173.

61 Soenen, H., Berghmans, H., Winter, H.H. and Overbergh, N. (1997) *Polymer*, **38**, 5653–5660.

62 Soenen, H., Liskova, A., Reynders, K., Berghmans, H., Winter, H.H. and Overbergh, N. (1997) *Polymer*, **38**, 5661–5665.

63 Kleppinger, R., Mischenko, N., Reynaers, H.L. and Koch, M.H.J. (1999) *J. Polym. Sci., B: Polym. Phys.*, **37**, 1833–1840.

64 Kleppinger, R., Mischenko, N., Theunissen, E., Reynaers, H.L., Koch, M.H.J., Almdal, K. and Mortensen, K. (1997) *Macromolecules*, **30**, 7012–7014.

65 Reynders, K., Mischenko, N., Kleppinger, R., Reynaers, H., Koch, M.H.J. and Mortensen, K. (1997) *J. Appl. Crystallogr.*, **30**, 684–689.

66 Kleppinger, R., Reynders, K., Mischenko, N., Overbergh, N., Koch, M.H.J., Mortensen, K. and Reynaers, H. (1997) *Macromolecules*, **30**, 7008–7011.

67 Mortensen, K., Theunissen, E., Kleppinger, R., Almdal, K. and Reynaers, H. (2002) *Macromolecules*, **35**, 7773–7781.

68 Reynders, K., Mischenko, N., Mortensen, K., Overbergh, N. and Reynaers, H. (1995) *Macromolecules*, **28**, 8699–8701.

69 Prasman, E. and Thomas, E.L. (1998) *J. Polym. Sci., B: Polym. Phys.*, **36**, 1625–1636.

70 Mischenko, N., Reynders, K., Mortensen, K., Overberg, N. and Reynaers, H. (1996) *J. Polym. Sci., B: Polym. Phys.*, **34**, 2739–2745.

71 Krishnan, A.S. (2010) Morphological, property and phase studies of multicomponent triblock copolymer networks, PhD thesis, North Carolina State University.

72 Sharma, R.V. and Sharma, K.C. (1977) *Physica A*, **89**, 213–218.

73 Kleppinger, R., van Es, M., Mischenko, N., Koch, M.H.J. and Reynaers, H. (1998) *Macromolecules*, **31**, 5805–5809.

74 Flory, P.J. (1953) *Principles of Polymer Chemistry*, Cornell University Press, Ithaca.

75 Mooney, M. (1940) *J. Appl. Phys.*, **11**, 582–592.

76 Rivlin, R.S. (1948) *Phil. Trans. R. Soc. A*, **241**, 379–397.

77 Krishnamoorti, R. and Vaia, R.A. (2002) *Polymer Nanocomposites: Synthesis, Characterization, and Modeling*, American Chemical Society, Washington, DC.

78 van Maanen, G.J., Seeley, S.L., Capracotta, M.D., White, S.A., Bukovnik, R.R., Hartmann, J., Martin,

J.D. and Spontak, R.J. (2005) *Langmuir*, **21**, 3106–3115.

79 Drzal, P.L. and Shull, K.R. (2005) *Eur. Phys. J. E*, **17**, 477–483.

80 Jin, J. and Song, M. (2005) *Thermochim. Acta*, **438**, 95–101.

81 Mishra, J.K., Ryou, J.H., Kim, G.H., Hwang, K.J., Kim, I. and Ha, C.S. (2004) *Mater. Lett.*, **58**, 3481–3485.

82 Mishra, J.K., Kim, G.H., Kim, I., Chung, I.J. and Ha, C.S. (2004) *J. Polym. Sci., B: Polym. Phys.*, **42**, 2900–2908.

83 Paglicawan, M.A., Balasubramanian, M. and Kim, J.K. (2007) *Macromol. Symp.*, **249–250**, 601–609.

84 Spontak, R.J. and Patel, N.P. (2004) Phase behavior of block copolymer blends, in *Developments in Block Copolymer Science and Technology* (ed. I.W. Hamley), John Wiley & Sons, Inc., New York, pp. 159–212.

85 Leibler, L., Ajdari, A., Mourran, A., Coulon, G. and Chatenay, D. (1994) Spreading of homopolymers on copolymer modified surfaces, in *Ordering in Macromolecular Systems* (eds A. Teramoto, M. Kobayashi and T. Norisuye), Springer-Verlag, Berlin, p. 353.

86 Walker, T.A., Semler, J.J., Leonard, D.N., van Maanen, G.J., Bukovnik, R.R. and Spontak, R.J. (2002) *Langmuir*, **18**, 8266–8270.

87 Quintana, J.R., Jáñez, M.D. and Katime, I. (1996) *Langmuir*, **12**, 2196–2199.

88 Flosenzier, L.S., Rohlfing, J.H., Schwark, A.M. and Torkelson, J.M. (1990) *Polym. Eng. Sci.*, **30**, 49–58.

89 Flosenzier, L.S. and Torkelson, J.M. (1992) *Macromolecules*, **25**, 735–742.

90 Quintana, J.R., Díaz, E. and Katime, I. (1998) *Langmuir*, **14**, 1586–1589.

91 Mazard, C., Benyahia, L. and Tassin, J.F. (2003) *Polym. Int.*, **52**, 514–521.

92 Barbe, A., Bökamp, K., Kummerlowe, C., Sollmann, H., Vennemann, N. and Vinzelberg, S. (2005) *Polym. Eng. Sci.*, **45**, 1498–1507.

93 Jackson, N.R., Wilder, E.A., White, S.A., Bukovnik, R. and Spontak, R.J. (1999) *J. Polym. Sci., B: Polym. Phys.*, **37**, 1863–1872.

94 Sengupta, P. and Noordermeer, J.W.M. (2005) *Macromol. Rapid Commun.*, **26**, 542–547.

95 Kane, L., Norman, D.A., White, S.A., Matsen, M.W., Satkowski, M.M., Smith, S.D. and Spontak, R.J. (2001) *Macromol. Rapid Commun.*, **22**, 281–296.

96 Spontak, R.J., Wilder, E.A. and Smith, S.D. (2001) *Langmuir*, **17**, 2294–2297.

97 Wilder, E.A., White, S.A., Smith, S.D. and Spontak, R.J. (2003) Gel network development in AB, ABA and AB/ABA block copolymer solutions in a selective solvent, in *Polymer Gels: Fundamentals and Applications*, Vol. 833, American Chemical Society, pp. 248–261.

98 Zhulina, E.B. and Halperin, A. (1992) *Macromolecules*, **25**, 5730–5741.

99 Halperin, A. and Zhulina, E.B. (1991) *Europhys. Lett.*, **16**, 337–341.

100 King, M.R., White, S.A., Smith, S.D. and Spontak, R.J. (1999) *Langmuir*, **15**, 7886–7889.

101 Li, B.Q. and Ruckenstein, E. (1998) *Macromol. Theory Simul.*, **7**, 333–348.

102 Halperin, A. (1996) *J. Adhes.*, **58**, 1–13.

103 de Gennes, P.-G. and Prost, J. (1993) *The Physics of Liquid Crystals*, 2nd edn, Oxford University Press, Oxford, New York.

104 Li, M.H., Keller, P., Yang, J.Y. and Albouy, P.A. (2004) *Adv. Mater.*, **16**, 1922–1925.

105 Kempe, M.D., Scruggs, N.R., Verduzco, R., Lal, J. and Kornfield, J.A. (2004) *Nat. Mater.*, **3**, 177–182.

106 Kempe, M.D., Verduzco, R., Scruggs, N.R. and Kornfield, J.A. (2006) *Soft Matter*, **2**, 422–431.

107 Halperin, A. and Williams, D.R.M. (1994) *Phys. Rev. E*, **49**, R986–R989.

108 Welton, T. (1999) *Chem. Rev.*, **99**, 2071–2083.

109 Patel, N.P. and Spontak, R.J. (2004) *Macromolecules*, **37**, 2829–2838.

110 He, Y. and Lodge, T.P. (2007) *Chem. Commun.*, 2732–2734.

111 Spontak, R.J. and Smith, S.D. (2001) *J. Polym. Sci., B: Polym. Phys.*, **39**, 947–955.

112 Wu, L.F., Cochran, E.W., Lodge, T.P. and Bates, F.S. (2004) *Macromolecules*, **37**, 3360–3368.

113 Smith, S.D., Spontak, R.J., Satkowski, M.M., Ashraf, A., Heape, A.K. and Lin, J.S. (1994) *Polymer*, **35**, 4527–4536.

114 Bansil, R., Nie, H.F., Koňák, Č., Helmstedt, M. and Lal, J. (2002) *J. Polym. Sci., B: Polym. Phys.*, **40**, 2807–2816.

115 Nie, H.F., Li, M.H., Bansil, R., Koňák, Č., Helmstedt, M. and Lal, J. (2004) *Polymer*, **45**, 8791–8799.

116 Gindy, M.E., Prud'homme, R.K. and Panagiotopoulos, A.Z. (2008) *J. Chem. Phys.*, **128**, 164906.

117 Florez, S., Muñoz, M.E. and Santamaria, A. (2005) *J. Rheol.*, **49**, 313–325.

118 Quintana, J.R., Jáñez, M.D., Hernáez, E.Z., García, A. and Katime, I. (1998) *Macromolecules*, **31**, 6865–6870.

119 Terech, P. and Weiss, R.G. (1997) *Chem. Rev.*, **97**, 3133–3159.

120 George, M. and Weiss, R.G. (2006) *Acc. Chem. Res.*, **39**, 489–497.

121 Sangeetha, N.M. and Maitra, U. (2005) *Chem. Soc. Rev.*, **34**, 821–836.

122 Lim, G.S., Jung, B.M., Lee, S.J., Song, H.H., Kim, C. and Chang, J.Y. (2007) *Chem. Mater.*, **19**, 460–467.

123 Hanabusa, K., Yamada, M., Kimura, M. and Shirai, H. (1996) *Angew. Chem. Int. Ed. Engl.*, **35**, 1949–1951.

124 Žinic, M., Vögtle, F. and Fages, F. (2005) *Top. Curr. Chem.*, **256**, 39–76.

125 Etter, M.C. (1991) *J. Phys. Chem.*, **95**, 4601–4610.

126 Zaworotko, M.J. (1994) *Chem. Soc. Rev.*, **23**, 283–288.

127 Meléndez, R.E., Carr, A.J., Linton, B.R. and Hamilton, A.D. (2000) Controlling hydrogen bonding: from molecular recognition to organogelation, in *Molecular Self-assembly Organic versus Inorganic Approaches*, Vol. 96, Springer, Berlin, pp. 31–61.

128 Dastidar, P. (2008) *Chem. Soc. Rev.*, **37**, 2699–2715.

129 Abdallah, D.J. and Weiss, R.G. (2000) *Adv. Mater.*, **12**, 1237–1247.

130 Leiserowitz, L. and Schmidt, G.M.J. (1969) *J. Chem. Soc. A*, 2372–2382.

131 Fan, E.K., Yang, J., Geib, S.J., Stoner, T.C., Hopkins, M.D. and Hamilton, A.D. (1995) *J. Chem. Soc., Chem. Commun.*, 1251–1252.

132 Hanabusa, K., Tanaka, R., Suzuki, M., Kimura, M. and Shirai, H. (1997) *Adv. Mater.*, **9**, 1095–1097.

133 Ohishi, T., Sugi, R., Yokoyama, A. and Yokozawa, T. (2008) *Macromolecules*, **41**, 9683–9691.

134 Lockwood, N.A., van Tankeren, R. and Mayo, K.H. (2002) *Biomacromolecules*, **3**, 1225–1232.

135 George, M., Snyder, S.L., Terech, P. and Weiss, R.G. (2005) *Langmuir*, **21**, 9970–9977.

136 van Esch, J.H. and Feringa, B.L. (2000) *Angew. Chem. Int. Ed.*, **39**, 2263–2266.

137 Fages, F., Vögtle, F. and Žinic, M. (2005) *Top. Curr. Chem.*, **256**, 77–131.

138 Carr, A.J., Meléndez, R.E., Geib, S.J. and Hamilton, A.D. (1998) *Tetrahedron Lett.*, **39**, 7447–7450.

139 van der Laan, S., Feringa, B.L., Kellogg, R.M. and van Esch, J. (2002) *Langmuir*, **18**, 7136–7140.

140 de Loos, M., van Esch, J., Kellogg, R.M. and Feringa, B.L. (2001) *Angew. Chem. Int. Ed.*, **40**, 613–616.

141 Schoonbeek, F.S., van Esch, J.H., Wegewijs, B., Rep, D.B.A., de Haas, M.P., Klapwijk, T.M., Kellogg, R.M. and Feringa, B.L. (1999) *Angew. Chem. Int. Ed.*, **38**, 1393–1397.

142 Thierry, A., Straupé, C., Wittmann, J.C. and Lotz, B. (2006) *Macromol. Symp.*, **241**, 103–110.

143 Wilder, E.A., Spontak, R.J. and Hall, C.K. (2003) *Mol. Phys.*, **101**, 3017–3027.

144 Wilder, E.A., Hall, C.K., Khan, S.A. and Spontak, R.J. (2003) *Langmuir*, **19**, 6004–6013.

145 Yamasaki, S., Ohashi, Y., Tsutsumi, H. and Tsujii, K. (1995) *Bull. Chem. Soc. Jpn.*, **68**, 146–151.

146 Smith, T.L., Masilamani, D., Bui, L.K., Khanna, Y.P., Bray, R.G., Hammond, W.B., Curran, S., Belles, J.J. and Bindercastelli, S. (1994) *Macromolecules*, **27**, 3147–3155.

147 Wilder, E.A., Hall, C.K., Khan, S.A. and Spontak, R.J. (2002) *Recent Res. Dev. Mat. Sci.*, **3**, 93–115.

148 Tan, C.H., Su, L.H., Lu, R., Xue, P.H., Bao, C.Y., Liu, X.L. and Zhao, Y.Y. (2006) *J. Mol. Liq.*, **124**, 32–36.

149 Bai, B.L., Wang, H.T., Xin, H., Zhang, F.L., Long, B.H., Zhang, X.B., Qu, S.N.

and Li, M. (2007) *New J. Chem.*, **31**, 401–408.

150 Aoki, M., Murata, K. and Shinkai, S. (1991) *Chem. Lett.*, 1715–1718.

151 Xing, B.A., Choi, M.F., Zhou, Z.Y. and Xu, B. (2002) *Langmuir*, **18**, 9654–9658.

152 Lin, Y.C. and Weiss, R.G. (1987) *Macromolecules*, **20**, 414–417.

153 Lin, Y., Kachar, B. and Weiss, R.G. (1989) *J. Am. Chem. Soc.*, **111**, 5542–5551.

154 Song, X.D., Geiger, C., Farahat, M., Perlstein, J. and Whitten, D.G. (1997) *J. Am. Chem. Soc.*, **119**, 12481–12491.

155 Song, J.H. and Sailor, M.J. (1997) *J. Am. Chem. Soc.*, **119**, 7381–7385.

156 Shoji, Y., Tashiro, K. and Aida, T. (2004) *J. Am. Chem. Soc.*, **126**, 6570–6571.

157 Murata, K., Aoki, M., Suzuki, T., Harada, T., Kawabata, H., Komori, T., Ohseto, F., Ueda, K. and Shinkai, S. (1994) *J. Am. Chem. Soc.*, **116**, 6664–6676.

158 Llusar, M. and Sanchez, C. (2008) *Chem. Mater.*, **20**, 782–820.

159 Kawano, S., Fujita, N. and Shinkai, S. (2004) *J. Am. Chem. Soc.*, **126**, 8592–8593.

160 Xue, M., Liu, K.Q., Peng, J.X., Zhang, Q.H. and Fang, Y. (2008) *J. Colloid Interface Sci.*, **327**, 94–101.

161 Lu, L.D., Cocker, T.M., Bachman, R.E. and Weiss, R.G. (2000) *Langmuir*, **16**, 20–34.

162 Clavier, G., Mistry, M., Fages, F. and Pozzo, J.L. (1999) *Tetrahedron Lett.*, **40**, 9021–9024.

163 Pozzo, J.L., Desvergne, J.P., Clavier, G.M., Bouas-Laurent, H., Jones, P.G. and Perlstein, J. (2001) *J. Chem. Soc., Perkin Trans.*, **2**, 824–826.

164 Terech, P., Clavier, G., Bouas-Laurent, H., Desvergne, J.P., Demé, B. and Pozzo, J.L. (2006) *J. Colloid Interface Sci.*, **302**, 633–642.

165 Ku, C.Y., Nostro, P.L. and Chen, S.H. (1997) *J. Phys. Chem. B*, **101**, 908–914.

166 Abdallah, D.J., Lu, L.D. and Weiss, R.G. (1999) *Chem. Mater.*, **11**, 2907–2911.

167 Nostro, P.L. and Chen, S.H. (1993) *J. Phys. Chem.*, **97**, 6535–6540.

168 Hozumi, A. and Takai, O. (1997) *Thin Solid Films*, **303**, 222–225.

169 Yamanaka, M., Sada, K., Miyata, M., Hanabusa, K. and Nakano, K. (2006) *Chem. Commun.*, 2248–2250.

170 Selinger, J.V., MacKintosh, F.C. and Schnur, J.M. (1996) *Phys. Rev. E*, **53**, 3804–3818.

171 Fuhrhop, J.H., Spiroski, D. and Boettcher, C. (1993) *J. Am. Chem. Soc.*, **115**, 1600–1601.

172 Huang, B.Q., Hirst, A.R., Smith, D.K., Castelletto, V. and Hamley, I.W. (2005) *J. Am. Chem. Soc.*, **127**, 7130–7139.

173 Das, A.K. and Banerjee, A. (2006) *Macromol. Symp.*, **241**, 14–22.

174 Maji, S.K., Malik, S., Drew, M.G.B., Nandi, A.K. and Banerjee, A. (2003) *Tetrahedron Lett.*, **44**, 4103–4107.

175 Hirst, A.R., Huang, B.Q., Castelletto, V., Hamley, I.W. and Smith, D.K. (2007) *Chem. Eur. J.*, **13**, 2180–2188.

176 George, M. and Weiss, R.G. (2001) *J. Am. Chem. Soc.*, **123**, 10393–10394.

177 Li, Y.G., Liu, K.Q., Liu, J., Peng, J.X., Feng, X.L. and Fang, Y. (2006) *Langmuir*, **22**, 7016–7020.

178 Suzuki, M., Nakajima, Y., Yumoto, M., Kimura, M., Shirai, H. and Hanabusa, K. (2004) *Org. Biomol. Chem.*, **2**, 1155–1159.

179 Zhu, G.Y. and Dordick, J.S. (2006) *Chem. Mater.*, **18**, 5988–5995.

180 Kato, T. (2002) *Science*, **295**, 2414–2418.

181 Mizoshita, N., Hanabusa, K. and Kato, T. (1999) *Adv. Mater.*, **11**, 392–394.

182 Kato, T., Kutsuna, T., Hanabusa, K. and Ukon, M. (1998) *Adv. Mater.*, **10**, 606–608.

183 Yabuuchi, K., Rowan, A.E., Nolte, R.J.M. and Kato, T. (2000) *Chem. Mater.*, **12**, 440–443.

184 Kato, T., Hirota, N., Fujishima, A. and Fréchet, J.M.J. (1996) *J. Polym. Sci., Part A: Polym. Chem.*, **34**, 57–62.

185 Deindörfer, P., Eremin, A., Stannarius, R., Davis, R. and Zentel, R. (2006) *Soft Matter*, **2**, 693–698.

186 Mizoshita, N., Hanabusa, K. and Kato, T. (2003) *Adv. Funct. Mater.*, **13**, 313–317.

187 Suzuki, Y., Mizoshita, N., Hanabusa, K. and Kato, T. (2003) *J. Mater. Chem.*, **13**, 2870–2874.

188 Pakula, T. and Zentel, R. (1991) *Makromol. Chem.*, **192**, 2401–2410.

189 Jones, R. (2002) *Soft Condensed Matter*, Oxford University Press, New York.

190 Hamley, I.W. (2007) *Introduction to Soft Matter*, John Wiley & Sons, Ltd, Chichester.

191 Ozin, G.A., Arsenault, A.C. and Cademartiri, L. (2009) *Nanochemistry: A Chemical Approach to Nanomaterials*, 2nd edn, Royal Society of Chemistry, Cambridge.

192 Shankar, R., Ghosh, T.K. and Spontak, R.J. (2007) *Soft Matter*, **3**, 1116–1129.

193 Shankar, R., Ghosh, T.K. and Spontak, R.J. (2007) *Macromol. Rapid Commun.*, **28**, 1142–1147.

194 George, M. and Weiss, R.G. (2003) *Chem. Mater.*, **15**, 2879–2888.

27
Self-assembly of Linear Polypeptide-based Block Copolymers

Sébastien Lecommandoux, Harm-Anton Klok, and Helmut Schlaad

27.1
Introduction

Now that the fundamental principles that underlie the self-assembly of block copolymers have been addressed in numerous theoretical and experimental studies, these materials are finding increasing interest in several nanotechnology applications, such as nanostructured membranes, templates for nanoparticle synthesis and high-density information storage [1–3]. The self-assembly of block copolymers composed of two (A and B) or more (A, B, C, ...) chemically incompatible, amorphous segments is determined by the interplay of two competitive processes [4, 5]. On the one hand, in order to avoid unfavorable monomer contacts, the blocks segregate and try to minimize the interfacial area. Minimization of the interfacial area, however, involves chain stretching, which is entropically unfavorable. It is the interplay between these two processes that determines the final block copolymer morphology. Depending on the volume fractions of the respective blocks, amorphous AB diblock copolymers can form lamellar, hexagonal, spherical and gyroid structures. More complex morphologies can also be generated, but these require alternative strategies. One possibility is to increase the number of different blocks [6]. Other strategies to create more complex block copolymer nanostructures include the introduction of more rigid or (liquid) crystalline blocks [7, 8]. Often, such conformationally restricted segments introduce additional secondary interactions, such as electrostatic, hydrogen bonding or π–π interactions, which have an impact on the block copolymer self-assembly. In particular, rod–coil type block copolymers composed of a rigid (crystalline) block and a flexible (amorphous) segment have attracted increased interest over the past years [9, 10]. Morphological studies on rod–coil block copolymers have revealed several unconventional nanoscale structures, which were previously unknown for purely amorphous block copolymers. These findings underline the potential of manipulating chain conformation and interchain interactions to further engineer block copolymer self-assembly.

The self-assembly of amphiphilic block copolymers in solution, driven by the incompatibility of constituents, into ordered structures in the sub-micrometer

Advanced Nanomaterials. Edited by Kurt E. Geckeler and Hiroyuki Nishide
Copyright © 2010 WILEY-VCH Verlag GmbH & Co. KGaA, Weinheim
ISBN: 978-3-527-31794-3

range is a current topic in colloid and materials science [11–15]. The basic structures of diblock copolymers in solution are, in the order of decreasing curvature, spherical and cylindrical micelles and vesicles; the curvature of the core–corona interface is essentially determined by the volume fractions of comonomers and environmental factors (solvent, ionic strength, etc.) [16]. Polymer vesicles, often also referred to as "polymersomes", are particularly interesting as mimetics for biological membranes [17–20]. Deviation from this conventional aggregation behavior and appearance of more complex superstructures occur, as in biological systems, when specific non-covalent interactions, chirality and secondary structure effects come into play [21, 22]. Particularly interesting are block copolymers that combine advantageous features of synthetic polymers (solubility, processability, rubber elasticity, etc.) with those of polypeptides or polysaccharides (secondary structure, functionality, biocompatibility, etc.).

This chapter discusses the solid-state and solution structures, organization and properties of polypeptide-based block copolymers. Most of the block copolymers studied so far are composed of a synthetic block and a peptide segment and are an interesting class of materials, both from a structural and a functional point of view [23, 24]. Peptide sequences can adopt ordered conformations, such as α-helices or β-strands (Figure 27.1). In the former case, this leads to block copolymers with rod–coil character. Peptide sequences with a β-strand conformation can undergo intermolecular hydrogen bonding, which also offers additional means to direct nanoscale structure formation compared with purely amorphous block copolymers. Combining peptide sequences and synthetic polymers, however, is not only interesting to enhance control over nanoscale structure formation, but can also result in materials that can interface with biology. Such biomimetic hybrid polymers or molecular chimeras [25] may produce sophisticated superstructures

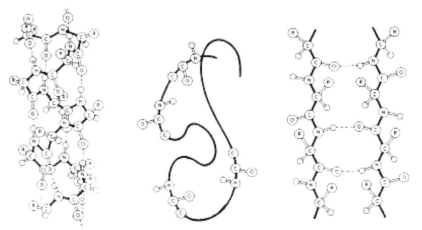

Figure 27.1 Illustration of the basic secondary structure motifs of polypeptides: (a) α-helix, (b) random coil and (c) antiparallel β-sheet.

with new materials properties. "Smart" materials based on polypeptides may reversibly change the conformation and, along with it, properties in response to an environmental stimulus, such as a change in pH or temperature [26]. Also, polypeptide block copolymers may be used as model systems to study generic self-assembly processes in natural proteins. Obviously, such materials would be of great potential interest for a variety of biomedical and bioanalytical applications.

27.2
Solution Self-assembly of Polypeptide-based Block Copolymers

27.2.1
Aggregation of Polypeptide-based Block Copolymers

27.2.1.1 Polypeptide Hybrid Block Copolymers
Corona-forming Polypeptides Block copolymers with soluble and thus corona-forming polypeptide segments include linear diblock and triblock copolymer samples. In most cases, studies on the aggregation behavior were carried out in aqueous solutions with samples consisting of a soft polybutadicne (PB) or polyiso-prene (PI) (exhibiting a glass transition temperature, T_g, below the freezing point of water) and an α-helical poly(L-glutamate) (PLGlu) or poly(L-lysine) (PLLys) segment.

The first study, reported in 1979 by Nakajima *et al.* [27], dealt with the structure of aggregates of symmetric triblock copolymers consisting of a coiled *trans*-1,4-PB middle-block and two α-helical poly(γ-benzyl L-glutamate) outer-blocks, PBLGlu$_{53-188}$-*b*-PB$_{64}$-*b*-PBLGlu$_{53-188}$ (subscripts denote number averages of repeat units, P_n), so-called "once-broken rods", in chloroform. The shape and dimensions of the aggregates in solution were calculated on the basis of simple thermodynamic considerations by taking into account chain conformation and the interfacial free energy. Predictions were found to be in good agreement with the structures (of solvent-cast films) observed by transmission electron microscopy (TEM). Depending on the composition of the copolymer, aggregates had a spherical, cylindrical or lamellar structure with a characteristic size of about 25–45 nm. Similar data were also obtained for block copolymers based on poly(γ-methyl L-glutamate) (PMLGlu) and poly(N^{ε}-benzyloxycarbonyl L-lysine) (PZLLys) [28]. These results suggest that the aggregation of rod–coil block copolymers might be treated in the same way as conformationally isotropic samples [29], provided that the rigid segment is dissolved in the continuous phase.

Schlaad and coworkers [30] and Lecommandoux and coworkers [31] investigated the aggregates of PB$_{27-119}$-*b*-PLGlu$_{20-175}$ in aqueous saline solution by dynamic and static light scattering (DLS/SLS), small-angle neutron scattering (SANS) and TEM. Copolymers were found to form spherical micelles with a hydrodynamic radius of $R_h < 40$ nm (70–75 mol-% glutamate) or unilamellar vesicles with $R_h = 50$–90 nm

(17–54 mol-% glutamate); cylindrical micelles have not been observed so far. Against all expectations, however, Klok and Lecommandoux and coworkers [32, 33] earlier reported not micelles but vesicles for a PB_{40}-b-$PLGlu_{100}$ containing 71 mol-% glutamate.

DLS and SANS showed that any pH-induced changes of the secondary structure of poly(L-glutamate) from a random coil (pH > 6) to an α-helix (pH < 5) (CD spectroscopy) did not have a severe impact on the morphology (curvature) of the aggregates. SANS further suggested that aggregation numbers remained the same [31], despite equilibration of the sample and a dynamic exchange of polymer chains between aggregates [30]. Coiled and α-helical polypeptide chains seem to have similar spatial requirements at the core-corona interface (see also [27]). However, as the contour length of an all-*trans* polypeptide chain is more than twice that of an α-helix, in particular the hydrodynamic size of the aggregates might decrease when decreasing the pH of the solution. A decrease of the hydrodynamic radius by 20% or less could be observed for PB_{48}-b-$PLGlu_{56-145}$ [31] but, however, not for PB_{27-119}-b-$PLGlu_{24-64}$ [30]. PI_{49}-b-$PLLys_{123}$ micelles in saline, also reported by Lecommandoux and coworkers [34], exhibited a hydrodynamic radius of $R_h \approx 44$ nm at pH 6 (coil) and of 23 nm at pH 11 (helix) (DLS), which corresponds to a decrease in size by almost 50%. Interestingly, although the polypeptide segment is of nearly the same length, the effect seen for PI_{49}-b-$PLLys_{123}$ micelles is much larger than that for PB_{48}-b-$PLGlu_{114}$ (≈8%). A possible explanation might be that the PLGlu helices are disrupted [28] and/or folded and hence less stretched as PLLys helices.

It is worth noting that the coronae of micelles of PB_{48}-b-$PLGlu_{114}$ and PI_{49}-b-$PLLys_{178}$ could be stabilized using 2,2′-(ethylenedioxy)bisethylamine and glutaric dialdehyde as cross-linking agents, respectively (Lecommandoux *et al.* [35]). The size and morphology of the aggregates were not affected by the chemical modification reaction.

Ouchi and coworkers [36] studied the aggregation of poly(L-lactide)-*block*-poly(aspartic acid), $PLL_{95,270}$-b-$PAsp_{47-270}$, in pure water and in 0.2 M phosphate buffer solution at pH 4.4–8.6. Irrespective of the chemical composition of the copolymer (21–74 mol-% aspartic acid), however, only spherical aggregates with $R_h = 10$–80 nm could be observed [DLS and scanning force microscopy (SFM)]. The aggregation behavior of the samples was rationalized in terms of a balance between hydrophobic interactions in the PLL core and electrostatic repulsion and hydrogen-bridging interactions in the PAsp corona. That the aggregates might be non-equilibrium structures being kinetically trapped in a "frozen" state due to semi-crystallinity of PLL chains was not considered. Metastability could be an explanation for the exclusive formation of spherical micelles in addition to the seemingly arbitrary changes of the size of aggregates in buffer solutions at different pH.

Klok and coworkers [37] used DLS/SLS, SANS and analytical ultracentrifugation (AUC) for the analysis of aggregates formed by $PS_{8,10}$-b-$PLLys_{9-72}$ (PS = polystyrene) in dilute aqueous solution at neutral pH; at this pH, PLLys was in a random coil conformation. They observed, however, cylindrical micelles regardless of the length of the PLLys segment. A conclusive explanation of this unexpected aggregation behavior could not be given.

Core-forming Polypeptides   Aggregates with an insoluble polypeptide core have been prepared with block or random copolymers having linear or branched architecture. Most studies focused on aqueous systems (designated for use as drug carriers in biomedical applications), and the only inverse systems investigated to date are PB-*b*-PLGlu in dilute tetrahydrofuran (THF) and CH_2Cl_2 solution and PS-*b*-PZLLys in CCl_4.

Harada and Kataoka [38–40] were the first to investigate the formation of polyion complex (PIC) micelles in an aqueous milieu from a pair of oppositely charged linear polypeptide block copolymers, namely of PEG_{113}-*b*-$PAsp_{18,78}$ polyanions and PEG_{113}-*b*-$PLLys_{18,78}$ polycations [PEG = poly(ethylene glycol)]. Complexation studies were carried out at pH 7.29, where both block copolymers had the same degree of ionization ($\alpha = 0.967$) and were thus double-hydrophilic in nature and did not form aggregates in water. Mixing of the copolymers at a 1:1 ratio of amino acid residues resulted in the formation of stable and monodispersed spherical core-shell assemblies of 30 nm in diameter (DLS). Another interesting feature connected with PIC micelles is that of "chain-length recognition" [40]. PIC micelles are exclusively formed by matched pairs of chains with the same block lengths of polyanions and polycations, even in mixtures with different block lengths. The key determinants in this recognition process are considered to be the strict phase separation between the PIC core and the PEO corona, requiring regular alignment of the molecular junctions at the core–corona interface, and the charge stoichiometry (neutralization).

Yonese and coworkers [41] studied the aggregation behavior of PEG_{113}-*b*-$PMLGlu_{20,50}$ and lactose-modified PEG_{75}-*b*-$PMLGlu_{32}$ in water. As shown by DLS, the copolymers formed large aggregates with a hydrodynamic radius of $R_h \approx 250$ nm. Contrary to what was claimed by these workers, it seems more likely that these aggregates were vesicles rather than spherical micelles. Key in the aggregation behavior might be the association of α-helical PMLGlu segments, as evidenced by CD spectroscopy, promoting the formation of plane bi-layers which then close into vesicles [15]. Further systematic studies on this system and detailed analysis of structures are lacking.

Closely related to this system are PEO_{272}-*b*-$PBLGlu_{38–418}$ and $PNIPAAm_{203}$-*b*-$PBLGlu_{39–123}$ [PEO = poly(ethylene oxide), PNIPAAm = poly(*N*-isopropylacrylamide)] described by Cho and coworkers [42, 43]. The aqueous polymer solutions, prepared by the dialysis of organic solutions against water, contained large spherical aggregates ($R_h \approx 250$ nm) with a broad size distribution (DLS). Although the size suggested a vesicular structure of the aggregates, aggregation numbers ($Z < 100$, method of determination not specified) were far below the values of several thousands typically being reported for polymer vesicles [18]. It is also worth noting that the PNIPAAm chains exhibit LCST (lower critical solution temperature) behavior. However, raising the temperature to the LCST ($\approx 34\,°C$) had no serious impact on the size of the aggregates.

Dong and coworkers described symmetric triblock copolymers with a glyco methacrylate middle block and two outer poly(L-alanine) (PLAla) or PBLGlu blocks [44, 45]. The aggregates formed in dilute aqueous solution were spherical in shape and were 200–700 nm in diameter (TEM). TEM further revealed a compact structure of the aggregates as with multi-lamellar vesicles. The dimensions of the particles, however, were found to decrease with increasing concentration of the copolymer.

Naka *et al.* [46] studied the aggregation behavior of poly(acetyliminoethylene)-*block*-poly(L-phenylalanine), $PAEI_{41}$-*b*-$PLPhe_{4,8}$, in a 0.05 M phosphate buffer at pH 7. Aggregates were observed despite the very low number of hydrophobic L-phenylalanine units, and the size of which was in the order of R_h = 425 nm (DLS). The seemingly high tendency of these polymers to form aggregates was attributed to the establishment of hydrogen bridges between the amino acid units, as shown by IR spectroscopy, in addition to hydrophobic interactions. Visualization of the aggregates with TEM strongly suggested the presence of coacervates or large clusters of small micelles but no vesicles.

The existence of vesicles could be demonstrated for PS_{258}-*b*-$PZLLys_{57}$ in dilute CCl_4 solution (Losik and Schlaad [47]). Scanning electron microscopy (SEM) showed collapsed hollow spheres of about 300–600 nm in diameter, indicative of vesicles, and also sheet-like structures, supposedly bi-layers that are not yet closed to vesicles (Figure 27.2A) [15]. The preference for a lamellar structure might be, as in the previous examples, attributed to a stiffening of the core by the 2D-arrangement of crystallizable PZLLys α-helices. PS_{258}-*b*-$PZLLys_{109}$, on the other hand, was found to form large compact fibrils being hundreds of nanometers in diameter and several tens of microns in length (Figure 27.2B); these aggregates might be cylindrical multi-lamellar vesicles. However, the processes involved in the formation of these structures are not yet known.

Likewise, Lecommandoux and coworkers [31] found vesicles for PB_{48}-*b*-$PLGlu_{20}$ in THF and in CH_2Cl_2 solution (R_h = 106–108 nm, DLS/SLS). The formation of

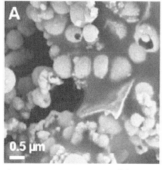

Figure 27.2 SEM images of the aggregates formed by (A) PS_{258}-*b*-$PZLLys_{57}$ and (B) PS_{258}-*b*-$PZLLys_{109}$ in dilute CCl_4 solution; specimens were prepared by shock-freezing a 0.2 wt-% polymer solution with liquid nitrogen and subsequent freeze-drying [47].

vesicles rather than micelles was attributed to the α-helical rod-like secondary structure of the insoluble PLGlu that forms a planar interface.

27.2.1.2 Block Copolypeptides

Besides the polypeptide hybrid block copolymers described earlier, a few purely peptide-based amphiphiles and block/random copolymers (copolypeptides) exist. In the latter case, both the core and corona of aggregates consist of a polypeptide. All of the studies reported so far have dealt with aggregation in aqueous media.

Doi and coworkers [48, 49] observed the spontaneous formation of aggregates of $PMLGlu_{10}$ with a phosphate (P) head group in water. Immediately after sonication, the freshly prepared solution contained globular assemblies (diameter: 50–100 nm, TEM), which after 1 h transformed into fibrous aggregates, promoted by intermolecular hydrogen bonding between peptide chains. After one day, these fibrils assembled into a twisted ribbon-like aggregate (TEM). As its thickness was ≈ 4 nm, which is close to the contour length of $PMLGlu_{10}$-P in a β-sheet conformation (CD and FT-IR spectroscopy), these workers concluded that the formation of the ribbon was driven by a stacking of anti-parallel β-sheets via hydrophobic interactions (see above) [22].

Lecommandoux and coworkers [50] showed that zwitterionic $PLGlu_{15}$-b-$PLLys_{15}$ in water can self-assemble into unilamellar vesicles with a hydrodynamic radius of greater than 100 nm (Figure 27.3). A change in the pH from 3 to 12 induced an inversion of the structure of the membrane (NMR) and was accompanied by an increase in the size of vesicles from 110 to 175 nm (DLS).

Non-ionic block copolypeptides made of L-leucine and ethylene glycol modified L-lysine residues, $PLLeu_{10-75}$-b-$PELLys_{60-200}$, were described by Deming and coworkers [51]. The copolymers adopted a rod-like conformation, due to the strong tendency of both segments to form ct-helices, as confirmed by CD spectroscopy. The self-assembled structures observed in aqueous solutions included (sub-) micrometer vesicles, sheet-like membranes and irregular aggregates. Here again, it was shown that the vesicle formation is related to the systematic presence of the polypeptide in a rod-like conformation in the hydrophobic part of the membrane, inducing a low interfacial curvature and as a result a hollow structure.

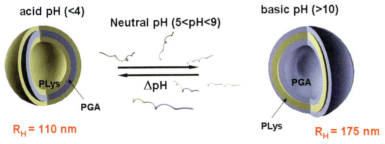

Figure 27.3 Schematic representation of the self-assembly of zwitterionic $PLGlu_{15}$-b-$PLLys_{15}$ in water into unilamellar vesicles [50].

Meyrueix and coworkers [52] performed the selective precipitation of PLLeu$_{180}$-b-PLGlu$_{180}$ and obtained nanoparticles, which could be purified and further suspended in water or in a 0.15 M phosphate saline buffer at pH 7.4. The colloidal dispersions were stable, due to the electrosteric stabilization of the particles by poly(sodium L-glutamate) brushes, containing spherical or cylindrical micelles, besides the large hexagonally-shaped platelets with a diameter of about 200 nm (TEM). Different shapes of particles were due to the heterogeneity of copolymer chains with respect to chemical composition (NMR): glutamate-rich chains formed micelles and leucine-rich ones formed platelets. CD spectroscopy and X-ray diffraction suggested that the core of platelets consisted of crystalline, helical PLLeu segments, and the structural driving force was thus related to the formation of leucine zippers in a three-dimensional array.

27.2.2
Polypeptide-based Hydrogels

Protein-based hydrogels are used for many applications, ranging from food and cosmetic thickeners to support matrices for drug delivery and tissue replacement. These materials are usually prepared using proteins extracted from natural resources, which can give rise to inconsistent properties unsuitable for medical applications.

Recently, Deming and coworkers [53–56] designed and synthesized diblock copolypeptide amphiphiles containing charged and hydrophobic segments. It was found and demonstrated that gelation depends not only on the amphiphilic nature of the polypeptides, but also on chain conformation, meaning α-helix, β-strand or random coil. Specific rheological measurements were performed to evidence the self-assembly process responsible for gelation [55]: the rod-like helical secondary structure of enantiomerically pure PLLeu blocks is instrumental for gelation at polypeptide concentrations as low as 0.25 wt-%. The hydrophilic polyelectrolyte segments have stretched coil configurations and stabilize the twisted fibrillar assemblies by forming a corona around the hydrophobic cores (Figure 27.4).

Interestingly, these hydrogels can retain their mechanical strength up to temperatures of about 90 °C and recover rapidly after stress. This new mode of assembly was found to give rise to polypeptide hydrogels with a unique combination of properties, such as heat stability and injectability, making them attractive for applications in foods, personal care products and medicine. In this context, their potential application as tissue engineering scaffolds has been recently studied [57].

27.2.3
Organic/Inorganic Hybrid Structures

Recently, polypeptide-based copolymers have also been used for the stabilization or synthesis of inorganic species. As a first example, Stucky and coworkers [58] used block copolypeptides to direct the self-assembly of silica into spherical and columnar morphologies at room temperature and neutral pH.

a)

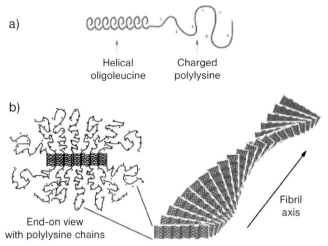

Helical
oligoleucine

Charged
polylysine

b)

End-on view
with polylysine chains

Fibril
axis

Figure 27.4 Drawings showing (a) representation of a block copolypeptide chain and (b) proposed packing of block copolypeptide amphiphiles into twisted fibrillar tapes, with helices packed perpendicular to the fibril axes. Polylysine chains were omitted from the fibril drawing for clarity (reprinted from [55] with permission of The American Chemical Society).

Stucky, Deming and colleagues [59] also designed a double-hydrophilic block copolypeptide poly{N^ε-2[2-(2-methoxyethoxy)ethoxy]acetyl L-lysine}$_{100}$-b-poly(sodium L-aspartate)$_{30}$ (PELLys-b-PNaLAsp) that can direct the crystallization of calcium carbonate into microspheres. They incorporated PLAsp in the diblock because domains of anionic aspartate residues are known to nucleate calcium carbonate crystallization. This effect is believed to be caused by matching interactions between aspartate and the atomic spacing of certain crystal faces in the growing mineral.

Also worth mentioning is the application of polypeptide-based copolymers in the production of magnetic nanocomposite materials. Lecommandoux et $al.$ [60] obtained stable dispersions of super-paramagnetic micelles and vesicles by combining an aqueous solution of PB$_{48}$-b-PLGlu$_{56-145}$ with a ferrofluid consisting of maghemite (γ-Fe$_2$O$_3$) nanoparticles. Incorporation of one mass equivalent of ferrofluid into the hydrophobic core of aggregates did not alter their morphology, as deduced from SLS and SANS data, but caused a substantial increase in the outer diameter by a factor of 6 (DLS). Interestingly, the hybrid vesicles underwent deformation under a magnetic field, as shown by 2D-SANS experiments. Held and coworkers [61] earlier reported that monodisperse, highly crystalline maghemite nanoparticles in organic solvents could be transferred into an aqueous medium using tetramethylammonium hydroxide stabilized at neutral pH. Combination of the aqueous maghemite solution with PELLys$_{100}$-b-PAsp$_{30}$ led to the formation of uniform clusters comprising approximately 20 nanoparticles (Figure 27.5).

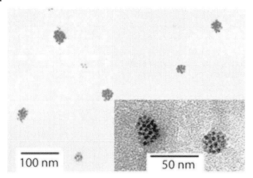

Figure 27.5 TEM images of clusters of maghemite nanoparticles deposited from dispersions in water in the presence of PELLys$_{100}$-*b*-PAsp$_{30}$ (reprinted from [61] with permission of The American Chemical Society).

27.3
Solid-state Structures of Polypeptide-based Block Copolymers

27.3.1
Diblock Copolymers

27.3.1.1 Polydiene-based Diblock Copolymers
One of the first reports on the nanoscale solid-state structure of peptide–synthetic hybrid block copolymers was published by Gallot and coworkers [62] in 1976. In this publication, the solid-state structure of a series of PB-*b*-PBLGlu and PB-*b*-PHLGln diblock copolymers was investigated using a combination of techniques, including infrared and CD spectroscopy, X-ray scattering and TEM. The block copolymers covered a broad composition range with peptide contents ranging from 19 to 75%. Interestingly, SAXS revealed a well-ordered lamellar superstructure characterized by up to four higher-order Bragg spacings for all of the investigated samples. The lamellar superstructure was confirmed by electron microscopy experiments, which were carried out on OsO$_4$-stained specimens. The intersheet spacings determined from the electron micrographs were in good agreement with the diffraction data. Wide-angle X-ray scattering (WAXS) experiments indicated that the α-helical peptide blocks were assembled in a hexagonal array. For a number of block copolymer samples it was found that the calculated length of the peptide helix was larger than the thickness of the polypeptide layer. To accommodate this difference, it was proposed that the peptide helices were folded in the peptide layer. The lamellar structure consists of plane, parallel equidistant sheets. Each sheet is obtained by superposition of two layers: (1) the PB chains in a more or less random coil conformation and (2) the α-helical polypeptide blocks in a hexagonal array of folded chains. The hexagonal-in-lamellar structure was also found for PB-*b*-PZLLys and PB-*b*-PLLys block copolymers by the same

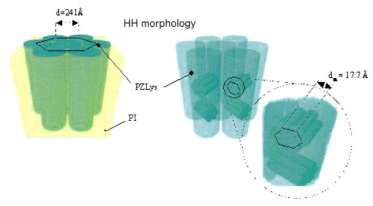

Figure 27.6 Schematic model representation of the hexagonal in hexagonal (HH) morphology obtained for PI$_{49}$-*b*-PZLLys$_{178}$ copolymer cast from dioxane solution.

workers [63–66]. In the case of the PB-PLLys block copolymers, no periodic arrangement of the PLLys chains in the peptide layer was found. This is due to the fact that the polypeptide segments in these block copolymers are not exclusively α-helical but are composed roughly of 50% random coil, 35% α-helix and 15% β-strand domains [63].

More recently, the solid-state nanoscale structure of PI-*b*-PZLLys diblock copolymers was reported [34]. Diblock copolymers composed of a P1 block with a number-average degree of polymerization of 49 and a PZLLys block containing 61–178 amino acid residues were investigated with dynamic mechanical analysis and X-ray scattering. For the PI$_{49}$-*b*-PZLLys$_{35}$, PI$_{49}$-*b*-PZLLys$_{61}$ and PI$_{49}$-*b*-PZLLys$_{92}$, the X-ray scattering data were in agreement with a hexagonal-in-lamellar morphology. Interestingly, for PI$_{49}$-*b*-PZLLys$_{92}$ the lamellar spacing was found to decrease when the samples were prepared from dioxane instead of THF/N,N-dimethylformamide (DMF) and suggested folding of the peptide helices. For PI$_{49}$-*b*-PZLLys$_{123}$ and PI$_{49}$-*b*-PZLLys$_{178}$ a hexagonal-in-hexagonal structure was found. This morphology is illustrated in Figure 27.6. This structure is unprecedented for polydiene-based peptide hybrid block copolymers, but has also been found for low molecular weight PS-*b*-PBLGlu copolymers [73].

27.3.1.2 Polystyrene-based Diblock Copolymers

In an early study, Gallot and coworkers [64] reported on the bulk nanoscale structure of PS-*b*-PZLLys diblock copolymers, which were based on a PS block with a number-average molecular weight of $M_n = 37$ kg mol^{-1} and had peptide contents ranging from 18 to 80 mol-%. X-ray scattering patterns of dry samples that had been evaporated from dioxane showed two sets of signals, characteristic of a hexagonal-in-lamellar superstructure. At very low angles, Bragg spacings characteristic of a layered superstructure were found, whereas at somewhat larger angles,

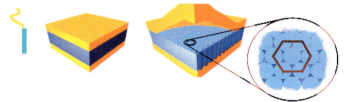

Figure 27.7 Schematic model representation of the hexagonal-in-lamellar (HL) morphology obtained for polypeptide-based rod–coil diblock copolymers cast from solutions.

there was a second set of reflections pointing towards a hexagonal arrangement of peptide helices. For several samples, the calculated length of the peptide helix was larger than the peptide layer thickness as determined from the X-ray data. In these cases, it was proposed that the helical PZLLys chains were folded in the peptide layer. Thus, the bulk nanoscale structure of the PS-*b*-PZLLys copolymers can be described in terms of the same hexagonal-in-lamellar model as was also proposed for the PB-based block copolymer described earlier (Figure 27.7).

Removal of the of the side-chain protective groups of the peptide segment resulted in PS-*b*-PLLys diblock copolymers [63]. These copolymers were not water soluble, but formed mesomorphic gels at water contents of less than 50%. The X-ray scattering patterns indicated a lamellar superstructure, both in the gel state and the dry samples. In contrast to the side-chain protected block copolymers, no evidence for a periodic arrangement of the peptide chains was found. This is not too surprising considering that IR spectra indicated that roughly 50% of the peptide blocks have a random coil conformation, 35% an α-helical secondary structure and 15% a β-strand conformation.

Along the same lines, Douy and Gallot [66] also studied the bulk nanoscale organization of PS-*b*-PBLGlu. For block copolymers composed of a PS block with $M_n = 25 \, \text{kg mol}^{-1}$ and containing 31–94 mol-% peptide, the same hexagonal-in-lamellar morphology as described above for the PS-*b*-PZLLys was found. The biocompatibility of PS-*b*-PBLGlu copolymers has been discussed in two publications [67, 68]. Mori *et al.* [68] studied diblock copolymers composed of a PS block with a number-average degree of polymerization of 87 and PBLGlu segments with number-average degrees of polymerization of 23, 52 or 83. Thrombus formation was assessed by exposing films of the diblock copolymers and the corresponding homopolypeptides to fresh canine blood. It was found that thrombus formation on the diblock copolymer films was reduced compared with the corresponding homopolymers. For the block copolymers, thrombus formation decreased with decreasing PBLGlu block length. Also, adsorption of plasma proteins such as bovine serum albumine, bovine γ-globulin and bovine plasma fibrinogen was reduced on the block copolymers compared with PS homopolymer.

The characterization of the solid-state nanoscale organization of PS-polypeptide hybrid block copolymers has recently been refined in a series of publications by

Schlaad and coworkers [69–71]. In a first report, three PS-*b*-PZLLys diblock copoly-mers with peptide volume fractions of 0.48, 0.74 and 0.82 were investigated [69]. SAXS patterns recorded from DMF cast films confirmed the hexagonal-in-lamellar morphology published earlier by Gallot and coworkers [64]. In their paper, Schlaad and coworkers went a step further and analyzed their SAXS data using the interface-distribution concept and the curvature-interface formalism. These evalu-ation techniques suggested that the bulk nanoscale structure of the PS-*b*-PZLLys diblock copolymers does not consist of plain but of undulated lamellae. The concept of the interface-distribution function and the curvature-interface formal-ism were also applied to compare the solid-state structures of two virtually identical PS based diblock copolymers; PS_{52}-*b*-$PZLLys_{111}$ ($\phi_{peptide} = 0.82$) and PS_{52}-*b*-$PBLGlu_{104}$ ($\phi_{peptide} = 0.79$) [70]. Analysis of the SAXS data obtained on DMF-cast films indi-cated a hexagonal-in-undulated (or zigzag) lamellar morphology for both block copolymers. However, the X-ray data also revealed two striking differences between the samples. The first difference concerns the thickness of the layers, which are a factor of three smaller for PS_{52}-*b*-$PBLGlu_{104}$ as compared with PS_{52}-*b*-$PZLLys_{111}$. Whereas the PZLLys helices are fully stretched, the PBLGlu helices are folded twice in the layers. As peptide folding increases the area per chain at the PS-PBLGlu interface, the thickness of the PS layers also has to decrease in order to cover the increased interfacial area. The second difference concerns the packing of the peptide helices. For the PZLLys-based diblock copolymer it was estimated that about 180 peptide helices form an ordered domain. The level of ordering, however, was considerably lower for the peptide blocks of PS_{52}-*b*-$PBLGlu_{104}$ and only ≈80 helices were estimated to form a single hexagonally ordered domain.

In addition, the influence of the polydispersity of the polypeptide block on the solid-state morphology of PS-*b*-PZLLys diblock copolymers has also been studied [71]. To this end, a series of five diblock copolymers was prepared from an identical ω-amino-polystyrene macroinitiator ($P_n = 52$; polydispersity index, PDI = 1.03). The peptide content in these diblock copolymers varied between 0.43 and 0.68 and the PDI ranged from 1.03 to 1.64. Evaluation of the SAXS data with the interface-distribution function and the curvature-interface formalism confirmed, as expected, the hexagonal-in-undulated (or zigzag) lamellar solid-state morphology. Fractiona-tion of the peptide helices according to their length leads (locally) to the formation of an almost plane, parallel lamellar interface, which is disrupted by kinks (undula-tions). The curvature at the PS-PZLLys interface, however, was found to be strongly dependent on the chain length distribution of the peptide block. Block copolymers with the smallest molecular weight distribution produced lamellar structures with the least curvature. Increasing the chain length distribution of the peptide block (block copolymers with PDI ≈ 1.25) leads to larger fluctuations in the thickness of the PZLLys layers, which increases the number of kinks and the curvature at the lamellar interface. At even larger polydispersities (PDI ≈ 1.64), however, the number of kinks decreases again. With increasing polydispersity of the peptide block, the thickness fluctuations become larger and larger, as does the interfacial area. At a certain point, at sufficiently high polydispersity, the system tries to compensate for the increased interfacial tension and minimizes the number of kinks (Figure 27.8).

Ludwigs *et al.* [72] used SFM to investigate the formation of hierarchical structures of PS_{52}-*b*-$PBLGlu_{104}$ in thin films. Thin films with a thickness of ≈4 and 40 nm were prepared by spin-coating of dilute polymer solutions on silicon substrates and were subsequently annealed in saturated THF vapor to achieve a controlled crystallization of the α-helical PBLGlu. On the smallest length-scale, the structure was found to be built of short ribbons or lamellae of interdigitated polymer chains. PBLGlu helices were fully stretched in thin films, in contrast to what has been observed in the 3D organized bulk mesophase (see above). Depending on the time of solvent annealing, different ordered structures on the micrometer length-scale could be observed (Figure 27.9).

The examples discussed so far have all involved relatively high molecular weight diblock copolymers. In these cases, the molecular weight of the polypeptide block is usually sufficiently high so that it forms a stable α-helix and the common hexagonal-in-lamellar morphology is found. The situation changes, however, when the molecular weight of the block copolymers is significantly decreased. The influence of molecular weight on the solid-state organization of polystyrene-based peptide–synthetic hybrid block copolymers has been studied for a series of low-molecular weight PS-*b*-PBLGlu and PS-*b*-PZLLys [73, 74]. These diblock copolymers consisted of a short PS block with $P_n ≈ 10$, a polypeptide block containing ≈10 to 80 amino acid repeat units and were characterized by means of variable temperature FT-IR spectroscopy and X-ray scattering. These experiments allowed the construction of "phase diagrams", which are shown in Figure 27.10. The phase diagrams reveal a number of interesting features. At temperatures below 200 °C and for sufficiently long polypeptide blocks, a hexagonal arrangement of the diblock copolymers was found, analogous to the hexagonal-in-lamellar morphology of the high-molecular weight analogues. Upon decreasing the length of the peptide block, however, several novel solid-state structures were discovered. For very short peptide block lengths (PS_{10}-*b*-$PBLGlu_{10}$, PS_{10}-*b*-$PZLLys_{20}$, PS_{10}-*b*-$PZLLys_{40}$ and PS_{10}-*b*-$PZLLys_{60}$) a lamellar supramolecular structure was found. This is

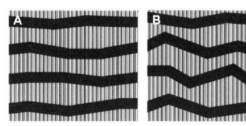
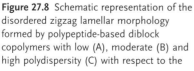
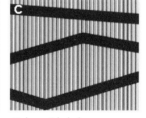

Figure 27.8 Schematic representation of the disordered zigzag lamellar morphology formed by polypeptide-based diblock copolymers with low (A), moderate (B) and high polydispersity (C) with respect to the length of helices. Polypeptide helices are represented as cylinders, and polyvinyl sheets are depicted in black (reprinted from [71] with permission of The American Chemical Society).

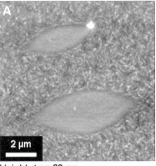

Height, $\Delta z = 60$ nm

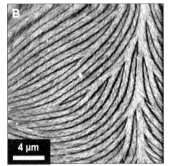

Height, $\Delta z = 25$ nm

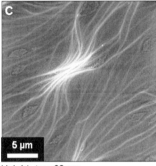

Height, $\Delta z = 80$ nm

Figure 27.9 SFM height images of a film of PS_{52}-b-$PBLGlu_{104}$ obtained by spin-coating from a 5 mg mL^{-1} THF solution and subsequent exposure to saturated THF vapor for 3.5 (A), 22.5 (B) and 42 h (C) (reprinted from [72] with permission of The American Chemical Society).

due to the fact that for such short peptide block lengths, a substantial fraction of the peptide blocks adopts a β-strand secondary structure. Self-assembly of these diblock copolymers in a β-sheet type fashion results in the lamellar structures observed by SAXS. For PS_{10}-b-$PBLGlu_{20}$ a peculiar and until then unprecedented structure was found. This structure that consisted of hexagonally packed diblock copolymer molecules, which are organized in a hexagonal superlattice, has been referred to as the double hexagonal or hexagonal-in-hexagonal morphology. Apart from several unconventional solid-state nanoscale structures, another factor that distinguishes the phase diagrams in Figure 27.10 from those of most conventional, conformationally isotropic block copolymers is the influence of temperature. For a number of diblock copolymers, increasing the temperature above 200 °C results in a change from a hexagonal-in-hexagonal (PS_{10}-b-$PBLGlu_{20}$) or hexagonal (PS_{10}-b-$PBLGlu_{40}$, PS_{10}-b-$PZLLys_{80}$) to a lamellar morphology. FT-IR spectroscopy experiments suggested that these morphological transitions are induced by an increase in the fraction of peptide blocks that have a β-strand conformation.

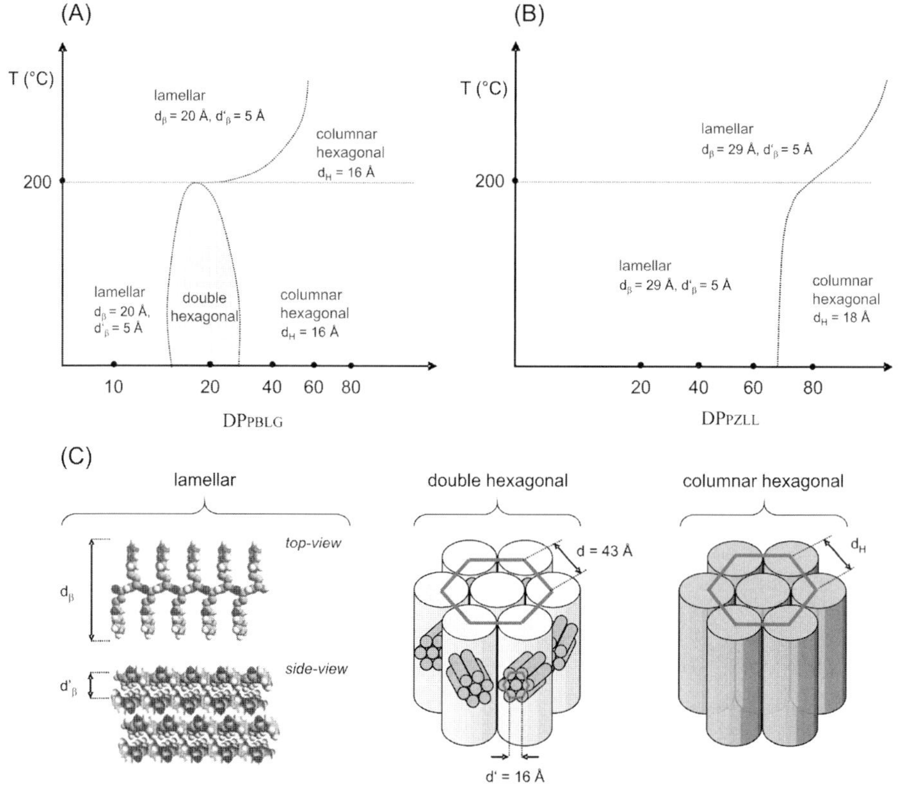

(A)

(B)

(C)

Figure 27.10 Phase diagrams describing the solid-state nanoscale structure of (A) PS-b-PBLGlu and (B) PS-b-PZLLys diblock copolymers; (C) illustration of the lamellar, double hexagonal and hexagonal morphologies found for the low molecular weight hybrid block copolymers (reprinted from [73, 74] with permission of The American Chemical Society).

27.3.1.3 Polyether-based Diblock Copolymers

PEG–polypeptide block copolymers are of particular interest, from both a structural and a functional point of view. Unlike the hybrid block copolymers discussed in the previous paragraphs, which were based on amorphous synthetic polymers, PEG is a semi-crystalline polymer. In addition to microphase separation and the tendency of the peptide blocks towards aggregation, crystallization of PEG introduces an additional factor that can influence the structure formation of these hybrid block copolymers. Ma and coworkers [75] have investigated the solid-state structure and properties of three PEG-b-PAla copolymers that were prepared from a PEG macroinitiator with $M_n = 2 \, kg \, mol^{-1}$. The diblock copolymers contained 39.8, 49.6 and 65.5 mol-% alanine. From FT-IR spectra and DSC measurements, these workers proposed a microphase-separated bulk structure.

AB diblock and ABA triblock copolymers composed of PEG as the A block and random coil segments of poly(D,L-valine-co-D,L-leucine) as the B block(s) were investigated by Cho and coworkers [76]. DSC experiments revealed PEG crystallization and showed that the PEG melting temperature was decreased compared with that of the PEG homopolymer. TEM micrographs suggested a larnellar microphase-separated structure for one of the triblock copolymer samples.

27.3.1.4 Polyester-based Diblock Copolymers

One of the first studies focusing on the solid-state properties of polypeptide–polyester synthetic hybrid block copolymers was reported by Jérôme and coworkers [77]. DSC experiments on a poly(ε-caprolactone)$_{50}$-*block*-poly(γ-benzyl L-glutamate)$_{40}$-(PCL$_{50}$-*b*-PBLGlu$_{40}$) diblock copolymer revealed two endotherms. The first endotherm was found at 60 °C and is due to the melting of the PCL. The second endotherm, which was located at 110 °C, was, mistakenly, interpreted as the melting transition of PBLGlu. This transition, however, is not a melting transition, but instead reflects the conformational transition of the PBLGlu helix from a 7/2 to an 18/5 helical structure. Although no further structural investigations were carried out, the observation of two separate endotherms occurring at temperatures identical to the transitions found for the respective homopolymers was a first indication for the existence of a microphase-separated structure. Similar results were reported by Chen and coworkers [78] who investigated the thermal properties of a series of PCL-*b*-PBLGlu copolymers composed of PCL blocks containing 13–51 repeat peptide segment units with and 22–52 amino acid repeat units.

Caillol *et al.* [79] have studied the solid-state structure and properties of a series of PLL-*b*-PBLGlu [PLL = poly(L-lactide)] copolymers. The PLL block in these copolymers contained 10–40 repeat units and the peptide segments were composed of 20–100 repeat units. DSC thermograms of the block copolymers revealed three transitions corresponding to the T_g of PLL ($\approx$50 °C), the 7/2 to 18/5 helix transition of PBLGlu ($\approx$100 °C) and the melting temperature of PLL ($\approx$160 °C), respectively. This observation was already providing a first hint towards a microphase-separated bulk morphology. SAXS experiments, which were performed at 100 °C, indicated the existence of hexagonally ordered assemblies of α-helical PBLGlu chains. With decreasing glutamate content, the peaks corresponding to this hexagonal organization decreased in intensity and another scattering peak appeared, which was ascribed to a lamellar assembly of PBLGlu chains with a β-strand secondary structure. Increasing the temperature to 200 °C not only resulted in melting of PLL, but also led to a decrease in intensity of the diffraction peaks corresponding to the hexagonally ordered α-helical PBLGlu segments and an increase in the fraction of PBLGlu segments that are ordered in a lamellar β-strand fashion.

27.3.1.5 Diblock Copolypeptides

A step forward in the design of hierarchically ordered structures with biofunctionality has been the subject of recent reports on the synthesis of block copolymers based on polypeptides. In the first such report [80], organo-nickel initiators rather than amines were used to avoid the unwanted α-amino acid N-carboxyanhydrides (NCA) side reactions, which had, for more than 50 years, hampered the formation of well-defined copolypeptides. This approach gave rise to various peptidic-based

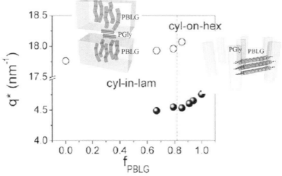

Figure 27.11 Dependence of the WAXS peak positions on the PBLGlu (PBLG) volume fraction, corresponding to the distance between PBLGlu α-helices (filled circles) and PGly β-sheets (open circles). The interhelix (inter-sheet) distance increases (decreases) with increasing polypeptide volume fraction. The vertical line separates the two nanodomain morphologies.

block copolymers that have mainly been studied in solution (see previous section). A second approach addressed the side reaction problem directly by using amines in combination with high-vacuum techniques, to ensure the necessary conditions for the living polymerization of NCAs: PBLGlu-*b*-PGly (PGly = polyglycine) were prepared for the first time with this methodology [81].

Despite these important synthetic efforts, the solid-state morphology of purely peptidic block copolymers is largely unexplored. Hadjichristidis and coworkers [81] recently investigated the self-assembly of a series of narrow polydispersity PBLGlu-*b*-PGly diblock copolymers within the composition range $0.67 < f_{BLGlu} < 0.97$ and the temperature range $303 < T < 433\,K$. SAXS and WAXS, ^{13}C NMR and DSC were used for the structure investigation coupled with dielectric spectroscopy for both the peptide secondary structure and the associated dynamics. These techniques not only provided insight into the nanophase morphology but also gave information about the type and persistence of peptide secondary structures. Particular evidence has been found for hexagonal-in-lamellar and cylinder-on-hexagonal nanostructures (Figure 27.11). The thermodynamic confinement of the blocks within the nanodomains and the disparity in their packing efficiency results in multiple chain folding of the PGly secondary structure that effectively stabilizes a lamellar morphology for high f_{BLGlu}. Nanoscale confinement proved to be important in controlling the persistence length of secondary peptide motifs.

27.3.2
Triblock Copolymers

27.3.2.1 Polydiene-based Triblock Copolymers
Whereas Gallot and coworkers have mainly studied the solid-state organization of PB-based diblock copolymers, Nakajima *et al.* concentrated on ABA-type hybrid block copolymers containing PB as the B component ("once-broken rods"). In a

first series of publications, the structure and properties of PBLGlu-b-PB-b-PBLGlu triblock copolymers containing 7.5–32.5 mol-% (= 3.0–14.3 vol-%) PB were investigated [82–84]. Infrared spectroscopy and WAXS experiments on films of the triblock copolymers indicated that the PBLGlu blocks were predominantly α-helical. From the WAXS experiments, it was concluded that the PBLGlu blocks assembled into different structures, depending on the type of solvent that was used to cast the films. In benzene cast films, the peptide helices were relatively poorly ordered, similar to the so-called form A morphology of PBLGlu [85]. In contrast, the PBLGlu segments in films cast from CHCl₃ were well ordered and contained paracrystalline and mesomorphic regions. Based on TEM, a cylindrical microstructure was proposed for a triblock copolymer containing 8 vol-% PB. Electron micrographs for other samples were not reported, but based on volume fraction considerations it was predicted that triblock copolymers containing 12 and 14 vol-% PB would form either cylindrical or lamellar superstructures [83]. Interestingly, copolymers having the same composition but polypeptide segments made of either enantiomerically pure or racemic γ-benzyl glutamate exhibited not only different secondary structures (α-helix or random coil, respectively; FTIR and WAXS) but also different superstructures (TEM). A cylindrical or lamellar morphology was proposed in the first case and a more spherical superstructure in the second [86].

Further support for the microphase-separated structure of the PBLGlu-b-PB-b-PBLGlu triblock copolymers was obtained from dynamic mechanical spectroscopy and water permeability experiments [84]. The temperature dependence of the dynamic modulus and the loss modulus could be explained well by assuming a microphase-separated structure. Furthermore, the hydraulic permeability of water through membranes prepared from the copolymers was approximately three orders of magnitude larger compared with a pure PBLGlu membrane. The hydraulic water permeability was found to increase with increasing PB content in the block copolymers. This was explained in terms of microphase-separated structure and the presence of an interfacial zone that separates the ordered domains formed by the α-helical PBLGlu chains from the unordered PB phase (Figure 27.12). The interfacial zone consists of amino acid residues that are located close to the N-terminus of the peptide block and in the vicinity of the PB segment. The amino acid residues in the interfacial zone do not form regular secondary structures. As the amide groups of the peptide chains in the interfacial zone are not involved in intramolecular hydrogen bonding, they are able to bind water molecules. Consequently, increasing the interfacial zone, e. g., by increasing the PB content, leads to an increase in the water permeability.

The bulk and surface structure of solvent-cast films from a series of PBLGlu-b-PB-b-PBLGlu triblock copolymers with much higher PB contents (50–80 mol-%) than the samples discussed above have been described by Gallot and coworkers [87]. The organization of these copolymers was compared with that of three other triblock copolymers with approximately the same PB content (≈50 mol-%) but which were composed of poly($N^ε$-trifluoroacetyl L-lysine) (PTLLys), poly(N^5-hydroxyethyl L-glutamine) (PHLGln) or polysarcosine (PSar) as the peptide block. For any

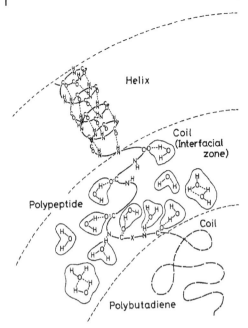

Figure 27.12 Hydrogen-bonded water and water clusters in an interfacial zone formed by unordered peptide chains that separate the PB domains from the helical PBLGlu phase in films of ABA triblock copolymers.

sample investigated, X-ray scattering experiments indicated a hexagonal-in-lamellar bulk morphology. X-ray photoelectron spectroscopy (XPS) measurements revealed that for the triblock copolymers with hydrophobic peptide blocks, i. e., PBLGlu or PTLLys, the surface composition was identical with that in the bulk of the sample. In contrast, the surfaces of films prepared from the triblock copolymers with the more hydrophilic peptide segments, i.e., PHLGln or PSar, were PB enriched. Furthermore, the XPS data suggested that the lamellar superstructures formed by the triblock copolymers were perpendicular to the air–polymer interface.

In addition, the solid-state organization and properties of PZLLys-b-PB-b-PZL-Lys triblock copolymers have been investigated. Nakajima and coworkers [88, 89] have studied copolymers composed of a central PB block with $M_n = 3.6 \, \text{kg} \, \text{mol}^{-1}$ and PB contents ranging from 12 to 52 mol-%. WAXS patterns obtained from solution-cast triblock copolymer films were in agreement with the α-helical secondary structure of the peptide blocks. The bulk microphase-separated structure of the five different block copolymer samples could be successfully characterized by means of TEM. For the samples with the largest PB volume fraction (56 and 65 vol-%), a lamellar superstructure was found. However, the electron micrographs suggested cylindrical and spherical microphase-separated structures for triblock copolymers with smaller PB volume fractions.

Other polybutadiene-based ABA type triblock copolymers that have been investigated include PMLGlu-*b*-PB-*b*-PMLGlu and PMGlu-*b*-PB-*b*-PMGlu [PMGlu = poly
(γ-methyl D,L-glutamate)] [86, 90]. Infrared spectroscopy experiments on solvent-
cast films indicated that the incorporation of 50% of the D-isomer disrupts the
α-helical secondary structure and induces a random coil conformation in significant portions of the peptide blocks. From the infrared spectra and WAXS experiments, it was estimated that the helix content of a PMGlu homopolypeptide was
about 60% of that of the corresponding PMLGlu. TEM images of OsO$_4$ stained
samples provided evidence for the microphase-separated solid-state structure.
Interestingly, different morphologies were observed when comparing the images
of PMLGlu-*b*-PB-*b*-PMLGlu and PMGlu-*b*-PB-*b*-PMGlu samples with the same PB
content (≈30 mol-%). A cylindrical morphology was proposed for the first and a
spherical structure for the second [86] (see above). The difference in morphology
was ascribed to the less regular secondary structure of the peptide block in the
case of the D,L-triblock copolymer, which prevents a highly ordered organization
of the peptide domains and facilitates the formation of spherical PB domains. The
ATRIR spectra further showed that adsorption of bovine serum albumine (BSA)
and bovine fibrinogen (BF) did not lead to denaturation. From these observations,
these workers concluded that the surfaces of the PMGlu-*b*-PB-*b*-PMGlu membranes interact only weakly or reversibly with these plasma proteins and it was
predicted that this may also lead to a good overall biocompatibility.

The solid-state structure and properties of PELGlu-*b*-PB-*b*-PELGlu [PELGlu =
poly(γ-ethyl L-glutamate)] triblock copolymers containing 31.5–94.5 mol-% (= 17–
88 vol-%) PELGlu have been studied using the same techniques as described
above for the other triblock copolymers [91, 92]. The secondary structure of the
PELGlu blocks was found to be predominantly α-helical and the helix content
in the triblock copolymers decreased from 95 to 60% upon decreasing the
peptide content from 95 to 61%. Interestingly, the WAXS data suggested that
the PELGlu helices were packed in a pseudohexagonal, i.e., monoclinic, arrangement instead of the hexagonal structure observed for most of the other investigated peptide–synthetic hybrid block copolymers. TEM experiments on OsO$_4$
stained films indicated a microphase-separated structure. Based on the electron
micrographs, a spherical microphase-separated structure was proposed for the
copolymer containing 17 vol-% PB, while cylindrical and lamellar morphologies
were suggested for triblock copolymers containing 28 and 44 vol-%, respectively,
68 and 88 vol-% PB. The biocompatibility of the PELGlu-*b*-PB-*b*-PELGlu triblock
copolymers was assessed by coating the samples onto a polyester mesh fiber
cloth, which was subsequently subcutaneously implanted in mongrel dogs for
four weeks. It was found that the foreign body reaction and degradation of the
PELGlu-*b*-PB-*b*-PELGlu samples were less pronounced as compared with
PMLGlu-*b*-PB-*b*-PMLGlu, PBLGlu-*b*-PB-*b*-PBLGlu and PZLLys-*b*-PB-*b*-PZLLys
triblock copolymers.

The bulk nanoscale structure of a series of PBLGlu-*b*-PI-*b*-PBLGlu copolymers
containing 37.4–81.1 mol-% PBLGlu was studied by means of infrared spectroscopy, WAXS, dynamic mechanical analysis and electron microscopy [93]. Based

on the electron micrographs, a cylindrical morphology was proposed for triblock copolymers containing 74.6 and 81.1 mol-% PBLGlu. Water permeability measurements also supported the microphase-separated bulk morphology [94]. Further insight into the bulk morphology of the PBLGlu-*b*-PI-*b*-PBLGlu triblock copolymers was obtained from pulsed proton NMR experiments [95]. The NMR signals of the block copolymers were composed of three components with different spin–spin relaxation times (T_2). The three different T_2 values were attributed to the microphaseseparated structure, which consists of three regions (the ordered helical peptide domains, the unordered interfacial peptide region and the rubbery PI phase) with different molecular mobility. The spin–lattice relaxation times (T_1) that were obtained provided insight into the domain sizes, which were in good agreement with the results from TEM. The surface structure of CHCl$_3$-cast films was studied by XPS and contact angle measurements [96]. It was found that the chemical composition of the microphase-separated films at the surface was different from that in the bulk. The PI content at the film surface was higher than that in the bulk. Water contact angle measurements indicated that the block copolymer films were wetted easier than the respective homopolymers for the same reasons as the previous samples.

Treatment of a PBLGlu-*b*-PI-*b*-PBLGlu film with a mixture of 3-amino-1-propanol and 1,8-octamethylenediamine led to the formation of hydrophilic, cross-linked PHLGln-*b*-PI-*b*-PHLGln membranes being obtained [97]. The swelling ratio of these membranes in pseudoextracellular fluid (PECF) was found to decrease with increasing PI content and increasing cross-link density. Tensile tests in PECF revealed that the triblock copolymer membranes had a larger Young's modulus, increased tensile strength and elongation at breaking compared with membranes prepared from PBLGlu homopolymer. Enzymatic degradation experiments using papain showed that the triblock copolymer films were more resistant towards degradation than the corresponding homopolypeptide membranes. The half-times for sample degradation increased with decreasing peptide content, which was in agreement with the swelling behavior of the membranes.

27.3.2.2 Polystyrene-based Triblock Copolymers

Tanaka and coworkers studied ABA type triblock copolymers composed of a central PS block flanked by two polypeptide segments (PBLGlu, PZLLys or PSar) [98]. TEM of a CHCl$_3$-cast film of PBLGlu$_{25}$-*b*-PS$_{165}$-*b*-PBLGlu$_{25}$ that was stained with phosphotungstic acid revealed a lamellar phase separated structure. In contrast, no microphase separation was observed in a film of PSar$_{73}$-*b*-PS$_{421}$-*b*-PSar$_{73}$. These workers proposed that the different block copolymer morphologies could be related to the different secondary structure of the peptide block; while the PBLGlu segments are predominantly helical, the PSar may not form any regular secondary structure. Fibrinogen adsorption on the block copolymer films was studied with ATR-IR spectroscopy and compared with that on the corresponding homopolymer films [98]. It was found that fibrinogen adsorption on PS and PSar homopolymer films and on PSar-*b*-PS-*b*-PSar triblock copolymer films led to denaturation of the protein. In contrast, protein adsorption on the microphase-separated PBLGlu-*b*-PS-*b*-PBLGlu surfaces was reported to stabilize the protein's secondary structure. Blood clotting tests suggested that thrombus formation was retarded compared with the respective homopolymers.

Samyn and coworkers [99] extended the investigations of ABA triblock copolymers and studied the solid-state organization of three different PBLGlu-*b*-PS-*b*-PBLGlu copolymers containing 34, 55 and 92 wt-% PBLGlu. TEM micrographs of ultramicrotomed and RuO_4 stained specimens and SAXS experiments indicated a lamellar morphology for the copolymers with 34 and 55 wt-% PBLGlu. The sample containing 92 wt-% PBLGlu did not form a lamellar structure. WAXS patterns yielded d-spacings reflecting the intermolecular distance between neighboring peptide α-helices. Ion permeability measurements on dioxane-cast films indicated that the bulk morphology influences the membrane properties [100]. The membranes prepared from the lamellae forming 34 and 55 wt-% PBLGlu containing triblock copolymers showed cation selectivity. In contrast, the membrane prepared from the triblock copolymer containing 92 wt-% PBLGlu did not show such selectivity. It was proposed that uptake of cations into the triblock copolymer membranes was facilitated by the interactions between the cations and the ester functions in the block copolymers. The difference in selectivity was explained in terms of the interfacial zone (as discussed earlier), which separates the PS and PBLGlu domains only in the films generated by the former two triblock copolymers.

27.3.2.3 Polysiloxane-based Triblock Copolymers

Imanishi and coworkers [101] have studied the structure, antithrombogenicity and oxygen permeability of ABA triblock copolymers composed of poly(dimethylsiloxane) (PDMS) as the B block and PBLGlu, PBGlu [poly(γ-benzyl D,L-glutamate)], PZLLys or PSar as the A block. Several series of triblock copolymers were prepared using bifunctional PDMS macroinitiators and targeting various peptide block lengths. TEM images of DMF-cast films provided evidence for a microphaseseparated morphology for PZLLys$_{49}$-*b*-PDMS$_{400}$-*b*-PZLLys$_{49}$ and PZLLys$_{91,160}$-PDMS$_{256}$-PZLLys$_{91,160}$. The images revealed a spherical morphology composed of PDMS islands in a PZLLys matrix. The formation of these spherical domains was attributed to the solvent that was used for sample preparation. While DMF is a good solvent for PZLLys, it is a poor solvent for PDMS. In a separate publication, the same workers also described non-spherical microphase-separated structures [102]. In CH_2Cl_2-cast films of a triblock copolymer with a very high PDMS content (PBLGlu$_{48}$-*b*-PDMS$_{508}$-*b*-PBLGlu$_{48}$, 83 mol-% PDMS) more extended, rod-like PBLGlu aggregates in a matrix of PDMS were observed. The TEM experiments also provided insight into the effects of peptide secondary structure and the nature of the casting solvent on the thin film morphology [101].

Thin films of PBLGlu$_{42}$-*b*-PDMS$_{148}$-*b*-PBLGlu$_{42}$ prepared from DMF showed a spherical morphology. Changing the solvent from DMF (a good solvent for PBLGlu) to CH_2Cl_2 (a fairly non-selective solvent) resulted in coarsening of the microphase-separated structures. PBGlu$_{42}$-*b*-PDMS$_{148}$-*b*-PBGlu$_{42}$ films prepared from CH_2Cl_2 also showed a microphase-separated structure in which spherical

PDMS domains were embedded in a PBGlu matrix. The dimensions of the spherical domains, however, were much smaller than those observed by TEM. These different morphologies reflect the influence of the peptide secondary structure on the block copolymer self-assembly; PBLGlu$_{42}$ adopts an α-helical conformation and PBGlu$_{42}$ a random coil conformation. Studies on adsorption/denaturation of proteins and oxygen permeation measurements from these triblock copolymers also tend to describe the relationship between the film morphology and these properties. In addition, a detailed study of the gas permeation properties of PBLGlu-*b*-PDMS-*b*-PBLGlu films cast from CH$_2$Cl$_2$ and DMF solution with PDMS contents ranging from 46 to 83 mol-% has been reported [103] and revealed that the oxygen permeability of the triblock copolymer films in water was found to increase exponentially with increasing PDMS content, in agreement with a microphase-separated morphology of the membranes. Similar results were reported by Kugo *et al.*, who studied oxygen and nitrogen transport across PBLGlu-*b*-PDMS-*b*-PBLGlu triblock copolymers containing 63–81 mol-% PBLGlu [104].

27.3.2.4 Polyether-based Triblock Copolymers

Inoue and coworkers [105, 106] studied the adhesion behavior of rat lymphocytes on solvent-cast films of PBLGlu-*b*-PEG-*b*-PBLGlu triblock copolymers. The triblock copolymers were prepared from α,ω-bis-amino functionalized PEG macroinitiators with molecular weights of 1.0 and 4.0 kg mol^{-1} and had PEG contents varying from 11 to 33 wt-%. Rat lymphocyte adhesivity was found to decrease with increasing PEG content. At the same PEG content, the adhesivity of the triblock copolymers based on the macroinitiator with a molecular weight of 4 kg mol^{-1} was lower than that of samples based on the macroinitiator with 1 kg mol^{-1}. In addition to overall lymphocyte adhesivity, these workers also studied the adhesion of specific subpopulations: B-cells and T-cells. All triblock copolymers showed a preference towards B-cells. These experiments, however, revealed that the observed differences in cell adhesion behavior were neither due to differences in the conformation of the peptide blocks, nor could they be attributed to differences in surface hydrophilicity. It was therefore proposed that the observed effects were caused by differences in the higher order surface structures, i.e., in terms of the microphase-separated morphology and/or PEG crystallinity.

Kugo *et al.* [107] studied the solid-state conformation of the peptide segment of a series of PBLGlu-*b*-PEG-*b*-PBLGlu copolymers containing a PEG segment with a molecular weight of 4 kg mol^{-1} and 36–86 mol-% PBLGlu. FT-IR spectroscopy experiments on CHCl$_3$-cast films revealed that the PBLGlu blocks, which had degrees of polymerization of 25–276, had an α-helical secondary structure. The helix content of the triblock copolymer containing PBLGlu$_{276}$ blocks was found to be similar to that of the PBLGlu homopolymer. Swelling the triblock copolymer films with water resulted in a decrease in helix content, as indicated by the CD spectra. This decrease in helicity was attributed to competition of water clusters to form hydrogen bonds with the peptide backbone. The effect was even more pronounced when pseudo-extracellular fluid was used instead of water.

A first detailed study of the solid-state nanoscale structure of peptide–PEG hybrid block copolymers was published by Cho *et al.* [108]. They investigated thin, $CHCl_3$-cast films of PBLGlu-*b*-PEG-*b*-PBLGlu copolymers, which were composed of a PEG block of $2 \, kg \, mol^{-1}$ and contained 25–76 mol-% PBLGlu. TEM micrographs of RuO_4-stained specimens revealed a lamellar morphology for triblock copolymers containing 25–64 mol-% PBLGlu. The microphase-separated structure was proposed to consist of chain folded, crystalline PEG domains and helical PBLGlu domains (IR). WAXS patterns were consistent with the ordered, crystalline-like solid-state modification C of PBLGlu. In contrast, in films cast from benzene, the peptide blocks only formed poorly ordered arrays. The sensitivity of the organization of the PBLGlu blocks towards the nature of the casting solvent is identical with the behavior of the PBLGlu homopolymer.

In a separate study, the enzymatic degradation behavior of PBLGlu-*b*-PEG-*b*-PBLGlu triblock copolymers was investigated [109]. The rate of degradation was found to increase with increasing PEG content in the triblock copolymers from 1.4 to 3.1 to 13.6 mol-%. A similar dependence on PEG content was observed for the level of swelling. Exposure of the triblock copolymer samples to a PBS solution without the enzyme did not result in measurable weight loss, indicating that hydrolytic degradation did not take place.

While the data reported by Cho *et al.* described the structure and organization of thin solvent cast films of PBLGlu-*b*-PEG-*b*-PBLGlu, Floudas *et al.* have extensively studied the bulk nanoscale organization of these materials [110]. To this end, a series of triblock copolymers with PBLGlu volume fractions (f_{PBLGlu}) ranging from 0.07 to 0.89 was investigated using SAXS/WAXS, polarizing optical microscopy (POM), DSC and FT-IR spectroscopy. For triblock copolymers with $f_{PBLGlu} \leq 0.25$, PEG crystallization was observed, however, with significant undercooling. Triblock copolymers with $f_{PBLGlu} \geq 0.43$ did not show PEG crystallization. SAXS experiments, which were carried out at 373 K, i.e., above the melting point of PEG, also revealed a different behavior for triblock copolymers with small and large PBLGlu volume fractions. For triblock copolymers with $f_{PBLGlu} \geq 0.43$ only a weakly phase separated structure was found, whereas for samples with $f_{PBLGlu} \leq 0.25$ the SAXS data clearly indicated a microphase-separated structure. WAXS patterns showed that in the microphase-separated state the PEG phase was semi-crystalline and the peptide phase consisted of hexagonally ordered assemblies of PBLGlu α-helices that coexisted with β-sheet structures. For triblock copolymers with $f_{PBLGlu} \geq 0.43$, PEG is amorphous and interspersed with aggregates of α-helical PBLGlu segments and unordered peptide chains. These different bulk structures are illustrated schematically in Figure 27.13. This figure illustrates how the competing interactions that promote the bulk self-assembly of the PBLGlu-*b*-PEG-*b*-PBLGlu triblock copolymers lead to the formation of hexagonally ordered structures covering different length scales. At the smallest length scale, hydrogen-bonding interactions stabilize peptide secondary structures (α-helices and β-strands) and PEG chain folding occurs. On the next higher level, peptide α-helices and β-strands form hexagonal assemblies and β-sheet structures, respectively. Finally,

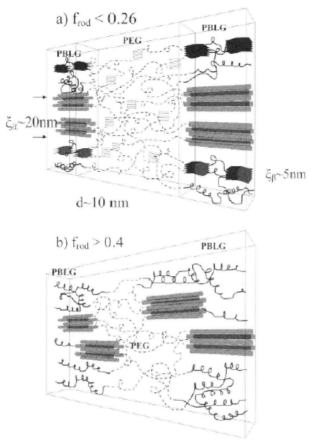

Figure 27.13 Highly schematic model of the phase state in the PBLGlu-*b*-PEG-*b*-PBLGlu triblock copolymers. (a) Phase state corresponding to low peptide volume fractions depicting a microphase-separated copolymer consisting of all the peptide and PEG secondary structures. (b) Phase state corresponding to $f_{rod} > 0.4$ depicting phase mixing resulting in the appearance of only one (α-helical) secondary structure (reprinted from [110] with permission of The American Chemical Society).

the mutual incompatibility of the peptide and PEG block leads to microphase separation.

Additional insight into the solid-state nanoscale organization of the triblock copolymers just discussed was obtained by combining SAXS/WAXS with various microscopic techniques (TEM and AFM) [111]. A "broken lamellar" morphology was observed in the TEM micrographs of PBLGlu$_{58}$-*b*-PEG$_{90}$-*b*-PBLGlu$_{58}$ (f_{PBLGlu} = 0.58). Annealing converted this metastable structure into a nonuniform microphase-separated pattern, which was proposed to consist of "pucklike" PEG domains in a PBLGlu matrix. For PBLGlu$_{105}$-*b*-PEG$_{90}$-*b*-PBLGlu$_{105}$ (f_{PBLGlu} = 0.67), a lamellar morphology was found in the as-cast film, which was transformed into a "broken lamellar" structure upon annealing. Based on these results, a morphology map was constructed.

In addition to PBLGlu-*b*-PEG-*b*-PBLGlu, PZLLys-*b*-PEG-*b*-PZLLys triblock copolymers have also been studied. Cho *et al.* [112] reported on the solid-state structure of a series of PZLLys-*b*-PEG-*b*-PZLLys composed of a PEG block with $M_n = 2\,kg\,mol^{-1}$ and PZLLys contents of 25.2, 49.9 and 83.0 mol-% (= 68, 86 and 98 vol-%). Infrared spectra of $CHCl_3$-cast films were in agreement with a helical secondary structure of the peptide blocks. DSC experiments provided a first hint for the existence of a microphase-separated structure and revealed two T_g values for all samples. The higher T_g was very close to that of the PZLLys homopolymer and the lower T_g approximately 20 °C higher than that of PEG homopolymer. A PEG melting transition was not observed. These results were interpreted in terms of a microphase-separated structure with hard, crystalline PZLLys domains and soft, amorphous PEG segments. The presence of a microphase-separated structure was confirmed by TEM micrographs of RuO_4-stained thin films.

Akashi and coworkers [113] reported on the solid-state nanoscale structure of ABA type triblock copolymers composed of a central PEG block flanked by two poly(β-benzyl L-aspartate) (PBLAsp) blocks. The molecular weight of the central PEG block was 11 or $20\,kg\,mol^{-1}$ and the degrees of polymerization of the peptide blocks ranged from 12 to 32. WAXS and POM studies on CH_2Cl_2-cast films showed PEG crystallization in all samples. The intensity of the crystalline PEG reflection peak, however, was found to decrease with increasing length of the PBLAsp block. The observation of PEG crystallization was interpreted as a first indication for microphase separation. In addition to the PEG signal, the WAXS patterns also contained reflections at $2\Theta = 5.9°$ (= 15 Å), which were assigned to a hexagonally packed array of PBLAsp helices, a result confirmed by FT-IR spectroscopy. In the SAXS patterns of PBLAsp$_{25}$-*b*-PEG$_{250}$-*b*-PBLAsp$_{25}$, PBLAsp$_{25}$-*b*-PEG$_{454}$-*b*-PBLAsp$_{25}$ and PBLAsp$_{32}$-*b*-PEG$_{454}$-*b*-PBLAsp$_{32}$ broad and weak diffraction peaks were observed, indicating the formation phase separated structures. Thermal analysis of the triblock copolymers, however, revealed several interesting properties. DSC experiments showed that the melting temperature of the crystalline PEG domains decreased linearly with increasing PBLAsp content, reflecting the strong influence of the peptide segments on PEG crystallization. More interestingly, these workers found that heating the as-cast films above 333 K and cooling down to 303 K, converted a certain fraction of the α-helical PBLAsp chains into β-strands and was accompanied by a decrease in PEG crystallinity. On a macroscopic level, this led to increased strength and elasticity of the films.

Cho *et al.* [114] have studied triblock copolymers composed of a middle block of poly(propylene glycol) (PPG) with a molecular weight of $2\,kg\,mol^{-1}$ flanked by two PBLGlu segments. Three triblock copolymer samples were investigated with PPG contents of 17.0, 26.0 and 60.0 mol-%, respectively. According to infrared spectra that were recorded from $CHCl_3$-cast films, the PBLGlu blocks possessed an α-helical secondary structure. WAXS patterns revealed a 12.5 Å interhelical spacing and were in agreement with a solid-state modification C of PBLGlu. No further details on the solid-state nanoscale structure and the possibility of microphase separation were reported. Platelet adhesion on glass beads coated with block copolymers containing 33 or 47 mol-% PPG was reduced compared with beads modified with PMLGlu homopolymer or block copolymers with 70.0 mol-% PPG. These differences were attributed to differences in surface composition and morphology, which, unfortunately, were not discussed further. Finally, Hayashi *et al.* [115] have reported on PMLGlu-*b*-PTHF-*b*-PMLGlu [PTHF = poly(tetrahydrofuran)] triblock copolymers that were prepared from an amino-functionalized PTHF macroinitiator with a molecular weight of 9.6 kg mol^{-1}. Three different triblock copolymers were studied with PMLGlu contents and degrees of polymerization of 86.8 mol-%/460, 89.7 mol-%/605 and 91.3 mol-%/730, respectively. With respect to the solidstate structure and organization, only infrared spectra and WAXS data were discussed.

27.3.2.5 Miscellaneous

The structure and properties of an ABA triblock copolymer composed of a poly(ether urethane urea) (PEUU) B block with M_n = 15.8 kg mol^{-1} and two PBLGlu$_{29}$ B blocks were described by Ito *et al.* [116]. DSC thermograms showed a single endotherm located between the T_g of the PEUU and the 7/2 to 18/5 helix transition of PBLGlu, suggesting that there was no phase separation. Platelet adhesion on DMF-cast films of the triblock copolymer was significantly reduced compared with the respective homopolymers and a PEUU/PBLGlu blend. Thrombus formation on block copolymer films was found to be ≈50% less compared with a glass surface. However, no significant difference in antithrombogenicity between PEUU, PBLGlu, their blend and the block copolymer was observed.

Another, very early study focused on two PBLGlu-*b*-PBAN-*b*-PBLGlu [PBAN = poly(butadiene-*co*-acrylonitrile)] triblock copolymers [117]. These block copolymers were composed of a PBAN block with M_n = 3.4 kg mol^{-1} and two PBLGLu segments containing either 80 or 160 repeat units. TEM micrographs of OsO4-stained films cast from dioxane, which is a selective solvent for PBLGlu, revealed a lamellar morphology. When the non-selective solvent CHCl$_3$ was used for the preparation of the TEM specimens, the images were more homogeneous and phase separation was less distinct. This suggests that, depending on the solvent conditions, the PBAN block can affect PBLGlu secondary structure.

Electro- and photoactive peptide–synthetic hybrid triblock copolymers have been prepared using bis(benzyl amine)-terminated poly(9,9-dihexylfluorene-2,7-diyl) (PHF) as a macroinitiator for the ring-opening polymerization of BLGlu-NCA [118]. The electroactive and photoactive properties of the triblock copolymers were similar to those of the PHF homopolymer, indicating that the introduction of the PBLGlu segments did not interfere with charge injection and transport and other material properties. FT-IR spectra of CHCl$_3$-cast films of PBLGlu$_{23}$-*b*-PHF$_{15}$-*b*-PBLGlu$_{23}$ indicated an α-helical secondary structure. The FT-IR spectra of PBLGlu$_{16}$-*b*-PHF$_{28}$-*b*-PBLGlu$_{16}$ displayed, however, an additional peak at 1630 cm^{-1}, indicating the coexistence of α-helical and β-strand conformations. The thin film morphologies of the triblock copolymers were investigated with AFM using different casting solvents (Figure 27.14). When 2,2,2-trifluoroacetic acid (TFA)–

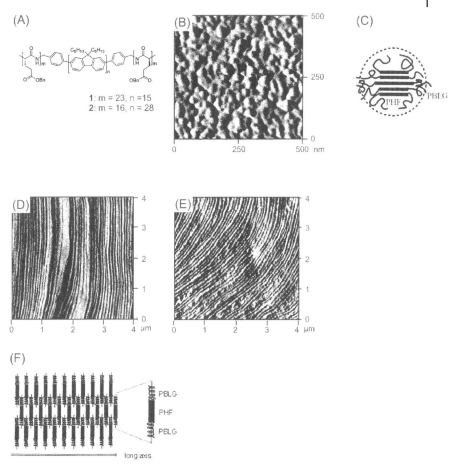

Figure 27.14 (A) Chemical structure of PBLGlu-*b*-PHF-*b*-PBLGlu triblock copolymers **1** and **2**; (B) AFM image of a thin film of **1** cast from TFA–CHCl₃ 30/70 (v/v); (C) schematic representation of the spherical nanostructures that can be observed in (B); (D) and (E) AFM images of triblock copolymers **1** (D) and **2** (E) cast from TFA–CHCl₃ 3/97 (v/v); (F) model proposed for the self-assembly of **1** and **2** in the fibrillar structures shown in (D) and (E) (reprinted from [118] with permission of The American Chemical Society).

CHCl₃ 30/70 (v/v) was used for the preparation of samples, globular or spherical aggregates with diameters of about 32 and 40 nm were observed for PBLGlu$_{23}$-*b*-PHF$_{15}$-*b*-PBLGlu$_{23}$ and PBLGlu$_{16}$-*b*-PHF$_{28}$-*b*-PBLGlu$_{16}$, respectively. In this solvent mixture, the PBLGlu chains have a random coil conformation and the triblock copolymers were proposed to form spherical nanostructures composed of a PHF core and a PBLGlu shell. When the solvent was changed to TFA–CHCl₃ 3/97 (v/v), the SFM images revealed parallel fibrillar structures being 79 ± 25 nm (PBLGlu$_{23}$-*b*-PHF$_{15}$-*b*-PBLGlu$_{23}$) and 83 ± 17 nm (PBLGlu$_{16}$-*b*-PHF$_{28}$-*b*-PBLGlu$_{16}$) in width and 4–10 μm in length. Under these conditions, PBLGlu adopts an α-helical secondary structure. As the widths of the fibrils were much larger than the extended length of the triblock copolymers, a side-by-side antiparallel stacking was proposed to explain the fibril formation.

27.4
Summary and Outlook

In summary, the first part of this chapter described and discussed the phase behavior of biomimetic polypeptide-based copolymers in solution with respect to the occurrence of secondary structure effects. Evidently, incorporation of crystallizable polypeptide segments inside the core of an aggregate has an impact on the curvature of the core–corona interface and promotes the formation of fibrils or vesicles, or other flat superstructures. Spherical micelles are not usually observed. Copolymers with soluble polypeptide segments, on the other hand, seem to behave as conventional block copolymers. A pH-induced change of the conformation of coronal polypeptide chains only affects the size of aggregates, not their shape. It is evident that the biomimetic approach using polypeptide hybrid polymers is very successful in the creation of novel superstructures with hierarchical order. However, although begun about 30 years ago in the mid-1970s, this field is still in a premature state. Most systematic studies on aggregation in solution have been reported only during the last five to ten years. A comprehensive picture of the processes involved in the formation of hierarchical structures is still lacking. The application potential of polypeptide copolymers has also not been exhausted. Most studies deal with ordinary micelles for the controlled delivery of drugs or genes. Not much attention has been, for whatever reason, paid to gel structures and other colloidal systems, such as emulsions, polymer latexes and inorganic–organic hybrid nanoparticles.

The solid-state structure, organization and properties of peptide–synthetic hybrid block copolymers were discussed in the second part. The most notable difference between peptide-based block copolymers and their fully synthetic and amorphous analogues is their hierarchical solid-state organization. In contrast to most synthetic amorphous block copolymers, which typically exhibit structural order only over a single length scale, peptide-based block copolymers can form hierarchically organized nanoscale structures that cover several different length scales. At the smallest length scale, peptide sequences fold into regular secondary structures, such as α-helices or β-strands. On the next higher level, peptide α-helices and β-strands can assemble into hexagonal superstructures and β-sheets, respectively. Finally, phase separation between the peptide and synthetic blocks leads to the formation of ordered domains with the largest characteristic length scales. For a large number of peptide-based block copolymers, lamellar phase separated morphologies have been observed. These lamellar structures, however, are often found over a much broader range of compositions compared with regular, fully amorphous diblock copolymers. This behavior, as in solution, can be explained easily considering the flat interface that is generated from the rod–rod packing. In addition to these more conventional morphologies, structural investigations on peptide-based hybrid block copolymers have also led to the discovery of various novel phase separated structures, which were not previously known for fully amorphous diblock copolymers. Both observations reflect the fact that the solid-state structure formation of peptide hybrid block copolymers is not solely dictated by phase separation, as is the case for amorphous diblock copolymers, but is also influenced by other factors such as intra- and intermolecular hydrogen bonding and chain conformation. While much of the early interest in peptide–synthetic hybrid block copolymers was driven by their potential use as membrane materials or for the development of antithrombogenic surfaces, more recent studies revealed that these materials can also have interesting mechanical properties [113].

The major drawback of most of the block copolymers discussed in this chapter is that they have been prepared via the conventional amine-initiated NCA polymerization. The polymerization of NCAs under these conditions does not allow very accurate control over polymer chain length, results in rather broad molecular weight distributions and is also not very useful for preparing defined block copolypeptides [119, 120]. It is obvious that these limitations possibly restrict further engineering of the structure and organization of peptide–synthetic hybrid block copolymers and could also hamper the exploration of their full practical potential. Over recent years, however, a number of alternative NCA polymerization strategies have been developed, which provide enhanced control over polypeptide chain length and chain length distribution and also allow access to defined block copolypeptides [121–123]. In the last 5 years, a great deal of effort has been focused on controlling the synthesis and understanding the structural behavior of these polypeptide-based block copolymers. These systems, which have been known for a long time, are currently gaining more and more attention due to the possibilities of making highly ordered materials on the nano- to micrometer-length scale and bio-compatible aggregates that respond to external stimuli. The door to new innovations in materials science is now open.

This chapter has been published previously in:
Lazzari, Massimo/Liu, Guojun/Lecommandoux, Sebastién (eds.)
Block Copolymers in Nanoscience
2006
ISBN-13: 978-3-527-31309-9-Wiley-VCH, Weinheim

References

1 Park, C., Yoon, J. and Thomas, E.L. (2003) *Polymer*, **44**, 6725.

2 Hamley, I.W. (2003) *Nanotechnology*, **14**, R39.

3 Lazzari, M. and López-Quintela, M.A. (2003) *Adv. Mater.*, **15**, 1583.

4 Bates, F.S. and Fredrickson, G.H. (1990) *Annu. Rev. Phys. Chem.*, **41**, 525.

5 Hamley, I.W. (1998) *The physics of block copolymers*, Oxford University Press, Oxford, New York, Tokyo.

6 Bates, F.S. and Fredrickson, G.H. (1999) *Phys. Today*, **52**, 32.

7 Walther, M. and Finkelmann, H. (1996) *Progr. Polym. Sci.*, **21**, 951.

8 Mao, G. and Ober, C.K. (1997) *Acta Polym.*, **48**, 405.

9 Klok, H.-A. and Lecommandoux, S. (2001) *Adv. Mater.*, **13**, 1217.

10 Lee, M., Cho, B.K. and Zin, W.C. (2001) *Chem. Rev.*, **101**, 3869.

11 Förster, S. and Antonietti, M. (1998) *Adv. Mater.*, **10**, 195.

12 Bates, F.S. and Fredrickson, G.H. (1999) *Phys. Today*, **52**, 32.

13 Cölfen H. (2001) *Macromol. Rapid. Commun.*, **22**, 219.

14 Förster, S. and Konrad, M. (2003) *J. Mater. Chem.*, **13**, 2671.

15 Antonietti, M. and Förster, S. (2003) *Adv. Mater.*, **15**, 1323.

16 Choucair, A. and Eisenberg, A. (2003) *Eur. Phys. J. E*, **10**, 37.

17 Discher, B.M., Won, Y.-Y., Ege, D.S., Lee, J.C.-M., Bates, F.S., Discher, D.E. and Hammer, D.A. (1999) *Science*, **284**, 1143.

18 Discher, D.E. and Eisenberg, A. (2002) *Science*, **297**, 967.

19 Taubert, A., Napoli, A. and Meier, W. (2004) *Curr. Opin. Chem. Biol.*, **8**, 598.

20 Kita-Tokarczyk, K., Grumelard, J., Haefele, T. and Meier, W. (2005) *Polymer*, **46**, 3540.

21 Cornelissen, J.J.L.M., Rowan, A.E., Nolte, R.J.M. and Sommerdijk, N.A.J.M. (2001) *Chem. Rev.*, **101**, 4039.

22 Löwik, D.W.P.M. and van Hest, J.C.M. (2004) *Chem. Soc. Rev.*, **33**, 234.

23 Vandermeulen, G.W.M. and Klok, H.-A. (2004) *Macromol. Biosci.*, **4**, 383.

24 Klok, H.-A. (2005) *J. Polym. Sci. Part A Polym. Chem.*, **43**, 1.

25 Schlaad, H. and Antonietti, M. (2003) *Eur. Phys. J. E*, **10**, 17.

26 Rodriguez-Hernandez, J., Chécot, F., Gnanou, Y. and Lecommandoux, S. (2005) *Progr. Polym. Sci.*, **30**, 691.

27 Nakajima, A., Kugo, K. and Hayashi, T. (1979) *Macromolecules*, **12**, 844.

28 Hayashi, T. (1985), in *Developments in block copolymers*, ed. Goodman, I., Elsevier Applied Science Publishers, London, p. 109.

29 Förster, S., Zisenis, M., Wenz, E. and Antonietti, M. (1996) *J. Chem. Phys.*, **104**, 9956.

30 Kukula, H., Schlaad, H., Antonietti, M. and Förster, S. (2002) *J. Am. Chem. Soc.*, **124**, 1658.

31 Chécot, F., Brûlet, A., Oberdisse, J., Gnanou, Y., Mondain-Monval, O. and Lecommandoux, S. (2005) *Langmuir*, **21**, 4308.

32 Chécot, F., Lecommandoux, S., Gnanou, Y. and Klok, H.-A. (2002) *Angew. Chem., Int. Ed. Engl.*, **41**, 1340.

33 Chécot, F., Lecommandoux, S., Klok, H.-A. and Gnanou, Y. (2003) *Eur. Phys. J. E*, **10**, 25.

34 Babin, J., Rodríguez-Hernández, J., Lecommandoux, S., Klok, H.-A. and Achard, M.-F. (2005) *Faraday Discuss.*, **128**, 179.

35 Rodríguez-Hernández, J., Babin, J., Zappone, B. and Lecommandoux, S. (2005) *Biomacromolecules*, **6**, 2213.

36 Arimura, H., Ohya, Y. and Ouchi, T. (2005) *Biomacromolecules*, **6**, 720.

37 Lübbert, A., Castelletto, V., Hamley, I.W., Nuhn, H., Scholl, M., Bourdillon, L., Wandrey, C. and Klok, H.-A. (2005) *Langmuir*, **21**, 6582.

38 Harada, A. and Kataoka, K. (1995) *Macromolecules*, **28**, 5294.

39 Harada, A. and Kataoka, K. (1997) *J. Macromol. Sci.-Pure Appl. Chem.*, A**34**, 2119.

40 Harada, A. and Kataoka, K. (1999) *Science*, **283**, 65.

41 Toyotama, A., Kugimiya, S-i, Yamanaka, J. and Yonese, M. (2001) *Chem. Pharm. Bull.*, **49**, 169.

42 Cheon, J.-B., Jeong, Y.-I. and Cho, C.-S. (1998) *Korea Polym. J.*, **6**, 34.

43 Cheon, J.-B., Jeong, Y.-I. and Cho, C.-S. (1999) *Polymer*, **40**, 2041.

44 Dong, C.-M., Sun, X.-L., Faucher, K.M., Apkarian, R.P. and Chaikof, E.L. (2004) *Biomacromolecules*, **5**, 224.

45 Dong, C.-M., Faucher, K.M. and Chaikof, E.L. (2004) *J. Polym. Sci., Part A: Polym. Chem.*, **42**, 5754.

46 Naka, K., Yamashita, R., Nakamura, T., Ohki, A. and Maeda, S. (1997) *Macromol. Chem. Phys.*, **198**, 89.

47 Losik, M. and Schlaad, H. unpublished results.

48 Doi, T., Kinoshita, T., Kamiya, H., Tsujita, Y. and Yoshimizu, H. (2000) *Chem. Lett.* 262.

49 Doi, T., Kinoshita, T., Kamiya, H., Washizu, S., Tsujita, Y. and Yoshimizu, H. (2001) *Polym. J.*, **33**, 160.

50 Rodríguez-Hernández, J. and Lecommandoux, S. (2005) *J. Am. Chem. Soc.*, **127**, 2026.

51 Bellomo, E.G., Wyrsta, M.D., Pakstis, L., Pochan, D.J. and Deming, T.J. (2004) *Nat. Mater.*, **3**, 244.

52 Constancis, A., Meyrueix, R., Bryson, N., Huille, S., Grosselin, J-M, Gulik-Krzywicki, T. and Soula, G. (1999) *J. Colloid. Interf. Sci.*, **217**, 357.

53 Deming, T.J. (2005) *Soft Matter*, **1**, 28.

54 Nowak, A.P., Breedveld, V., Pakstis, L., Ozbas, B., Pine, D.J., Pochan, D. and Deming, T.J. (2002) *Nature (London)*, **417**, 424.

55 Breedveld, V., Nowak, A.P., Sato, J., Deming, T.J. and Pine, D.J. (2004) *Macromolecules*, **37**, 3943.

56 Pochan, D.J., Pakstis, L., Ozbas, B., Nowak, A.P. and Deming, T.J. (2002) *Macromolecules*, **35**, 5358.

57 Pakstis, L.M., Ozbas, B., Hales, K.D., Nowak, A.P., Deming, T.J. and Pochan, D. (2004) *Biomacromolecules*, **5**, 312.

58 Cha, J.N., Stucky, G.D., Morse, D.E. and Deming, T.J. (2000) *Nature (London)*, **403**, 289.

59 Euliss, L.E., Trnka, T.M., Deming, T.J. and Stucky, G.D. (2004) *Chem. Commun.*, 1736.

60 Lecommandoux, S., Sandre, O., Chécot, F., Rodríguez-Hernández, J. and Perzynski, R. (2005) *Adv. Mater.*, **17**, 712.

61 Euliss, L.E., Grancharov, S.G., O'Brien, S., Deming, T.J., Stucky, G.D., Murray, CB. and Held, G.A. (2003) *Nano Lett.*, **3**, 1489.

62 Perly, B., Douy, A. and Gallot, B. (1976) *Makromol. Chem.*, **177**, 2569.

63 Billot, J.-P., Douy, A. and Gallot, B. (1976) *Makromol. Chem.*, **177**, 1889.

64 Billot, J.-P., Douy, A. and Gallot, B. (1977) *Makromol. Chem.*, **178**, 1641.

65 Douy, A. and Gallot, B. (1977) *Polym. Eng. Sci.*, **17**, 523.

66 Douy, A. and Gallot, B. (1982) *Polymer*, **23**, 1039.

67 Gallot, B., Douy, A., Hayany, H. and Vigneron, C. (1983) *Polym. Sci. Technol.*, **23**, 247.

68 Mori, A., Ito, Y., Sisido, M. and Imanishi, Y. (1986) *Biomaterials*, **7**, 386.

69 Schlaad, H., Kukula, H., Smarsly, B., Antonietti, M. and Pakula, T. (2002) *Polymer*, **43**, 5321.

70 Losik, M., Kubowicz, S., Smarsly, B. and Schlaad H. (2004) *Eur. Phys. J. E*, **15**, 407.

71 Schlaad, H., Smarsly, B. and Losik, M. (2004) *Macromolecules*, **37**, 2210.

72 Ludwigs, S., Krausch, G., Reiter, G., Losik, M., Antonietti, M. and Schlaad, H. (2005) *Macromolecules*, **38**, 7532.

73 Klok, H.-A., Langenwalter, J.F. and Lecommandoux, S. (2000) *Macromolecules*, **33**, 7819.

74 Lecommandoux, S., Achard, M.-F., Langenwalter, J.F. and Klok, H.-A. (2001) *Macromolecules*, **34**, 9100.

75 Zhang, G., Ma, J., Li, Y. and Wang, Y. (2003) *J. Biomater. Sci. Polym. Edn.*, **14**, 1389.

76 Cho, I., Kim, J.-B. and Jung, H.-J. (2003) *Polymer*, **44**, 5497.

77 Degée, P., Dubois, P., Jérôme, R. and Theyssié, P. (1993) *J. Polym. Sci. Part A Polym. Chem.*, **31**, 275.

78 Rong, G., Deng, M., Deng, C., Tang, Z., Piao, L., Chen, X. and Jing, X. (2003) *Biomacromolecules*, **4**, 1800.

79 Caillol, S., Lecommandoux, S., Mingotaud, A.-F., Schappacher, M., Soum, A., Bryson, N. and Meyrueix, R. (2003) *Macromolecules*, **36**, 1118.

80 Deming, T.J. (1997) *J. Am. Chem. Soc.*, **119**, 2759.

81 Papadopoulos, P., Floudas, G., Schnell, I., Aliferis, T., Iatrou, H. and Hadjichristidis, N. (2005) *Biomacromolecules*, **6**, 2352.

82 Nakajima, A., Hayashi, T., Kugo, K. and Shinoda, K. (1979) *Macromolecules*, **12**, 840.

83 Nakajima, A., Kugo, K. and Hayashi, T. (1979) *Macromolecules*, **12**, 844.

84 Nakajima, A., Kugo, K. and Hayashi, T. (1979) *Polymer J.*, **11**, 995.

85 McKinnon, A.J. and Tobolsky, A.V. (1968) *J. Phys. Chem.*, **72**, 1157.

86 Hayashi, T., Chen, G.-W. and Nakajima, A. (1984) *Polymer J.*, **16**, 739.

87 Gervais, M., Douy, A., Gallot, B. and Erre, R. (1988) *Polymer*, **29**, 1779.

88 Kugo, K., Hayashi, T. and Nakajima, A. (1982) *Polymer J.*, **14**, 391.

89 Kugo, K., Hata, Y., Hayashi, T. and Nakajima, A. (1982) *Polymer J.*, **14**, 401.

90 Kugo, K., Murashima, M., Hayashi, T. and Nakajima, A. (1983) *Polymer J.*, **15**, 267.

91 Chen, G.-W., Hayashi, T. and Nakajima, A. (1981) *Polymer J.*, **13**, 433.

92 Sato, H., Nakajima, A., Hayashi, T., Chen, G.-W. and Noishiki, Y. (1985) *J. Biomed. Mater. Res.*, **19**, 1135.

93 Yoda, R., Komatsuzaki, S., Nakanishi, E. and Hayashi, T. (1995) *Eur. Polym. J.*, **31**, 335.

94 Yoda, R., Komatsuzaki, S. and Hayashi, T. (1996) *Eur. Polym. J.*, **32**, 233.

95 Yoda, R., Shimoda, M., Komatsuzaki, S., Hayashi, T. and Nishi, T. (1997) *Eur. Polym. J.*, **33**, 815.

96 Yoda, R., Komatsuzaki, S. and Hayashi, T. (1995) *Biomaterials*, **16**, 1203.

97 Yoda, R., Komatsuzaki, S., Nakanishi, E., Kawaguchi, H. and Hayashi, T. (1994) *Biomaterials*, **15**, 944.

98 Imanishi, Y., Tanaka, M. and Bamford, C.H. (1985) *Int. J. Biol. Macromol.*, **7**, 89.

99 Janssen, K., Van Beylen, M., Samyn, C., Scherrenberg, R. and Reynaers, H. (1990) *Makromol. Chem.*, **191**, 2777.

100 Janssen, K., Van Beylen, M., Samyn, C. and Van Driessche, W. (1989) *Makromol. Chem. Rapid. Commun.*, **10**, 457.

101 Kumaki, T., Sisido, M. and Imanishi, Y. (1985) *J. Biomed. Mater. Res.*, **19**, 785.

102 Kang, I.-K., Ito, Y., Sisido, M. and Imanishi, Y. (1988) *Biomaterials*, **9**, 138.

103 Kang, I.-K., Ito, Y., Sisido, M. and Imanishi, Y. (1988) *Biomaterials*, **9**, 349.

104 Kugo, K., Nishioka, H. and Nishino, J. (1987) *Chem. Express*, **2**, 21.

105 Nishimura, T., Sato, Y., Yokoyama, M., Okuya, M., Inoue, S., Kataoka, K., Okano, T. and Sakurai, Y. (1984) *Makromol. Chem.*, **185**, 2109.

106 Yokoyama, M., Nakahashi, T., Nishimura, T., Maeda, M., Inoue, S., Kataoka, K. and Sakurai, Y. (1986) *J. Biomed. Mater. Res.*, **20**, 867.

107 Kugo, K., Ohji, A., Uno, T. and Nishino, J. (1987) *Polymer J.*, **19**, 375.

108 Cho, C.-S., Kim, S.-W. and Komoto, T. (1990) *Makromol. Chem.*, **191**, 981.

109 Cho, C.-S. and Kim, S.U. (1988) *J. Control. Release*, **7**, 283.

110 Floudas, G., Papadopoulos, P., Klok, H.-A., Vandermeulen, G.W.M. and Rodríguez-Hernandez, J. (2003) *Macromolecules*, **36**, 3673.

111 Parras, P., Castelletto, V., Hamley, I.W. and Klok, H.-A. (2005) *Soft Matter*, **1**, 284.

112 Cho, C.S., Jo, B.-W., Kwon, J.-K. and Komoto, T. (1994) *Macromol. Chem. Phys.*, **195**, 2195.

113 Tanaka, S., Ogura, A., Kaneko, T., Murata, Y. and Akashi, M. (2004) *Macromolecules*, **37**, 1370.

114 Cho, C.-S., Kim, S.-W., Sung, Y.-K. and Kim, K.-Y. (1988) *Makromol. Chem.*, **189**, 1505.

115 Hayashi, T., Kugo, K. and Nakajima, A. (1984) *Cont. Topics Polym. Sci.*, **4**, 685.

116 Ito, Y., Miyashita, K., Kashiwagi, T. and Imanishi, Y. (1993) *Biomat. Artif. Cells Immob. Biotechnol.*, **21**, 571.

117 Barenberg, S., Anderson, J.M. and Geil, P.H. (1981) *Int. J. Biol. Macromol.*, **3**, 82.

118 Kong, X., Jenekhe, SA. (2004) *Macromolecules*, **37**, 8180.

119 Kricheldorf, H.R. (1987) *α-Aminoacid-N-carboxyanhydrides and related heterocycles*, Springer-Verlag, Berlin, Heidelberg, New York.

120 Deming, T.J. (2000) *J. Polym. Sci. A Polym. Chem.*, **38**, 3011.

121 Deming, T.J. (1997) *Nature (London)*, **390**, 386.

122 Dimitrov, I. and Schlaad, H. (2003) *Chem. Commun.*, 2944.

123 Aliferis, T., Iatrou, H. and Hadjichristidis, N. (2004) *Biomacromolecules*, **5**, 1653.

28
Structural DNA Nanotechnology: Information-Guided Self-Assembly

Yonggang Ke, Yan Liu, and Hao Yan

28.1
Introduction

Although the Watson–Crick double helical model [1] of DNA has imparted major impacts on modern biology for more than 50 years, the significance of this simple–yet very elegant–model is not limited to one particular area. In 1982 [2], Ned Seeman proposed the building of nanostructures from DNA, an idea which led to the origination of the field now known as "structural DNA nanotechnology." During the past decade, this field has witnessed much significant progress, and in this chapter we will discuss the basic concepts and major research directions of structural DNA nanotechnology, the important progress that has been made in recent years, and some future perspectives of the field.

DNA, which serves as the genetic information carrier in most organisms on Earth, is also an ideal candidate for structural nanotechnology, which targets at controlling and organizing matter at the nanometer scale. First, DNA is a nanometer-scale object itself, with a diameter of 2 nm and a helical repeat of 10–10.5 nucleotide pairs, or ~3.5 nm for the common B-form DNA. Second, DNA hybridization is highly predictable because of the well-known Watson–Crick base-pairing that guanine (G) pairs with cytosine (C), and adenine (A) with thymine (T). Third, whilst single-stranded DNA is quite flexible, the DNA duplex is more rigid and has a persistence length of approximately 50 nm. It is this combined flexibility and rigidity that permits the design of DNA structures to form different geometric shapes. Furthermore, as a results of advances in modern chemistry and molecular biology, DNA molecules with any designed lengths, sequences, and a variety of functionalities can now be synthesized conveniently, and also manipulated by using the wide range of enzymes that are available for the cleavage, ligation and amplification of DNA. All of these facilities have provided scientists with an extreme degree of control for building DNA nanostructures and maneuvering DNA nanomachines.

Topologically speaking, DNA duplex is a one-dimensional (1-D) molecule. In order to create a two-dimensional (2-D) or three-dimensional (3-D) structure, it is

Advanced Nanomaterials. Edited by Kurt E. Geckeler and Hiroyuki Nishide
Copyright © 2010 WILEY-VCH Verlag GmbH & Co. KGaA, Weinheim
ISBN: 978-3-527-31794-3

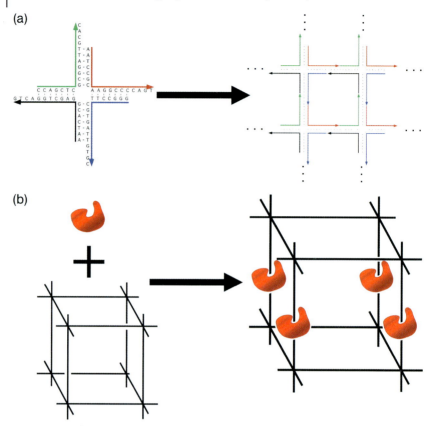

Figure 28.1 Ned Seeman's original proposal of construction of a periodic DNA array and its application. (a) Four-arm junction DNA tiles with sticky ends are connected together to form a 2-D periodic array through self-assembly process; (b) A 3-D DNA lattice templated protein array could be used for X-ray crystallography.

necessary to use branched objects. For example, Seeman proposed techniques that would allow DNA strands to assemble into branched junctions that could further self-assemble into periodic, 2-D arrays [3] (Figure 28.1a).

The basic building block here is the branched four-arm junction consisting of four single-stranded DNA oligonucleotides. To enable DNA junction building blocks to form higher ordered objects and lattices, single v stranded DNA overhangs called "*sticky ends*" are used to bring DNA junction molecules together. These sticky ends carries base sequences that are complementary to each other; for example, the red sticky end is complementary to the black end, and the green end is complementary to the blue end (Figure 28.1a). As a result, sticky end cohesion will cause the individual four-arm junctions to be "glued" together to form the 2-D array. Yet, this simple scheme illustrated a powerful method for building

DNA nanostructures: first, to design the branched DNA building blocks or "tiles", and then to assemble them together using sticky end cohesion.

Although other cohesions have also been used for mortaring DNA nanostructures–for example, paranemic crossovers (PX) cohesion [4] and edge-sharing [5]–sticky end cohesion is by far the most extensively used in DNA nanostructure design. Studies of crystal structures have revealed that the duplex formed by complementary sticky ends has an exactly identical structure to B-DNA [6], a feature which allows designers to predict and control the relative orientations of the DNA tiles. It is interesting to note that for sticky ends that are N-bases long, the number of unique sticky ends can be up to 4^N, which provides a library of programmable molecular interactions.

Ideally, the DNA sequences of a DNA nanostructure should be designed to achieve the highest stability, so that all other less-stable competitive structures will be less likely to form. On a practical basis, it is necessary to choose a set of optimized sequences that minimizes sequence symmetry at the branch points [7] so as to prevent the branch migration of DNA strands.

As originally proposed by Seeman, one potential application of structural DNA nanotechnology would be to use the highly ordered self-assembling DNA scaffolds to organize other types of macromolecule into 3-D crystals. Moreover, if the macromolecule could be attached to a 3-D DNA nanostructure (Figure 28.1b), the self-assembly of DNA would facilitate the organization of macromolecules into a periodic lattice, the periodicity and parameters of which could be well defined. This would in turn facilitate macromolecule structural analysis using X-ray diffraction.

Besides the above-described potential application, the DNA nanostructure might also be used as a scaffold to organize different nanomaterials. An example of this is the organization of nanoelectronic components into an addressable fashion, leading to the construction of DNA-templated nanoelectronics/nanophotonics devices. Indeed, recent developments in the use of 1-D and 2-D DNA nanostructures to template nanoparticles into rationally designed patterns have paved the way towards this goal.

28.2
Periodic DNA Nanoarrays

Seeman's original proposal has inspired many research groups to construct a variety of DNA tiles with different sizes and geometries, and assemble them into 2-D periodic arrays (Figure 28.2). For example, Mao *et al.* designed and constructed DNA parallelograms that would grow into micrometer-sized 2-D arrays (Figure 28.2a) [8].

A series of DNA tile molecules termed double crossover [9] (DX) DNA molecules were originally created by Seeman, and subsequently utilized in many of the later studies. The DX tile consists of two parallel DNA helices, joined together by two crossovers through strand exchange. Winfree *et al.* successfully built DNA 2-D

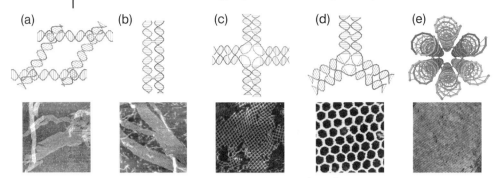

Figure 28.2 DNA tiles and their 2-D periodic arrays formed through sticky end cohesion. (a) A DNA parallelogram tile consisting of four Holliday junction structures; (b) A double crossover (DX) tile; (c) A 4 × 4 cross-shaped tile. Each arm of the cross is a four-arm DNA junction; (d) A three-point-star tile; (e) A six-helix-bundle tile.

arrays through the self-assembly of DX tiles (Figure 28.2b) [10]. A similar design strategy, learned from the DX tile construction, was then used by different groups when designing DNA tiles containing multiple parallel DNA helices; these included triple crossover tiles [11], and four-, eight-, and 12-helix tiles [12, 13].

Yan *et al.* used four four-arm DNA junctions to design a cross-shaped tile, named "4 × 4" tile [14]. In this design, four four-arm junctions are tethered together by a long central strand with a dT_4 loop in between (Figure 28.2c); the 4 × 4 tiles then self-assembled into a 2-D square lattice. By removing one arm from, or adding two more arms to the 4 × 4 tile, Mao's group were able to construct a three-point-star [15] and a six-point star tile [16] that could self-assemble into 2-D lattices with hexagonal and triangular cavities (Figure 28.2d).

Another family of tiles are tube-like tiles, including three-helix [17], six-helix [18] (Figure 28.2e), and eight-helix bundles [19]. With different sticky end designs, these tubes tile can be assembled to either 1-D tubes, 2-D arrays, or even 3-D lattices.

28.3
Finite-Sized and Addressable DNA Nanoarrays

For the purpose of DNA-directed macromolecule crystallization, the periodic arrays represent an excellent choice. However, to build a functional DNA nanoelectronic device, it must be possible to control the size of a DNA array and to attach functional species at particular locations on the array. Such needs have driven research teams to develop methods for building finite-sized and addressable arrays.

Park *et al.* [20] reported the construction of a square-shaped, addressable array consisting of sixteen 4 × 4 tiles (Figure 28.3a) through an hierarchical assembly

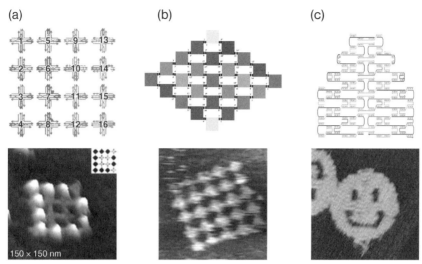

Figure 28.3 DNA finite-sized and addressable arrays. (a) A 4 × 4 tile array consisting of 16 distinctive 4 × 4 tiles. This is a fully addressable array on the level of individual tiles. Streptavidin protein was attached onto certain DNA tiles to display a letter "D"; (b) A 5 × 5 tile array consisting of 25 eight-helix tiles. The number of distinctive tiles was reduced from 25 to 13 by taking advantage of the array's C_2 symmetry; (c) Schematic of Rothemund's scaffolded DNA self-assembly and atomic force microscopy image of a nanometer-scale "smiley face." The black strand is the scaffold DNA that is folded by other short "staple" strands into the designed shape.

process. In order to demonstrate that the 16-tile array was fully addressable on the level of individual tiles, Park and colleagues first functionalized a few tiles on the array with biotin groups, and then attached streptavidin molecules to the array. In this way, the protein molecules could be organized to display the letters, "D", "N" and "A."

It is costly to make every tile in an array unique; moreover, the more complex the system, the greater the chances of self-assembly errors occurring. In order to build finite-sized arrays in a cost-efficient way, Liu *et al.* [21] demonstrated a strategy to utilize the geometric symmetry of the array to reduce the number of unique tiles. For a N-tile finite-size array with C_m symmetry, the number of unique tiles required is N/m, if N/m is an integral number, or $Int(N/m) + 1$, if N/m is a non-integral number. Consequently, Liu and coworkers demonstrated two 25-tile array examples with C_2 and C_4 fold symmetry. The 5 × 5 array with C_2 symmetry required 13 unique tiles instead of 25 (Figure 28.3b), while the 5 × 5 array with C_4 symmetry required only seven unique cross-shaped tiles instead of 25.

When comparing these two finite-size addressable arrays, it is possible to understand the dilemma that research groups often face when designing a DNA structure. On one hand, the addressability of an array can be increased by introducing more unique sequences/tiles into the system, although a low yield and a high error rate will be expected. On the other hand, the symmetry can be utilized to reduce complexity of the system so as to achieve a high yield and a low assembly error rate, but the high addressability would be lost. Thus, depending on the purpose of a DNA nanostructure, it is important for a designer to identify a good balance between the complexity and addressability of the system.

Another important approach when building an addressable DNA nanoarray is "nucleated DNA self-assembly." This method uses a long natural or synthetic DNA strand, which serves as a scaffold, to direct the DNA strands or tiles self-assembly. Yan *et al.* demonstrated the use of this technique by efficiently assembling DX tiles together into barcode-patterned lattices [22]. In 2006, Paul Rothemund reported an exciting breakthrough, in which he used more than 200 short "staple" DNA strands to fold 7249 base long, single-stranded M13 viral DNA into 2-D arrays ("DNA Origami") with a variety of shapes [23]. Every staple has its unique sequence, and can only hybridize with predefined parts of M13 according to a predetermined folding path (Figure 28.3c). The array resulted is fully addressable at the position of each individual staple strand. In theory, if the scaffold strand is long enough, it is possible to design any arbitrarily shaped 2-D DNA array and to functionalize staples at any position on that array.

28.4
DNA Polyhedron Cages

One branch of structural DNA nanotechnology focuses on the construction of 3-D DNA "cages." Potentially, DNA cages can be used for applications such as target-specific drug delivery or nanoparticle site-specific functionalization. For delivery purposes, the cages can be made to encapsulate cargos during their self-assembly and to display target recognition tags outside the cages. Because DNA duplex is linear, it is not surprising that most of the DNA cages built to date are polyhedral, using DNA duplex(es) as straight edges and branching point(s) at the vertices. When Chen and Seeman assembled the first DNA polyhedron, a cube consisting of ten DNA strands [24], they used endonucleases to cleave specific edges of the cube to prove the tube topology with polyacrylamide gel electrophoresis. Many years later, several groups have only recently developed a number of methods to build DNA polyhedral cages and more direct ways to prove their formation (Figure 28.4).

Goodman *et al.* built a DNA tetrahedron by hybridizing four single-stranded DNAs together in a one-step annealing process (Figure 28.4a) [25, 26]. In a later study, the same group built hairpin loops into the tetrahedron and demonstrated that the edge of this cage could be opened/closed by the addition of "fuel DNA" strands [27]. This mechanism would allow cargos to diffuse into the structure at the open stage, and to be captured in the close stage. Shih *et al.* used a 1.7 kilobase and five short, single-stranded DNAs to construct an octahedron (Figure 28.4b) [28] where the edges of the octahedron were double crossovers (DX) or paranemic crossovers (PX) DNA motifs. One noteworthy feature of this octahedron was that the 1.7 kilobase DNA was formed through polymerase chain reaction (PCR), which

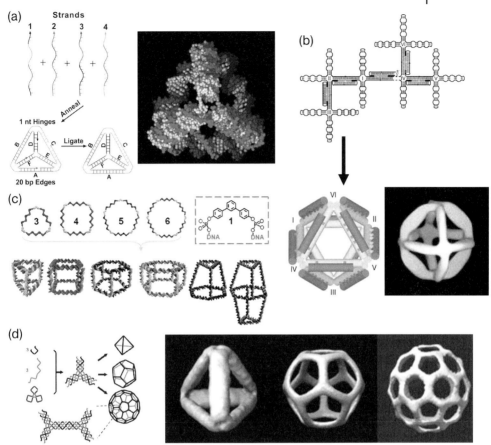

Figure 28.4 DNA polyhedron cages. (a) DNA tetrahedron: schematic drawing and a physical model; (b) DNA octahedron model and 3-D reconstruction image from cryo-electron microscopy (EM) experiments; (c) A series DNA polyhedra. Organic molecules are incorporated into the DNA circular strands to help the structures form. (d) DNA tetrahedron, dodecahedron, and buckyball assembled from three-point-star tiles. A 3-D reconstruction of the cryo-EM images of these polyhedra is shown on the right.

in turn means that the DNA could be easily amplified to produce large amounts of the DNA octahedron.

Aldaye *et al.* recently developed a new approach for building different-shaped DNA cages [29]. They introduced organic molecules into the circular single-stranded DNA during the DNA synthesis process (Figure 28.4c). This allowed the building of a series of DNA polygons with repeated sequences at each edge and branched organic molecules at the corners. Linker strands were then used to bring two or three polygons together to form the DNA cages.

Mao's group reported the design and formation of tetrahedron, dodecahedron, and buckyballs through hierarchical DNA assembly (Figure 28.4d) [30]. All three polyhedra shared the same basic building unit, namely the three-point-star tile

[15], although at low concentrations (50–75 nM) the tile tended to form discrete 3-D polyhedra rather than a 2-D array. Another variable in such a design was the length of the single-stranded loops at the center part of the tile, which provided a variable flexibility to the tile and controlled the angle at which the arms could bend out of the plane. Longer loops seemed to provide a higher flexibility at the center of tiles and allowed them to form polyhedra that required larger bending angles at the vertexes. For example, five-base loops were found to promote the formation of tetrahedrons, while three-base loops were suitable for the formation of dodeca-hedrons. By linking two of the three-point-star tiles, a tile with four arms was created that could self assemble into a buckyball-shaped DNA structure. In a later study, a five-point-star tile was shown to self-assemble into an icosahedron, using a similar design/assembly strategy [31].

28.5
DNA Nanostructure-Directed Nanomaterial Assembly

As noted at the start of this chapter, one central task of DNA nanotechnology is to control functional materials at the nanometer scale, and the DNA templated self-assembly of nanomaterials represents an important direction towards this goal. Until now, several groups have demonstrated the assembly of nanometer-scale materials such as metallic nanoparticles, quantum dots (QDs) and proteins on DNA nanostructures (Figure 28.5). The recognition events used to capture these functional materials included protein–protein/small molecule/aptamer (*in vitro* selected short DNA or RNA that can specifically bind to certain protein) interactions, functionalized metallic nanoparticle–DNA interactions, and DNA–DNA hybridization.

By using the well-known interaction between streptavidin and biotin, streptavi-din protein molecules can be organized onto 2-D DNA nanoarrays with a control-led periodicity and spacing [32]. The streptavidin–biotin interaction has also been shown to be useful for organizing nanoparticles. For example, Sharma *et al.* suc-cessfully built a patterned QD array by reacting streptavidin-coated QDs with biotin groups on the 2-D DNA nanoarray [33]. By attaching single/multiple copies of DNA onto the Au nanoparticle surface [34–36], Au nanoparticles could be rationally organized onto self-assembled DNA nanoarrays through DNA–DNA hybridization. Recently, Yan's group demonstrated that different-sized Au nano-particles could be used to control the conformation of DNA nanotubes (Figure 28.5d) [37].

Liu *et al.* incorporated thrombin-binding aptamers into a linear, three-helix DNA tile array and used aptamer/protein recognition to capture thrombin proteins into periodic arrays [38]. Recently, Rinker *et al.* [39] were able to take one step further by displaying two different thrombin aptamers, each target at a distinctive site of the protein, on a DNA array. The distances between the two aptamers were sys-tematically tuned such that a ~5.7 nm spacing led to an optimal "multivalent binding," which has much higher binding affinity than does any of the single aptamers (Figure 28.5c).

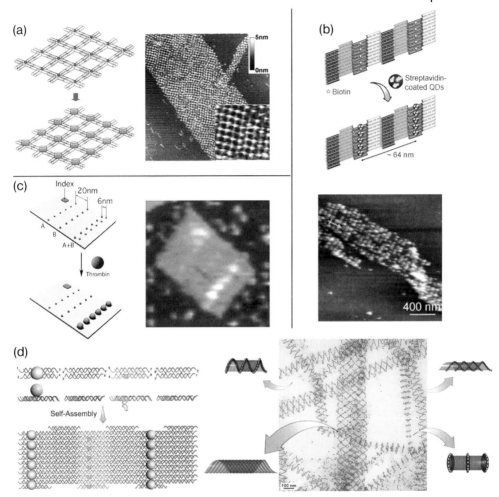

Figure 28.5 DNA nanostructure-directed nanomaterial assembly. (a) Streptavidin assembly on DNA 4 × 4 periodic 2-D array; (b) Quantum dot assembly on a double crossover (DX) array; (c) Thrombin line assembled on M13 viral DNA scaffolded 2-D array; (d) Gold nanoparticles were assembled on a DX 2-D array and forced the array to form tubes with different spiral patterns.

28.6
Concluding Remarks

The field of structural DNA nanotechnology has witnessed numerous break-throughs during the past decade, although only a small fraction of the prominent studies are outlined here. In that time, the complexity of the DNA structures has grown dramatically, from less than 100 nucleotides (four-arm DNA junction) to more than 10 000 nucleotides ("DNA Origami"), with series of 3-D objects having been built and characterized. Today, the capability is available to attach proteins, metallic and semi-conducting nanoparticles onto DNA arrays so as to create a variety of patterns of these materials. An increasing knowledge of the rules to design DNA structures has led to the building of more-complex DNA self-assemblies, and with fewer errors. With such progress, there is much optimism that DNA nanotechnology can potentially be applied to build DNA nanostructure-based nanocircuits and to control chemical/biochemical reactions in an ordered fashion that mimics enzyme cascade reactions. Nonetheless, many challenges remain. Although some functional materials have been successfully patterned by DNA arrays, the key ability is still lacking to create well-controlled, multicomponent nanoarchitectures. Whilst the current success with DNA 3-D objects has been limited to the building of polyhedron cages for purposes such as protein encapsulation and drug delivery, the task remains to create a universal strategy for building highly ordered 3-D structures. The "designer DNA" nanostructures produced, with their controlled geometry and topology, might find their use in biological applications such as interfacing cellular components through DNA scaffolds. Yet, much remains to be done in studying the biocompatibility, delivery, and stability of DNA nanostructures inside living systems. With the field of nanotechnology having successfully "borrowed" DNA from biological systems, it will not be surprising to see DNA nanotechnology contribute to the *in vivo* applications of nanotechnology in the future.

Acknowledgments

These studies were supported by grants from the National Science Foundation (NSF), the Army Research Office (ARO), and the Technology and Research Initiative Fund from Arizona State University to Y.L., and by grants from NSF, ARO, Air Force Office of Scientific Research, Office of Naval Research, and the National Institute of Health to H.Y.

References

1 Watson, J.D. and Crick, F.H.C. (1953) A structure for deoxyribose nucleic acid. *Nature*, **171**, 737–738.

2 Seeman, N.C. (1982) Nucleic acid junctions and lattices. *J. Theor. Biol.*, **99**, 237–247.

3 Holliday, R. (1964) A mechanism for gene conversion in fungi. *Genet. Res.*, **5**, 282–304.

4 Zhang, X., Yan, H., Shen, Z. and Seeman, N.C. (2002) Paranemic cohesion of topologically-closed DNA molecules. *J. Am. Chem. Soc.*, **124**, 12940–12941.

5 Yan, H. and Seeman, N.C. (2003) Edge-sharing motifs in structural DNA nanotechnology. *J. Supramol. Chem.*, **1**, 229–237.

6 Qiu, H., Dewan, J.C. and Seeman, N.C. (1997) A DNA decamer with a sticky end: the crystal structure of d-CGACGATCGT. *J. Mol. Biol.*, **267**, 881–898.

7 Seeman, N.C. (1990) De novo design of sequences for nucleic acid structural engineering. *J. Biomol. Struct. Dyn.*, **8**, 573–581.

8 Mao, C., Sun, W. and Seeman, N.C. (1999) Designed two-dimensional DNA Holliday junction arrays visualized by atomic force microscopy. *J. Am. Chem. Soc.*, **121**, 5437–5443.

9 Fu, T.J. and Seeman, N.C. (1993) DNA double-crossover molecules. *Biochemistry*, **32**, 3211–3220.

10 Winfree, E., Liu, F., Wenzler, L.A. and Seeman, N.C. (1998) Design and self-assembly of two-dimensional DNA crystals. *Nature*, **394**, 539–544.

11 LaBean, T.H., Yan, H., Kopatsch, J., Liu, F., Winfree, E., Reif, J.H. and Seeman, N.C. (2000) Construction, analysis, ligation, and self-assembly of DNA triple crossover complexes. *J. Am. Chem. Soc.*, **122**, 1848–1860.

12 Reishus, D., Shaw, B., Brun, Y., Chelyapov, N. and Adleman, L. (2005) Self-assembly of DNA double-double crossover complexes into high-density, doubly connected, planar structures. *J. Am. Chem. Soc.*, **127**, 17590–17591.

13 Ke, Y., Liu, Y., Zhang, J. and Yan, H. (2006) A study of DNA tube formation mechanisms using 4-, 8-, and 12-helix DNA nanostructures. *J. Am. Chem. Soc.*, **128**, 4414–4421.

14 Yan, H., Park, S.H., Finkelstein, G., Reif, J.H. and LaBean, T.H. (2003) DNA-templated self-assembly of protein arrays and highly conductive nanowires. *Science*, **301**, 1882–1884.

15 He, Y., Chen, Y., Liu, H., Ribble, A.E. and Mao, C. (2005) Self-assembly of hexagonal DNA two-dimensional (2D) arrays. *J. Am. Chem. Soc.*, **127**, 12202–12203.

16 He, Y., Tian, Y., Ribble, A.E. and Mao, C. (2006) Highly connected two-dimensional crystals of DNA six-point-stars. *J. Am. Chem. Soc.*, **128**, 15978–15979.

17 Park, S.H., Barish, R., Li, H., Reif, J.H., Finkelstein, G., Yan, H. and LaBean, T.H. (2005) Three-helix bundle DNA tiles self-assemble into 2D lattice or 1D templates for silver nanowires. *Nano Lett.*, **5**, 693–696.

18 Mathieu, F., Liao, S., Kopatsch, J., Wang, T., Mao, C. and Seeman, N.C. (2005) Six-helix bundles designed from DNA. *Nano Lett.*, **5**, 661–665.

19 Kuzuya, A., Wang, R., Sha, R. and Seeman, N.C. (2007) Six-helix and eight-helix DNA nanotubes assembled from half-tubes. *Nano Lett.*, **7**, 1757–1763.

20 Park, S.H., Pistol, C., Ahn, S.J., Reif, J.H., Lebeck, A.R., Dwyer, C. and LaBean, T.H. (2006) Finite-size, fully addressable DNA tile lattices formed by hierarchical assembly procedures. *Angew. Chem. Int. Ed. Engl.*, **45**, 735–739.

21 Liu, Y., Ke, Y. and Yan, H. (2005) Self-assembly of symmetric finite-size DNA nanoarrays. *J. Am. Chem. Soc.*, **127**, 17140–17141.

22 Yan, H., LaBean, T.H., Feng, L. and Reif, J.H. (2003) Directed nucleation assembly of DNA tile complexes for barcode-patterned lattices. *Proc. Natl Acad. Sci. USA*, **100**, 8103–8108.

23 Rothemund, P.W.K. (2006) Folding DNA to create nanoscale shapes and patterns. *Nature*, **440**, 297–302.

24 Chen, J. and Seeman, N.C. (1991) The electrophoretic properties of a DNA cube and its substructure catenanes. *Electrophoresis (Weinheim)*, **12**, 607–611.

25 Goodman, R.P., Berry, R.M. and Turberfield, A.J. (2004) The single-step synthesis of a DNA tetrahedron. *Chem. Commun.*, 1372–1373.

26 Goodman, R.P., Schaap, I.A.T., Tardin, C.F., Erben, C.M., Berry, R.M., Schmidt, C.F. and Turberfield, A.J. (2005) Rapid chiral assembly of rigid DNA building blocks for molecular nanofabrication. *Science*, **310**, 1661–1665.

27 Goodman, R.P., Heilemann, M., Doose, S., Erben, C.M., Kapanidis, A.N. and Turberfield, A.J. (2008) Reconfigurable, braced, three-dimensional DNA nanostructures. *Nat. Nanotechnol.*, **3**, 93–96.

28 Shih, W.M., Quispe, J.D. and Joyce, G.F. (2004) A 1.7-kilobase single-stranded DNA that folds into a nanoscale octahedron. *Nature*, **427**, 618–621.

29 Aldaye, F.A. and Sleiman, H.F. (2009) Modular access to structurally switchable 3D discrete DNA assemblies. *J. Am. Chem. Soc.*, **129**, 13376–13377.

30 He, Y., Ye, T., Su, M., Zhang, C., Ribble, A.E., Jiang, W. and Mao, C. (2008) Hierarchical self-assembly of DNA into

symmetric supramolecular polyhedra. *Nature*, **452**, 198–201.

31 Zhang, C., Su, M., He, Y., Zhao, X., Fang, P., Ribble, A.E., Jiang, W. and Mao, C. (2008) Conformational flexibility facilitates self-assembly of complex DNA nanostructures. *Proc. Natl Acad. Sci. USA*, **105**, 10665–10669.

32 Park, S.H., Yin, P., Liu, Y., Reif, J.H., LaBean, T.H. and Yan, H. (2005) Programmable DNA self-assemblies for nanoscale organization of ligands and proteins. *Nano Lett.*, **5**, 729–733.

33 Sharma, J., Ke, Y., Lin, C., Chhabra, R., Wang, Q., Nangreave, J., Liu, Y. and Yan, H. (2008) DNA-tile-directed self-assembly of quantum dots into two-dimensional nanopatterns. *Angew. Chem. Int. Ed. Engl.*, **47**, 5157–5159.

34 Sharma, J., Chhabra, R., Liu, Y., Ke, Y. and Yan, H. (2006) DNA-templated self-assembly of two-dimensional and periodical gold nanoparticle arrays. *Angew. Chem. Int. Ed. Engl.*, **45**, 730–735.

35 Sharma, J., Chhabra, R., Andersen, C.S., Gothelf, K.V., Yan, H. and Liu, Y. (2008) Toward reliable gold nanoparticle patterning on self-assembled DNA nanoscaffold. *J. Am. Chem. Soc.*, **130**, 7820–7821.

36 Sharma, J., Chhabra, R., Yan, H. and Liu, Y. (2008) A facile in situ generation of dithiocarbamate ligands for stable gold nanoparticle-oligonucleotide conjugates. *Chem. Commun.*, 2140–2142.

37 Sharma, J., Chhabra, R., Cheng, A., Brownbell, J., Liu, Y. and Yan, H. (2009) Control of self-assembly of DNA tubules through integration of gold nanoparticles. *Science*, **323**, 112–116.

38 Liu, Y., Lin, C., Li, H. and Yan, H. (2005) Aptamer-directed self-assembly of protein arrays on a DNA nanostructure. *Angew. Chem. Int. Ed. Engl.*, **44**, 4333–4338.

39 Rinker, S., Ke, Y., Liu, Y. and Yan, H. (2008) Self-assembled DNA Nanostructures for distance dependent multivalent ligand-protein binding. *Nat. Nanotechnol.*, **3**, 418–422.

Index

Advanced Nanomaterials. Edited by Kurt E. Geckeler and Hiroyuki Nishide
Copyright © 2010 WILEY-VCH Verlag GmbH & Co. KGaA, Weinheim
ISBN: 978-3-527-31794-3